Particulate Products

Particle Technology Series

Volume 19

Many materials exist in the form of a disperse system, for example powders, pastes, slurries, emulsions and aerosols, with size ranging from granular all the way down to the nanoscale. The study of such systems necessarily underlies many technologies/products and it can be regarded as a separate subject concerned with the manufacture, characterization and manipulation of such systems. The series does not aspire to define and confine the subject without duplication, but rather to provide a good home for any book which has a contribution to make to the record of both the theory and applications of the subject. We hope that engineers and scientists who concern themselves with disperse systems will use these books and that those who become expert will contribute further to the series.

The Springer Particle Technology Series is a continuation of the Kluwer Particle Technology Series, and the successor to the Chapman & Hall Powder Technology Series.

For further volumes:
http://www.springer.com/series/6433

Henk G. Merkus • Gabriel M.H. Meesters
Editors

Particulate Products

Tailoring Properties for Optimal Performance

Springer

Editors
Henk G. Merkus
Retired Associate Professor
Delft University of Technology
Pijnacker, The Netherlands

Gabriel M.H. Meesters
DSM Food Specialities
Delft University of Technology
Delft, The Netherlands

ISSN 1567-827X
ISBN 978-3-319-00713-7 ISBN 978-3-319-00714-4 (eBook)
DOI 10.1007/978-3-319-00714-4
Springer Cham Heidelberg New York Dordrecht London

Library of Congress Control Number: 2013954059

Printed on acid-free paper

Springer is part of Springer Science+Business Media (www.springer.com)

Preface

For a long time, chemical engineering focused on the characterization, understanding, and designing of chemical processes for the manufacture of commodity chemicals. These processes typically contain several unit operations, which are continuously operated. The products generally consist of a single phase and their quality is usually defined in terms of, e.g., purity, hardness, calorific value, boiling point/range, and/or melting point/range. Typical examples are processes in the oil industry for production of gasoline, etc. As a result of the extensive chemical research in the twentieth century, knowledge and understanding of the general relationships between chemical structure of components and their properties significantly improved, and more complex products arose by formulation with different ingredients, each of which serving specific performance goals. A good example is the development of detergents from classical soaps via synthetic detergents and powders to detergent pearls, containing active material, enzymes, whiteners, etc.

Gradually, the focus changed to consumer products, often produced in batch processes, and product complexity increased through the presence of different phases. Also, particulate materials were more and more incorporated in these products. Nowadays, particulate products make up about 80 % of all chemical products. Here, performance properties no longer solely depend on chemical composition and structure but also on particle size, size distribution, and particle shape (and resulting microstructures). Although the field of particle technology has existed for several decades, in which several basic principles have been elucidated and many instrumental measurement techniques developed, it still offers many major challenges. Catalysts are a good example. Particulate properties, such as size, pore structure, and surface area, not only influence activity (reaction rate) and selectivity (ability to form the desired product) but also mechanical strength (attrition), pressure drop, deactivation, and regeneration possibilities. Concrete and paints are other examples, as can be seen in Chaps. 7, 12 and 13.

Consequently, good design of such products is not only science but also still an art. Besides knowledge and understanding of the basic relationships, a lot of creativity is required to combine component properties and technological capabilities into a commercial product that is desired or accepted by the market.

Design and development of these complex products is a complex process in itself. Thus, several books have been written on its basic principles as well as on how this process can be managed in an optimum manner. They typically contain some molecular structure – property relationships for chemical components and some historic cases of product innovation in addition to a general management scheme. However, very limited insight is given so far for particulate products.

Therefore, we invited experts in different product fields to write a chapter in this book that presents the state of the art for each specific type of complex particulate product. We are very pleased that many of them accepted this invitation and indebted for their contribution in which they share their experience. Their names and background has been listed separately. We hope and trust that the leads in one product field will promote advances in other fields.

Henk G. Merkus
Gabriel M.H. Meesters

Acknowledgment

The editors gratefully acknowledge Mr. Maurice Wedd for his contribution to clear English texts in this book.

Contents

Author Information

Dr. M. Cristina Bonferoni graduated in 1984 in chemistry and pharmaceutical Technology, in 1991 followed by a Ph.D. degree. She became assistant professor in 1993 and is associate professor since 2001. Her research interests concern preformulation and formulation of conventional and especially of controlled release formulations. Early work involved the characterization of hydrophilic polymers used in oral matrix tablets. Swelling/erosion and diffusion/dissolution processes involved in hydrophilic matrices have been studied. Rheological measurements (viscosity and viscoelasticity) under different conditions of pH and ionic strength were related to the performance of the polymer in hydrophilic matrices. Dr. Bonferoni's study of the rheological properties of mucoadhesive interface has contributed to the understanding of the mechanism of adhesion and the rationale choice of mucoadhesive polymers. Polymeric micelles for topical administration of poorly soluble drugs and the development of polymeric vehicles for the delivery of hemoderivatives are her most recent research interests . Dr. Bonferoni's research work resulted in 140 papers in scientific journals, 11 book chapters, 6 patents, and more than 200 contributions to scientific meetings.

Prof. Dr. Carla M. Caramella is professor of pharmaceutical technology and biopharmacy at the Faculty of Pharmacy, University of Pavia, with teaching responsibilities both in undergraduate and Ph.D. programs. She has been responsible for the Erasmus/Socrates programs for the Faculty since 1998. She has served as dean of the Faculty of Pharmacy in the years 2003–2006. From 2002 to 2005 she has acted as Director of the Interuniversity Consortium TEFARCOInnova. She serves in the EAB of *European Journal of Pharmaceutical Sciences*, *Journal of Pharmaceutical Sciences*, *Pharmaceutical Development and Technology*, *AAPS PharmSciTech*, *Journal of Drug Delivery Science and Technology* and *Journal of Drug Delivery*. She has been appointed as an expert by the Italian Agency for Medicines for the European Directorate of Quality of Medicines (Certification Division) and for EMA. In 2001, she was elected AAPS Fellow. She has been awarded by AFI (Association of Industrial Pharmacists) in 2002, in 2004, and in 2010 for her cooperation with the Association in the organization of scientific events. In 2008, she became the first recipient of the prestigious "Ralph Shangraw Memorial Prize" created by the IPEC America Foundation. She has been the principal investigator of research project funded by the Italian Ministry of Education, by bank foundations, and by regional institutions. Her research interests are in the field of biopharmaceutics and drug formulation, including preformulation studies and controlled release dosage forms. Her present research interests are focused on the therapy of mucosal and skin disorders. She has published about 180 papers in journals, has co-authored 5 book chapters and 13 patents.

Dr. Menno B. Claase received his master's degree in chemistry with honors at the University of Nijmegen. Subsequently he worked in the Polymer Chemistry and Biomaterials Group of Professor Jan Feijen at the University Twente, where he received his Ph.D. in 2004.

Immediately after, he started as head of R&D at BYK-Cera, Deventer, the Netherlands. There he was responsible for the development of wax additives for the coating and printing ink industry, especially the development and upscaling of micronized waxes.

In March 2007, he started working for DSM Powder Coating Resins as R&D manager of the powder coating application laboratory. Currently he holds the position of R&D manager application, Powder, Can and Coil at DSM Coating Resins.

Dr. David Fairhurst is president of Colloid Consultants Ltd., a company that offers consulting services (from exploratory research projects through formulation and characterization to product production) in the fields of cosmetics, personal care and pharmaceutics, and in all general areas of emulsion and dispersion technology and the methods to characterize them.

Dr. Franca Ferrari has received her Pharm. Sci. degree in pharmaceutical chemistry and technology at the University of Pavia in 1981 and Ph.D. in pharmaceutical analysis and technology, at the same university, in 1984. She became assistant professor in 1984 and associate professor in 1998 at the School of Pharmacy, University of Pavia, where she is presently active both in undergraduate and graduate education. From 2002 to 2010 she has been the coordinator of the II level Masters Course in Preformulation, Pharmaceutical Development and Control of Medicines. Her scientific expertise is in the field of dosage form technology and biopharmaceutical implications and includes: study of tablet disintegrants and of disintegration process, characterization of hydrophilic polymers (swelling and water uptake phenomena), characterization of semisolids by means of viscosity and viscoelastic measurements, evaluation and study of mucoadhesion mechanisms, and evaluation and study of absorption enhancement mechanisms. Dr. Ferrari's current research projects are in the field of the design and evaluation of drug delivery systems and include buccal and vaginal semisolid (gels, in situ gelling systems) and solid (films, matrices) formulations, micro- and nanoparticulate systems for ophthalmic delivery, and sponge-like dressings for cutaneous and mucosal application. Her research work resulted in

128 contributions published in scientific journals, 4 patents, 12 book chapters, and 250 contributions to scientific meetings.

Dr. Huan He was born in Zhejiang Province, China. He was awarded his bachelor's and master's degrees in civil engineering with honors from Beijing University of Technology in 2002 and 2005, respectively. He got his Ph.D. from Delft University of Technology in the Netherlands in 2010. Then, he started to work as a postdoctoral fellow in Laboratory of Building Materials of GeMMe group of University of Liège in Belgium. Recently, he obtained a position at the College of Civil Engineering and Architecture of the Beijing University of Technology in China. His main interests include road and bridge engineering, characterization, and numerical modeling of concrete materials, supplementary cementitious materials, etc. He is a RILEM TC-SCM committee member and serves as a reviewer for several international journals such as *Cement and Concrete Research*, and *Image Analysis and Stereology*.

Dr. Stéphen L.A. Hennart is a French citizen from Normandy. He graduated from the National Institute for Applied Sciences in Rouen, France, with a chemical engineering degree and from the University of Rouen with a Master of Science degree in organic chemistry. He performed his thesis at the Department of Research and Development, DSM Food Specialties, Delft, the Netherlands, with the support of the Delft University of Technology and was co-funded by the Marie Curie actions of the European Commission and DSM Food Specialties BV. The main research topic of his thesis entitled "Sub-micron grinding of a food product" is the optimization of preservatives in food coatings by means of particle size reduction..

After graduating, Stephen worked as Product Application Specialist with DSM Food Specialties, specializing on food safety and food quality with focus on antibiotic residues in milk and dairy products. Three years later, he decided to change company and country and started as Global Product Manager for Veterinary Diagnostics within the Qiagen Group in Hilden, Germany. Simultaneously, he takes responsibility for ELISA and PCR assays for the detection of animal diseases and for business development in France and Latin America.

Anthony J. Hickey is a distinguished fellow (appointed June 2012) and a senior research pharmacologist at Research Triangle Institute. Dr. Hickey has more than 30 years of academic and research experience in pulmonary biology, aerosol physics, powder dynamics, pharmacokinetics and drug disposition, formulation design, and device development. Since joining RTI in 2011, he has conducted research related to pulmonary drug and vaccine delivery for tuberculosis treatment and therapy. Additionally, Dr. Hickey is an adjunct professor of biomedical engineering at the University of North Carolina at Chapel Hill School of Medicine, emeritus professor of molecular pharmaceutics at the University of North Carolina at Chapel Hill Eshelman School of Pharmacy, and founder and president of Cirrus Pharmaceuticals, Inc. He has received many awards and honors and is an active member of several professional societies and committees, including the American Association for the Advancement of Science, American Association for Aerosol

Research and the Society of Toxicology. He is a fellow of the Society of Biology, the American Association of Pharmaceutical Scientists and the American Association for the Advancement of Science. Dr. Hickey holds 17 patents and has authored numerous books, book chapters, and more than 160 journal articles.

Georgia K. Hinkley received a Bachelor of Science degree in biology and chemistry from Florida State University and is currently an Alumni Fellow with the Center for Environmental & Human Toxicology at the University of Florida. Her research focuses on factors influencing the agglomeration state of nanomaterials in the body and the effect of agglomeration on bioavailability. Her research also includes study of the maternal transfer and distribution of silver nanoparticles and its implications.

Dr. Saul M. Lemkowitz studied chemical engineering (B.Sc.) at Rutgers University in the United States. After working at Allied Chemical Europe in the Netherlands, he studied chemical engineering at Delft University of Technology (M.Sc., Ph.D.). He remained at Delft University, researching dust explosions and teaching explosion safety (later both with Prof. Hans Pasman). For almost 30 years, he has also taught explosion safety to industry. While formally retired, Dr. Lemkowitz still actively teaches at Delft University and to industry.

Dr. Gabriel M.H. Meesters has a B.Sc. and M.Sc. in chemical engineering with a major in BioProcessTechnology from the Delft University of Technology. He also has a Ph.D. in particle technology also from the same university. He worked at biotechnology companies like Gist-Brocades in the Netherlands, as well as for Genencor International and currently at DSM in research and development in the Netherlands. In all these functions he was working on formulation and product development. Since 1996, he has held a part-time position at the Delft University of Technology, as assistant professor at the faculty of Applied Sciences, first in the Particle Technology group, later the Nano-Structured Materials Group and currently in the Product and Process Engineering group. He has supervised over 15 Ph.D. students and more than 50 M.Sc. students. He has published around 60 refereed papers and holds around 15 patents and patent applications. He (co-)organized several international conferences in the field of particle technology and was president of the World Congress on Particle Technology in 2010.

Drs. Henk G. Merkus graduated in physical organic chemistry at the University of Amsterdam. He worked several years at the Royal Dutch Shell Laboratories in Amsterdam on research in the field of detergents and industrial chemicals, followed by development work on thermal wax cracking for production of $C_2 - C_{14}$ olefins and on acid-catalyzed synthesis of carboxylic acids from $C_3 - C_6$ olefins. Then, he made the change to analytical chemistry, involving both measurements and method development with a large variety of techniques and methods, first at Shell's process development department and later in the chemical engineering department of Delft University of Technology. Gradually, his analytical horizon widened: first surface area and porosity measurements were added to chemical analysis, later followed by particle size analysis. He is author of the book *Particle Size Measurements* (2009,

Springer) and many journal articles. Moreover, he participates in standardization activities regarding particle size measurement, both in the Dutch NEN and the ISO.

Dr. Tosko A. Misev has received his B.Sc. in chemical technology and Ph.D. in polymer technology from the University of Cyril and Methodius in Skopje, Macedonia, and M.Sc. in polymer technology from the University of Zagreb in Croatia. He started as Resin Plant Chemist at the chemical and pharmaceutical industry Alkaloid and then joined the Resin Business Group of DSM in the Netherlands, where he held the positions of R&D Manager Powder Coating Resins and R&D Director of the Coating Resin Group and R&D Vice President of DSM Desotech in the USA, a world leader in radiation curing coatings for optical fibers. In 2004, he returned to the Netherlands for the position of R&D Director of DSM Coating Resins until his retirement in 2011. His scientific activities primarily focused on resins for powder coatings and resulted in one book on powder coatings, co-authorship on two other books on the same subject, more than 20 publications and conference papers, and 21 patents.

Dr. Makio Naito received B.Sc., M.Sc., and Ph.D. degrees in chemical engineering from Nagoya University, Japan, in 1980, 1982, and 1987, respectively. He was with Hosokawa Micron Corp. from 1982 to 1993 and engaged in the R&D of powder processing technology. He joined Japan Fine Ceramics Center (JFCC), Nagoya, Japan, in 1993, where he focused on powder characterization and powder processing technology in ceramics manufacturing. He was Vice Director of JFCC from 2000 to 2002, and then became a professor at the Joining and Welding Research Institute (JWRI), Osaka University, Japan, in 2002. He continued focusing on important studies on innovative powder and nanoparticle processing to develop advanced materials about energy and environmental issues. He was promoted to be the Director of Smart Processing Research Center, JWRI in 2007, and is also the Vice Director of JWRI from 2009. In addition, he has served as a Director of Hosokawa Micron Corp. from 2005. His publications cover a wide range of studies in the fields related to the advanced materials. He has authored or coauthored more than 500 technical articles, including more than 200 refereed journal articles. He has contributed to 52 books including 13 books as an editor. He has received many awards, and is a Fellow of The American Ceramic Society.

Prof. Dr. Ir. Hans J. Pasman graduated in chemical technology in 1961 at the Delft University of Technology and received the degree of Doctor in Technical Sciences working for Shell in 1964. He joined the Netherlands Organization for Applied Research TNO and worked in analytical chemistry, thermal stability of substances and explosiveness and investigated industrial accidents. Later he initiated gas and dust explosion research, risk analysis, and in particular consequence analysis methodology and a whole variety of defense-related subjects such as test methods, operations research, and systems analysis. He went through the ranks of TNO as research coordinator, institute director, and director of marketing and programs. At the same time, he was active as chairman of the Loss Prevention Working Party of the European Federation of Chemical Engineering organizing symposia, and co-founding the European Process Safety Centre. In 1998, he became professor of Chemical Risk Management at the Delft University of Technology and after retirement in 2007 research professor at the Mary Kay O'Connor Process Safety Center of the Artie McFerrin Chemical Engineering Department at the Texas A&M University.

Prof. Dr. Stephen M. Roberts received a Ph.D. from the University of Utah College of Medicine and subsequently completed a National Institutes of Health (NIH) individual postdoctoral fellowship in pharmacokinetics at SUNY Buffalo. He is currently professor at the University of Florida with joint appointments in the College of Veterinary Medicine, College of Medicine, and College of Public Health and Health Professions. He also serves as director of the Center for Environmental & Human Toxicology. His teaching responsibilities at the university include graduate courses in toxicology and risk assessment. He has served on science advisory boards for the US Environmental Protection Agency, US Department of Health and Human Services, and the US Food and Drug Administration and has an active research program on mechanisms of toxicity and toxicokinetics. His current research projects include assessment of behavior, uptake, and biological activity of nanoparticles. He previously chaired the Nanotechnology Working Group of the National Toxicology Program (U.S.) and currently serves as an Associate Editor for the journal *Nanotoxicology*.

Dr. Silvia Rossi graduated in pharmaceutical chemistry and technology at the School of Pharmacy, University of Pavia, in 1988. In 1995, she got her Ph.D. degree. She became assistant professor in 1998 and associate professor in 2006 at the above-mentioned School of Pharmacy, where she is presently active both in undergraduate and graduate education. Her scientific expertise is in the field of dosage form technology and biopharmaceutical implications and includes: tablet disintegrants, characterization of hydrophilic polymers (swelling and water uptake phenomena), characterization of semisolid by means of viscosity and viscoelastic measurements, mucoadhesion evaluation and study of the mechanisms, and absorption enhancement mechanisms. Her current research projects are in the field of the design and evaluation of drug delivery systems and include buccal and vaginal semisolid (gels, in situ gelling systems) and solid (films, matrices) formulations, micro- and nanoparticulate systems for ophthalmic delivery, sponge-like dressings for cutaneous application, and formulations containing hemo-components for the treatment of mucosal and skin lesions. Her research work has resulted in 86 papers in scientific journals, 10 book chapters, and more than 220 contributions to scientific meetings.

Dr. Giuseppina Sandri graduated in pharmaceutical chemistry and technology at Faculty of Pharmacy of University of Pavia in 1999. She had got her Ph.D. in pharmaceutical chemistry and technology in the technological field at Department of Pharmaceutical Chemistry of University of Pavia in 2003. Since December 2008, she is assistant professor at the Department of Drug Sciences of University of Pavia. She is member of board of teachers of Ph.D. students on Biopharmaceutics-Pharmacokinetics and of Master students on "Preformulation, pharmaceutical development, and control of medications". Her present research activity is focused on the development of systems (wound dressings, powders for cutaneous application, contact lenses) to deliver hemoderivatives to treat chronic lesions (mucositis, skin, and corneal ulcers). Her research activity has resulted in the publication of 42 research papers in international journals, 7 review articles in international journals, 5 book chapters, and 3 patents. Also, more than 100 proceedings have been published. She serves as reviewer for NRF (National Research Foundation) South Africa and for several scientific journals.

Dr. Elke Scholten received her M.Sc. in physical and colloid chemistry from the University of Utrecht. She received a Ph.D. from Wageningen University, the Netherlands, where she focused on protein-polysaccharide mixtures and related interfacial phenomena. She then joined the Chemical Engineering group at MIT, USA, from 2006 to 2008, where she worked with spinning and spraying techniques to create fibers and coatings. Then, she joined AKZO Nobel as a research scientist where she worked in the field of waterborne paints. In 2009, she returned to Wageningen University, where she currently holds a position as assistant professor in the Food Physics group. She mainly focuses on physical phenomena in food products, with a special interest in emulsions, gels, and fracture phenomena. One of her interests is molecular gastronomy.

Prof. Dr. Piet Stroeven was employed at Delft University of Technology from 1963 until his retirement in 2002, in a broad range of concrete technologies. Since then, he still is active as guest professor in the same university. His research experience includes a wide variety of subjects within the concrete field, such as shell structures, damage mechanics, optical methods for stress analysis, quantitative image analysis, alternative binders, fiber-reinforced concrete, and discrete element method. He has been awarded by ACI/CANMET (2001) and Wroclaw University of Technology (1999), and is appointed as honorary professor at Beijing Jiaotong University (1997). He has written about 1000 publications in journals, books, conference papers, reports, and reviews.

Dr. Paul Vercoulen received a master's in chemical engineering at the Delft University of Technology followed by a Ph.D. in powder technology at the same university. He started his industrial career at DSM in the area of Fine Chemicals. He held various positions within R&D and New Business Development of different DSM businesses and is currently (since 2011) responsible for the R&D and Innovation efforts of the Powder, Can & Coil Coatings Resins unit of DSM, located in Zwolle, the Netherlands.

Dr. Zhen Xu obtained his Ph.D. in 2010 in the Eshelman School of Pharmacy, University of North Carolina. His doctoral research in Dr. Anthony Hickey's lab focused on dry powder aerosol formulations for the treatment of asthma, with an effort on elucidating the heterogeneous microparticle interactions in the solid-state, and achieving cost-effective formulation screening and performance prediction. He is currently a postdoctoral fellow in the School of Pharmacy, University of

Maryland, contributing in the field of aerosol device technologies, using valvedii holding chamber and facemask systems for pediatric pulmonary delivery. He also earned a M.S. degree in chemistry previously at Michigan State University, where he worked on small molecule drug design and synthesis from the structural richness of carbohydrates.

Chapter 1
Introduction

Henk G. Merkus and Gabriel M.H. Meesters

When you can measure what you are talking about, and express it in numbers, you know something of your subject.
But if you cannot measure it, when you cannot express it in numbers, your knowledge is of a meager and unsatisfactory kind.

Lord Kelvin.

Abstract There is a great variety of industrial particulate products. Quality and performance of such products as well as their production processing depend upon the characteristics of the particulate ingredients and of product processing. These characteristics should be selected with due care in direct relation to product quality aspects. Measured data should be capable to discriminate between different product qualities, and the measurement quality should be sufficient. The last decades have shown a vast expansion and differentiation of measurement techniques as well as strong improvement of their quality. The design and selection of optimum particle characteristics for a variety of commercial products is the subject of this book. Its goal is to improve future product development. This chapter gives an introduction on product types, relevance of particle characteristics for product performance, schemes and statistical tools for optimum product development and design, and preferred particle characteristics in relation to product quality.

H.G. Merkus (✉)
Retired Associate Professor, Delft University of Technology, Park Berkenoord 30, Pijnacker 2641 CZ, The Netherlands
e-mail: henkmerkus@hetnet.nl

G.M.H. Meesters
DSM Food Specialities, Delft University of Technology, Alexander Fleminglaan 1, 2613 AX Delft 1, Delft 2600 MA, The Netherlands
e-mail: gabrie.meesters@DSM.com

H.G. Merkus and G.M.H. Meesters (eds.), *Particulate Products: Tailoring Properties for Optimal Performance*, Particle Technology Series 19,
DOI 10.1007/978-3-319-00714-4_1, © Springer International Publishing Switzerland 2014

1.1 Objective of This Book

Particulate products make up around 80 % of all chemical products. Particle characteristics play an essential role in their quality and performance. The design of such products is the subject of this book. It can be used for improving existing products as well as for designing new products. Often, particulate products are composed of several ingredients. The particulate ingredients are not only solid particles, but they include all entities that have a difference in phase with their surroundings. Thus, products for consideration include dry powder mixtures, suspensions, emulsions and creams (liquid droplets in an immiscible liquid), sprays (droplets in gas) and foams (gas bubbles in a liquid or a solid material), etc. The other ingredients in the final products serve goals related to consistency, stability, taste, smell, etc.

The main objective of this book is to guide optimum design of particle characteristics of particulate ingredients in commercial products in relation to the required performance characteristics, both during processing and for the final product. Such designs start with good identification of the various performance aspects and understanding of their relationships with particle characteristics, viz. particle size distributions (PSD), particle shape, porosity and/or zeta-potential. The choice of optimum characteristic parameters requires such knowledge, in relation to desired performance tasks as well as for adequate specifications of both ingredients and products. Since very little seems to be published on the basic parameter choices to be made for particulate products, we asked various experts to write a chapter on this subject for a variety of important particulate products. We trust that their views will provide the available insight in the different, essential characteristics of composed particulate products. Their texts clearly show the different approaches taken in different product areas. This may lead to better understanding of the challenges in all product fields.

A further requirement for good product quality is that the particle parameters can be measured with sufficient precision, sensitivity and resolution. In addition to the information given for the products, a separate Chap. 3 deals with some major techniques for measurement of the parameters. It includes the quality of sampling, dispersion and measurement.

In the next sections of this introductory chapter, the relevance of particle characteristics for product behavior and methods for product design and development are illustrated.

1.2 Relevance of Particle Characteristics for Product Quality

Product quality can be identified in different ways. A simple definition is "*Conformance to product specifications*". Such specifications, however, are far from always sufficient to cover all performance aspects. Also, broader definitions have been formulated, which can be combined in:

"*The totality of features and characteristics of a product that bear on its ability to satisfy stated or implied expectations, needs and requirements of*

Table 1.1 Examples of quality aspects caused by particle characteristics

Product stability and 'structure' in dispersions
Dissolution rate, reactivity
Drying rate, efficiency
Rheological properties of powders, emulsions and suspensions: powder flowability, dosability, fluidization, viscosity, (non-) Newtonian behavior
Air dispersability, inhalation possibility
Filtration rate, removal possibilities
Pharmaceutical efficacy and side effects
Safety, toxicity, dusting, explosion sensitivity/severity
Catalytic activity and selectivity
Concrete/ceramic strength
Taste, mouthfeel, skin feel
Gloss, hiding power, transparency
Abrasive/polishing quality
Packing density and product porosity

customers in exchange for monetary considerations". Features and characteristics depend on product type. Typical examples are: performance, reliability, safety, toxicity, medical effects, appearance, taste, nutritional properties, texture, shelf life, etc.

The general concept is nicely covered in the following poem ([25] plus a little extension):

Quality is . . . giving a customer what he wants today,
at a price he is pleased to pay,
at a cost we can contain,
again, and again, and again;
and when more desires develop for the future
giving him something even better tomorrow.

There is a large variety of commercial products that have a particulate nature or contain particulate ingredients. Different industries produce such products. Examples for simple products or ingredients are abrasives, coal, filter aids, pigments, sand and sugar. Examples for composed products are polishers, agro-products, ceramics, chocolate, concrete, emulsions (in both dilute and concentrated form, e.g. milk, margarine and sunscreen), ice cream, medical inhalers, paint (both powder coatings and suspensions), powders for pharmaceutical tablets, sprays, toners, etc. Their chemical composition and possible specific (crystal) structure are primarily responsible for the basic material properties, such as hardness, color, taste, solubility, efficacy and safety. Particle characteristics take a second position for product quality. They are involved in many quality aspects, as illustrated in Table 1.1. A third important effect often comes from structural effects in dispersions resulting from attractive or repulsive forces between particles.

Particulate ingredients are either produced through a process of *size reduction* (crushing/milling/breakage/attrition, spraying, emulsification, de-agglomeration) or *size enlargement* (precipitation, crystallisation, polymerisation,

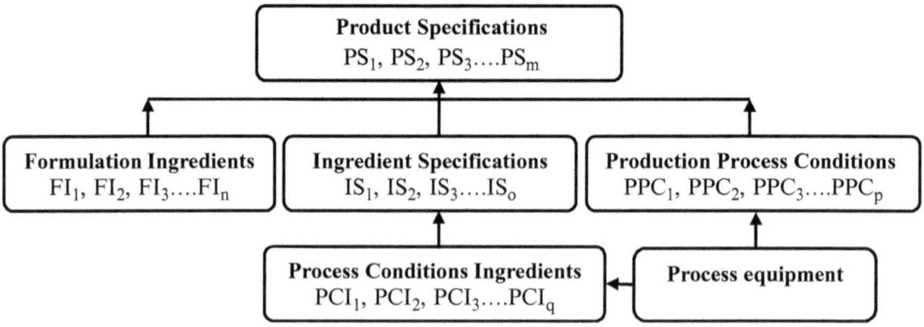

Fig. 1.1 Schematic of dependency of product specifications (*PS*) from formulation ingredients (*FI*), ingredient specifications (*IS*), process equipment, process conditions ingredients (*PCI*) and production process conditions (*PPC*)

compaction/granulation, extrusion, spray-drying). Often, a *classification* step in (hydro-) cyclones or by sieving/screening is integrated in the production process, for adequate control of the particle size distribution.

In order to obtain the desired quality at lowest possible costs, most products are formulated. This means that they contain several ingredients, each of which has a defined function or purpose, sometimes related to improved quality, sometimes to cost reduction. In addition to particles, there are e.g. wetting agents, flavoring agents and thickeners. Product quality and specifications are related to type, concentration and specifications of all ingredients as well as on (the control of) their production processes. Also, the choice of adequate process equipment is important. A general block scheme that relates the specifications of the end product to formulation ingredients, ingredient specifications, conditions of production processes and production equipment involved is given in Fig. 1.1.

Some of the general relationships between the quality of ingredients and product are known either from theory or from experimental results. Chapter 2 gives an overview of available relationships. For many products, however, adequate quantitative relationships are not (yet) available. There are several reasons for this lack of knowledge:

1. Particulate properties depend upon the chemical composition and (crystal) structure of the particles and, thus, upon their purity and route for preparation. Specific relationships are often required which may or may not be known.
2. Some ingredients cause synergistic or antagonistic effects within different product quality aspects, which complicates the relationships.
3. Some product properties are described in terms that do not yet allow an easy, direct translation to particle characteristics. Examples are sensory characteristics in 'soft' customer desires, e.g. 'easy to swallow', 'mouthfeel', 'crispy', 'bite', 'creamy texture' and 'soft skin feeling'. Here, individual preferences and psychological factors play an important role in human perception.
4. The width of the PSD affects product properties. This is far from always incorporated in the PSD parameters chosen.
5. The results of size distributions of non-spherical particles may strongly depend on the technique used for determination. Also the product performance may

depend upon particle shape. Then again, application of a new ingredient or measurement technique causes that a new relationship has to be established.
6. Moreover, compromises must be found for particle characteristics in view of contradictory effects coming from different quality aspects or particle characteristics. Different products typically ask for different compromises.

Product specifications are meant to guarantee that product quality/performance corresponds to expectation. At present, many specifications seem to come from subjective judgment based on experience. To our opinion, only a (semi)quantitative (mathematical) model that accounts for all effects, gives an optimum basis for setting the specification. This applies equally to the design and specification of both single and composed products. This is a major challenge for the future.

1.3 Product Development and Design

At this moment, there seem to be three main lines for development of new, complex products, viz. Technology Push, Market Pull and Quality Function Deployment/ House of Quality [22, 32]. Their main differences are the driving force for the development of the new product and the way of organizing and planning the development activities.

Technology Push is the conventional method for product design. Typically, all activities are done by a single responsible person or a small expert team within the same company. It is often based on the skills of that person or team. The individual or team may or may not consult the marketing people, R&D experts, production engineers and/or others, to define and develop the best products for that company. This should be done in relation to the company's strategic goals, while taking into account available and newly developed technologies. Thus, it reflects the optimum possibilities for new products in the view of the producer company to serve the market. The iPad is an example of pushed products. Test panels may be used to assess product quality. In this method, strict rules for planning and development are not prescribed and, often, decisions are made implicitly. For products having clear performance requirements, the method is adequate, provided that the above weak points are addressed.

Market Pull is the opposite approach for product design. Here, some group of 'customers' requests for a product with new properties. Examples are low-calorific foods and the bicycle with electric support. Often, its basis is improvement of an existing product, based on customer desires. However, for many products customer desires are too often, poorly defined.

Quality Function Deployment (QFD) or its basic element, *House of Quality* (HOQ), is a fairly new approach. It offers a good compromise between Technology Push and Market Pull. A production company usually takes the initiative. A new aspect is that all activities for the development of a new product are explicitly described and planned. This includes collection of all customer desires, including 'soft' ones, and their translation into measurable characteristics. Typically, customer input has a (much) stronger influence on product development than in the

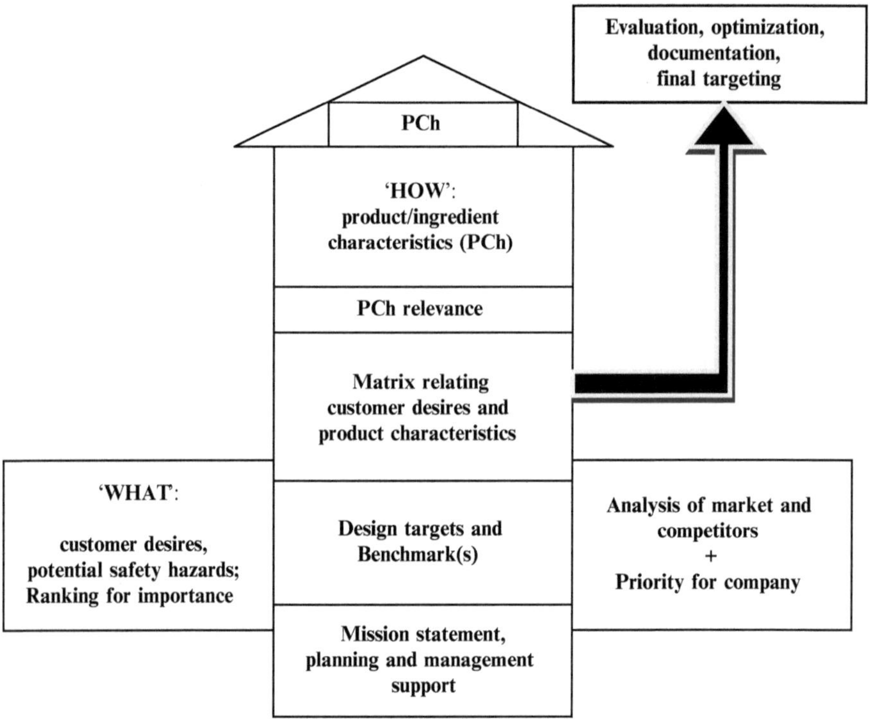

Fig. 1.2 Basic elements in House of Quality. (*PCh,* product characteristics)

Technology Push method [4, 5, 9, 11, 19, 21, 23, 27, 29, 31, 33]. In comparison to the Market Pull method, customer needs are more explicitly translated into measurable attributes. Consequently, it is useful in development of new products that the customer input as well as valuable chemical and engineering knowledge are combined. Note that different planning schemes are used, in relation to the type of product to be developed. A general scheme for the House of Quality method is presented in Fig. 1.2.

The basic steps are (for more details, see [9, 30]):

1. Prepare a clear, concise mission statement for the product and its development. This should include a short description of the product, its critical advantages over available products and its importance (priority) of the product for the producer company. The Mission statement should also include costs and the organization of the development, business goals for development time, profits and market share targets as well as assumptions and constraints related to the development.
2. Assure management support, plan all activities and set up a multi-functional expert team, in which as much expertise of colleagues as possible (marketing and R&D) is brought together for brainstorming, discussions, evaluation and decisions. Have a project leader appointed to lead the discussions effectively and to promote that planning schedules are met.

3. Collect information from potential customers on their qualitative desires (needs, wants and benefits) for the stated product through interviews of individual or groups of users/customers. Translate these into short 'quality tools' (the WHATs or NEEDS or Voice of the Customer). Make sure that the expert team defines these tools as clearly as possible. For interpretation and translation of 'soft' customer desires, product engineers and trained user panels may be helpful.
4. Identify potential hazards of both product and ingredients with respect to human and environmental safety and integrate these aspects in the group of customer needs.
5. Give a ranking to all customer needs in view of their importance and list them in order of importance (CN_1, CN_2, ..., CN_n). Three levels of ranking will generally suffice, e.g. 1 (low) – 3 (medium) – 9 (high) [9, 23, 26]; often these levels are indicated by specific symbols or colors to allow a rapid overview. Here too, trained user panels may be helpful.
6. Investigate and judge the market situation and competition. Select one or two (competitive or own) products to act as 'benchmarks' for quality.
7. Translate all customer desires, including 'soft' ones, and any hazards into measurable characteristic product parameters (the HOWs or Voice of the Company).
8. Investigate and describe the positive (synergistic) and negative (antagonistic) interrelationships between these characteristic product parameters and design trade-offs (in a matrix in the roof of the House).
9. Investigate and set up quantitative relationships between customer desires, potential hazards, importance for producer and characteristic parameters of product and/or ingredients and give them a weighting (in the central matrix, the Relationship Room).
10. Investigate the feasibility of (additional) ingredients and production processes to meet the desired characteristic product parameters and estimate the production costs.
11. Choose the best product and manufacturing process, and define the operational goals for the characteristic product parameters, both values and margins.
12. Evaluate the results in terms of potential, risks and regulatory issues for the producer company in the development team, and set targets. Address intellectual property in terms of patents and/or secrecy.
13. Define targets, benchmarks and specifications for product and/or ingredients and equipment and conditions for the production process.
14. Document all considerations and results well for eventual future use.

A major advantage of this approach is that it provides good planning opportunities as well as an objective overview of customer desires and market potential in relation to production possibilities and requirements. For example, the relation between 'soft' customer needs and product/ingredient characteristics, such as PSD parameters, is addressed explicitly. Another advantage is that it contains a weighting for priorities and costs. Hence, it provides a repository for product planning information. However, full application of all aspects of Quality Function Deployment/House of Quality requires a lot of time and money, although mathematical procedures for effective judgment are available. Therefore, often only its basic concept is used to improve planning and analysis of product development as well as for reference in subjective judgments. Based on adequate knowledge and planning, QFD has proven in several fields of application (e.g. ships, electronics, cars and food) to decrease both cycle time and costs for product development, despite the substantial expert time needed initially for collection, discussion and translation of customer information [4, 5, 11, 23, 29].

The application of QFD can be extended through the sequential use of similar schemes for planning, process design, production and process control, in a Planning Matrix, a Design Matrix, a Manufacturing/Operating Matrix and a Control Matrix [9, 26, 29].

It should be noted that the present QFD schemes seem to be used primarily for improvement of existing products and not for development of highly innovative products. A reason may be that customer needs are not yet clear for such products and that production processes and/or product composition still have to be invented. Here, the Technology Push approach with an open mind for applications seems to be more appropriate.

QFD/HOQ design schemes are often rather abstract and seem to be mainly directed towards management. For the design and manufacture of our complex type of products, the core activities for chemists and engineers are expressed in the following ten-step template [12, 34]:

I. Prepare a clear, concise **mission statement** for the product and its development.
II. Identify **needs** for the envisaged product and translate them into measurable attributes and (preliminary) specifications.
III. Generate **product ideas** (ingredient characteristics, ingredients and products) that may satisfy these needs.
IV. Determine (or estimate) **quantitative relationships** between product performance parameters and product composition and ingredient characteristics.
V. **Select** the most promising product ideas, in relation to performance and quality as well as to potential safety hazards.
VI. Generate **ideas** for **manufacturing** processes of both ingredients and product.
VII. Investigate and set up **quantitative relationships** between product performance and the manufacturing process(es) and process conditions.
VIII. **Select** the best manufacturing process possibilities (batch/continuous, flow sheet, equipment and operating conditions), in relation to both least complexity and costs.
IX. **Evaluate** most promising ideas and **select** the best combination of product and process, plus final specifications and process control parameters.
X. **Set up** equipment and **prepare** a test portion of the product.

A review for the management can be envisaged to get support and to enable a go/no-go decision for the next steps, e.g. at the end of steps I, V, VIII and IX of this template. Then, also costs and human and environmental safety can be considered. The steps in this template fit well within the QFD scheme, but they are more explicitly defined. In step II, product users are interviewed and their needs are translated into measurable attributes and specifications. The emphasis in steps III to V rests on specialized chemists. On the basis of knowledge, experience, literature and experiments, they should select the most promising ingredients and products. The same holds for specialized engineers in steps VI to VIII for the manufacturing equipment and process conditions. Note that an existing plant may be the starting point to ideas for a new product line. Optimum decisions in step IX depend on the total group of chemists and engineers. Reference [34] provides examples and background for some creams and pastes. References [12, 22, 32] give numerous examples for a wide variety of chemical products, illustrating the choices to be made in the design steps. Scientific and empiric backgrounds are discussed as well as differences in the various steps for different types of products.

All approaches for design of composed chemical products show three similarities:

- They require basic data that are necessary for optimum design of the new product
- They require basic data that are necessary for optimum design of the manufacturing process
- They use similar ways for optimal application of these data in the design.

The basic product data are measurable, characteristic (particulate and other) features that relate best to assumed or expressed customer desires. For ingredients, they include the influence of wetting and stabilizing agents, fillers, viscosity improvers, taste enhancers, etc. Their basic relationships may come either from theory or from experiments. Chapter 2 gives examples of existing relationships.

The basic process data usually relate to equipment, such as crystallizers, emulsifiers, homogenizers, (colloid) mills, mixers and sterilizers, where the influence of mechanical forces, shear forces and cooling/heating dominate in the quality of the product.

1.4 Particle Characteristics for Product Quality and Process Control

Relevant particle characteristics for particulate product performance include aspects of the particle size distribution (PSD), particle shape, particulate concentration, porosity and/or zeta-potential. Sometimes, the relationship between performance and particle characteristics are known from theory or literature. Often, however, this is not the case and research is necessary to find both best characteristics and functional relationships [15]. Characteristic parameters are preferred when their relationships with product quality and performance are based on both good correlation and understanding of their physical background. The parameters that are regarded essential must be laid down in specifications for ingredients and product, which are meant to distinguish good and poor products. Thus, they should correlate well with product quality and have adequate measurement possibilities. Note that redundancy of specifications may lead to unnecessary costs for measurement and contradictory results for quality. For control of the production process, sometimes, alternative parameters are used in relation to process parameters. The relevant size parameters depend on the product as well as on the precision with which they can be measured.

Following PSD parameters for particulate ingredients seem logical choices from the product point of view [20]:

- Mean size of the distribution, weighted according to number, area, volume, etc., depending on application and theoretical background. The mean value has the advantage that the contributions of *all* particles to the performance are taken into account.
- Size of the largest particle if this value is essential to product quality. Sometimes, the D_{90} or the fraction of particles larger than a stated size is used instead, in view of easier determination.[1]
- Fraction of particles smaller than a stated size (e.g. 45 μm) or the D_{10} if these smaller particles are essential to product quality[1].
- The width of the size distribution, in addition to one or some PSD parameters chosen, expressed as ratio D_{90}/D_{10}, D_{84}/D_{16} or the width parameter of a modeled distribution[1].
- Stated volume fractions in given size classes to minimize voids, e.g. in sand for concrete (see Chap. 7).

It is again remarked that particle size measurement of non-spherical particles, which are normal in industrial products, may yield distributions of equivalent sphere diameters that depend on the physical principle applied in the technique. Thus, different techniques generally yield different PSD results, the more so when the particles' aspect ratios significantly differ from 1 (for more information see Chap. 3). This weakens, of course, the physical basis for relationships between PSD characteristics and product performance aspects.

Note that PSD's are sometimes characterized by model equations, which contain a size location parameter as well as a distribution width parameter. Typical examples are: normal, log-normal, Rosin-Rammler (or Weibull) and Gaudin-Schumann distributions. Such model parameters, however, seem to be more useful for identification and explanation of particulate production processes and for process control purposes in existing equipment than for relating PSD parameters

[1] In all cases an indication has to be given for the type of weighting of the value, viz. by number, area or volume, etc. and for the measurement technique.

to product performance. For, often, real size distributions do not follow these equations over their full range.

Note further that the median (D_{50}) of a PSD is *not* suitable to qualify products, since it is a statistical parameter that has no basic relationship to product quality. For, it does neither reflect information about the width of the PSD nor about the presence of outlying particles, which may damage product quality. In combination with some other characteristic PSD parameters, however, the D_{50}[1] can be suitable for both process control and for testing measurement precision and bias of instruments with standard samples.

Particle shape characteristics may involve [20]:

- Macro-shape features, related to the 3-dimensional form of the particles, as expressed e.g. in aspect ratio
- Meso-shape features, related to the general aspects of roundness and angularity
- Micro-shape features, related to rugosity and smoothness as well as porosity and other structural heterogeneities.

At this moment, typical shape characterization concentrates on only one or two of the main shape features. Typical examples are aspect ratio, angularity, porosity and surface area. Also, modeling is sometimes used. Examples are (1) evaluation of the contours of two-dimensional particle projections in a so-called Fourier series, specific coefficients of which being used in correlation with specific performance features, and (2) expression in terms of fractal dimension [20]. So far, shape characteristics are mostly used for qualitative characterization of product performance. Only porosity features are used quantitatively.

For product specifications as well as process control, techniques and procedures for sampling, dispersion and measurement should be applied that have a known, fit-for-purpose quality. This means, for example, that both precision and sensitivity for the selected parameters must be adequate to distinguish between different product qualities. All procedures should be laid down in written form, for important products preferably in international standards (ISO, EN, ASTM, etc.). This simplifies international trade.

The following aspects should be addressed during development of particulate products:

1. Product composition, basic data and application(s)
2. Consumer desires for product quality and translation into measurable attributes
3. Potential hazards and regulations with respect to human and environmental safety
4. Critical behavior/performance aspects in quantitative terms for differentiation between good and poor products
5. Production process(es) and equipment used for ingredients and product, e.g. crystallization, (colloid) milling, mixing, polymerization, precipitation
6. Role of ingredients and/or production process conditions for product behavior/performance, in qualitative relationships as well as quantitative property functions for product and process(es)

7. Selected critical particle characteristics of the particulate ingredients for product behavior/performance
8. Selected particle/product parameters for process control
9. Specifications for particle and other characteristics of product/ingredients, incl. defined confidence intervals and occurrence of outliers
10. Preferred techniques and methods for sampling, dispersion and measurement in relation to specifications.

1.5 Statistical Tools for Product and Process Development

Statistical methods can be very helpful in the optimization steps of both product and process design. About half a century ago, application of statistics was more or less limited to reduction of large data sets into a few numbers, e.g. mean, median, mode, variance, standard deviation and confidence interval. With the arrival of powerful computers, applications greatly expanded. For example, regression analyses to fit data to a model and tests for significance, hypotheses, outliers and variance can now be performed easily. Also, computer simulations can be executed with the models obtained, to find optimum products or process conditions.

Caution: Advanced statistical analysis should only be applied to measurement data that has a proven pedigree as to its quality and fitness for analysis. Statistical analysis of poor data may provide false indications leading to expensive and unnecessary further work.

In the past, *classical statistical* methods were most often applied. They only use measured results and assume that this data is subject to random errors and, thus, have probability distributions. Nowadays, *Bayesian statistics* offer a nice alternative. Here, both causes (observations) and effects (parameters) are assumed to have probability distributions. Moreover, prior knowledge is included in a mathematically elegant manner [1, 8, 10, 18]. The advantage over classical statistics is that this approach allows estimation of probabilities for hypotheses, e.g. for model selection, and of unknown parameters in model equations together with confidence intervals. Although various applications of statistics in chemistry and engineering are published, wide application is still far away. Hopefully, some examples will be included in the chapters on specific products, so that further application is promoted. Of course, adequate knowledge is required to choose the optimum statistical tools.

Statistical methods can be applied to set up of experiments, for data interpretation, for selection of key parameters and models and/or for hypothesis testing. A variety of methods is available, e.g. Factorial Design, Principle Component Analysis/Factor Analysis and Linear/Non-linear Regression/Artificial Neural Networks. Both classical statistics and Bayesian statistics can be used. For optimization of relationships, often a method of least squares is applied. Significance can be tested with the aid of e.g. the method of Analysis of Variance.

Note that statistical correlation between two variables does not automatically imply a functional relationship. There may be a hidden factor (confounder) that is responsible for the correlation between the two [14]. The correlation between presence of storks and delivery of babies is an example. Always check for the presence of confounders. Understanding the relationships from physical laws helps the exclusion of confounders and extends the range of validity for the relationships found.

Factorial Design is meant to find optimum quality of results with a minimum of experimental effort by systematic planning of experiments and statistical evaluation of their results [2, 16, 17]. The typical goal is to find out how changes in controlled and independent input variables (e.g. temperature, concentration, particle size, duration of treatment, within a defined, limited data input range) affect the dependent output variables (e.g. conversion, selectivity, strength, hiding power and taste).

There are three steps in Factorial Design. If a large number of input variables are anticipated, some *Screening or Fractional Factorial Design* is necessary. Here, prior knowledge and a limited number of experiments are used to find the most significant input variables. Thus, some combinations of variables are omitted in these preliminary experiments.

If the number of significant variables is reduced to a maximum of about 4–6 variables, a *Full Factorial Design* is employed, in which the influence of the input variables and their interactions with the output variables are investigated by changing the input variables at two levels (i.e. two discrete values for each variable), viz. 'high' and 'low'. If there are N variables, 2^N experiments are required. Note that the number of experiments strongly increases with N. For example, 2^6 is already 64! This type of factorial design implies that all relationships between input parameters and output parameters are presupposed to be linear or sufficiently linear over the range of changes in the input factors. The statistical analysis of the results allows the presentation of an empirical equation that relates the result with the variables and their interactions with the significance of each term. Note that all experiments should be performed in random order, to reduce the effect that bias due to the order of experimentation may have on the results. If the parameter levels used in the design are in good balance, then the Factorial Design has a built-in replication that helps mitigate the impact of random errors on the model estimates. Some replication measurements should be done to estimate precision and bias. The assumed linearity of the relationships can be checked through inclusion of some extra mid-point measurements.

If the relationship between input and output variables is expected or found to be non-linear in the range of interest, then, more experiments are to be executed in a third step, in a so-called *Response Surface Design*. At least 3 value settings are then required to reveal the curvature of the 'interrelation surface'. Here, a predefined model is not necessary, but the number of data points limits the assessment of non-linearity.

The strategic approach for Factorial Design is now listed.

1. Collect information on any relationships between input and output variables.
2. Define objectives for product quality output.
3. Select the relevant ranges for the input variables.

4. Select the appropriate type of Factorial Design in view of the estimated number of input variables.
5. Run the designed experiments in random order.
6. Analyze and interpret the experimental results, to select the significant input variables or/and to find the model equations.
7. Go to the next stage of design with more experiments if the objectives are not met.
8. If the objectives are met, confirm the established equations and apply the results to find the required optimum product property.
9. Document all results and considerations for eventual future use.

Principle Component Analysis (PCA) is used to extract the **key variables** in large data sets, containing information on relationships between different aspects of a product (e.g. measured data of one or more of its performance aspects versus composition), through statistical analysis [3, 13, 20, 28]. In PCA, a set of measured data, which contain some correlation, is transformed – without loss of information – into linear combinations of the original variables. This leads to a set of uncorrelated (but abstract) variables, the so-called principal components (PC's), which, in fact, are the eigenvectors of the variance-covariance or correlation matrix. These PC's are derived in decreasing order of significance. In consequence the first PC accounts for the greatest significance of the data set in the relationship (greatest contribution to the variance of the data set), the second PC for the second greatest significance and so on. If the degree of correlation in the data set is high, then only the first few PC's account for significant variance of the data. The remaining components may often be considered as noise. The relative contributions of the PC's to the variance of the data set are often named PC scores. So-called loadings can be calculated that give the relative contributions of the original variables to each PC. The most significant PC's must be examined to find out which PC or combination of PC's contains the desired information for the product. Reference [13] gives some nice examples of PCA applications. Note that the first PC is not always relevant, since it may be related to e.g. concentration fluctuations.

Factor Analysis is comparable to PCA. The difference is that, based upon prior knowledge, a smaller number of significant PC's (called Factors) of the data set is selected before further calculation. Scores and loadings here have the same meaning as in case of the PC's. A drawback of both methods is that the PC's or Factors have an abstract nature, the interpretation of which in terms of properties or applications may be difficult. For proper selection and interpretation of relevant PC's or Factors, sufficient knowledge and an insight into the relationships between input and output data is necessary.

Linear and non-linear regression can be applied to quantify (complex) relationships between input parameters (e.g. PSD characteristics) and output parameters (e.g. tasty, creamy, gritty). These regression methods typically use pre-formulated linear, polynomial or exponential functions. They require a pre-formulated equation (model) for the relationship [6, 14]. The validity of the model equations can be tested, e.g. through the correlation coefficient. Note that it may be useful to reduce the complexity of relationships in order to reduce the number of parameters to be assessed.

Artificial Neural Networks (ANN) provide a special way of non-linear regression [6]. They are inspired by neural nets in living organisms. A typical network consists of an input layer of neurons, an intermediate 'hidden' layer (where non-linear transformations are performed) and an output layer, with weighting factors for the strengths of the connections between the different layers. Here too, the initial relationships in the hidden layer predominantly depend on experience and prior knowledge. The advantage of ANN is that the weighting factors can be adjusted by 'learning' through the use of known relationships between input data and output data. The number of connections between layers can be minimized. For the learning stage least squares criteria are often used, which may account for measurement errors in input and output data and uncertainty effects in the model structure. In ANN, no pre-formulated model equation is required.

Analysis of Variance (ANOVA) is one of the methods for significance and hypothesis testing [20]. Other test methods are the *t*-test, F-test, Grubb's test and Dixon's or Q-test. They are all meant to decide whether a measured value does or does not belong to a defined population of values, at a defined level of confidence (often 95 % probability). In all methods the problem is posed in terms of an a-priori hypothesis that a difference is not significant. Statistical analysis of the experimental results then deduces the probability of the occurrence of that difference. When this probability is considered low, it is concluded that the difference is significant. It can be used to test the significance of the influence of input parameters (e.g. product characteristics) on output parameters (e.g. product performance), or of differences in measured characteristic product parameters of different samples [20, 27].

The final design step is choosing the best equipment (if existing equipment is not the starting point) and operating conditions for manufacture of the product. Here, similar statistical method lines can be followed as for product design. Typically, fairly simple processes are applied, such as (colloid) milling, mixing and emulsification.

After having established – and preferably understood – the various equations relating product performance; its ingredients; composition and its manufacturing process to optimum basic parameters, the next phase in product development regards how to use these basic data for the design of the best product at minimum costs. To this end, the equations, if determined separately, have to be given relative weights for product performance, quality, costs, human safety and environmental impact by the expert team involved. For ingredients and relatively simple products, the weighted relationships can be overviewed subjectively by an expert or expert team to find the best solution. For more complex products, where many relationships exist with sometimes opposite outcomes for some parameters, such subjective consideration becomes increasingly difficult. In these cases adequate weighting of all the relevant equations for product performance, etc. together with implementing them in a mathematical model (the product property function [24]) becomes necessary. The model is, for example, set up as a matrix, as found e.g. in the central matrix of QFD. Usually, a computer with appropriate software is required. *Artificial Neural Networks*, if used in the earlier phase, can be useful. *Bayesian statistics* offer a nice alternative to classical statistics. Computer simulations can be valuable to find optimum solutions.

In our computer age, it is possible to apply a rich variety of methodologies coming from statistics, mathematics and the various decision sciences without being restrained by the complexity and size of the various product designs. In addition, computerized libraries and electronic databases are available. Adequate statistical knowledge is required to choose the optimum tools.

Of course, product chemists and process engineers should apply existing knowledge to interpret results, verify the validity of the resulting equations and identify the necessity and type of any additional or alternative ingredients.

The above statistical methods can also be applied for optimization of existing production processes, through directly relating customer desires to the process parameters. An example of ANN is given for ice cream in [7]. This direct approach has the advantage that the number of error sources is minimized and, thus, the confidence intervals of the control parameters optimized. A drawback for the design of new products is that it does not address the influence of specific (e.g. particulate) characteristics of both ingredients and product and, thus, excludes available knowledge in this field.

The essential elements in the overall development process of new or improved particulate products appear to be:

1. Explicit, clear definition of desired performance qualities.
2. Effective organization and efficient planning of activities.
3. Explicit, careful selection and definition of the desired performance parameters and of related characteristic (particle PSD, shape, concentration, etc.) parameters that can be measured with sufficient, stated quality.
4. Sufficient acquisition of quantitative knowledge that relates these parameters (for use in product design and specifications, and in process control).
5. Adequate selection and optimization of production process(es) and process control parameters.
6. Optimum set up of experiments, statistical evaluation of R&D results and weighting of relationships between performance quality parameters, characteristic PSD, etc. parameters, formulation, costs, human safety and environmental impact.
7. Final choice of best product, its specifications, process equipment, operating conditions and process control parameters.
8. Safeguarding of intellectual property of product.
9. Full documentation of all information, experimental results, considerations, conclusions and decisions.

1.6 Set Up of the Book

The main objective of this book is to guide optimum design of particle characteristics of particulate ingredients in commercial products in relation to the required performance characteristics, both during processing and for the final product. Chapter 1 describes the general aspects of the design of (particulate) products in a systematic way. It is followed by Chap. 2 on basic relationships between behavior and performance aspects of particulate products and particle characteristics (size, shape, particulate concentration, porosity, surface area and zeta-potential). Chapter 3 gives an overview of quality aspects and limitations of the most popular measurement techniques for particle characteristics. Since explosion hazards and toxicity to humans and environment may be involved in the production and application of particulate materials, Chaps. 4 and 5 deal with these subjects. The next chapters describe – for a variety of selected products – which information is required and how it is used to reach the desired product quality, viz. ceramics (Chap. 6), concrete (Chap. 7), chocolate (Chap. 8), ice cream (Chap. 9), medical dry powder inhalers (Chap. 10), pharmaceutical nanopowders (Chap. 11), pigment and paint dispersions (Chap. 12), powder coatings (Chap. 13), sunscreen (Chap. 14), and cheese coatings (Chap. 15). Although these product fields differ strongly, some of the basic characteristics of the processing stage or the final product quality can be compared. They provide examples for product design as it stands today.

1.7 Definitions, Abbreviations and Symbols

Aspect ratio	Ratio of maximum to minimum Feret diameter of a particle (also often used for ratio of length to breadth for fairly short particles, when breadth is about equal to thickness)
Equivalent sphere	Sphere (or collection of spheres) that has the same property as the observed particle(s) in relation to a given measurement principle
Mean/average size	Arithmetic mean particle size for a given population of particles, weighted according to number, area, volume, etc.
Median size	Particle size at the point in a cumulative size distribution, where 50 % of the particles is smaller and 50 % larger.
ANN	Artificial neural networks
ANOVA	Analysis of variance
ASTM	American Society for Testing and Materials
CAD	Computer aided design
CN	Customer needs
D	Particle size (equivalent diameter from a defined measurement technique)
$D_{50;3}$	Median size (50th percentile) of a volume-based PSD; other percentiles of the cumulative distribution are indicated by related first subscripts. The second subscript relates to the basis of the PSD (viz. number ($_{;0}$), volume ($_{;3}$), etc.)
$D_{90;0}$	Nineteith percentile of a number-based PSD
EN	Written standard of European community
FI	Formulation ingredients
HOQ	House of quality
IS	Ingredient specifications
ISO	International standards organization

(continued)

(continued)

PC	Principle component
PCh	Product characteristics
PCI	Process conditions ingredients
PPC	Production process conditions
PS	Product specifications
PSD	Particle size distribution
QFD	Quality function deployment

References

1. Ando, T.: Bayesian Model Selection and Statistical Modeling. Chapman & Hall/CRC, Boca Raton (2010)
2. Antony, J.: Design of Experiments for Engineers and Scientists. Butterworth-Heinemann, Amsterdam (2008)
3. Azzam, M.I.S., Jutan, A., Rodriguez, E.: Particle size characterization of a jet mill product via principal component analysis. Part. Sci. Technol. **6**, 95–104 (1988)
4. Bech, A.C., Hansen, M., Wienberg, L.: Application of house of quality in translation of consumer needs into sensory attributes measurable by descriptive sensory analysis. Food Qual. Prefer. **8**, 329–348 (1997).
5. Benner, M., Linneman, A.R., Jongen, W.M.F., Folstar, P.: Quality function deployment (QFD) – can it be used to develop food products? Food Qual. Prefer. **14**, 327–339 (2003)
6. Bhadeshia, H.K.D.H.: Review: Neural networks in materials science. ISIJ Int. **39**, 966–979 (1999)
7. Bongers, P.M.M.: Model of the product properties for process synthesis. Comp. Aided Chem. Eng. **25**, 55–60 (2008); ESCAPE 18, 18th Eur. Symp. Comp. Aided Process Eng.; B. Braunschweig, X. Joulia (eds.); 2008 Elsevier BV/Ltd
8. Chang, M.: Modern Issues and Methods in Biostatistics. Springer, New York (2011)
9. Cohen, L.: Quality Function Deployment: How to Make QFD Work for You. Addison-Wesley, Reading (1995)
10. Congdon, P.: Bayesian Statistical Modelling. Wiley, New York (2001)
11. Costa, A.I.A., Dekker, M., Jongen, W.M.F.: Quality function deployment in the food industry: a review. Trends Food Sci. Technol. **11**, 306–314 (2001)
12. Cussler, E.L., Moggridge, G.D.: Chemical Product Design. Cambridge Univ. Press, New York (2011)
13. Davis, J.C.: Statistics and Data Analysis in Geology. Wiley, New York (1973)
14. Freedman, D.A.: Statistical Models: Theory and Practice. Cambridge Univ. Press, New York (2009)
15. Gani, R., Abildskov, J. In: Bröckel, U., Meier, W., Wagner, G. (eds.) Product Design and Engineering, vol. I, Basic Technologies. Wiley-VCH, Weinheim (2007)
16. Hendrix, C.D.: What every technologist should know about experimental design. Chem. Tech. 167–174 (1979)
17. Hockman, K.K., Berengut, D.: Design of experiments. Chem. Eng. 142–147 (1995)
18. Koch, K.R.: Introduction to Bayesian Statistics. Springer, New York (2007)
19. Linneman, A.R., Benner, M., Verkerk, R., van Boekel, M.A.J.S.: Consumer-driven food product development. Trends Food Sci. Technol. **17**, 184–190 (2006)
20. Merkus, H.G.: Particle Size Measurements; Fundamentals, Practice, Quality. Springer, Dordrecht (2009)
21. Moskowitz, H., Hartmann, J.: Consumer research: creating a solid base for innovative strategies. Trends Food Sci. Technol. **19**, 581–589 (2008)

22. Ng, K.M., Gani, R., Dam-Johansen, K. (eds.): Chemical Product Design: Towards a Perspective through Case Studies. Elsevier, Boston (2007)
23. Park, T., Kim, K.-J.: Determination of an optimal set of design requirements using house of quality. J. Oper. Manage. **16**, 569–581 (1998)
24. Polke, R., Schäfer, M. Chapter 5, Characterization of disperse systems. In: Bröckel, U., Meier, W., Wagner, G. (eds.) Product Design and Engineering, vol. I, Basic Technolgies. Wiley-VCH, Weinheim (2007)
25. Seaver, M. (ed.): Gower Handbook of Quality Management. Gower, Burlington (2003)
26. Seider, W.D., Seader, J.D., Lewin, D.R., Widagdo, S.: Product and Process Design Principles. Wiley, New York (2010)
27. Summers, D.C.S.: Quality. Prentice Hall, Boston (2010)
28. Taylor, M.K., Ginsburg, J., Hickey, A.J., Gheyas, F.: Composite method to quantify powder flow as a screening method in early tablet or capsule formulation development. AAPS Pharm. Sci. Tech. **1**, art.18 (2000)
29. Temponi, C., Yen, J., Tiao, W.A.: House of quality: a fuzzy logic - based requirements analysis using the house of quality. Eur. J. Oper. Res. **117**, 340–354 (1999)
30. Ulrich, K.T., Eppinger, S.D.: Product Design and Development. McGraw-Hill Comp., Ney York (2008)
31. Vairaktarakis, G.L.: Optimization tools for design and marketing of new/improved products. J. Oper. Manage. **17**, 645–663 (1999)
32. Wei, J.: Product Engineering: Molecular Structure and Properties. Oxford Univ. Press, New York (2007)
33. Wesselingh, J.A., Kill, S., Vigild, M.E.: Design and Development of Biological, Chemical, Food and Pharmaceutical Products. Wiley, Hoboken (2007)
34. Wibowo, C., Ng, K.M.: Product – oriented process synthesis and development: creams and pastes. AIChE J. **47**, 2746–2767 (2001)

Chapter 2
Basic Information for Design of Particulate Products

Henk G. Merkus

Good correlation between parameters is fine,
Some added understanding is better.

Abstract In this chapter some basic features of particulate products are examined, which are relevant for adequate design and best performance. Particle packing and product porosity is considered first. Closest packing is reached within size distributions, where small particles fill the voids in between the larger particles. Smallest void fractions in monosized powders are reported to be about 40 % (related solids content 60 %); the ideal void fraction is 26 %. For polydisperse powders, especially those containing polymodal size distributions minimum void fractions may be considerably smaller and maximum packing densities larger. Powder flow and cohesivity strongly depend upon particle size and shape. Various empirical values for the characterization of flowability are presented. In general, it can be said that dry powders having a particle size above 50 μm and regular particle shape, flow easily. When the particles are much smaller and/or when they have a large aspect ratio, they tend not to flow at all: they are called cohesive. The fluidization of powders follows a similar trend where granular particles in the 20–200 μm size range fluidize readily and smaller particles together with ones of high aspect ratio, can cause difficulties. Geldart has categorized powders as cohesive, aeratable, bubbling and spouting; the latter two words mean that particles larger than about 200 μm show another 'fluidization' behavior. Equations for fluid flow through packed beds are developed by different authors. The equation of Carman and Kozeny is used in filtration. The viscous behavior of emulsions and suspensions is strongly related to particulate concentration. At low concentrations, the viscosity shows a Newtonian behavior; it is independent of particle size and

H.G. Merkus (✉)
Retired Associate Professor, Delft University of Technology, Park Berkenoord 30, Pijnacker 2641 CZ, The Netherlands
e-mail: henkmerkus@hetnet.nl

H.G. Merkus and G.M.H. Meesters (eds.), *Particulate Products: Tailoring Properties for Optimal Performance*, Particle Technology Series 19,
DOI 10.1007/978-3-319-00714-4_2,

relates linearly with concentration. At high concentrations, various types of non-Newtonian behavior exist. Viscosity depends on particle size (distribution), but is no longer linearly related to particulate concentration. Differences between solids concentration and maximum packing density of the particles play an important role. Moreover, rheological behavior may be time-dependent through changing particle-particle or molecular interactions. Adequate non-Newtonian behavior is essential for good performance during production and for the final quality of various products, such as chocolate, ice cream and paint dispersions. Long-term stability of dispersions often relates to the zeta-potential. Absolute values of at least about 50 mV can cause stable dispersions. At low zeta-potential values, the particles tend to agglomerate or flocculate. Dispersion stability may be improved by application of steric stabilizers. In paint, color, scattering, transparency and hiding power of dispersed particles are also strongly related to particle size and shape. Several equations are provided that are useful in the design of pigment mixtures. These aspects play a role in human appreciation of paints, foods and creams. Sensorial characteristics are important in foods, sweets, beverages and cosmetics. Some terminology is provided. Interpretation of some of the terminology is subjective. This is due to the characteristics not only having a physical background but may also relate to particle size and shape, psychological or cultural aspects. Finally, this chapter deals with the adsorption and diffusion of components into the pores of adsorbents and catalysts. In such cases the diffusion rate is strongly linked to the pore size distribution and the effective molecular size of the diffusing component.

2.1 Introduction

Particle characteristics play an important role in the quality aspects of particulate products. In general, they have an effect on e.g. particle packing, sensorial aspects, transparency and hiding power as well as on adsorption, catalysis and separations. In the case of dry powders, there is a relationship with powder flow and fluidization. For emulsions and suspensions, they affect rheological properties in a positive or negative way. These effects and existing relationships will be treated in the following paragraphs.

2.2 Particle Packing and Product Porosity

Particle packing is of fundamental importance for powders. It has influence with respect to the mass of powder in a given volume and thus the dose of drug in a tablet. It also influences their flow properties, which in turn affects powder operations like the flow from hoppers. Moreover, a higher degree of packing causes less shrinkage during sintering and better product strength in ceramics and concrete. Particle packing is usually expressed as (effective) powder density, product porosity or void fraction. A higher degree of packing means a higher powder density and less voids between the particles, and vice versa. The void fraction ε is defined by:

$$\varepsilon = \frac{V_V}{V_B} = 1 - \frac{V_P}{V_B} = 1 - \frac{\rho_B}{\rho_e} = 1 - \frac{M}{\rho_e V_B} \tag{2.1}$$

where:

ε = void fraction in powder bed (powder bed porosity)
V_V = volume of voids
V_B = volume of powder bed
V_P = volume of particles in bed
ρ_B = effective bulk density of powder bed
ρ_e = envelope particle density (including all intra-particle pores)
M = mass of powder bed

The degree of particle packing is related to the number and area of contacts between particles and is governed by the attractive and the frictional forces between the particles. The way that a packing is prepared has an important influence on the degree of packing. For example, tapping (vibration) usually enables particles in a loose powder to take positions in closer proximity. Tapped powder density can be considerably larger than the density in loose packing. Particle size, size distribution and particle shape may have a substantial influence (see below).

The closest packing of monosized spheres contains about 26 % voids, regardless of particle size [8, 51]. Thus, about 74 % of the overall volume is taken by the particles. This maximum is reached in a regular, rhombohedral particle arrangement, where each sphere touches 12 neighbors. It can only be reached by careful building of the particle arrangement, but never in general powders. In the case of large, monosized, smooth spheres, friction between particles causes the packing density to be considerably lower than the above maximum, and thus powder porosity is larger. Maximum density for random packing, obtained by careful vibration, appears to be 63.7 % v/v [69]. Less careful mixing, imperfect spheres and surface roughness increasing packing resistance, results in still lower packing densities with void fractions larger than 40 % v/v [8, 77]. The influence of particle size becomes important for sizes below about 20 μm. Decreasing particle size leads to stronger attractive forces (mainly Van der Waals- type for dry powders) between particles, also causing more difficult particle rearrangements. Typical porosity of dry particle beds increases for particles smaller than about 20 μm. Note that presence of moisture strongly increases adhesive forces between particles.

In the case of free-flowing powders (size distribution width, D_{90}/D_{10} 1.5–10 and particle size D_{90} above about 20 μm), the inter-particle voids may be filled, in principle, with smaller particles. In practice, a minimum void fraction of about 25 % v/v can be attained under these conditions (75 % v/v solids) [8, 51, 77]. Optimum polymodal mixtures may result in still higher densities for random packing, up to about 85 % v/v. Size ratios as well as relative quantities of the modes should be attuned in such a way that the finer particles exactly fill the voids left by the larger particles [19, 55].

If a high degree of packing, approaching the maximum, is desirable as in ceramics or concrete, it may be reached through the addition of a liquid to form a paste, which reduces the inter-particle friction.

Small amounts of specific solid ingredients may give the same result as well as reducing the liquid content (see e.g. Sect. 2.3 and Chap. 7).

As stated above, the density of particle packing in powders decreases progressively with increasing ratio of attractive forces between particles and inter-particle friction to the gravity force for particles. The attractive forces and inter-particle friction depend upon surface area (size squared) and particle shape. The gravitational force depends upon mass (size cubed). Therefore, powder cohesivity increases with decreasing particle size and increasing surface roughness. An empirical relationship between powder porosity and particle size, has been reported for the gravitational packing (both poured and tapped) for irregularly shaped fused alumina powders, having a particle size greater than 1 μm [87].

$$\varepsilon = \varepsilon_0 + (1 - \varepsilon_0)\exp\left(-aD_{50;3}^{b}\right) \tag{2.2}$$

where:

ε = powder bed porosity (void fraction in powder bed)
ε_0 = limiting porosity (in either tap density or some effective bed density)
$D_{50;3}$ = median particle size of volume-based size distribution
a = powder packing factor (depends on way of packing)
b = particle shape factor

For cohesive powders, where attractive forces between the particles are large, the degree and ease of packing depends on the size and strength of existing agglomerates. This size can be expressed in an equivalent packing size [87]. A power-law relation was determined between the ratio of equivalent packing size[1] $r_{i,j}$ between components i and j and the corresponding equivalent volume diameter $R_{i,j}$ for bimodal mixtures:

$$r_{i,j} = R_{i,j}^{p} \tag{2.3}$$

where:

p = powder packing exponent (this depends on way of packing and strength of attractive forces)

Note that the constants in the above equations depend on the interactive forces between the particles and, thus, depend on the product composition.

[1] The equivalent packing size represents the size of agglomerates, which act as (composed) particles during packing.

The effect of particle shape on the degree of packing becomes significant when the particles have a flat or rough surface or a large aspect ratio, such as with plates and fibers. Such particles tend to be more cohesive than spheres or irregular particles. One reason is the increased surface for inter-particle contacts, which leads to increased attractive forces. A rough particle surface causes larger friction between particles. Fibers have a larger difficulty for free movement due to their length being prone to causing entanglement. Extreme examples are fibers of glass or asbestos. Thin-cut potato crisps with their curved platelet structure, provides a nice example of the influence of particle shape. The irregular structure of the conventional type causes an extremely loose packing. The structure of a newer type (Kellogg's Pringles) where all crisps have the same shape, on the other hand, allows a very close (platelet-like) packing, but only after careful construction of the packing. Flat mica platelets are capable of extremely close packing as realized by nature.

2.3 Powder Flow and Fluidization

Flowability, cohesivity and fluidization are important characteristics of dry particulate materials. Critical examples are emptying of hoppers, mixing or dosing of powders, or use of particulate materials in catalytic processes. A particle size of at least about 20–50 μm and regular particle shape are required for good flow behavior. Note that the presence of moisture in powders may also cause poor flow properties through hydrogen bonding or capillary forces [17].

A. **Powder Flow**

Dry powders show large differences in flow behavior. Some flow very easily; others flow not at all: they are called cohesive. The differences depend upon particle size, shape and surface roughness. Although there has been considerable progress in understanding bulk powder behavior, much work needs to be done to develop the functional relationships between particles and bulk powders with respect to flowability [5, 20, 51, 54].

Differences in flowability/cohesivity of dry powders can easily be seen by comparing powdered sugar used for icing cakes (particle size about 1 μm) and crystal sugar (particle size about 400 μm). The former behaves as a cohesive powder; the latter is free-flowing, which makes flow and dosing much easier. The same holds for detergents (dry powder versus compacted pearls). In the general case of dry powders, the differences are caused by Van der Waals forces, which are related to the surface-to-volume ratio of the particles. For hygroscopic powders, such as detergents, flow properties are often deteriorated by the presence of moisture, due to their hygroscopic nature.

The surface-to-volume ratio increases with decreasing particle size, which in turn causes an increased inter-particle attraction. If the powder has undergone some compaction, by its own weight and/or by vibration or tapping, then the number of contacts between the particles increases. This can not only lead to stronger cohesivity but at its limit significantly increases the mechanical strength of the compressed powder. For particles smaller than about 20 μm and having an aspect ratio of about 1, the attractive forces cause agglomeration of the particles. This means that such powders are cohesive and do not flow easily. If de-agglomeration of such particles is needed, strong shear forces are required, until the point where particle bonding in the clumps is too strong. Sometimes, a small amount of sub-micrometre particles (e.g. fumed silica) can be added to improve flowability. For particles larger than about 50 μm and having an aspect ratio of about 1, the attractive forces are insignificant and the resistance to flow is only caused by friction. Then, these powders are free-flowing. Segregation in a batch of particles, according to size may be expected. Increased aspect ratios, such as in the case of fibers and plates, also causes decreased flowability, as the number of contact points between particles is increased.

In practice, the Carr Compressibility Index *CI* or the related Hausner Ratio *H* are often used to indicate the both compressibility and flowability [9, 81]:

$$CI = 100\frac{V_f - V_t}{V_f} = 100 \times \left(1 - \frac{\rho_f}{\rho_t}\right) = 100 \times (1 - 1/H) \tag{2.4}$$

and:

$$H = \frac{V_f}{V_t} = \frac{\rho_t}{\rho_f} \tag{2.5}$$

where:

CI = Carr Compressibility Index (%)
H = Hausner Ratio
V_f = freely settled volume per unit mass of powder
V_t = tapped volume per unit mass of powder
ρ_f = freely settled bulk density of powder
ρ_t = tapped bulk density of powder

A simple, popular test for flowability is the measurement of the angle of repose of a cone-like heap of powder after dropping it onto a fixed base plate from a funnel at some elevation. The angle of spatula, the Jenike and similar shear tests, the flow rate through an orifice (the latter only for free flowing materials) and avalanching tests are also used [21, 36, 37, 45, 46, 53, 82]. For the mass flow from a hopper, the wall friction and the corresponding critical angle also influences the flowability [52]. Some values for classification of the flow properties of powders are given in Table 2.1.

Table 2.1 Classification of flow properties of powders

Flow property	Carr index (%)	Hausner ratio	Angle of repose, degrees
Excellent	≤ 10	1.00–1.11	25–30
Good	11–15	1.12–1.18	31–35
Fair; aid not needed	16–20	1.19–1.25	36–40
Passable; borderline	21–25	1.26–1.34	41–45
Poor; must agitate, vibrate	26–31	1.35–1.45	46–55
Very poor	32–37	1.46–1.59	56–65
Very, very poor	≥ 38	≥ 1.60	> 66

For example, Carr Index < 15, Hausner Ratio < 1.18 and Angle of Repose $< 35°$ are considered as indications for good flowability, compared with Carr Index > 25, Hausner Ratio > 1.34 and Angle of Repose $> 45°$ for poor flowability.

Compression (consolidation) of a powder, which already may occur during storage under its own weight, will decrease flowability, as the number of point contacts and Van der Waals forces increase. The effect of consolidation can be tested in shear cells.

Note that small amounts of moisture increases cohesivity of powders significantly [53]. Typically, flowability gets worse with increasing moisture content due to formation of liquid bridges between particles. At a certain moisture content, maximum cohesivity is reached. At still higher moisture contents, the liquid starts acting as a glidant and flowability increases. When all voids are filled, the powder reaches the state of a paste or a slurry.

Taylor et al. [81] have applied principal component analysis to reveal a flowability index for pharmaceutical powders based on measurements of critical orifice, compressibility and angle of repose. They conclude that a combined index provides a better characterization of powder flow than the individual tests. This is due to the fact that the different tests challenge different aspects of the complex flow behavior of powders.

Extensive research has not yet lead to an adequate understanding and basic relationships between the flowability of powders and the size and shape characteristics of the constituting particles.

B. **Fluidization**

When a fluid is passed upwards through a packed bed of particles that is unrestrained at its upper surface, there is an increase of the frictional resistance and pressure loss in the fluid with increasing flow rate. At some point, the upward drag force of the fluid on the particles becomes equal to the mass of the particles in the bed and the pressure drop remains constant with increasing flow. Often some excess pressure is required to free the particles from adhesion or interlocking. Four groups of powders can be distinguished, as shown in the classical Geldart diagram. The distinction has been found empirically for powders when exposed to dry air at ambient pressure and temperature (Fig. 2.1) [20, 84].

Type C powders, with particle size smaller than about 20–80 μm (depending on density difference ρ_P-ρ_F of particles and fluid) are cohesive. Such powders cannot be fluidized by a stream of fluid. When subjected to a gas flow, the gas passes in channels through the powder or lifts slugs of particles. For these powders, the attractive forces between particles are much greater than particle mass [84].

Type A powders with larger particles, sizing from tens to hundreds micrometers, are aeratable. They can be fluidized in an upward gas flow above minimum fluidization velocities. In these powders, the attractive forces between the particles are in the same order of magnitude as their mass. Upon aeration of the powder, the powder bed expands considerably without bubbles being formed. A fluidized bed behaves like a liquid, in which the particles are subject to weak attractive forces [21]. This type of

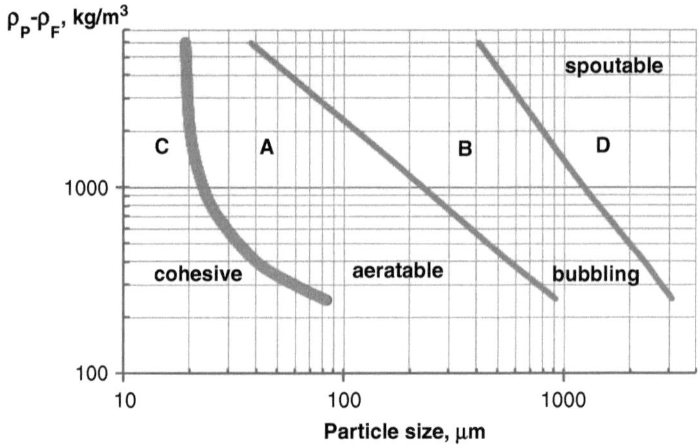

Fig. 2.1 Geldart fluidization diagram (Adapted from [20])

behavior is important e.g. in catalytic reactions where it provides optimum contact between gaseous reactants and solid catalyst particles, in addition to adequate temperature control.

Type B powders, again larger particles start forming bubbles as soon as the lifting force of the fluid velocity exceeds the particle mass. Particle mass is much greater here than inter-particle forces. The bubbles coalesce as they rise through the bed. When they take up the size of the cross section of the bed, they may cause 'slugging', i.e. lift a section of the bed contents over a short distance until the particles drop down again.

Type D powders contain still larger particles. Bubbles are also formed here but the required fluid velocity is so high and the particles so heavy, that these 'bubbles' bypass the solids in so-called spouts. These spouts are fluid jets, which entrain some particles upwards and let them fall back again later at the sides of the jets.

In all cases, particles are blown out of the bed at very high fluid velocities, when the lifting forces are larger than particle mass.

For the fluidization regime, it can be derived that [21, 64, 70]:

$$\frac{\Delta P}{H_b} = \left(1 - \varepsilon_{fl}\right)(\rho_P - \rho_F)g \tag{2.6}$$

where:

ΔP = pressure drop over bed
H_b = bed height
ε_{fl} = void fraction during fluidization
ρ_P = effective density of particles
ρ_F = density of fluid
g = gravitational acceleration constant

Several empirical relationships have been derived for the minimum fluidization gas velocity in relation to particle size [64].

The understanding of fluidization behavior has significantly increased in the last decades and methods for improvement for cohesive powders have been published [84]. For example, some powders can sometimes be fluidized through weakening of the attractive inter-particle forces, i.e. by addition of a small amount of nano-particles that adhere to the particulate surface, or through modification of temperature, fluid viscosity or pressure. Additionally, mechanical vibration or application of an external electric or magnetic field have been applied. It has also been found that some nano-particles aggregate/agglomerate to form clusters having sizes of tens of micrometers and low density, which can be fluidized.

2.4 Fluid Flow Through Particulate Beds

Darcy has found that the volumetric flow rate of a liquid in a channel is proportional to the pressure drop and inversely proportional to the liquid viscosity under laminar flow conditions [1, 8, 22, 64]:

$$\Phi_v = \frac{\Delta P}{\eta_L \cdot K} \tag{2.7}$$

where:

Φ_v = volumetric flow rate
ΔP = applied pressure drop
η_L = liquid viscosity
K = resistance constant depending on channel geometry

Later, Hagen and Poiseuille derived (independently) a similar but more specific equation for laminar flow of fluids through circular capillaries:

$$< u >= \frac{d^2}{32\eta_F} \frac{\Delta P}{L} \tag{2.8}$$

where:

$<u>$ = mean linear fluid velocity
d = diameter of capillary
η_F = fluid viscosity
L = length of capillary

The volumetric flow rate Φ_v can be derived from $< u >$ since:

$$\Phi_v = \frac{\pi \cdot d^2 \cdot < u >}{4} \tag{2.9}$$

Later again, Carman and Kozeny worked out an equation for laminar fluid flow through a non-compressible, randomly packed fixed bed of solid, monosized, spherical particles by considering the bed pores equivalent to a set of parallel capillaries. They assumed the particle surface of the bed, which relates to the particle diameter squared, per unit volume of pore space to be the important factor for bed permeability:

$$< u_s >= \frac{D^2 \cdot \varepsilon^3}{180\eta_F(1-\varepsilon)^2} \frac{\Delta P}{H_b} \qquad (2.10)$$

where:

$<u_s>$ = mean linear superficial fluid velocity
D = spherical particle diameter
ε = bed porosity (void fraction)
H_b = bed height

Note that the constant factor 180 is an approximation and includes factors relating to particle shape and channel tortuosity. The value of this 'constant' may depend somewhat on particle size range and bed porosity.

In the case of non-spherical particles in a particle size distribution, particle diameter D should be replaced by the surface weighted (or Sauter) mean diameter $D_{3,2}$ in Eq. 2.10. The Kozeny-Carman equation is most often used for estimating conditions in industrial filtration.

In the above equations, viscosity only depends on the type of fluid and the temperature.

2.5 Rheology of Emulsions and Suspensions

The rheology of industrial emulsions and suspensions is of particular interest in relation to processability, stability and applicability of such products. Moreover, some sensorial aspects relate to rheological behavior (see Sect. 2.8).

The Newtonian flow model, which is used to describe the flow of a gas, a liquid or a dilute dispersion, is the most well-known. The Eqs. 2.7, 2.8, and 2.10 are special cases of the Newtonian flow. In this model, the required force (shear stress) to maintain flow is proportional to the shear (strain[2]) rate, with the viscosity as the proportionality constant. In other words, the viscosity is a measure of the resistance of a fluid to flow [22, 25, 28, 49, 51, 80]:

[2] For solids, strain can be seen as deformation.

$$\tau = \eta_L \left(\frac{d\gamma}{dt} \right) \tag{2.11}$$

where:

τ = shear stress (force per unit area)
η_L = liquid (Newtonian) viscosity
$(d\gamma/dt)$ = shear rate (or shear strain rate)

The rheological behavior of particulate dispersions depends on the particulate concentration and – at high particulate concentrations – also on the particle size, size distribution (PSD) and particle shape.

(A). **Particulate concentration**

The viscosity of dilute dispersions of spherical particles increases with the concentration of particles [14, 22, 28, 48, 49, 51, 55, 65, 71, 72]:

$$\eta_d = \eta_L(1 + 2.5\Phi) \tag{2.12}$$

or:

$$\eta_{rel} = \eta_d / \eta_L = 1 + 2.5\Phi \tag{2.13}$$

where:

η_d = viscosity of dispersion
η_L = viscosity of pure liquid
η_{rel} = relative viscosity of dispersion
Φ = effective volume fraction of dispersed phase (involving the hydrodynamic diameter)

The constant factor 2.5 in the equations is named the Einstein constant [15]. Other names for this constant are intrinsic viscosity or limiting viscosity number of particles, i.e. their contribution to the viscosity of a dispersion. This linear relationship only holds for dilute dispersions of rigid spheres, which have the same size and no interaction. In such cases, inter-particle distance must be large and concentration low (less than about 5 % v/v). Moreover, the proportionality constant of 2.5 in Eqs. 2.12 and 2.13 may have to be adapted somewhat in relation to particle shape, for shapes that deviate significantly from spherical. No dependency on particle size has been found in the absence of particle-particle interaction.

Electrical interactive forces between particles may lead to an increased viscosity in comparison to Eqs. 2.12 and 2.13, through the electro-viscous effect. In such cases the intrinsic viscosity (above: 2.5) may have to be doubled, even at concentrations below 5 % [28].

In most dispersions having an intermediate concentration range (up till about 30 % v/v), particulate concentration oversteps the limits for a linear relationship as particle-particle interactions may occur. For the case of rigid, non-interacting spheres, an extended form of the Einstein equation has been derived, known as the Brinkmann-Roscoe equation [6, 16, 66]:

$$\eta_{rel} = (1 - \Phi)^{-2.5} \tag{2.14}$$

Note that, for small particulate concentrations, Eq. 2.14 is similar to Eq. 2.13.

More complex equations have been proposed, but all appear to lack general validity [63, 71, 72]. In the absence of a general theory, the relation between the relative viscosity and particulate concentration is often approximated by a polynomial [13, 28, 48, 71, 72, 76]:

$$\eta_{rel} = 1 + c_1 \cdot \Phi + c_2 \cdot \Phi^2 + c_3 \cdot \Phi^3 + \ldots. \tag{2.15}$$

Here, c_1, c_2, c_3 are proportionality constants of an estimated polynomial, in which the squared term may be assumed to account for binary particle interactions, the cubed term for ternary interactions, etc.

In general, the viscosity not only depends on particle size and particle size distribution but also upon the type of fluid, electrolyte, surfactant, zeta-potential of particles, particulate concentration, temperature and interactive forces (Van der Waals or electrostatic). The relationship with particle size is an inverse one: smaller particles cause higher viscosity at the same concentration, due to higher number per unit volume and larger surface area. This dependency is expressed in the constants c_1, c_2, etc., which strongly depend upon the components in the product and the conditions of the application.

For colloid dispersions close to maximum packing (see Sect. 2.2), several equations relate the viscosity to the relative fraction of particulates in comparison to the maximum packing. This fraction also depends upon the size distribution, as indicated earlier. One example is provided by the Krieger-Dougherty equation [41]:

$$\eta_{rel} = \left(1 - \frac{\Phi_e}{\Phi_m}\right)^{-c\Phi_m} \tag{2.16}$$

where:

Φ_e = effective particulate fraction
Φ_m = particulate fraction of maximum packing
c = constant (intrinsic viscosity; often close to the Einstein value of 2.5)

Note that the effective fraction of particulates in dispersions is often increased by adsorbed molecules (dispersant and liquid) and by electric and/or steric repulsion between particles.

The general case of concentrated industrial mixtures is even more complex, as they often show non-Newtonian behavior, i.e. the dependency of viscosity on shear rate and exposure time. Here too, the difference between actual particulate concentration and the concentration in densest packing determines the rheological behavior. In fact, electric and steric effects, relevant to particle-particle interactions, as well as additives are often used deliberately to cause some kind of network 'structure' in the product. In these circumstances a sensitivity exponent has to be added to the shear rate ($d\gamma/dt$). This is often indicated as [22, 23, 39, 50, 51, 55]:

$$\eta_m = \eta_0 \left(\frac{d\gamma}{dt} \right)^{m-1} \tag{2.17}$$

where:

η_m = apparent viscosity of complex mixture
η_0 = inherent viscosity without shear (also called consistency index)
m = shear rate sensitivity exponent (also called flow index)

The equation indicates the non-linear relationship between shear stress and shear rate in non-Newtonian rheological behavior, meaning that the apparent viscosity depends upon the shear rate. This behavior is caused by the fact that three-dimensional network 'structures' are present that may break down and rearrange during shear, depending upon the shear rate, particulate concentration and inter-particle forces. Note that emulsion droplets may also deform or coalesce during shear, forming non-spherical or larger droplets. Paint is a good example of the positive effects of the non-linear relationship between viscosity and shear rate of concentrated suspensions during its application. The viscosity should be low enough for good mixing and painting with a brush (when applying a fairly small shear rate), but high enough to remain on the surface without dripping (in the absence of shear). It will be clear that Newtonian flow behavior is characterized by $m = 1$. Several cases can be distinguished (see Fig. 2.2).

Pseudoplastic or shear thinning fluids occur most often (at small shear rates). They show an apparent viscosity that decreases with increasing shear rate (or in Eq. 2.17: $m < 1$). Suspensions of paper pulp and pigments, yoghurt, ketchup, mayonnaise and polymeric solutions (e.g. Xanthan gum) are typical examples. Such behavior occurs when the inter-particle distance is equal to or smaller than the particle size and the particle-particle interactions are medium strong. The resulting three-dimensional network 'structures' break down gradually at increasing shear rates to form two-dimensional arrangements, which have a lower viscosity.

This characteristic is advantageous to processability, stability and applicability of the product (e.g. while mixing or pouring) [3, 4].

Dilatant or shear thickening fluids show an apparent viscosity that increases with increasing shear rate (or in Eq. 2.17: $m > 1$). This behavior indicates that increasing shear rates cause less freedom for movement of the particles, so that fluid flow becomes more difficult. Typical examples are ceramic pastes, (concentrated) suspensions of starch and beach sand. In fact, most products show this behavior at high shear rates, if their particulate concentration is close to maximum packing. High shear rates cause the formation of three-dimensional particle clusters, within which fluid flow is strongly limited. For example, mixing or pumping may require

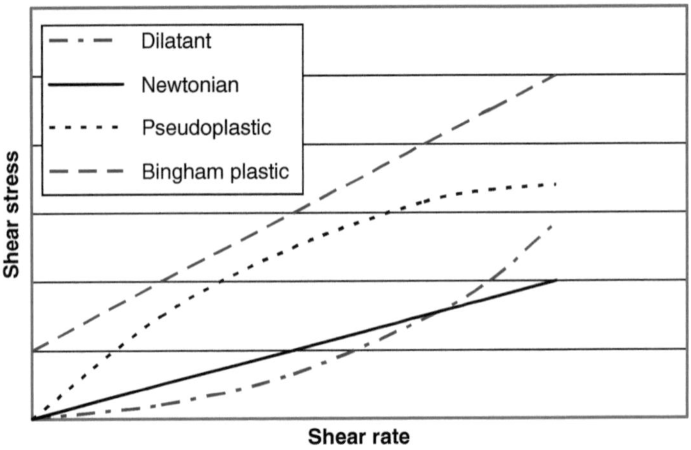

Fig. 2.2 Different types of fluids, where properties are independent of time under shear

much energy or even becomes problematic [3, 85]. Dilatancy, i.e. increase in total volume, often accompanies shear thickening.

Optimum size distributions that promote an increase of maximum particle packing may be investigated to lessen problems (see Sect. 2.5.B) [10, 55, 58, 75]. On the other hand, the same characteristic may be used to improve the performance of composite materials such as shock-absorptive skis and soft body armor [85].

For *Bingham plastic* fluids, such as molten chocolate, aqueous suspensions of rock or clays and sewage sludge, the relating line between shear stress and shear rate does not pass through the origin, since a certain minimum stress (the yield stress) is required to overcome particulate structures. This yield stress is added as a constant to the right hand side of Eq. 2.17 as can be seen in Eq. 2.18:

$$\eta_m = \tau_0 + \eta_0 \left(\frac{d\gamma}{dt} \right)^{m-1} \tag{2.18}$$

At higher shear rates, the relationship between shear stress and shear rate may be linear. Most often, however, Bingham plasticity occurs in combination with shear-thinning or shear-thickening.

The typical viscosity behavior against shear rate – after removal of relaxation and time-dependency – of many concentrated particulate products can be schematically illustrated as in Fig. 2.3 [7, 16, 41].

Typically, four regions can be indicated, for which both the presence and exact structure depends on product composition:

I. Region of zero-shear, Newtonian viscosity (viscosity often typified by η_0)
II. Region of shear-thinning (more or less power law, i.e. linear in double logarithmic plot)

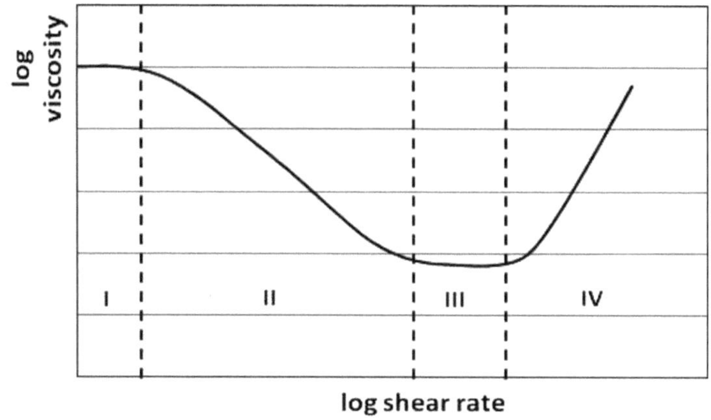

Fig. 2.3 Schematic viscosity – shear rate relationship (relaxation and time-dependency removed)

III. Second region of Newtonian behavior (viscosity often typified by η_∞)
IV. Region of shear-thickening.

Region I corresponds to medium-high particulate concentrations, which – at low shear rates – show no or low dependency of shear rate. In this case, Brownian motion has a dominant influence upon viscosity.

Region II shows shear-thinning behavior, which relates to some kind of alignment of particles. At increasing shear rates, a decrease in viscosity is observed. Here, shear forces between particles dominate the viscosity; this often occurs at high particulate concentrations.

Region III indicates a second period of Newtonian behavior that may occur at high particulate concentrations when shear forces have rearranged the particulate structure to one forming a minimum viscosity.

Finally, region IV corresponds to very high particulate concentrations (near to closest packing). In this state only few particles, or parts of complete particle structures, have the possibility to move. The high shear rate further decrease the possibilities for movement of the mixture.

For some dispersions, the rheological behavior is time-dependent, where behavior depends upon the duration of the shear.

Thixotropic fluids are the first example. They possess a 3-dimensional 'structure' due to inter-particle interactions, which breaks down fairly slowly as a function of time under shear. As a result of this breakdown, shear stress decreases. In other words: the longer the fluid undergoes stress, the lower its viscosity. In the absence of shear, the structure builds up again, which also proceeds fairly slowly. Examples are molten chocolate, some clays, drilling mud and some paints (at high particulate concentrations).

Rheoplastic fluids are the opposite case (also called negative thixotropy). They show an apparent viscosity increase with time when shaken or tapped. Here, some structure builds up during shear. Examples are aqueous sols of bentonite, vanadium pentoxide and gypsum.

In order to obtain typical results in both cases, viscosity measurements are usually executed in steady-state experiments, in which thixotropic effects are counteracted by pretreatment (pre-shearing) of the measurement sample for some (standardized) time period.

Finally, there are *viscoelastic* products. They are intermediate between viscous and elastic products as they show both characteristics [79]. This behavior occurs if the characteristic stress relaxation time τ of the product is comparable to the experimental time t. It can be expressed in terms of the dimensionless Deborah number D_e which relates to both times:

$$D_e = \tau/t \tag{2.19}$$

In case of viscoelastic behavior, D_e equals about 1. Viscous behavior occurs when $D_e << 1$, or $\tau << t$. Elastic behavior is the other extreme; then, $D_e >> 1$, or $\tau >> t$. Elastic objects stretch/deform under stress, but recover their original dimensions when stress is stopped (see a spring as the example). Viscoelastic products, when imposing stress, instantaneously stretch/deform, followed by a slower viscous response. When stress is removed, elastic recovery occurs, followed by a relatively slower (often exponential) viscous recovery. Typically, no full recovery to the initial situation is reached.

Viscoelasticity of products is typically determined through measurement of their response (compliance) in dynamic techniques, viz. from so-called creep curves after sudden application of stress or by oscillatory techniques [79]. The former technique leads to an instantaneous elastic modulus *G(0)*, in addition to a viscous component. Through studying the relaxation after removing the stress, the time-dependency of the elastic modulus G can be evaluated. The response of viscoelastic fluids to oscillatory stress application yields a complex modulus G*, from which the so-called storage (elastic) modulus G' and the so-called loss (viscous) modulus G" can be derived, the latter being the imaginary component of the complex modulus.

Ice cream (cf Chap. 9), blood and polymer solutions are examples of viscoelastic products. An application of viscoelasticity ensures optimum droplet formation from low-viscosity liquids by ink jets. Here, it plays a role by avoiding satellite droplets at high frequencies/speeds [83].

(B). **Particle size distribution**

For concentrated dispersions, the particle size distribution (PSD) influences the rheological behavior in addition to particulate concentration. PSD width and multimodality are important as they are decisive for the maximum packing density of the particles: highest packing densities can be reached in carefully designed multimodal mixtures (see Sect. 2.2). At a given (high) particulate concentration, such increase of the maximum packing density may lead to an enormous decrease of the (apparent) viscosity of the dispersion. This difference between actual and maximum packing density is often used in equations for the rheological behavior (see above). For such concentrated dispersions, steric effects and (temporal) structures of particles as a result of particle-particle interactions play an important role, especially in liquids of (medium) low viscosity. In general, these interactions relate to the particles' surface area. Thus, the Sauter mean diameter, $D_{3,2}$, is often regarded as the most significant parameter to characterize these relationships. This is especially true if strong electrostatic interactions exist between particles [55].

(C). **Particle shape**

For non-dilute solid concentrations, particle shape influences rheological behavior, as it decreases maximum packing density. The effects increase with increasing aspect ratio and are very important for fibers and plates [55]. For the general case of commercial dispersions, the effects are still significant. Note that emulsion droplets may deform from spherical to elongated shapes, which also influences viscosity.

Given all the possibilities of complex rheological behavior mentioned and the absence of a basic theory, it will be clear that adequate measurements, as well as knowledge and understanding of the influence of particulate characteristics on the behavior, are necessary for good design of products and their production processes. This has become ever clearer, since the complexity of many products has increased significantly over the last decades and thus, composition has become critical. Particle size, size distribution, width and modality, particle shape, zeta-potential, particulate concentration as well as added ingredients play their role. Consequently, the influences on rheological behavior are often highly product-specific. All of these parameters need to be optimized, for rheological behavior as well as for other performance aspects, to reach best products in individual cases.

2.6 Emulsion and Suspension Stability

Industrial emulsions and suspensions are generally required to be stable over an appreciable period of time. This means that sedimentation, flocking, agglomeration, coagulation and coalescence should be prevented. In products, this requires a typical particle size smaller than 1–30 μm, depending on particle density and adequate stabilization. In such cases, sedimentation or creaming is virtually absent. Dispersions containing larger sizes also exist. For these larger particle sizes, high-viscosity liquids, thickening agents or high particulate concentrations usually prevent or limit sedimentation effects. Two stabilizing factors can be distinguished in the general case of dispersions [28, 32, 39, 54, 60, 61]:

(a). Repulsive forces between the particles by electric charges and/or steric hindrance induced at the surface of the particles that prevent particles from coming close together. At high particulate concentrations, when the particles are forced to close proximity, e.g. in pastes, stabilizing structures may result from these repulsive forces.

(b). Brownian motion (particle diffusion), if its velocity is much larger than the rate of settling or creaming (at low particulate concentrations).

Table 2.2 Stability behavior of colloidal dispersions

Zeta-potential, mV	Stability behavior
$< \|10\|$	Rapid coagulation/flocculation
$\|10\|–\|30\|$	Instable
$\|30\|–\|60\|$	Moderate/good stability
$> \|60\|$	Excellent stability

Ionic stabilization of aqueous suspensions results from ionization of hydroxyl or oxide groups at the particle surface. This may be through the influence of the pH, and/or by adsorption of polyvalent ions at this surface. Typically, there is also adsorption of some water molecules in the boundary layer. Emulsions are often stabilized by ionic or non-ionic surfactants. These adsorbed species are strongly bonded to the surface so that they remain attached to the particle as the particle moves. Note that this boundary layer represents a real distance but is conceptual, since the adsorbed species, at some small molecular distance from the true surface, exchange relatively fast with species in the dispersion medium and the surface of shear varies slightly with time and location. Thus, the boundary layer and the surface of shear have a statistical nature. Steric hindrance may give another type of stabilization. It is effective if large polymeric molecules are adsorbed at the surface, which prevent other particles from approaching. Ionic and steric effects are often combined for the stabilization of industrial particulate dispersions. The hydrodynamic particle size – i.e. the particle size at the surface of shear when the dispersed particle is moving in a liquid – is usually somewhat larger than the size of the solid particles as it includes the adsorbed layer of molecules. The differences are typically in the range 1–100 nm, depending on electrolyte concentration and dielectric constant. The electric charge at the shear plane of the moving particle is called zeta-potential (ζ-potential). It is considered to be the relevant parameter for describing the interactions between dispersed particles and the stability of the dispersion. Note that the zeta-potential depends strongly on surface properties of particulate material as well as on the ionic concentration and pH of the dispersion. An increase of the electrolyte concentration of the dispersion fluid, causes a decrease of the zeta-potential. Typical dispersion stability characteristics are provided in Table 2.2 (for aqueous media of low viscosity).

The table shows that long-time stability of dispersions requires a zeta-potential larger than about plus or minus 50 mV, causing strong inter-particle repulsion. A zeta-potential around zero will lead to particle clusters through agglomeration, coagulation or flocculation. A low zeta-potential if often arranged to deliberately induce aggregation and flocculation in order to facilitate filtration or sedimentation. It can also be used in cases that all or specific colloidal particles are to be removed from a dispersion.

Flotation, on the other hand, requires a minimum absolute zeta-potential of about 40 mV. This involves the adsorption of anionic or cationic surfactants (with their polar heads) to convert hydrophilic oxidic particle surfaces, into hydrophobic ones (with their non-polar hydrocarbon tails). This, in turn, facilitates the attachment of air bubbles and, thus, particle removal by flotation attached to the froth formed.

At medium/high particulate concentrations, the above particle interaction effects often cause an increase of the apparent viscosity of a suspension or even some kind of 'structure' for the particulate product, by which means the particles experience a considerable resistance to movement. This resistance is sometimes enlarged by the presence of thickening additives in the mixture (see further Sect. 2.5).

The rate of settling or creaming at viscous flow conditions (i.e. Reynolds number for settling particles < 0.25) in dilute liquid dispersions in relation to particle size is given by Stokes' law [54]:

$$v = \frac{H_s}{t} = \frac{(\rho_P - \rho_L) \cdot g \cdot D_{St}^2}{18 \cdot \eta_L} \tag{2.20}$$

where:

v = terminal settling velocity of a particle
H_s = settling height
t = time for settling over height H
ρ_P = effective particle density[3]
ρ_L = liquid density
g = gravitational acceleration constant
D_{St} = equivalent Stokes' diameter of a particle
η_L = liquid viscosity

This Eq. 2.20 is derived for viscous creep flow conditions, which assume that the drag coefficient C_D is inversely related to Reynolds number Re through:

$$C_D = 24/Re \tag{2.21}$$

The settling rate at higher Reynolds number becomes slower than predicted by Stokes' law through an increased drag force [1]. With an increase in Reynolds number, the relationship between the drag coefficient C_D and the Reynolds number Re becomes more complex. Several approaches for solutions in the 'intermediate' region of $0.25 < \mathrm{Re} < 1{,}000$ have been reported, in the form of both equations and tables.

Concentrated suspensions also cause the sedimentation to become slower than the above velocity predicted by Stokes' law [54]. In this case, the liquid density has to be replaced by an apparent density of the suspension, which relates to the particulate concentration (see further Sect. 2.5). Furthermore, some 'porosity' factor for the solids fraction may have to be added in the equation, which reaches a maximum value of one for dilute suspensions. This 'porosity' factor is analogous to that of powders.

[3] The effective particle density to be used in Stokes' law is best determined by a method where a known weight of dispersed particles displaces a measured volume of suspending liquid. This method regards the influence of pores in the same way as in sedimentation. An absolute density value, determined by gas pyknometer, is not appropriate in the presence of pores.

The Stokes-Einstein equation gives the relation between (hydrodynamic) particle size in colloids and their diffusion coefficient in the absence of particle-particle interaction [54]:

$$\mathbb{D} = \frac{k_B T}{3\pi\eta_L D_h} \tag{2.22}$$

where:

$\mathbb{D}$ = diffusion coefficient
k_B = Boltzmann constant
T = absolute temperature
η_L = liquid viscosity
D_h = (effective) hydrodynamic particle size

In the general case of dilute dispersions of low viscosity, the competition between Brownian motion and sedimentation is decisive for dispersion stability. In aqueous media or other media of low viscosity, Brownian motion typically becomes significant for particle sizes below about 1 μm (assuming particle density is low). In such media, sedimentation becomes very slow below about 0.5 μm. Of course, the exact sizes depend on the density of the particles and the apparent viscosity of the media. Note that attractive or repulsive forces between particles occurring at intermediate or high concentrations may limit Brownian motion significantly.

Many industrial particulate products contain some kind of 'structure' for stabilization, due to their particulate concentration or other reasons. Such 'structures' generally dominate the effects of Brownian motion, creaming and sedimentation (see Sect. 2.5).

For settling of aerosols, Eq. 2.20 is slightly modified [26]:

$$v = \frac{H_s}{t} = \frac{\rho_P \cdot g \cdot C_c \cdot D_{St}^2}{18 \cdot \eta_A} \tag{2.23}$$

In this equation, the relatively small air density is omitted, C_c represents the Cunningham slip correction factor and η_A the air viscosity. C_c accounts for the 'slip' between the particle and the air and reduces the value of the Stokes' drag force. It relates to the Knudsen number and becomes important if the particle size approaches the mean free path of the air molecules. It depends upon both particle size (below about 10 μm) and air pressure. Values for C_c have been tabulated [26].

The aerodynamic particle size (D_{ae}) is commonly used for aerosol particles instead of the Stokes' diameter. It assumes a particle density of 1,000 kg/m^3. Thus:

$$D_{ae} = D_{St} \cdot \sqrt{\rho_P} \tag{2.24}$$

2.7 Color, Opacity, Gloss and Transparency

Visible, white light is composed of light having different colors, which have wavelengths from about 380 nm (violet) to 780 nm (red), named the visible spectrum (see Fig. 2.4). This spectrum can be made visible e.g. by a prism, as refraction depends upon the wavelength (Fig. 2.5). In a rainbow, it is caused by refraction of sunlight through the rain drops. In this spectrum, there are three so-called primary colors, viz. blue, green and red. Other colors are formed by combination of these three, as we learn as children when we start producing colored drawings (Fig. 2.6). Note that the human perception of colors depends upon the relative spectral sensitivity of the cones in the retina of our eyes, the so-called luminous efficiency function.

Many materials have the property of absorbing light at specific wavelengths, where the unabsorbed fraction of the incident light is transmitted or reflected. Thus, these materials appear to be colored. The same phenomenon causes these materials to give another color impression if the spectrum of light, falling upon them, for example from public lighting at night, differs from sunlight. Color may be regarded as an impression of the combined intensities of light at different wavelengths, reaching our eyes and interpreted by our brain.

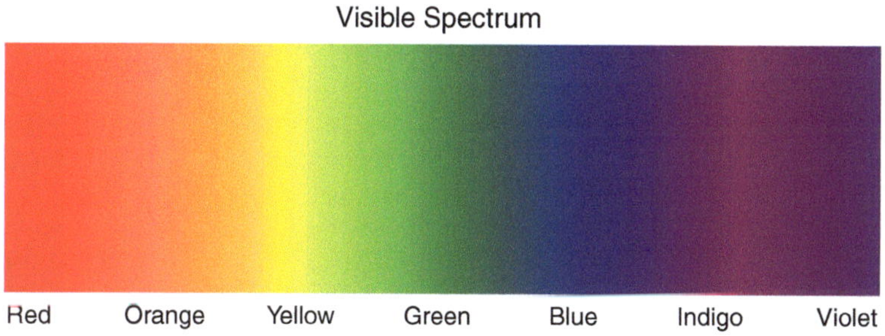

Fig. 2.4 Visible spectrum

Fig. 2.5 Refraction of white light in a prism

Fig. 2.6 Color composition

There are two types of colorant substances, viz. dyes and pigments. Dyes are soluble in the substrate, in which they are dispersed; pigments are insoluble. Both types are typically used to improve the esthetic appeal of objects. The colorant effect of dyes comes from molecular structure, viz. the number of conjugated double bonds. The color of pigments is primarily governed by their chemical composition and crystal structure (see also Chap. 12). These properties determine the refractive index of the particles and, thus, their absorption, reflection and scattering at the different wavelengths of visible white light. Particle size and shape are important as well, in view of their relationship to the degree of scattering of light by the particles [50, 54].

The effects of light absorbance in relation to concentration and path length are quantified in the equation of Beer-Lambert-Bouguer:

$$A = \ln(I/I_0) = \varepsilon_\lambda \cdot c \cdot l \tag{2.25}$$

where

A = absorbance (or light extinction)
I_0 = intensity of incident light of given wavelength
I = intensity of transmitted light at given wavelength
ε_λ = molecular extinction coefficient (at a given wavelength)
c = concentration
l = path length

The gloss of surfaces is related to a uniform reflection of light (angle of incoming light equals angle of reflected light; parallel rays in cause parallel rays out). Rough surfaces or the presence of particles or pores causes reduction of gloss (matting) as the reflection becomes more diffuse.

The opacity of pigments is caused by light scattering and absorption by the composing surface molecules and/or optical heterogeneities (e.g. particles). For e.g. white paints, containing small TiO_2 particles, light scattering is the dominant

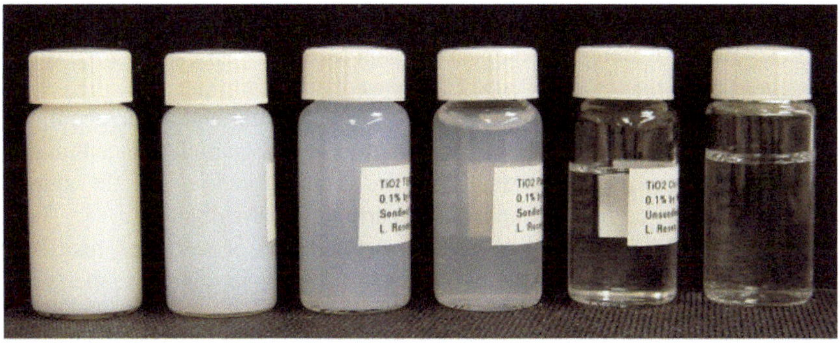

$D_{50;3}$	199 nm	87 nm	44 nm	26 nm	3 nm	pure water
$D_{90;3}$	333 nm	204 nm	74 nm	45 nm	6 nm	

Fig. 2.7 Influence of TiO_2 particle size on turbidity (0.1 % w/w in water) [68]

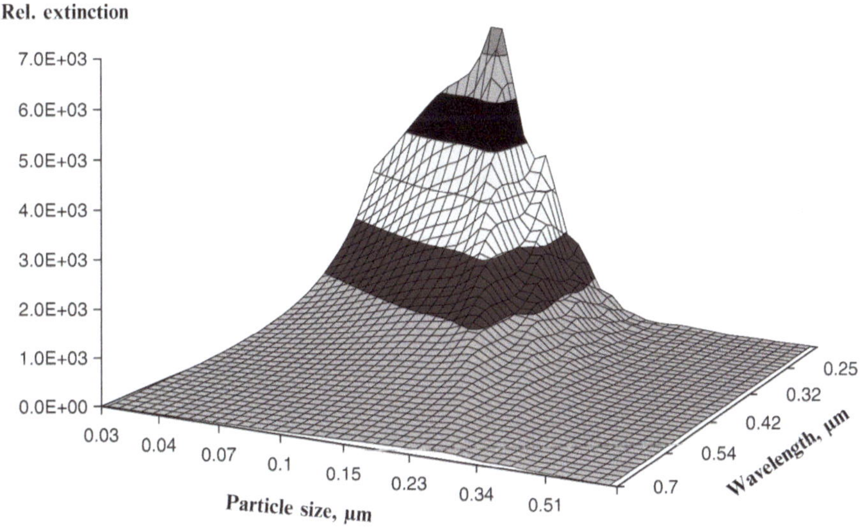

Fig. 2.8 Relative extinction for equal volumes of particles of different size; particle RI = 2.4 – 0 i, in water RI = 1.33. Courtesy M. Wedd

cause for opacity (see also Chap. 12). At the optimum particle size (for rutile about 200–300 nm), the scattering efficiency (ratio of optical cross section to actual particle cross section) is about 5, against about 2 for larger particles than about 10 μm. For particles smaller than about 100 nm, the scattering power decreases with the 6th power of particle size. The scattering power of particles also relates to the inverse of wavelength to the 4th power. Therefore, scattering power strongly increases with decreasing wavelength. This assists some dispersions to become fairly or even fully transparent at visible wavelengths but permit absorption of the dangerous UV wavelengths as needed for sunscreens. Figure 2.7 nicely illustrates the influence of particle size on turbidity for aqueous TiO_2 dispersions. Such very small particle sizes may be attractive in e.g. sunscreens (see Chap. 14). Figure 2.8

exemplifies the relative extinction at different wavelengths for equal volumes of particles having different sizes in suspension (at RI of particles = 2.4). Note that particle volume relates to the cube (3rd power) of particle size.

Particle size and shape affect the quality of e.g. paint and sunscreen in tinting strength, hiding power and transparency. And at low particle concentration, the scattering differences are used in both the laser diffraction and the ultrasound extinction technique for measurement of particle size distributions.

Computer codes have been derived for calculation of scattering matrices that describe the angular light scattering patterns of spheres in relation to their size and optical properties. Various theories are used as basis for these codes. Mie theory is the general theory. It describes the general relationship between the particle size of spheres, refractive index of the particles, the wavelength of the light source, the angular pattern of scattered light intensities, and the scattering efficiencies for each size. Rayleigh theory is also used but is limited to particles smaller than about 100 nm. It assumes particles to be point sources and calculates that the scattered light intensities decrease with the 6th power of particle size (thus, 2nd power per unit of volume of such particles) and are equal in all directions. Fraunhofer theory is exclusively based upon the diffraction at the particle contours and is an approximation that is generally valid for particles larger than about 50 μm, though fully opaque particles allow a lower limit (down to about 2 μm). Its main advantage is that it does not require values for the refractive index and that it allows for much simpler calculations [54] at the cost of reduced accuracy of the quantity of small particles. Most computer codes for light scattering calculations are limited to low concentrations of spherical particles, since they assume that light is scattered only once by a particle (single scattering). When the concentration is increased, the scattered light may be scattered again by other particles (multiple or diffuse scattering). The fine-structure of the scattering patterns decreases by the presence of particles of different size as well as by multiple scattering. As a consequence, calculations become increasingly complex. Modern computers, however, have proven to be capable of performing such calculations through using e.g. so-called T-matrix codes for evaluation of the basic Maxwell equations [56, 57, 89]. These codes can also be used for calculation of scattering patterns of non-spherical and heterogeneous particles.

In order to reach the desired color and hiding power for e.g. paint applications, pigments are usually mixed. The equations given by Kubelka-Munk and Duncan are often used in the design of such mixtures.

Both color and opacity can be derived from the measured diffuse reflectance curves with respect to wavelength, through the Kubelka-Munk theory [24, 27, 38, 43, 44]:

$$\left(\frac{K}{S}\right)_{\lambda} = \frac{\left(1 - R_{\lambda,\infty}\right)^2}{2R_{\lambda,\infty}} = f(R) \tag{2.26}$$

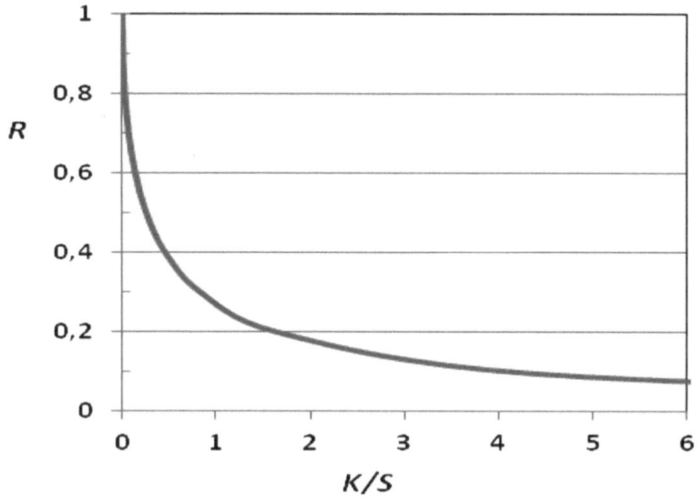

Fig. 2.9 Relationship between R and K/S

where:

K = absorption coefficient of the K-M theory
S = scattering coefficient of the K-M theory
$R_{\lambda,\infty}$ = measured reflectance at given wavelength and infinitely thick layer
λ = wavelength

Although the theory has a phenomenological origin, its absorption and scattering coefficients were later proven to relate to the single-scattering and extinction coefficients. In K is twice the extinction coefficient of the Beer-Lambert-Bouguer equation (Eq. 2.25), even at high pigment concentrations. S has a more complex relationship with the Mie scattering coefficient, since it also depends on pigment concentration [38, 59, 62, 67].

The K-M equation is quite helpful since its results, illustrated in Fig. 2.9, inform us that reflectance will decrease at increasing K/S, for example if we add a strongly absorbing pigment such as carbon black to the system. On the other hand, reflectance increases if we increase S/K, e.g. by addition of a strongly scattering white pigment.

As illustrated above, pigments are most often used in combinations to provide the required color and hiding power. Duncan has demonstrated that the contributions of scattering and absorption by individual constituents in a mixture can be summed [12, 27, 38]:

$$\left(\frac{K_{mix}}{S_{mix}}\right)_\lambda = \left(\frac{c_1K_1 + c_2K_2 + c_3K_3 + \ldots}{c_1S_1 + c_2S_2 + c_3S_3 + \ldots}\right)_\lambda \tag{2.27}$$

Equation 2.27 is often used to design color as well as hiding power of pigment mixtures.

The desired color is the result of mixing various colored pigments at appropriate concentrations, with the hiding power achieved through the addition of white pigment (see further Chap. 12).

Pigment application for hiding power is most effective when particle size is about half the dominant wavelength of the light (about 200–400 nm); then, a least amount of pigment is required for the given task of protection and obliterating the color of the substrate. Particle size for maximum hiding power for spheres having a high RI can be calculated with the approximating equation given by Weber [86]:

$$D_{sc\max} = \frac{2\lambda}{\pi(n_P - n_L)} \tag{2.28}$$

where:

D_{scmax} = particle size at maximum scattering power
λ = wavelength incident light
n_P = refractive index particle
n_L = refractive index liquid

For transparent applications, much smaller particles – in the nano-size range – are used; these have much less capacity for scattering light (see above, Rayleigh theory and Figs. 2.7 and 2.8). Nano-particles may also be used for catalytic applications, where maximum catalytic surface area per unit mass is desired.

2.8 Sensorial Characteristics

Sensorial characteristics are important in our human appreciation of foods, sweets, beverages and cosmetics [18, 40, 73, 78]. Taste, for example, relates to a combination of aroma, texture, odor, visual and auditive characteristics more or less independent of product type (solid, semi-solid or fluid) and use. Bitter, salt, sour, sweet and umami (savoury) are presently considered as the five basic tastes [29, 47]. Some examples of popular terminology in relation to specific products are presented in Table 2.3. A lexicon of terms and definitions has been published by ASTM [2].

Most consumer products of this kind consist of two or more phases. The sensorial properties have both psychological and physical aspects. The terminology is semi-quantitative. The psychological aspects are individual. For example, some people like bitter chocolate, others like milk chocolate. Sense of taste also depends to some extent on the country of origin culture, for example appreciation or dislike of some special kinds of vegetables, cheeses or spice. Physical aspects can be subdivided in mechanical, geometrical and other characteristics, in addition to aroma. Mechanical aspects correspond to some kind of texture, geometrical ones to the particulate nature.

A wide variety of causes may influence the physical characteristics of products such as rheological behavior, chewiness, etc. Particle size and size distribution

Table 2.3 Examples of popular terminology for taste characteristics in products

Product type	Characteristics	Product examples
Solid	Crisp, brittle, crumbly, powdery Creamy, gritty, grainy, coarse Tender, chewy, tough, plastic Moist, dry, sticky, soggy	Apples, biscuits, carrots, chocolate, cornflakes, gum, meat, sweets
Semi-solid	Pasty, mealy, coherent Moist, dry, sticky, soggy Lumpy, smooth	Butter, cheese, creams, fish, margarine
Fluid	Thin, watery, viscous Creamy, fatty, greasy Sticky, tacky Bubbly, tingly	Beverages, margarine, mayonnaise, milk, soups

(of solid, liquid and air 'particles' dispersed in another phase), particle shape and particulate concentration are important geometrical aspects. The type and melting range of the ingredients (in relation to temperature of the environment or the human body), moisture content, etc. are included in other aspects. Note that the particle size and shape of ingredients may influence the sensorial properties of a product in addition to its processability (rheological behavior).

Specific products may show different or similar relationships between particle characteristics and performance. For example, they might wish to limit the degree of grittiness in our mouths, which depends on product composition and temperature. In chocolate we experience a change in mouth-feel from creamy to gritty when the $D_{90;3}$ of the sugar particles increases from about 20 to 30 μm and in ice cream above about 55 μm [30]. Over these size ranges, we are able to sense some individual solid particles in between our finger tips. Other similar effects are the influence of particle size on e.g. powder flow and suspension viscosity (see Sects. 2.3 and 2.5).

In view of its complex nature, taste is generally tested by trained panels, which use standard procedures and terminology [88]. Often, these tests are combined with measurements of e.g. viscosity, elasticity, hardness, turbidity, etc.

2.9 Adsorption and Catalysis

Porosity and pore size distribution of particles are essential factors for their ability to act as adsorbents, catalysts or ion exchangers in industrial processes. On the one hand, wide pores favor quick access and removal of components. On the other hand, narrow pores provide for a large surface area for adsorption or reaction. The minimum pore width for a given application is defined by the size of the molecules involved. Pore sizes are classified into three categories, viz. [74]:

- Macropores, with pore diameters larger than 50 nm
- Mesopores, with pore diameters in the range 2–50 nm
- Micropores, with pore diameters smaller than 2 nm.

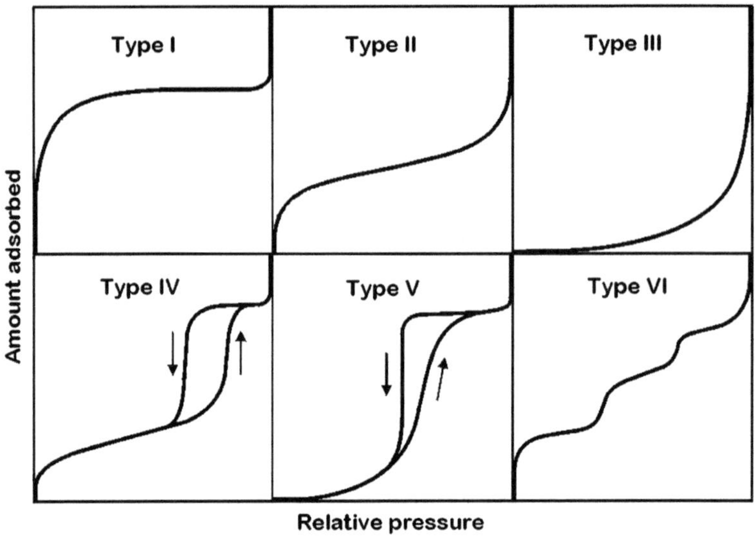

Fig. 2.10 Types of physisorption isotherms

Note that the size of many voids in between particles is in the same order of magnitude as the size of the particles. Although voids are usually larger than pores, some overlap in size occurs due to the irregular shape of the voids.

Mercury penetration and gas adsorption techniques (using physisorption with physical attractive forces between adsorbent and adsorbate) are most often used for determination of pore size distributions and specific surface area (see also Chap. 3). Mercury penetration technique employs the pressure required for the penetration of mercury into the pores and voids as a measure of pore size. The inverse relationship between pressure and pore size requires that higher pressures are needed for smaller pores/voids (see also Chap. 3).

In the gas adsorption technique, the amount of adsorbate for a known amount of adsorbent is determined at various relative pressures (P/P_0). For characterization, nitrogen is often used as adsorbate and measurement takes place at its boiling point temperature of about − 196 °C (77 K). For most adsorbates, the majority of physisorption isotherms may be grouped into six types, as shown in Fig. 2.10 [31, 33–35, 74]:

Type I isotherms show a limiting value at increasing relative pressures. This is reversible and typical for microporous solids, where the sorption is limited by the accessible micropore volume rather than the internal surface area.

Type II isotherms are also reversible and the normal form for non-porous and macroporous adsorbents. They represent unrestricted monolayer-multilayer adsorption. Within the relative pressure range of about 0.25–0.35 there is a point at which monolayer coverage is nearly complete and multilayer adsorption is about to begin. This is usually regarded to be the end point for application of the BET equation for surface area measurement (see Chap. 3).

Type III isotherms are also reversible and typical for systems where adsorbate-adsorbate interactions play a more important role than adsorbent-adsorbate interactions. This type is not common.

Type IV isotherms are the most common form for mesoporous sorbents. The initial part of the isotherm is attributed to monolayer-multilayer adsorption (as in type II), but at higher relative pressures they show a hysteresis loop, which is associated with capillary condensation in the mesopores. Different shapes of the hysteresis loops can be related to different specific pore structures. Similar to type II isotherms, relative pressures up to about 0.35 are used for surface area measurement (see Chap. 3).

Type V isotherms are typical for weak adsorbent-adsorbate interactions. They are uncommon and show relationship to type III isotherms.

Type VI isotherms show a stepwise increase of adsorbate at increasing relative pressures and are also uncommon. The step height represents the monolayer capacity for each adsorbed layer. The steps show stepwise multilayer adsorption, onto a uniform non-porous surface.

Within the measurement techniques for pore size distributions (cf. Chap. 3), a cylindrical pore shape is generally assumed. Other types of idealized pore shapes used are e.g. ink bottle pores (having a restricted opening), slit-shaped pores and wedge-shaped pores. In practice, however, pores often have an irregular shape.

Processes that include adsorption, catalysis and ion exchange, are governed by the presence of both specific chemical elements or groups and a large specific surface area. Specificity is often favored by using a specific particular pore architecture for connecting different pore sizes.

Adsorption occurs by physical attractive forces (physisorption), but specific chemical bonding may occur as well (e.g. between carbon monoxide and platinum; chemisorption). In industrial applications, adsorbents are typically applied to remove water or hazardous components from flue gas, waste water or process water (e.g. sulphur dioxide or sulphate). Specificity may be obtained by using adsorbents with a specific pore size, which limits entrance of the large species present. This is, for example, used in the separation of normal and branched paraffins in molecular sieves, where the former can enter the pores and the latter cannot ('molecular sieving phenomenon').

Ion exchange is used to remove ions that are considered to harm a process or the environment (e.g. calcium or heavy metals). It requires the presence of specific, harmless ions (e.g. sodium) at the surface that can be exchanged with the harmful ions. Here too, some specificity can be obtained by using ion exchangers with a specific pore size, which favors entrance of the targeted ions.

Catalysis is a very complex process. It requires adequate adsorption of several reactants onto a surface as well as the ability of that surface to catalyze the rapid and selective conversion into the desired components. Solid catalysts require a large surface area of catalytic material (usually a highly dispersed metal on a large inert surface) as well as rapid access of reactants and rapid removal of produced components. Catalytic reactions are sometimes executed in fluid beds, sometimes

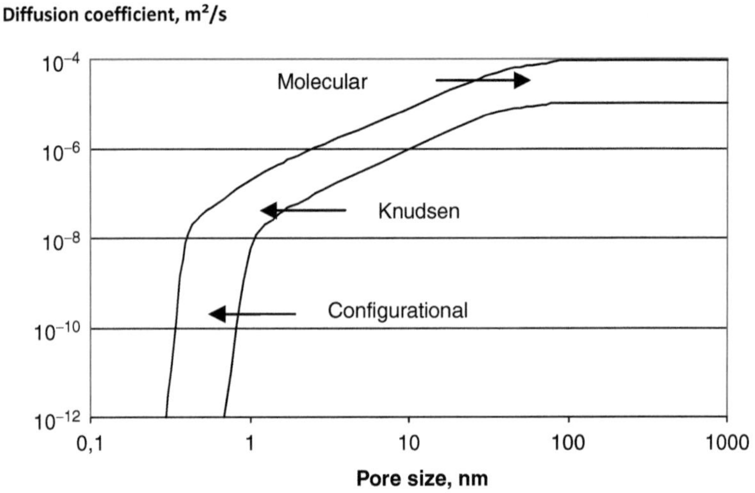

Fig. 2.11 Schematic of gas diffusion regions in pores

in fixed beds. Adequate behavior of the catalysts requires adequate engineering of particle size, pore size distribution and architecture in relation to the molecular size of all components involved.

The specific surface area of porous particles, is usually established from a gas adsorption measurement, at the point where a monolayer of adsorbed gas may be assumed (see above and Chap. 3). It includes any internal surface of pores. Specific surface area lies within the range of 100–1,000 m^2/g for porous particles. For non-porous, spherical particles, it inversely relates to particle size:

$$S = \frac{6\pi \cdot D^2}{\pi \cdot D^3 \cdot \rho} = \frac{6}{D \cdot \rho} \tag{2.29}$$

where:

S = specific surface area
D = spherical particle diameter
ρ = true particle density

From this equation it can be easily derived that solid, non-porous spheres with diameters in the range of 0.1–10 μm have a specific surface area of about 60–0.6 m^2/g. The above quoted value of 100 m^2/g is only reached for solid particles smaller than about 50 nm (calculated for unit density of 1,000 kg/m^3).

Diffusion (Brownian motion) is the driving force for transport of molecules in pores [11, 42]. Besides temperature, the effective molecular size, the pore size and adsorption determine the possibilities and the rate of transport. Four diffusion mechanisms can be distinguished in the following schematic way, which may act in conjunction (cf. Fig. 2.11).

- *Molecular diffusion* (also called *bulk or Fick diffusion)* occurs when the pore diameter is large in comparison to the mean free path length of the molecules. Molecules will diffuse independently of pore size at a rate related to their diffusion coefficient, which is inversely related to molecular size. Small molecules diffuse faster than larger ones. Generally, diffusion coefficients are in the range 10^{-4}–10^{-5} m^2/s.
- *Knudsen diffusion* prevails, when the mean free path length of the molecules is fairly large in comparison to the pore diameter. Frequent collisions of the molecules with the pore wall retard diffusion. The Knudsen diffusion coefficients are proportional to the pore diameter and inversely proportional to molecular weight of the diffusing component. Their value is generally in the range 10^{-5}–10^{-7} m^2/s.
- *Activated, configurational diffusion* occurs when the size of the molecules is almost the same as the pore diameter. Strong interaction between molecules and pore wall result in very slow transport. Generally, diffusion coefficients are smaller than 10^{-8} m^2/s.
- *Surface diffusion* is possible, when the molecules have a greater tendency for adsorption. In this case, transport mainly takes place in the 'liquid' adsorbate layer, which may strongly enhance diffusional flux in comparison to both Knudsen and configurational diffusion.

Note: Molecules, of course, cannot enter pores when they are larger than the pore size.

2.10 Definitions, Abbreviations and Symbols

Adsorbate	Enriched adsorptive at the surface of the adsorbent
Adsorbent	Solid that adsorbs the adsorptive
Adsorption	Enrichment of the adsorptive at a solid surface
Adsorptive	Gas or vapor (or a specific liquid component) to be adsorbed
Agglomerate	Assemblage of primary particles with intermediate attractive forces (sometimes named aggregate)
Agglomeration	Formation of agglomerates
Aggregate	Assemblage of primary particles with strong attractive forces (sometimes named agglomerate)
Aspect ratio	Ratio of maximum to minimum Feret diameter of a particle (in some literature, the inverse is used, i.e. ratio of minimum to maximum Feret diameter)
Coagulation	(active) formation of (large) agglomerates, often by addition of electrolyte or other specific chemical to a suspension
Coalescence	Collision of droplets in an emulsion, followed by merging to a larger droplet
Cohesivity	Stickiness of a powder, caused by strong attractive and/or frictional forces between individual particles
Colloid	Dispersion of particles having a size range of about 1 nm – 1 μm in a liquid

(continued)

(continued)

Cumulative size distribution	Distribution of the fraction of material smaller (undersize) or larger (oversize) than given particle sizes against particle size
Effective powder density	Density of powder bed measured at well defined conditions
Effective particle density	Particle density measured at well defined, optimum conditions for dispersion in a liquid that includes effects of any retained gas or liquid within the closed or open pores of the particle
Equivalent sphere	Sphere that has the same property as the observed particle in relation to a given measurement principle
Floc	Assemblage of loosely bound primary particles in a liquid
Flocculation	Formation of flocs, usually intentional by addition of a specific flocculation aiding chemical
Median size	Particle size at the point in a cumulative size distribution, where 50 % of the particles is smaller and 50 % larger
Particle size	Diameter of some defined equivalent sphere
Pore	Cavity or channel within an object, such as a particle
Powder bed porosity	Void fraction in the powder bed (packed or fluidized)
Principal component analysis	Mathematical procedure for resolving large data sets into orthogonal components, whose linear combinations approximate the original data to any desired degree of accuracy
Shear stress	Type of stress resulting from shear forces, coplanar with a material cross section
Strain	Deformation of a body due to stress
Stress	Force within a deformable body as a reaction to an external force (often also used for the external force)
Umami	Fifth basic taste (brothy, savory), imparted by glutamate and ribonucleotides [29, 47]
Void	Space between particles, usually in a powder
Zeta potential	electrostatic potential at the layer of shear of a particle suspended in a liquid
IFPRI	International Fine Particle Research Institute
ISO	International Standards Organisation
IUPAC	International Union of Pure and Applied Chemistry
PSD	Particle size distribution
a	Powder packing factor (depends on way of packing)(Eq. 2.2)
b	Particle shape factor for packing equation (Eq. 2.2)
c_1, c_2, c_3	Proportionality constants of an estimated polynomial (Eq. 2.15)
C_D	Drag coefficient
CI	Carr Compressibility Index (Eq. 2.4)
D	Particle size (diameter of equivalent sphere)
$D_{3,2}$	Sauter mean diameter, area-weighted mean size, mean value of surface area-weighted PSD
$D_{50;3}$	Median particle size of volume-based size distribution
$D_{90;3}$	90th percentile of cumulative, volume-based particle size distribution
D_{ae}	Aerodynamic particle size of aerosol particles
D_h	(Effective) hydrodynamic particle size
D_{St}	Equivalent Stokes' diameter of a particle
$\mathbb{D}$	Diffusion coefficient
g	Gravitational acceleration constant

(continued)

(continued)

H	Hausner Ratio (Eq. 2.5)
H_s	Settling height
H_b	Bed height
k_B	Boltzmann constant
K	Resistance constant depending on channel geometry (Eq. 2.7)
L	Length of capillary
m	Shear rate sensitivity exponent (also called flow index) (Eq. 2.17)
M	Mass of powder bed
p	Powder packing exponent (depends on way of packing) (Eq. 2.3)
P	Actual pressure
P_0	Saturation pressure
ΔP	Pressure drop over bed or tube
$r_{i,j}$	Ratio of equivalent packing sizes between components *i* and *j* of a bimodal mixture (Eq. 2.3)
$R_{i,j}$	Ratio of equivalent volume diameters between components *i* and *j* of a bimodal mixture (Eq. 2.3)
Re	Reynolds number
S	Specific surface area
t	Time for settling over height H_s
T	Absolute temperature
u_s	Linear fluid velocity
$<u_s>$	Mean linear superficial fluid velocity
v	Terminal settling velocity of a particle
V_B	Volume of powder bed
V_f	Freely settled volume per unit mass of powder
V_P	Volume of particles in bed
V_t	Tapped volume par unit mass of powder
V_V	Volume of voids
$(d\gamma/dt)$	Shear strain rate (or shear rate)
ε	Powder bed porosity (void fraction in powder bed)
ε_{fl}	Void fraction during fluidization
ε_0	Limiting porosity (in either tap density or some apparent density)
η_d	Viscosity of dispersion
η_F	Fluid viscosity
η_L	Viscosity of liquid (Newtonian)
η_m	Apparent viscosity of complex mixture
η_0	Inherent viscosity without shear
η_{rel}	Relative viscosity
Φ	Effective volume fraction of dispersed phase
Φ_v	Volumetric flow rate
ρ	True particle density
ρ_B	Effective bulk density
ρ_e	Envelope particle density (including all intra-particle pores)
ρ_f	Freely settled bulk density of powder
ρ_F	Density of fluid
ρ_L	Liquid density
ρ_P	Effective particle density
ρ_t	Tapped bulk density of powder
τ	Shear stress (force per unit area)

References

1. Allen, T.: Powder Sampling and Particle Size Determination. Elsevier, Boston (2003)
2. ASTM DS72: Lexicon for Sensory Evaluation of Aroma, Flavor, Texture and Appearance. American Society for Testing and Materials, Philadelphia (2011)
3. Barnes, H.A.: Shear-thickening ("Dilatancy") in suspensions of nonaggregating solid particles dispersed in newtonian liquids (review). J Rheol. **33**(2), 329–366 (1989)
4. Barnes, H.A.: Thixotropy – a review. J. Non-Newtonian Fluid Mech. **70**, 1–33 (1997)
5. Behringer, R., Louge, M., McElwaine, J., Mort, P., Pfeffer, R., Sundaresan, S.: Report of the IFPRI owder flow working group SAR30-08 (2005)
6. Brinkmann, H.C.: The viscosity of concentrated suspensions and solutions. J. Chem. Phys. **20**, 571 (1952)
7. Caldwell, D.H., Babbitt, H.E.: Flow of muds, sludges and suspensions in circular pipe. Ind. Eng. Chem. **33**, 249–256 (1941)
8. Carman, P.C.: Fundamental principles of industrial filtration. Chem. Eng. Res. Des. **16a**, 168–188 (1938)
9. Carr, R.L.: Evaluating flow properties of solids. Chem. Eng. **72**, 163–168 (1965)
10. Cheng, D.C.H., Kruszewski, A.P., Senior, J.R., Roberts, T.A.: The Effect of particle size distribution on the rheology of an industrial suspension. J. Materials sci. **25**, 353–373 (1990)
11. Coulson, J.M., Richardson, J.F., Backhurst, J.R., Harker, J.H.: Chemical Engineering, vol. I, Fluid Flow, Heat Transfer and Mass Transfer. Butterworth-Heinemann (1999)
12. Duncan, D.R.: The identification and estimation of pigments in pigmented compositions by reflectance spectrophotometry. J. Oil Color Chem. Assoc. **45**, 300–324 (1962)
13. Eilers, H.: Die Viskosität von Emulsionen Hochviskoser Stoffe als Funktion der Konzentration. Kolloid Zeitschrift **102**, 154–169 (1943)
14. Einstein, A.: Investigations on the Theory of Brownian Movement. Dover Publications, New York (1956)
15. Einstein, A. Ann. Physik Lpz. **19** (1906) 289 and **34** (1911) 591
16. Eley, R.R. In: ASTM Paint Testing Manual, pp. 333–368. ASTM Internat (1995)
17. Emery, E., Oliver, J., Pugsley, T., Sharma, J., Zhou, J.: Flowability of moist pharmaceuticals. Powder Technol. **189**, 409–415 (2009)
18. Engelen, L., van der Bilt, A.: Oral physiology and texture perception of semisolids. J. Text. Studies **39**, 83–113 (2008)
19. Farris, R.J.: Prediction of the viscosity of multimodal suspensions from unimodal viscosity data. Trans. Soc. Rheol. **12**, 281–301 (1968)
20. Geldart, D. (ed.): Gas Fluidization and Technology. Wiley, New York (1986)
21. Geldart, D. In: McGlinskey, D. (ed.), Characterization of Bulk Solids. Oxford, CRC-Blackwell (2005)
22. German, R.M., Park, S.J.: Handbook of Mathematical Relations in Particulate Materials Processing. Wiley, Hoboken (2008)
23. Green, D.W., Maloney, J.O. (eds.): Perry's Chemical Engineers Handbook. McGraw-Hill, New York (1984)
24. Hammond III, H.K., Kigle-Boeckler, G.: Gloss. In: ASTM Paint Testing Manual. ASTM International (1995)
25. Hiemenz, P.C., Rajagopalan, R.: Principles of Colloid and Surface Chemistry. Marcel Dekker, New York (1997)
26. Hinds, W.C.: Aerosol Technology: Properties, Behavior and Measurement of Airborne Particles. Wiley, New York (1999)
27. Howell, D.M.: The technology, formulation and application of powder coatings; series in surface coatings technology. In: Sanders, J.D. (ed.), vol. 1. Wiley, New York (2000)
28. Hunter, R.J.: Zeta Potential in Colloid Science; Principles and Applications. Academic, New York (1981)

29. Ikeda, K.J.: New seasonings. Tokyo Chem. Soc. **30**, 820–836 (1909) (Japanese); English translation in: Chem. Senses **27** (2002) 847–849
30. Imai, E., Saito, K., Hatakeyama, M., Hatae, K., Shimada, A.: Effect of physical properties of food particles on the degree of graininess perceived in the mouth. J. Text. Studies **30**, 59–88 (1999)
31. ISO 9277.: Determination of the Specific Surface Area of Solids by Gas Adsorption Using the BET Method; (in revision) International Standards Organisation (1995)
32. ISO 13099 (in prep.).: Methods for Zeta Potential Determination; International Standards Organisation
33. ISO 15901-1.: Evaluation of Pore Size Distributions of Solid Materials by Mercury Porosimetry. International Standards Organisation (2005)
34. ISO 15901-2.: Evaluation of Pore Size Distributions of Solid Materials by Gas Adsorption (and corrigendum). International Standards Organisation (2006/2007)
35. ISO 15901-3.: Analysis of Micro-pores by Gas Adsorption. International Standards Organisation (2007)
36. Jenike, A.W.: Gravity flow of bulk solids. Utah. Eng. Exp. Stn. Bull. **108**, 1–294 (1961); Univ. Utah, Salt Lake City
37. Jenike, A.W.: Storage and flow of solids. Utah. Eng. Exp. Stn. Bull. **123**, 1–194 (1964); Univ. Utah, Salt Lake City
38. Johnston, R.M.: Color theory (Ch. III-D-b). In: Patton, T.C. (ed.), Pigment Handbook, vol. III, Characterization and Physical Relationships. Wiley, London (1973)
39. Kissa, E.: Dispersions: Characterization, Testing and Measurement. Marcel Dekker, New York (1999)
40. Kramer, A., Szczesniak, A.S.: Texture Measurements of Foods. D. Reidel, Dordrecht (1973)
41. Krieger, I.M., Dougherty, T.J.: A mechanism for non-newtonian flow in suspensions of rigid spheres. Trans. Soc. Rheol. **3**, 137–152 (1959)
42. Krishna, R.: A unified approach to the modelling of intraparticle diffusion in adsorption processes. Gas Separ. Purif. **7**, 91–104 (1993)
43. Kubelka, P., Munk, F.: Ein Beitrag zur Optik der Farb Anstriche. Zeitschr. f. techn. Physik **12**, 593–601 (1931)
44. Kubelka, P.: New contributions to the optics of intensely light - scattering materials, Part I. J. Opt.Soc. Am. **38**, 448–457 (1948)
45. Lavoie, F., Cartilier, L., Thibert, R.: New methods characterizing avalanche behavior to determine powder flow. Pharm. Res. **19**, 887–893 (2002)
46. Lee, Y.S.L., Poynter, R., Podczek, F., Newton, J.M.: Development of a dual approach to assess powder flow from avalanching behavior. AAPS PharmSciTech. **1** (2000) art. 21
47. Lindemann, B., Ogiwara, Y., Ninomiya, Y.: The discovery of Umami. Chem. Senses **27**, 843–844 (2002)
48. Lyklema, J. (ed.): Fundamentals of Interface and Colloid Science. Particulate Colloids, vol. IV. Elsevier, London (2005)
49. Macosco, C.W.: Rheology: Principles, Measurement and Applications. VCH Publishers, New York (1994)
50. Marrion, A.: The Chemistry and Physics of Coatings. Royal Soc. Chem, Cambridge (2004)
51. Masuda, H., Higashitani, K., Yoshida, H.: Powder Technology; Fundamentals of Particles, Powder Beds and Particle Generation. CRC Press, Boca Raton (2007)
52. McGee, E.: Predicting powder flow. Pharm. Tech. (June, 2007)
53. McGlinskey, D. In: McGlinskey, D. (ed.), Characterization of bulk solids. CRC-Blackwell, Oxford (2005)
54. Merkus, H.G.: Particle Size Measurements; Fundamentals, Practice, Quality. Springer, Dordrecht (2009)
55. Metzner, A.B.: Rheology of suspensions in polymeric liquids. J. Rheol. **29**, 739–775 (1985)
56. Mischenko, M.I., Travis, L.D., Lacis, A.A.: Scattering, Absorption and Emission of Light by Small Particles. Cambridge Univ. Press, New York (2002)
57. Mischenko, M.I., Travis, L.D., Lacis, A.A.: Multiple Scattering of Light by Particles: Radiative Transfer and Coherent Backscattering. Cambridge Univ. Press, New York (2006)

58. Mongia, G., Ziegler, G.R.: The role of particle size distribution of suspended solids in defining the flow properties of milk chocolate. Int. J. Food Prop. **3**, 137–147 (2000)
59. Mudgett, P.S., Richards, L.W.: Multiple scattering calculations for technology. Appl. Optics **10**, 1485–1502 (1971)
60. Nelson, R.D.: Dispersion of powders in liquids. In: Kirk-Othmer Encyclopedia of Chemical Technology, vol. 19. Wiley, Chichester (1996)
61. Nelson, R.D.: Dispersion of Powders in Liquids. Elsevier, New York (1988)
62. Phillips, D.G., Billmeyer Jr., F.W., Coatings, J.: Predicting Reflectance of Color and Paint Films; IV Kubelka - Munk Scattering Coefficient. J. Coatings Technol. **48**^{616}, 30–36 (1976)
63. Qi, F., Tanner, R.I.: Random close packing and relative viscosity of multimodal suspensions. Rheol. Acta, 04 Oct. **2011** (online)
64. Rhodes, M.: Introduction to Particle Technology. Wiley, Chichester (1998)
65. Riley, F.L.: Structural Ceramics; Fundamentals and Case Studies. Cambridge Univ. Press, New York (2009)
66. Roscoe, R.: The viscosity of suspensions of rigid spheres. Brit. J. Appl. Phys. **3**, 267–269 (1952)
67. Schabbach, L.M., Bondioli, F., Ferrari, A.M., Petter, C.O., Fredel, M.C.: Colour in ceramic glazes: Efficiency of the Kubelka - Munk model in glazes with a black pigment and opacifier. J. Eur. Ceram. Soc. **29**, 2685–2690 (2009)
68. Scott, D.M.: Improving control of particulate processes via ultrasonic spectroscopy. In: APACT10: Advances in Process Analytics and Control Technology, Manchester, 28–30 April 2010
69. Scott, G.D., Kilgour, D.M.: The density of random close packing of spheres. J. Phys. D (Brit. J. Appl. Phys.) **2**, 863–866 (1969)
70. Seville, J.P.K., Tüzün, U., Clift, R.: Processing of Particulate Solids. Chapman and Hall, London (1997)
71. Sherman, P.: The viscosity of emulsions. Rheol. Acta **2**, 74–82 (1962)
72. Sherman, P.: Industrial Rheology. Academic, London (1970)
73. Sherman, P. (ed.): Food Texture and Rheology. Academic, New York (1979)
74. Sing, K.S.W., Everett, D.H., Haul, R.A.W., Moscou, L., Pierotti, R.A., Rouquérol, J., Siemieniewska, T.: Pure Appl. Chem **57,** 603–619 (1985) (© 1985 IUPAC)
75. Smith, P.A., Haber, R.A., Am, J.: The effect of particle packing in the filtration and rheology behavior of extended size distribution alumina suspensions. Ceram Soc. **78**, 1737–1744 (1995)
76. Strivens, T.A.: Introduction to rheology (Ch. 14). In: Lambourne, R., Strivens, T.A. (eds.) Paint and Surface Coatings, 2nd edn. Woodhead Publ, New York (1999)
77. Suzuki, M., Sato, H., Hasegawa, M., Hirota, M.: Effect of size distribution on tapping properties of fine powder. Powd. Technol. **118**, 53–57 (2001)
78. Szczesniak, A.S.: Texture is a sensory property. Food Qual. Pref. **13**, 215–225 (2002)
79. Tadros, T.F.: Colloids in Paints. Colloids and Interface Science Series, vol. 6. Wiley-VCH, Weinheim (2010)
80. Tanner, R.I., Walters, K.: Rheology: An Historical Perspective. Elsevier, New York (1998)
81. Taylor, M.K., Ginsburg, J., Hickley, A.J., Gheyas, F.: Composite method to quantify powder flow as a screening method in early tablet or capsule formulation development. AAPS PharmSciTech **1**(3), E18 (2000). doi:article 18
82. US Pharmacoeia 29-NF24 p.3017 <1174 > Powder Flow. http://www.pharmacopeia.cn/v29240/usp29nf24 s0_c1174.html
83. Vadillo, D.C., Tuladhar, T.R., Mulji, A.C., Mackley, M.R.: The rheological characterization of linear viscoelasticity for ink jet fluids using piezo axial vibrator and torsion resonator rheometers. J. Rheol. **54**, 781–795 (2010)
84. Valverde Millán, J.M.: Fluidization of Fine Powders. Particle Technology Series, vol. 18. Springer, Dordrecht (2013)
85. Wagner, N.J., Brady, J.F.: Shear thickening in colloidal dispersions. Phys. Today 27–32, (2009)

86. Weber, H.W.: Lichtstreuung and Teilchengröszenverteilung Kugelförmiger Teilchen. Kolloid Zeitschrift u. Zeitschift f. Polym. **188**, 40–44 (1962)
87. Yu, A.B.., Bridgewater, J., Burbidge, A.: On the modeling of the packing of fine particles. Powd. Technol. **92**, 185–194 (1997)
88. Ziegler, G.R., Mongia, G., Hollender, R.: The role of particle size distribution of suspended solids in defining the sensory properties of milk chocolate. Int. J. Food Prop. **4**^{2}, 353–370 (2001)
89. www.giss.nasa.gov/mmischenko/t_matrix.html

Chapter 3
Measurement of Particle Size, Shape, Porosity and Zeta-potential

Henk G. Merkus

Particle size analysis is not an objective in itself,
it is a means to an end,
i.e. the application of such knowledge to
some manufacturing process or performance of some product.

H. Heywood [13].

Abstract This chapter gives an overview of the most popular methods for measurement of particle size and shape. It covers microscopy and image analysis, laser diffraction, dynamic light scattering and gravitational and centrifugal sedimentation. In view of its capability for measurements at high particulate concentrations, the ultrasound attenuation technique is also included. Moreover, measurement techniques for porosity, surface area and pore size distributions and zeta potential are examined. This chapter also contains sections on sampling and dispersion of dry powders, since either of the two often limits the accuracy of the measured data.

3.1 Introduction

Performance and quality of particulate products depend upon specific characteristics of the particles, viz. size distribution parameters, shape, surface area and/or – for suspensions – zeta-potential. Adequate definition and measurement of the relevant characteristics is essential for good quality control of the products. Note that particles may include any difference in phase between the 'particles' and their surroundings. In addition to the particle characteristics in dry powders and suspensions, are those in

H.G. Merkus (✉)
Retired Associate Professor, Delft University of Technology, Park Berkenoord 30, Pijnacker 2641 CZ, The Netherlands
e-mail: henkmerkus@hetnet.nl

H.G. Merkus and G.M.H. Meesters (eds.), *Particulate Products: Tailoring Properties for Optimal Performance*, Particle Technology Series 19,
DOI 10.1007/978-3-319-00714-4_3,

emulsions (liquid droplets in an immiscible liquid) and foams (gas bubbles in a liquid or solid material) that may all be subjects for measurement.

Given the importance of a large variety of particulate products and the wide range of particle sizes, many different techniques have been developed for size measurement. A selection is presented in Sect. 3.4; a full overview is presented in an earlier book [31]. All of the techniques interpret some size-related feature of the particles as an equivalent particle size. This may be, for example, the diameter of a particle's projection, the size related to its sedimentation rate or an interpretation of the scattering signature of a group of particles. Thus, a series of 'equivalent' sizes exists in practice for non-spherical particles (see Sect. 3.4.1) [31]. This means that the 'equivalent' size distributions derived by different techniques may deviate from one another.

Moreover, the amounts of particles in the basis of the size distribution may be different. For example, microscopy yields size distribution results by counting the number of particles of specific measured sizes. This gives a number-basis for the size distribution. In other techniques, like laser diffraction, sedimentation and sieving, distributions have a more or less volumetric (or mass) basis. Since different techniques measure PSD's from a different point of view (principle), it can also be concluded that they may supplement each others information for a collection of particles. In this way, data sets can be strengthened.

For spheres, the mathematical conversion of a number-based PSD into one based on e.g. area or volume, and vice versa, is straightforward since their area and volume relate to the diameter squared and cubed, respectively. For particles having another shape, however, the situation is different. Their size is derived from equivalent spheres, the diameter of which depends upon the principle used for measurement. Thus, a measured PSD offers only one aspect of reality and different techniques may lead to different PSDs.

Also the basis of the measurement is important, since measurement errors are transferred in the conversion in a relative manner. This causes, for example, the amount of the smallest particles in a broad PSD to be identified at better precision on a number-basis than on a volume-basis, since large particle numbers only take a relatively small volume.

At the upper end of broad PSDs, relatively small particle numbers represent a significant mass/volume (see below). Here, number statistics relate the precision of the measured PSD parameters to the minimum sample mass. In practice, the sample mass applied for the analysis often limits the overall precision of parameters at the upper PSD end.

The particle sizes of particulate products can cover many orders of magnitude. At the small size end, they may be in the nanometer range. At the upper end, they go up to centimeters. Also the width of the size distributions differs. Industrial products usually show a medium or broad distribution; few particulate materials have a narrow or ultra-narrow size distribution.

Size distributions are often shown in a graphical manner that can take different forms. Most often, differential or cumulative curves are used [16, 31]. An example of a cumulative undersize particle size distribution is presented in Fig. 3.1, in the form of a number-based and a volume-based distribution. It presents a log-normal size distribution of spheres having a geometric standard deviation $s_g = 2$ and ln

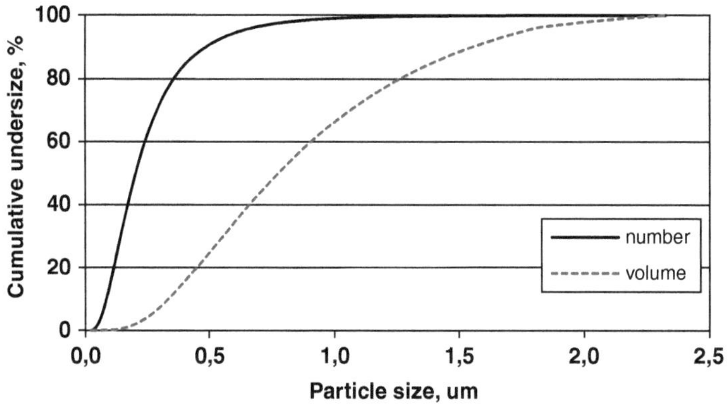

Fig. 3.1 Example of a cumulative size distribution

$(s_g) = 0.693$ around a mean size of 0.2 μm by number. Due to the width of this PSD example, there are large differences between both distribution types.

These cumulative size distributions allow easy estimation of percentile parameters such as the D_{10} (10 % undersize), the median (D_{50} at 50 %) and the D_{90} (90 % undersize), as well as the undersize or oversize fraction at a given size. Some number-based percentile features of this PSD are median diameter $D_{50;0} = 0.20$ μm, $D_{10;0} = 0.08$ μm, $D_{90;0} = 0.49$ μm and $D_{90;0}/D_{10;0} = 5.9$. More than 98 % n/n of the particles is smaller than 1 μm. Some volume-based characteristic features of this distribution are median diameter $D_{50;3} = 0.78$ μm, $D_{10;3} = 0.33$ μm, $D_{90;3} = 1.55$ μm and $D_{90;3}/D_{10;3} = 4.7$.[1] Note that the D_{90} and the D_{10} in a normal Gaussian distribution are separated by about 2.6 times the standard deviation of that distribution.

The differences between the two types of distributions are obvious. For example, the $D_{50;3}$ is about 4 times larger than the $D_{50;0}$. Moreover, only about 2 % v/v is smaller than the $D_{50;0}$; as much as 34 % v/v is larger than the $D_{99;0}$ and only 0.05 % n/n corresponds to the largest 5 % v/v (i.e. is larger than the $D_{95;3}$). These differences illustrate the need to be on your guard when reviewing particle size distribution data. Note that narrower distributions mean closer proximity between the two distribution types and vice versa.

A comparison of differential and cumulative curves is presented in Fig. 3.2. These figures show that resolution and sensitivity in terms of bimodality or shoulders are more easily identified in differential distribution curves than in cumulative ones. The cases shown have a fairly narrow overall size range of about 2:1. Here, the high resolution required for showing details is only possible with high-resolution techniques such as Electrical Sensing Zone, Image Analysis and Line-Start Disc Centrifuge.

Commercial products typically have a much broader size range, typically up to about 10:1 and they are analyzed by medium-resolution techniques, such as Laser Diffraction, Sieving and Ultrasound Attenuation. Also in these lower-resolution

[1] The second subscript value of the D represents its weighting: 0 for number, 1 for length, 2 for area and 3 for volume [ISO 9276].

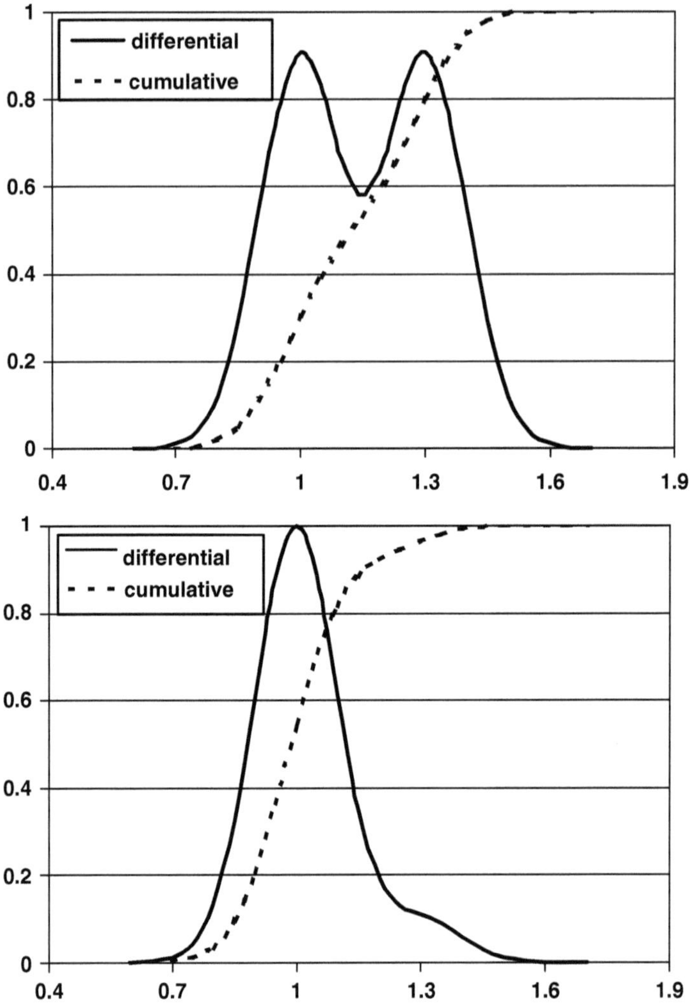

Fig. 3.2 Comparison of cumulative and differential distribution curves (sum of two normal distributions; modes at 1 and 1.3; $\sigma = 0.1$; 50/50 ratio *above* and 90/10 ratio *below*)

cases, differential curves show relatively more details than the cumulative curves. Note, however, that the measurement technique may introduce artefacts (e.g. broadening of the distribution or shoulders).

Fineness of particles and width of particle size distributions are defined differently in different fields of application. In the opinion of the author, it seems best to standardize [31]:

Fineness ($D_{90;0}$):		PSD width (D_{90}/D_{10} ratio)[2]:	
Nanoparticles	< 0.1 μm	Monosized	< 1.02 (ideally: 1.00)
Ultrafine	0.1–1 μm	Ultra narrow	1.02–1.05
Fine	1–10 μm	Narrow	1.05–1.5
Medium	10–1,000 μm	Medium	1.5–4
Coarse	1–10 mm	Broad	4–10
Very coarse	> 10 mm	Very broad	> 10

Adequate determination of PSD's in relation to product quality requires six basic elements, viz. stated quantitative analysis criteria, good sampling, good dispersion/dilution of the measurement sample, well-trained analysts, good instrumentation and complete reporting of the measurement results [31].

The *analysis criteria* relate stated parameters of the PSD or particle shape or surface area to stated aspects of the product performance. This is often necessary to assist consistent production or adequate quality for the ultimate product, in a quantitative manner. This means that product performance characteristics have been translated into measurable parameters with a stated precision (repeatability, reproducibility) and resolution.

Inadequate *sampling* often limits the quality of characterization of a batch of product. Good sampling requires a measurement sample that is representative for the total batch of product and contains sufficient particles for measuring the most critical parameter (e.g. largest size or few deviating particle shapes) at the desired precision (see further Sect. 3.2).

Good *dispersion/dilution* requires that the measured state of the particles agrees with the goals for measurement. Here, the dry powder dispersion must be distinguished from the dilution of concentrated dispersions. In the analysis of dispersed dry powders, the emphasis often lies in the PSD of the primary particles. For concentrated dispersions, the goal is to characterize the state of dispersion. Effective dispersion should not introduce a change of state and adequate care should be given that this goal is fully realized (see further Sect. 3.3).

Analysts who perform the measurements should be well trained. This means that they have been proven to be capable to perform the measurements, including laboratory sampling and dispersion/dilution, at the required quality and working according to written procedures. Furthermore, they should be capable to signal any deviation from the normal situation while doing this. Their qualification should be regularly tested and approved. Testing of analysts can be carried out with typical in-house products that have PSD parameters of known quality. Well characterized in-house products can be employed for comparative measurements to test for consistent results. Certified reference materials should be used for fundamental accuracy determinations of both apparatus and analysts.

[2] Note that NIST uses the word 'monodisperse system' for a collection of particles for which at least 90 % lies within a range of 5 % of the median size, and 'polydisperse system' for a particle collection, where more than 10 % differs at least 5 % from the median size [25].

Measurement techniques should be chosen that fulfill the quality requirements. The first quality requirement is for sufficient repeatability and reproducibility (precision), but also accuracy, resolution and sensitivity for discrimination between good and poor products is often important (see further Sect. 3.4).

Good *reporting of measurement results* means that it is done in accordance with given instructions, preferably in the form of a standard analysis sheet. Often, a short report gives only few basic measured parameters necessary for process or product control purposes. Extensive reports are useful in research and to investigate problems or unexpected results. All data should be stored in a computer memory or logbook that also contains data on instrument testing/qualification and that is accessible with authorization to both measurement personnel and users of the data. Minimum report requirements are: sample date and ID; analyst ID and date; instrument ID; measured characteristic values for size, shape and/or surface area plus their range for acceptance and potential analyst remarks on signaled deviations from normal. Maximum report contents are in addition to the above data: all measured values; sampling conditions; dispersion conditions; settings of both instrument and product parameters (e.g. RI, density) and sample concentration during measurement.

3.2 Sampling

A representative test sample is required for obtaining analysis results that are representative for the corresponding batch of product. The primary lessons for sampling are to always take a sample from a moving stream and to always take the sample from the full width of the moving stream. This ideal state is not always possible. For representative sampling of dry powders, it is usually necessary to take samples at different spots from any heaps or stock piles and/or at different times during product transport. The assembled sub-samples should be mixed carefully and then subdivided in an adequate laboratory riffler to obtain the test sample. For representative sampling of liquid dispersions, it is usually sufficient to mix them well and take the sample from the mixture whilst it is moving, thus preventing or at least limiting sedimentation or other types of systematic deviation (bias).

A fully representative sample of sufficient quantity is required for effective measurement. Only with such a representative sample can the properties of the particles in the sample be said to be characteristic of the full product. This sets two requirements to the sampling procedure of particulate products [31]:

(a). The product shall be randomly sampled, or the samples or sample increments shall be collected over the full volume of the lot of product in order to account for potential segregation (related to the 'segregation error') and to give each particle the same chance of being sampled according to its proportion of occurrence.

(b). The number of particles in the sample or sample increment shall be large enough for the precision required for the measured parameter (related to the 'fundamental error', which depends on number statistics).

Ad a: Fully random sampling of particles from a lot is virtually impossible. Neither is it possible to estimate the degree of segregation by a theoretical approach; only empirical estimation is possible. Therefore, the usual procedure is:

1. To collect and analyze samples of sufficient amount (in view of fundamental error) from different locations in the lot (or at different times during product transport or from the production process). The resulting set of measurements can be used to calculate the degree of segregation from the result variations.
2. To estimate the number of different locations (or times) required for the desired precision in view of occurring segregation.
3. To use samples from those locations (or times) as sample increments to be combined into a collective sample. (Note that the collective sample may be further split up in the laboratory to form the test sample).

The segregation error (degree of segregation) can be approximated by the overall error provided that the segregation error is significantly greater than the fundamental error and the errors from dispersion and measurement are relatively insignificant. This overall error can be calculated for any parameter y from size measurements of N different samples through:

$$s_y^2 = \frac{\Sigma(y_i - <y>)^2}{N-1} \quad (3.1)$$

where:

s_y = standard deviation of parameter y (overall)
y_i = i-th measurement of y
$<y>$ = mean measurement result for y
N = number of different samples taken for analysis of y

These calculations enable estimation of the minimum number of samples or sample increments to reach a defined standard deviation for any parameter of a size distribution.

Ad b: The minimum number of particles in a sample to reach a stated precision for a given characteristic PSD parameter can be calculated on the basis of number statistics. Large numbers of particles are required for high precision. Useful equations for this purpose in case of quasi-two-component mixtures arise from binomial statistics:

$$s_p^2 = \frac{p(1-p)}{n} \quad (3.2)$$

or from Poisson statistics:

$$s_n^2 = <n> \quad (3.3)$$

where:

s_p = standard deviation of p
p = number based proportion of particles in one of the two components of the mixture
$1\text{-}p$ = number based proportion of particles in the other component of the mixture
s_n = standard deviation of $<n>$
$<n>$ = mean number of particles measured in a size class

Note that the uncertainty in the number fraction of particles is calculated above. It has to be divided by the slope of the size distribution at the corresponding point for calculation of the uncertainty in the corresponding size parameter.

The above calculations give the minimum sample size for reaching a defined standard deviation for any parameter of the size distribution. It holds for ideal, random mixtures or for ideal, fully random sampling.

For further information and some examples, reference is made to [20, 31]. The latter also gives information on sampling methods and devices.

3.3 Dispersion

Most solid particulate ingredients consist in the form of dry powders. Examples are cement, pigments, sand, sugar, ceramics and pharmaceutical powders. For measurement of their particle size distribution, the particles should present as individuals, i.e. they should be adequately dispersed. Sometimes, dry powders are dispersed in air. Such dry dispersion of powders is typically only feasible if the particles are larger than about 10 μm. Below this size, bonding in agglomerates is often too strong. The advantage of dry dispersion is that no liquid is necessary in which the particles can dissolve or swell.

Most often, however, dispersions in a liquid are prepared which can be diluted to the required particulate concentration for analysis. To achieve the desired dispersion an adequate liquid (zero solubility or swelling for the solid), a suitable dispersing agent (adequate wetting of the solid and stabilization of the suspension) and adequate dispersion energy (to break-up agglomerates) are all necessary. The advantage of liquid dispersions is that they allow much easier confirmation that the required state of dispersion has been achieved. For visual confirmation, e.g. a microscope may be used.

The general goals of dispersion are [5, 14, 21, 31, 32]:

- The particles should be contained in the final dispersion as individual entities ('primary particles'), i.e. free of agglomerates, aggregates and flocs
- The individual primary particles should not undergo a change of size – by breakage, dissolution or swelling – during the dispersion or dilution process

- The particulate concentration should be adequate for measurement
- The dispersion state should not change during analysis or testing.

There are two fundamental steps for adequate dispersion of dry powders into a liquid, viz.:

- Wetting of the solid particles by the liquid and displacement of present air. This requires a liquid of suitable, low surface tension, either or not by the addition of a surface-active agent (surfactant)
- De-agglomeration of particle clusters through application of energy to the suspension by stirring, shaking or sonication.

Wetting of the surface of solid particles is the first step in their dispersion. The driving force for wetting is the difference in surface tension between the solid-vapor interface (γ_{SV}) and the solid–liquid interface (γ_{SL}). The resisting force is the energy required for increasing the surface area of the liquid drop. The balance between these forces is represented by the equation of Young-Dupré:

$$\gamma_{SV} - \gamma_{SL} = \gamma_{LV} \cos \Theta \quad (3.4)$$

Adequate wetting is only possible if the contact angle Θ between solid and liquid is smaller than 90° ($\cos \Theta > 0$) and the driving force ($\gamma_{SV} - \gamma_{SL}$) approaches γ_{LV}. Compare, for example, water droplets on cars that are nicely waxed with the wet surface in the absence of the wax.

After wetting, much less energy is required for de-agglomeration.

The ability of liquids to wet the particles is dictated by their surface tension, which must be smaller than that of the solid. Adequate wetting properties of a liquid are shown by a low contact angle between solid and liquid. Dispersions of poor quality often show low stability with time, where e.g. a sediment layer or floating particle clusters can be seen by the eye. For clusters of small particles, microscopic examination may be necessary.

Most industrial liquid based particulate products are concentrated dispersions, either during processing and/or as end product. However, very few techniques are capable of measuring concentrated dispersions. Therefore, most product dispersions are diluted before measurement of their particle size distribution. Particle characteristics should not alter during dilution. Note that most industrial emulsions and suspensions contain emulsifiers and/or stabilizers for adequate performance. Such particle dispersions lose their stability upon dilution with mere solvent, and flocs and/or agglomerates are formed. Supplementing these stabilizing components during dilution is, therefore, essential but no guarantee of avoiding flocculation and agglomeration.

More dispersion background is provided in the references [5, 14, 21, 31, 32].

3.4 Overview of Most Popular Techniques for Size Measurement

3.4.1 Introduction

The measurement techniques for particle size distributions (PSD) show two basic differences:

(a). The principle that is employed for measurement
(b). The way of quantification of the relative amounts of particles of given sizes.

Many different principles are used for size measurement, which can be grouped into fingerprint techniques, separation techniques and particle-ensemble techniques.

In the *fingerprint techniques*, each individual particle provides a size related to a signal that comes from some kind of detection forming a 'fingerprint'. The numbers obtained from these signals are classified into size classes, often after a process of calibration. Examples are microscopy and image analysis.

In the *separation techniques*, particles of different size are physically separated. The size fractions are quantified e.g. by weighing or extinction of electromagnetic radiation. Examples are sieving and sedimentation techniques.

In the *particle-ensemble techniques*, a set of signals arising from an ensemble of particles is mathematically converted to a best-fitting size distribution, using some kind of model for particle behavior. In the ensemble method both size and quantity information are derived from the same set of signals. Examples are dynamic light scattering and laser diffraction.

All techniques should yield, in principle, the same results for ensembles of spherical particles. For non-spherical particles, equivalent sizes are obtained, however the sizes reported depend upon the principle used for the determination. Note that all techniques, except microscopy, assume in their modelling that all particles have the same physical properties, e.g. density or refractive index.

Some examples of equivalent sizes are:

- Equivalent projected area diameter, i.e. diameter of a circle with the same area as the particle
- Equivalent settling diameter (or Stokes' diameter), i.e. diameter of a sphere with the same settling rate as the particle
- Equivalent sieve diameter, i.e. diameter of a particle with the same size as the sieve opening
- Equivalent surface area diameter, i.e. diameter of a sphere with the same surface area as the particle
- Equivalent volume diameter, i.e. diameter of a sphere with the same volume as the particle

Aerodynamic (for aerosols) and hydraulic particle size (for sediments) represent special types of Stokes' diameter. Particle density for these cases is assumed to be

1,000 and 2,650 kg/m^3, respectively. These standard values are useful for settling of particle mixtures that have unknown or different densities. If the test sample density is different from the standard value then the size distribution is only a relative guide.

Note that aerodynamic sizes are, sometimes, calculated from other equivalent sizes than from the Stokes' diameter.

An overview of all particle size measurement techniques is given by Merkus [31]. In this chapter, only a summarized background is presented on some modern and popular techniques, of microscopy/image analysis, laser diffraction, dynamic light scattering (DLS), sedimentation and ultrasound attenuation. More detailed information and information on other techniques is given in the same book [31]. Note that most techniques do not allow size distribution measurements in concentrated dispersions, although DLS may provide some 'structural' information. If PSD information is required in such dispersions, e.g. if dilution results in changes of the distribution, then only sieving and currently less popular techniques such as nuclear magnetic resonance, small angle X-ray scattering and ultrasound attenuation can be used.

3.4.2 Microscopy and Image Analysis

Microscopy

Microscopy allows the direct inspection of individual particles. It provides not only information on particle size but also on particle shape. The size scale is calibrated by means of suitable reference scales or materials; the quantity information is gathered by counting the particles in relation to size. Digital image analysis is used for size analysis (see section "Image Analysis"), but direct measurement by an operator is also carried out. The experience and quality of training of the operator are important. In the following section only optical microscopy and electron microscopy are described.

Optical microscopy of resting particles is most commonly used. It uses visible light and a lens system for magnification of the particles image. Maximum magnification is about 1,500 times. Such high magnification can only be achieved using the oil immersion method. Particles can be visualized at sizes larger than about 0.3 μm; quantitative size measurement is possible from about 3 μm onwards. For quantification, both manual procedures and image analysis can be used. Information about fine structure of particle shape requires larger particles, as the ability to recognize fine structure depends on the number of picture points available at the particle's contour. In 'manual' procedures for shape recognition, usually reference shapes are used for comparison.

Optical microscopy can also be used for moving particles, e.g. in a process stream. This is becoming increasingly popular as it provides direct process information. Measurement periods should then be short in comparison to the particle velocity in the stream in order to get sharp pictures (e.g. through using flash illumination and/or short shutter times). Moreover, concentration should be low to avoid overlapping of particles in the image. In this application, maximum magnification is about 5-fold lower than in the static case.

For static images requiring large magnification, **scanning or transmission electron microscopy** (SEM or TEM) is used, with or without image analysis. SEM allows magnifications up to about 100,000 times, TEM up to about 1,000,000. In electron microscopy, electron beams (wavelength about 0.04 nm) are used instead of light and electromagnetic lenses instead of optical ones. The technique requires substantial particle preparation involving vacuum deposition of a gold coating and depositing onto a suitable filter or grid.

Summary of Quality Aspects

An advantage of optical microscopy is that it allows a quick impression of particle size range, particle shape, and particle dispersion quality. This method also enables images to be recorded for further inspection and later reference.

Type of information. Typically, a number-based size distribution is obtained of equivalent area diameters or linear size parameters, related to the images shown. Shape information can be obtained as well.

Sample type. The measurements can be performed in emulsions, powders, suspensions and dry particles on a surface. The only requirement is that the refractive index of the particles differs from that of the surrounding medium.

The *overall size range* of optical microscopy is about 0.3–500 μm. In the range 0.3–3 μm only qualitative or semi-quantitative inspections are possible. For SEM the size range is about 10 nm–500 μm, for TEM about 1 nm–5 μm. At a single magnification, the range for quantitative measurement is about a factor 30 between the largest and the smallest particle.

The typical *particle number* per image for quantitative measurement is about 5–50. The number is limited by the requirement that individual particles should not show a significant degree of overlap and have a sufficient magnification. For qualitative inspection, the number per image may be greater. For measurement of size distributions, large numbers of particles – i.e. many images – have to be analyzed in view of repeatability.

The *measurement time* is typically about 1 min to several hours. It depends on whether qualitative or quantitative information is required and which quality is required for that information.

The *repeatability* of size measurement on a single particle is about 0.5–2 %. This precision requires adequate magnification to provide a suitable number of image pixel elements for the smallest particle and good operator performance [40]. For size distributions, repeatability also depends on the number of particles counted; reasonable precision requires sizing of (often many) thousands of particles governed by the range of particle sizes within the distribution.

Good *accuracy and* absence of *bias* strongly depend upon proper calibration of the magnification and upon proper focusing for sharp images. Overlap of particles may also lead to bias, if the combination is taken as a single particle. Finally, dirty lenses and insufficient contrast between particles and background together with lack of uniform illumination of the background, may cause biased results.

Resolution of microscopic measurement can be high, provided that large particle images are available created by a significant number of pixels.

Sensitivity. Particles that deviate from the typical population can be fairly easily recognized. Quantitative interpretation of such quality requires an experienced operator.

The *traceability* of microscope techniques is good, since it is very direct and certified standards for calibration are available.

On-/in-line capability. Most applications are off-line. On- and in-line application of optical microscopy is feasible with special equipment. In-line measurement is only possible for product streams, which have a low particulate concentration.

Note that non-representative sampling is often a major error source. One advantage is that optical microscopy can show the presence of agglomerates due to inadequate dispersion.

Image Analysis

Image analysis can be understood to mean two different things. In its basic form, it is a subjective, qualitative description of images by an observer, with or without comparison with standard images. The quality of results strongly depends on the experience and quality of the observer. In the early days of microscopy, a.o. Van Leeuwenhoek (1632–1723) reported on his objects by describing what he saw. Especially when standard references are used, this approach is still fast and powerful but only a limited number of particles are generally observed. Nowadays, it usually requires digital analysis of images by computer software. For very small particles, the images are usually produced after magnification in a microscope (optical, SEM, TEM). After threshold detection, digitization and restoration, the images are processed according to some code and algorithms employed to yield size and shape information. The quality of results depends strongly on the degree of magnification (i.e. the number of pixels per particle image [40]), the quality of the image (especially contrast and uniformity), the number of images included and the quality of the algorithms.

Summary of Quality Aspects

Type of information. A number-based size distribution is obtained of equivalent area diameters or linear size parameters, related to the particle images shown. Shape information can be obtained from the images as well, often as a ratio of linear size parameters.

Sample type. The measurements can be carried out in any images coming from emulsions, suspensions and – with limitations – aerosols.

The *overall size range* of quantitative image analysis is determined by the microscope technique used for magnification of the particle images.

The typical *particle number* per image for quantitative measurement is about 5–50. The number is limited by the requirement that individual particles should not show a significant degree of overlap as well as have a sufficient magnification. Thus, the statistical significance of size distribution results can only be obtained by measurement of a large number of images.

The *measurement time* is typically about 1–60 min. It strongly depends on the degree of automation for change of image fields.

The *repeatability* of size measurement on a single particle can be about 0.5–2 %. For this precision, adequate magnification is required. Typically, area-based size has better precision than linear size. For size distributions, repeatability also depends on the number of particles counted; reasonable precision requires sizing of thousands of particles.

Good *accuracy and* absence of *bias* strongly depend upon proper calibration and segmentation and upon the creation of sharp images. Overlap of particles may also lead to bias, if segmentation is insufficient and the combination is taken as a single particle.

Resolution and *sensitivity* of image analysis results depend strongly on the number of pixels in the particle images and on the number of particles measured.

The *traceability* of image analysis in combination with a microscope is good, since certified standards for calibration are available.

On-/in-line capability. Measurements can be done off-line, on-line and in-line, provided that sharp images of good quality are available.

3.4.3 Laser Diffraction

The laser diffraction technique uses a monochromatic light source, typically a laser, to illuminate a flowing collection of particles, either dispersed in air or in a transparent liquid within a specified concentration range. The angular light scattering pattern is measured by a series of light sensitive detector elements in forward and backward directions. This pattern is converted to a PSD by means of some optical model for light scattering of (usually spherical) particles. Various optical models can be chosen. The model based on the Mie theory which is exact for all sizes of spherical particle is most often used. Its application requires information on the real and the imaginary part of the refractive index of both particles and dispersion medium. The model based on Fraunhofer diffraction theory, is often used for larger and/or opaque particles. The Fraunhofer approximation does not require information on the refractive index. The technique can be used in the laboratory as well as in processes, provided that the particle concentration is neither too high nor too low. Above the maximum concentration allowed, multiple scattering occurs that yields biased size distributions. At too low concentrations, signal to noise ratio becomes too small. Note that, with some assumptions, some macroshape information may be obtained [7, 28, 29, 31].

Summary of Quality Aspects

Type of PSD information. A volume-based size distribution is obtained of diameters of spheres that show the same scattering pattern.

Sample type. The measurements can be implemented for emulsions, suspensions (in both air and liquid), sprays and aerosols.

The *overall size range* for laser diffraction instruments is about 0.1–10,000 μm. Per measurement the dynamic range is usually about a factor 1,000.

The *typical concentration range* is about 0.001–1 % v/v (5–30 % optical concentration). It depends on particle size.

The *measurement time* ranges from about 0.01 to 30 s.

The *repeatability* of laser diffraction results can be excellent (about 0.5 % relative), provided that representative samples are being analyzed in sufficient quantities and after adequate dispersion. Poor repeatability is often caused by the use of non-representative or poorly dispersed samples.

Biased results – for non-spherical particles meaning results that deviate systematically from standard results – may be caused by sub-standard use of this method, due to application of inadequate optical models, to input of incorrect refractive index parameters, to drifted detector sensitivity (fully eliminated in current apparatus) or to application of too high particulate concentrations. It is essential to describe the input data in operating procedures and to qualify the instrument and operator performance through regular measurement of a reference material, for which adequate data is available. Note that particle shape has a significant influence upon the reported particle size, since spheres are assumed in the models.

The *resolution* of the technique is usually medium. Size differences of about 10–40 % are typically taken for the limits of the size classes. This resolution may be improved to some extent in special cases with special hardware and software as it depends upon the size and number of detector elements and the degree of smoothing applied in the deconvolution procedure. The standard size classes are satisfactory for most applications.

The *sensitivity* of laser diffraction is satisfactory to medium (better than about 5 % m/m).

Traceability. Laser diffraction is a first principles technique but due to the additional uncertainty and errors induced by dispersion, sampling, material properties and operator parameter choices calibration is necessary to ensure traceability.

On-/in-line capability. Off-line as well as on- and in-line application is feasible. In-line measurement is possible for product streams, which have a suitable (low to medium) particulate concentration in a transparent medium.

3.4.4 Dynamic Light Scattering(DLS) or Photon Correlation Spectroscopy (PCS)

As in laser diffraction, an ensemble of particles, dispersed in a transparent liquid, scatters light from a laser beam. In DLS the variation of the scattered light intensity with time, at some defined angle, is measured.[3] The rate of change of this intensity is related to the diffusion coefficient of the particles, which in turn is related to hydrodynamic particle size by the Stokes-Einstein equation (see also Chap. 2).

$$\mathbb{D} = \frac{k_B T}{3\pi \eta_L D_h} \tag{3.5}$$

where:

$\mathbb{D}$ = diffusion coefficient
k_B = Boltzmann constant
T = absolute temperature
η_L = liquid viscosity
D_h = (apparent) hydrodynamic particle size

Several mathematical methods are used for conversion of the intensity-time relationship to a PSD. Signal processing can be carried out either with a digital Correlator or by Spectrum analysis. From either of the signal processors a further mathematical conversion is required. Such data is regarded as ill-conditioned and has a poor signal-to-noise ratio which limits the amount of PSD information that can be obtained. Usually, the polynomial 'cumulants' method is favored for data analysis. It leads to a mean size and a value for PSD width. Other methods claim to lead to a PSD, but often results are not stable.

There are two versions for measurement. The conventional technique operates usually at a fixed angle of 90°, or another specified angle, at very low concentration. New techniques, such as fiber-optics quasi-elastic light scattering (FOQELS) and diffusive wave spectroscopy, use back-scattered light and may operate at higher concentration. At higher concentrations particle-particle interactions often influence the particle movement and, thus, the sizing result. However, a correlation with product quality may be found provided that the particulate concentration is kept constant.

A variant of the technique can also be used for measurement of the zeta-potential (see Sect. 3.7).

Summary of Quality Aspects

Type of PSD information. Typically, a size distribution is obtained for diffusion coefficient related diameters, for which the quantity axis relates to scattered light intensity. For particles less than 100 nm scattering intensity reduces as the 6th power of particle diameter. In liquid dispersions, these hydrodynamic diameters may be (slightly) larger than the actual particle size as they include a layer of

[3] A new instrument measures Brownian motion of suspended particles directly, through microscopic inspection after illumination by a focused laser beam.

attached ions often needed to stabilize the dispersion. This additional layer of ions is described as the Stern layer [31].

Note: The extended layer caused by the attached ions increases the surface drag of the particle causing it to diffuse more slowly which is then reported by DLS as a larger hydrodynamic particle size.

Sample type. The measurements can be executed as emulsions, lipids, colloids and suspensions and – with substantial limitations – to aerosols.

The *overall size range* of the technique in liquids is about 0.005–1 μm. The lower size limit depends upon the intensity of the illuminating light source since the scattered light intensity for very small particles drops with the 6th power of particle size. It also depends upon the difference in the refractive index of the test sample particles compared with the suspending liquid. The upper size limit depends upon the density of the test material, i.e. the onset of sedimentation, together with the low number of large particles with their attendant long diffusion times. These combined influences may result in a practical upper size limit being well below 1 μm. Best results are obtained for fairly narrow size distributions.

The *typical concentration range* for measurement is around 10^{-2}–10^{-3} % v/v. In FOQELS and diffusive wave spectroscopy, higher concentrations may be used. Then, the resulting ‘size’ is dependent on particle-particle interactions.

The typical *measurement time* is about 0.1–5 min.

The *repeatability* of the mean DLS diameter is very good if the cumulants procedure is applied, viz. better than about 2 % relative. Repeatability of the polydispersity index is typically poor, viz. about 20 % relative. Precision, due to the intensity weighting, remains limited in general and quite variable if other information extraction procedures are used for obtaining size distributions.

The *bias* of the mean intensity weighted DLS diameter obtained in the cumulants procedure is, at optimum conditions, less than about 2 % relative. Note that bias is to be interpreted for non-spherical particles as systematic deviation from standard results by this technique. However, DLS is vulnerable to distortions and artefacts caused by applying different de-convolution procedures, increased particulate concentrations (due to particle-particle interactions and multiple scattering), too low concentrations (due to fluctuations of the number of particles in the measurement zone) and presence of one or few relatively large particles (‘dust’) or air bubbles. Validation of results is recommended at regular time intervals and whenever it seems necessary. It can ensure proper functioning of the instrument as well as competence of the operator.

The *resolution* of the technique is low. The cumulants procedure only delivers a mean DLS diameter together with a polydispersity index for PSD width. Special mathematical procedures such as CONTIN and MEM claim that they are capable to extract more PSD information. The results, however, have low resolution and are very sensitive to noise.

Sensitivity for small changes in the PSD is very poor due to the limited resolution. In fact, changes can usually only be identified through changes in the mean size or the polydispersity index.

Traceability of the DLS technique is good, since first principles are applied. Of course, this only holds within the limits of proper application.

On-/in-line capability. Off-line measurement is preferred, since any flow influences the diffusion. Stop-flow on-line measurement is only possible, if this is not the case. Furthermore, on-line measurement requires a product stream, which has a (very) low particulate concentration. Note that relative product performance parameters may be extracted at high concentrations, if the concentration is kept constant.

3.4.5 Ultrasound Attenuation

In this technique, the attenuation of ultrasound and/or retardation of ultrasound velocity by an ensemble of particles suspended in a liquid is measured at a series of frequencies. Similar to laser diffraction, the measured pattern is then converted to a PSD by a mathematical deconvolution through the use of a matrix that contains attenuation patterns per unit volume of particles in defined size classes. This matrix is either calculated from a theoretical model or obtained in an empirical way from measurements of known size fractions of the same material. In order to construct the model-based matrix, various properties of both particulate phase and dispersion medium have to be known. These relate to thermodynamic, mechanical and transport behavior. These models are sometimes simplified through an assumption of some kind of model size distribution (e.g., normal or log-normal). At increased concentration, particle–particle interactions and overlap of acoustical fields of different particles may become significant in relation to attenuation (depending on material properties). In such cases adaptation of the model is required to compensate for these effects, or specific parts of the ultrasound attenuation spectrum is used (in case of process control).

A new instrument operates as a particle counter. A single transducer is used as transmitter and receiver. Signals are classified by their amplitude and then converted to particle size.

Summary of Quality Aspects

Type of PSD information. A volume-based size distribution is obtained modeled upon spherical particles that shows the same ultrasound attenuation pattern.

Sample type . The measurements can be performed in emulsions and suspensions.

The *overall size range* for ultrasound attenuation is about 0.01–3,000 μm. This range can often be met without changing the instrument. The range of frequencies covered by an instrument determines the measurement range.

The *typical concentration range* is about 0.5–70 % (v/v). For the newly developed counting instrument it is about 1–1,000 ppmv.

The *measurement time* is about 1–10 min.

Repeatability of characteristic sizes in between D_{10} and D_{90} is typically within 3 % (coefficient of variation).

Bias of particle size distributions – for non-spherical particles meaning results that deviate systematically from standard results – may be caused by sub-standard use of this method. Typically, bias may be smaller than about 3 %. Application of inadequate models/model parameters in the matrix that relates extinction spectrum to PSD and particulate concentration or inaccurate calibration may lead to more severe bias. Regular qualification of instruments is recommended.

The *resolution* of the technique is low. Typically, particle size distributions are presented as log-normal distributions, i.e. as the mean size and the standard deviation of such distributions.

The *sensitivity* of the technique is, in relation to resolution, also low. Often, US attenuation at one or few specific frequencies is used for control purposes.

The direct *traceability* of sizing results is poor, due to the complex theory behind it and the deconvolution procedure. Both attenuation spectra and PSDs can be validated through use of appropriate reference materials at appropriate conditions.

On-/in-line capability. Measurements can be carried out off-line as well as on- and in-line. In-line measurement is possible for product streams, which have a medium to very high particulate concentration.

3.4.6 Gravitational and Centrifugal Sedimentation

Sedimentation is applied for particle sizing in suspensions within various types of instruments. The instruments differ, first in the force field that is applied for sedimentation, viz. gravity or a centrifugal field. A second difference is that either a line start or a homogenous start is applied. A third difference is the way of detection of particle quantity after size separation. All techniques require a known liquid viscosity, a constant temperature, absence of sources of vibration and knowledge of effective particle density.

A short overview of gravitational sedimentation and centrifugal sedimentation is given below.

Gravity Sedimentation

The terminal settling velocity of spherical particles in suspensions due to gravity is the basis for this technique. Viscous – creep flow conditions and dilute liquid dispersions are assumed (i.e. Reynolds number for settling particles < 0.25) and Stokes' law is applied to convert this velocity to particle size (see also Chap. 2).

$$v = \frac{H_s}{t} = \frac{(\rho_P - \rho_L) \cdot g \cdot D_{St}^2}{18 \cdot \eta_L} \tag{3.6}$$

where:

v = terminal settling velocity of a particle
H_s = settling height
t = time for settling over height H
ρ_P = effective particle density
ρ_L = liquid density
g = gravitational acceleration constant
D_{St} = equivalent Stokes' diameter of a particle
η_L = liquid viscosity

Effective particle density must be known for this conversion (in addition to liquid viscosity and liquid density). The word 'effective' means that it takes into consideration the effects coming from gas or liquid present within potential closed and open pores of the particles, which decrease their true solid density. Its measurement should be described in a written procedure that guarantees both optimum dispersion and pore filling. 'Hydraulic' particle size assumes a particle density of 2,650 kg/m^3. Generally, one starts with a homogenous suspension of particles. In the most simple (pipette) version of the technique, samples are taken at predefined times at a fixed or a variable height in the dispersion, and then dried and weighed. In the instrumental version, X-rays, visible light or a balance is used to obtain remaining particle mass concentration or settled mass reported against time.

Summary of Quality Aspects

Type of PSD information. A mass-based size distribution is obtained of Stokes' diameters. Note that the method of detection may set limits to the mass determination.

Sample type. The measurements are executed in suspensions, in which there is sufficient (positive) density difference between particles and medium.

The *overall size range* for gravity sedimentation in aqueous suspension is about 0.3–200 μm. For particles larger than about 65 μm or smaller ones having a high density, more viscous liquids other than water must be used in order to sustain viscous-creep flow conditions. When a major part of the particles is smaller than about 1 μm, centrifugal sedimentation is recommended.

The *typical concentration* for gravity sedimentation is smaller than about 0.2 % v/v, in order to avoid particle-particle interactions.

The *measurement time* is about 0.5–8 h, depending on the rate of settling and operational conditions.

At optimum conditions of settling rate and particulate concentration, the *repeatability* (standard deviation) of the results is better than 1 % w/w at the same Stokes' diameter. Precision of manual operations depends strongly on the skills of the operator.

Biased results are obtained when standard conditions are not met, e.g. if a too high concentration is used. Note that bias is to be interpreted for non-spherical particles as systematic deviation from standard results by this technique. At the lower size end, Brownian motion may shift the sizing results to lower sizes, since it counteracts sedimentation. Regular validation of both instrument and procedure with certified reference materials can expose this.

Size resolution is about 5–10 %; instrumental methods are better than manual ones.

The *sensitivity* of the technique for fractional mass is typically better than 2 % w/w. For sedimentation instruments it depends on the net extinction by the particles and the stability of the extinction signal.

Traceability of the technique is good, since first principles are applied and the total equipment can be calibrated and validated. Note that an equivalent sedimentation or Stokes' diameter is obtained.

On-/in-line capability. Measurements can only be carried out off-line.

Centrifugal Sedimentation

Centrifugal sedimentation uses the centrifugal force for more rapid settling (within the Reynolds number limit) of the particles in a liquid than occurs under gravitational conditions.[4] It is recommended when a major part of the particle distribution is smaller than about 1 μm, where the influence of Brownian motion becomes significant in gravity sedimentation. Centrifuge speeds of 500–24,000 rpm are used, where smaller particles require higher speeds. Two types of instruments are used, viz. the disc centrifuge and the cuvette centrifuge. A line start is applied in the disc, a homogenous start in the cuvette. White or monochromatic light or X-rays are used for detection of particle concentration against time at a given position. Here, an adapted Stokes' law is applied for conversion of settling rate to particle size.

$$D_{St} = \sqrt{\frac{18 \cdot \eta_L \cdot \ln(R_d/R_i)}{(\rho_P - \rho_L) \cdot \omega^2 \cdot t}} \tag{3.7}$$

where:

D_{St} = equivalent Stokes' (hydrodynamic) particle diameter
η_L = liquid viscosity
R_d = distance between detector and center of rotation
R_i = distance between injected sample layer and center of rotation
ρ_P = effective particle density
ρ_L = liquid density
ω = angular velocity of centrifuge ($2\pi N_c/60$), radians/s

[4] Also an aerosol centrifuge has existed for separation of aerosol particles according to size.

t = sedimentation time
N_c = centrifuge speed, rpm

Summary of Quality Aspects

Type of PSD information. A mass-based size distribution is obtained of Stokes' (hydrodynamic) diameters, which may include a layer of adsorbed dispersant molecules. The quantity axis depends on the technique of concentration measurement (light or X-rays).

Sample type. The measurements are carried out in suspensions, in which there is sufficient (positive) density difference between particles and medium.

The *size range* for centrifugal sedimentation in water is about 0.02–10 μm. This can only be met in a single analysis if the centrifuge speed is programmed with time.

The particulate *concentration* should be typically smaller than about 0.2 % v/v, in order to avoid particle-particle interactions.

The *measurement time* is about 10–20 min.

At optimum conditions the *repeatability* of results (standard deviation) is better than 1 % w/w at the same hydrodynamic diameter. Precision of manual operations depends strongly on the skills of the operator.

Biased results are obtained when analysis conditions systematically deviate from standard conditions. Note that bias is to be interpreted for non-spherical particles as systematic deviation from standard results by this technique. Regular validation of both instrument and procedure with certified reference materials can show this.

Resolution capability depends on the technique and conditions used. An optimum disc design allows a size resolution of about 5 %; the cuvette technique is slightly worse, about 10 %.

The *sensitivity* of the centrifugal sedimentation technique is usually better than 2 % w/w. It is worse for particles, which have a low extinction coefficient.

Traceability of the technique is good, since first principles are applied and the total equipment can be calibrated and the procedure validated.

On-/in- line capability. Measurements can only be carried out off-line.

3.5 Particle Shape and Its Measurement

3.5.1 Introduction

Besides particle size, particle shape may have a significant effect on the properties of a particulate product. This influence is sometimes dominant. This is, for example, the case for fibers, flakes and platelets in relation to flow, packing and scattering behavior.

Conceptually, particle shape is the pattern of all points on the boundary of a particle. Thus, it includes every aspect of external morphology of the particle and consequently exhibits a wide variety of aspects. Three scales of shape can be discriminated: macroscale, mesoscale and microscale:

The *macroscale* is related to the general 3-dimensional form of particles, for example the ratio of their main three dimensions as used in aspect ratio, elongation and flakiness.

The *mesoscale* regards the general aspects of the roundness and angularity of the particle's contour.

The *microscale* involves surface rugosity or smoothness as well as porosity and other structural heterogeneities.

An overview of shape characteristics and their measurement is given in [17, 31].

3.5.2 Shape Measurement

A *qualitative* or *semi-quantitative* description for particle shape on macroscale and mesoscale can be given through their evident properties on the eye. The US Pharmacopoeia [34] gives such descriptions for pharmaceutical particles; BS 2955 [6] provides similar descriptors (e.g. acicular/needle shape, angular, dendritic, plate, rod, rounded). These descriptions can be used to express an expectation of the behavior of powders and/or particles. For example, the flow behavior of a powder and the vulnerability of particles to breakage or attrition during processing. The usual procedure is to magnify a suitably dispersed set of particles and to compare the particle images by the eye, with a standard set of images. If this exercise is conducted manually, then it will be clear that good direct observation by the eye of many particles requires both experience and time.

A *quantitative* description for particle shape on the macro- and mesoscale arises after magnification by optical or electron microscopy, from image analysis (see Sect. 3.4.2). *Note that large magnifications are required to reach sufficient accuracy for all parameters involved in the shape parameters* [17].

The determination of parameters such as e.g. length, breadth, perimeter and contour angles is mostly used, which are then related to e.g. aspect ratio and angularity.

In the so-called R-theta method, the distance of the center of gravity of a particle's projection to its contour is measured in many directions (i.e. over a large number of directions) [4, 27, 31]. In this way, the contour is 'unrolled'. The results are then compared to a library of given shapes, or they are evaluated in a Fourier series:

$$F(\Theta) = A_0 + A_k \cdot \cos(k \cdot \Theta - \phi_k) \tag{3.8}$$

where:

$F(\Theta)$ = Fourier function in polar coordinates; $R/\langle R\rangle$ as a function of angle Θ
Θ = (polar) angle for measurement of R
A_0 = mean radius $\langle R\rangle$
A_k = Fourier coefficients (harmonic amplitudes in shape function)
R = distance between center of gravity and contour of particle projection
$\langle R\rangle$ = mean value of R (radius of circle with the same area as the particle projection)
k = harmonic number in Fourier shape function
φ_k = phase angle of the k-th harmonic

In this equation, the lower-order harmonics (lower frequencies) reflect the aspects of macroscopic shape, the higher-order elements surface angularity and rugosity. At increasing k, more detail of the contour is revealed (note that this is only possible if the number of measured data points is large).

Fractal analysis of the contour of the particle's projection offers another method for shape determination. Different length scales λ are then used to evaluate the contour length (perimeter, $P(\lambda)$). In a given contour, decreasing length scales lead to increased perimeters, as they are capable of showing more and more detail of the fine structure, provided that a picture of high resolution is available [26, 31]. A straight line on a double-logarithmic plot of $P(\lambda)$ against λ relates to a self-similar structure, whose slope is the so-called fractal dimension δ. Small values of δ relate to simple structures (e.g. linear shape in 2-dimensional images), large values to complex information.

Sometimes, shape information is derived from the ratio between sizing results coming from different techniques. For the so-called Waddell sphericity factor, shape is defined by the ratio of surface area of a sphere with the same volume as the particle, compared to the actual surface area. Shape information may also be derived from theoretical behavior, such as in measurement results from the angular scattering pattern in laser diffraction [7, 31, 39].

3.6 Porosity, Surface Area and Pore Size Distributions and Measurement

3.6.1 Introduction

Porosity, surface area and the pore size distribution are cardinal properties in the capacity and selectivity of particles for absorption, adsorption, catalysis and separation. These same properties may also affect the flow of dry powders. For solid products, large pores and inter-particle voids are often the instigator of cracking of the product.

3.6.2 Porosity and Measurement

Porosity of Dry Powder Beds

Rocks and powdered rock are similar in that they are both solids, but differ in the fact that the rock is typically in one piece but the powdered rock consists of many particles. In between these particles, empty spaces or voids exist. The porosity of dry powders in a bed or batch typically considers only the fraction of voids in between the particles. Thus, void fraction is easily derived from the difference between effective bulk density and particle density if the particles do not contain pores.

$$\varepsilon = \frac{V_V}{V_B} = 1 - \frac{V_P}{V_B} = 1 - \frac{\rho_B}{\rho_e} = 1 - \frac{M}{\rho_e V_B} \tag{3.9}$$

where:

ε = void fraction in powder bed (powder bed porosity)
V_V = volume of voids
V_B = volume of powder bed (=$V_P + V_V$)
V_P = volume of particles in bed
ρ_B = effective bulk density of powder bed
ρ_e = envelope particle density (including all intra-particle pores)
M = mass of powder bed

Two types of powder bed porosity (powder bulk density) can be distinguished, viz. porosity of loose powder packing (usually measured under standardized conditions), and porosity after tapping until a constant volume is achieved.

Sometimes, the particles contain pores. For the proper calculation of the void fraction, the volume of all pores should be included in the envelope particle density. Note that the sizes of pores and voids may show overlap.

Note: Solid materials, e.g. concrete or pottery, often exhibit porosity that results from voids in between particles being created as they are formed. If the porosity is significant and the pores large, then cracks can be formed weakening the structure. These pores, if open, may be measured through mercury penetration or similar techniques (see section "Mercury Penetration Technique").

Porosity of Particles

Porosity of particles may have different functions. If the pores are open and accessible, they can provide access to a large active surface area for e.g. catalytic reactions, ion exchange and adsorption. In other cases, pores may be closed and, thus, in-accessible. With closed pores the effective density will be (much) smaller than the true density of the material. This may result in e.g. light-weight materials with increased insulation properties.

Measurement of Void Volume and Total Pore Volume

The total volume of open pores and voids in a dry powder can be determined by **titration** with a non-viscous liquid that is capable of wetting the solid, e.g. water or ethanol for hydrophilic powders or linseed oil for pigments. Titration involves liquid being added dropwise to a known mass of powder in a small container, while mixing with a spatula [2, 3]. Initially, the open pores in the particles are filled and the powder's appearance remains dry. As soon as all open pores are filled, the appearance changes into wet. This point represents the volume of the open pores. Continued addition of liquid fills the voids and the particles start forming small lumps. The titration end point occurs when the lumps have formed a single 'ball' and liquid start smearing the wall of the container. The volume of liquid used at that point is an indicator of the total pore volume.

The total volume of open pores and voids in a dry powder can also be derived from the measurement results of pore size distributions (see Sect. 3.6.4).

The presence and volume of in-accessible pores can only be determined by comparing the effective particle density with the true material density or by visual inspection or image analysis of cross-sections of particles or materials.

3.6.3 Surface Area and Measurement

The surface area of solid particles reflects both the external surface and the surface of the accessible pore walls. The external surface of solid, non-porous, spherical particles can be easily derived from its inverse relationship with their diameter:

$$S = \frac{6}{D \cdot \rho} \tag{3.10}$$

where:

S = specific surface area (usually expressed in m^2/g adsorbent)
D = particle diameter
ρ = true particle density

The total surface area of porous particles is often much larger than the external surface. It is usually measured by the gas adsorption method. In this method, the fraction of gas molecules adsorbed at the surface at constant temperature is determined at various relative gas pressures (adsorption isotherm). Most often, nitrogen or argon is used as the adsorptive at liquid nitrogen temperature (about – 196 °C). At some point on the isotherm, a complete monolayer of adsorbed molecules (adsorbate) is assumed to have formed. At this point the corresponding surface area is calculated from the amount of adsorbate together with the known cross-sectional area of the gas molecules adsorbed [1, 18, 30]. Typically, the BET equation (Brunauer-Emmett-Teller) is used for evaluation:

$$\frac{P}{V(P_0 - P)} = \frac{1}{V_m \cdot C} + \frac{C-1}{V_m \cdot C}\frac{P}{P_0} \tag{3.11}$$

where:

P = actual pressure during measurement
P_0 = saturation pressure of adsorptive
V = volume of adsorbate at pressure P (usually expressed in mole, or cm^3 at standard T and P)
V_m = monolayer of adsorbate (usually expressed in mole, or cm^3 at standard T and P)
C = constant (related to adsorbent-adsorbate interaction energy; dependent on type of porous material)

This Eq. 3.11 results for suitable materials in straight line plots of $P/V(P_0 - P)$ against P/P_0. The slope of this line is $(C-1)/V_m \cdot C$, with the intercept being $1/(V_m \cdot C)$. They are calculated by linear regression or determined graphically. The BET equation usually holds for type II and type IV isotherms and relative pressures P/P_0 of 0.05–0.35. Subsequently, V_m is calculated from slope and intercept. and, the specific surface area S(BET) by using Avogadro's number, the cross-sectional area of the adsorbate molecule and the amount of adsorbent used in the determination. If V_m is expressed in moles, Eq. 3.12 can be used for the calculation:

$$S(BET) = V_m \cdot a_m \cdot N_A / M \tag{3.12}$$

where:

S(BET) = specific surface area according to the BET method (here in m^2/g)
V_m = monolayer of adsorbate (here in mole)
a_m = cross-sectional area of adsorbate (e.g. for N_2: $0.162 \cdot 10^{-18}$ m^2)
N_A = Avogadro's number ($6.022 \cdot 10^{23}$ mol^{-1})
M = mass of adsorbent (here in g)

For C-values much larger than 1, the BET plots can be assumed to go through the origin and single-point measurement may be done at a fixed relative pressure (P/P_0 usually about 0.3).

Microporous materials show very strong adsorption. They exhibit a Langmuir type of isotherm (type I), since the size of the pores restricts adsorption to one or a few layers of adsorbate.

Frequently volumetric methods are used for the determination of adsorption isotherms of gases. These are the same methods as used for pore size distributions (see Sect. 3.6.4).

The *repeatability* of surface area measurements is about 1 % relative ± 0.1 m^2/g.

The BET method can only be used if some inflection point at complete monolayer coverage of the surface is visible in the adsorption isotherm, i.e. if there is adequate attraction between the surface of the adsorbent and the adsorptive. This is not the case if the C-value is near to 1 or negative. Neither is the method applicable in the case of micropores, where pore volume rather than monolayer coverage sets limits to the adsorption.

Complete 'outgassing' of the samples is essential for good results.

Various standard reference materials are available for validation of instruments and measurement results.

3.6.4 Pore Size Distribution and Measurement

Porous materials typically show a distribution of pore sizes. They are classified in three categories, viz. [37]:

- Macropores, with pore diameters larger than 50 nm
- Mesopores, with pore diameters in the range 2–50 nm
- Micropores, with pore diameters smaller than 2 nm.

Pore size distributions are determined by either gas adsorption methods (section "Gas Adsorption Methods") or mercury penetration (section "Mercury Penetration Technique").

Gas Adsorption Methods

In the gas adsorption method, the fraction of gas molecules adsorbed at the surface at constant temperature is determined at various relative gas pressures in a so-called adsorption isotherm. Nitrogen and argon are most often used as adsorptive at liquid nitrogen temperature (about 77 K or – 196 ºC). The volumetric measurement method is most popular. Initially a known amount of sample is 'outgassed' (i.e. all adsorbed molecules and air removed) by high vacuum and/or elevated temperature and the dead space of the equipment filled with test sample determined (with the aid of a known volume of helium at known pressure). Starting again from vacuum, known volumes of adsorptive are added stepwise and the resulting pressure determined after adsorption equilibrium is attained, until saturation pressure and related full pore filling is reached. Subsequently, the desorption isotherm may be determined by gradual decrease of the pressure. Both the adsorption and the desorption isotherm can be used as the basis for calculation of pore size distributions. The choice depends on several factors, viz. the type of pore system, whether relatively small pores are the only entrance and exit for larger pores and the presence of a tensile strength effect. Often, the type of hysteresis loop provides information on the shape of the accessible pores. This data may also give an indication whether the adsorption or the desorption isotherm is to be preferred [12, 37]. Most often, cylindrical pores (or, in some specific cases, slit pores) are assumed at whose walls the adsorbate exists as a liquid layer. For these the Kelvin equation leads to calculation of pore size [1, 23, 24, 30, 36].

For cylindrical pores:

$$d_p = 2r + 2t = \frac{4V \cdot \gamma}{R \cdot T\ln(P_0/P)} + 2t \tag{3.13}$$

and for slit pores:

$$W = 2r + 2t = \frac{2V \cdot \gamma}{R \cdot T\ln(P_0/P)} + 2t \tag{3.14}$$

where:

d_p = diameter of cylindrical pore
W = width of slit pore
r = radius or half-width of pore after some space has been occupied by liquid layer t
t = thickness of adsorbed liquid layer at pore wall
V_l = molar volume of the liquid condensate
γ = surface tension of the liquid condensate
R = gas constant
T = absolute analysis temperature

The thickness t of the adsorbate layer can be derived in most cases from the 'common' t-curve, which gives the average layer thickness independent of relative pressure for a variety of materials. Typically, t is in the order of 0.5–4 nm.

Summary of Quality Aspects

The *size range* for measurement of pore diameters is about 0.5–200 nm.

The *repeatability* of the measurements is about 3 % relative.

Complete 'outgassing' of the samples is essential for good results. This is especially true when micropores are present, in which adsorption can be strong. Note that the presence of some adsorbates (e.g. water) at vulnerable, large surfaces at high temperatures may decrease surface area by sintering.

Mercury Penetration Technique

This technique uses the principle that a non-wetting liquid can enter the accessible pores of a particulate system only when forced by pressure. Mercury is one of the very few liquids that provides the required property of non-wetting, regardless of the pore wall material. Thus, the pore size is determined by forcing mercury into an evacuated sample under increasing pressure, while measuring the volume of mercury intruded as a function of pressure. Measurements occur at each stepwise increasing pressure. Sometimes, they are followed by measurements at decreasing pressures, for example to check whether the pores have changed through the influence of the pressure applied. Pore size is calculated by means of the Washburn equation, which assumes free access of the mercury to cylindrical pores [1, 22, 30]:

$$d_p = \frac{-4\gamma_m \cdot \cos\Theta}{P} \tag{3.15}$$

where

d_p = diameter of a cylindrical pore
γ_m = surface tension of mercury
Θ = contact angle of mercury at the pore surface
P = applied pressure

A pressure range of about 0.001–400 MPa is covered in instruments, which allows pore diameter measurements of about 800 μm to about 4 nm. The Washburn equation shows an inverse relationship between pore size and pressure. Thus, the large pores are intruded at low pressure; the small pores require high pressures. At pressures near ambient pressure, small pressure variations have a significant influence on pore size. Thus, size resolution is low here. Note that voids are included in the measurement.

If the access to wide pores or voids is constrained by narrow channels, as in e.g. ink bottle-type pores (having a narrow entrance and a wider body afterwards) and voids in between particles, then the mercury intrusion occurs at the higher pressure relating to the narrow entrance. In these cases the total intruded volume is only attributed to the size of the entrance, and the inner size of the pores or voids remains undetected. The inner void size is often important, for example in relation to the strength of concrete, where cracks find their origin in the larger voids (see Chap. 7). Intrusion of Wood's metal at elevated temperature instead of mercury, which has a similar contact angle, may be applied to circumvent the limitation of the entrance. This metal has a melting point of about 66 °C and, thus, is a solid at room temperature. This permits the making of product sections, followed by quantitative image analysis of their electron microscopic images (see Sect. 3.4.2). This method clearly show the void sizes [10, 38, 41].

Summary of Quality Aspects

The *repeatability* of the mercury intrusion measurements is about 3 % relative.

The maximum volume of intruded mercury gives the pore (and void) volume in the meso- and macropore range. If high pressures are applied, then it may also include volumetric changes of the sample induced by the pressure (e.g. for compressible materials, originating from elasticity or structural collapse).

Good removal of adsorbed species from the samples by 'outgassing' before analysis is essential for good results.

Note: Mercury is poisonous. Therefore, its use is restricted in most countries.

3.7 Zeta-potential and Its Measurement

3.7.1 Zeta-potential

The zeta-potential (ζ-potential) plays an important role in the stability of colloidal suspensions and emulsions. It is defined as the electrostatic potential at the surface of shear of a particle suspended in a liquid. The electric charges result from ionization of hydroxyl groups at the particle surface through the influence of the pH and/or by adsorption of ions at this surface. The distribution of charges is affected by the dielectric constant of the dispersion medium. In aqueous dispersions, there is also adsorption of some water molecules in the boundary layer. These adsorbed species are so strongly bonded to the surface that they go with the particle if this moves.[5] The electrostatic potential results in repulsive forces between particles, which limit their free movement and prevent them from coming in close vicinity (see also Sect. 2.6). At somewhat higher concentration, it may lead to some 'structure' in the dispersion.

3.7.2 Measurement

Zeta-potential is not directly measured but is calculated via the electrophoretic mobility. The relationship between zeta-potential and electrophoretic velocity in an electric field is provided by the following equation [9, 11, 14, 15, 19].

$$v_{el} = \frac{2\varepsilon_d \cdot \varsigma \cdot E \cdot F(\kappa a)}{3\eta} \tag{3.16}$$

where

v_{el} = electrophoretic velocity of dispersed particle
ε_d = dielectric constant of dispersion medium
ζ = zeta-potential of dispersed particle
E = strength of electric field applied
u_{el} = electrophoretic mobility of dispersed particle ($= v_{el}/E$)
a = particle radius
κ = inverse statistical thickness of double layer around particle
$F(\kappa a)$ = Henry's function (depends on particle shape)

[5] Note that the boundary layer represents a real distance but is conceptual, since the surface of shear varies slightly with time because there is a rapid exchange of adsorbed species and those in the dispersion medium. Thus, the surface of shear and the boundary layer have a statistical nature. This is the reason for the term 'statistical thickness of double layer'.

Henry's function $F(\kappa a)$ relates to the ratio of the particle radius and the statistical thickness of the double layer (or Debije Length; $DL = 1/\kappa$). Large values indicate a relatively thin double layer and vice versa. The value of $F(\kappa a)$, in aqueous dispersions of high electrolyte concentration is often taken to equal 1.5. The Helmholtz-Smoluchowski approximation which assumes the electric field to be uniform and parallel to the particle surface, is valid for $\kappa a > 200$. In non-aqueous dispersions, with a low-dielectric constant, $F(\kappa a)$ may be assumed to equal 1 (Hückel approximation, which disregards the deformation of the electrical field in the neighborhood of the particle; and valid for $\kappa a < 0.3$). Values in the intermediate range may be calculated from Henry's function [33]. Some values are presented in [15, 35]. Note that Rahaman [35] gives values for ($F(\kappa a)$–1).

The electrophoretic velocity can be measured in several ways:

(a). Directly under a microscope
(b). In dynamic light scattering instruments by measurement of the Doppler shift of the frequency of the scattered light (caused by the movement of the particles; LDE = Laser Doppler Electrophoresis) or by measurement of its phase shift (PALS = Phase Analysis Light Scattering)
(c). In electro-acoustic instruments by measurement of the Colloid Vibration Potential (CVP) or the Colloid Vibration Current (CVI) or the Electrokinetic Sonic Amplitude (ESA).

The latter two techniques (b and c) are more sensitive and thus do not require strong electric fields. Strong fields may cause Joule-Thompson heating and thus distort the measurement due to thermal convection. These two techniques are more suited at low values of the zeta-potential [8, 11, 15]. A further advantage of the electro-acoustics technique is that it allows measurement in concentrated dispersions. Thus, the original dispersion can be inspected and dilution of the dispersion, which may lead to unwanted agglomeration or flocculation, is avoided [11].

Adequate cell design and measurement locations in the cell are required to mitigate ion induced fluid flow caused by the cell wall potential.

Summary of Quality Aspects

The *size range* for zeta-potential measurement depends on the technique used, but is at best about 0.005–30 μm.

Repeatability of zeta-potential measurements is about 0.3 mV, that of electrophoretic mobility about 2 % (relative standard deviation).

The typical *particle concentration* range for diluted samples is about10^{-3}–10^{-6} % v/v. Dilution of concentrated samples must be done with due care in order to preserve the existing charge state of the particle surface. This can be done by imitating the dispersion medium or by using clear centrifuged or filtered medium from the original dispersion.

Note that strong electric fields (field larger than about 100 V/cm) may heat up the dispersion and cause thermal motion of particles that will mask electro-kinetic effects.

Typically, titration facilities are present in the instruments so that the effects of pH on zeta-potential can be determined.

3.8 Standardization and Qualification

The word standardization used in this section has two aspects, viz. application of written standard procedures as well as the use of standard reference materials. National and international written standards exist for many measurement techniques and for many industrial products. At present, international standards have preference in view of the intensity of international trade. Most well known are ISO (International Standardization Organization), EN (European Union) and ASTM (American Society for Testing and Materials). They aim at optimum use of measurement instruments as well as describing optimum methods for measurement of particulate products. In the preparation of these standards, many experts from a variety of disciplines have been active to deliver contributions, so that the result is combined wisdom and experience. Often, the starting point is development of standards by individual companies to maintain their measurement quality and, thus, product quality.

Standard reference materials are available from official institutes like NIST (USA), BCR (Europe), APPIE (Japan) and BAM (Germany). Various experts have contributed to optimum certification results for these materials. These materials provide the best possibilities for testing and qualification of instruments. In addition, commercial companies offer a large variety of reference materials, some of which have traceable characteristics. For qualification, i.e. testing of instruments and operators within a company, and for operator training, batches of well-tested own products may be used. The choice of reference material depends upon characteristics of typical product, as well as upon the object of testing.

It is strongly recommended to visit the web sites of the institutes and companies involved to obtain an overview of available material. It is equally recommended to use both types of standards in ones work to ascertain optimum quality of measurement results. Moreover, ISO 9000 qualifications can only be obtained by maintaining adequate control of measurement quality. More information is given in [31].

3.9 Definitions, Abbreviations and Symbols

Agglomerate	Assemblage of primary particles with intermediate attractive forces (sometimes named aggregate)
Aggregate	Assemblage of primary particles with strong attractive forces (sometimes named agglomerate)
Colloid Vibration Current	AC current between two relaxed electrodes placed in a dispersion, if the latter is subjected to an ultrasonic field
Colloid Vibration Potential	AC potential difference between two relaxed electrodes placed in a dispersion, if the latter is subjected to an ultrasonic field (counterpart of ESA)
Effective powder density	Density of powder bed measured at well defined conditions
Effective particle density	Particle density measured at well defined, optimum conditions for dispersion in a liquid that takes into consideration the effects of any retained gas or liquid within the closed or open pores of the particles
Electrokinetic Sonic Amplitude	Amplitude of the ultrasonic field created by an AC electric field in a dispersion (counterpart of CVP)
Electrophoresis	Movement of colloidal particles, dispersed in a liquid, under the action of an external electric field
Floc	Assemblage of loosely bound primary particles in a liquid
Particle	Discrete piece of material
Pore	Cavity or channel in an object, such as a particle
'Primary particle'	Basic particle, which cannot be separated unless by breakage
Test sample	Representative sample of adequate size that is used entirely for testing
Void	Space between particles, usually in a powder
Zeta-potential	Electrostatic potential at the surface of shear of a particle suspended in a liquid
AC	Alternating current
APPIE	Association of Powder Process Industry and Engineering, Japan
BAM	Bundes Anstalt für Materialforschung und –prüfung, Germany
BCR	Bureau Communautaire de Référence, European Union
CVI	Colloid Vibration Current
CVP	Colloid Vibration Potential
DL	Debije Length
ESA	Electrokinetic Sonic Amplitude
FOQELS	Fiber-optics quasi-elastic light scattering
ISO	International Standards Organisation
LDE	Laser Doppler Electrophoresis
NIST	National Institute of Standards and Technology, USA
PALS	Phase Analysis Light Scattering
PSD	Particle size distribution
a	Particle radius
A_0	Mean radius $< R>$
A_k	Fourier coefficients (harmonic amplitudes in shape function)
C	Constant in BET equation (dependent on type of porous material)
d_p	Diameter of a cylindrical pore

(continued)

(continued)

D	Particle size (diameter of equivalent sphere)
D_h	(apparent) hydrodynamic particle size
D_{St}	Equivalent Stokes' particle diameter
$D_{50;0}$	Median size of a number-based size distribution
$D_{90;0}$	90th percentile size of a number-based size distribution (other numbers represent other percentile sizes or other basis for the PSD)
$D_{50;3}$	Median size of a volume-based size distribution
$D_{3,2}$	Sauter mean diameter, area-weighted mean size, mean value of an area-based PSD
$D_{4,3}$	Volume-weighted mean size, mean value of a volume-based PSD
$\mathbb{D}$	Diffusion coefficient
E	Strength of electric field applied
$F(\kappa a)$	Henry's function (for ratio of particle radius to thickness of double layer)
$F(\Theta)$	Fourier function in polar coordinates; $R/<R>$ as a function of angle φ
k	Harmonic number in Fourier shape function
k_B	Boltzmann constant
M	Mass of powder bed
$<n>$	Mean number of particles measured in a size class
N	Number of different samples used for measurement of y
N	Centrifuge speed, rpm
p	Proportion by number of particles in one of the two components of the mixture
$1\text{-}p$	Proportion by number of particles in the other component of the mixture
P	Applied pressure
P	Actual pressure during measurement
P_0	Saturation pressure of adsorptive
$\%\ m/m$	Percent by mass
$\%\ n/n$	Percent by number
$\%\ v/v$	Percent by volume
$Ppmv$	Parts per million by volume
r	Radius or half-width of pore after some space has been occupied by liquid layer t
r_p	Radius of cylindrical pore
R	Gas constant
R	Distance between center of gravity and contour of particle projection
$<R>$	Mean value of R (radius of circle with the same area as the particle projection)
R_d	Distance between detector and center of rotation
R_i	Distance between injected sample layer and center of rotation
s_g	Geometric standard deviation, coming from measurements
s_n	Standard deviation of $<n>$, coming from measurements
s_p	Standard deviation of p, coming from measurements
s_y	Standard deviation of parameter y (overall), coming from measurements
S	Specific surface area
t	Sedimentation time
t	Thickness of adsorbed liquid layer at pore wall
T	Absolute analysis temperature

(continued)

(continued)

u_e	Electrophoretic mobility of dispersed particle (v_e/E)
v_e	Electrophoretic velocity of dispersed particle
V	Volume of adsorbate at pressure P (usually expressed in mole)
V_B	Volume of powder bed ($=V_P + V_V$)
V_l	Molar volume of the liquid condensate
V_m	Monolayer of adsorbate (usually expressed in mole)
V_P	Volume of particles in bed
V_V	Volume of voids
W	Width of slit pore
y	Arbitrary parameter resulting from measurement
y_i	i-th measurement of y
$<y>$	Mean measurement result for y
γ	Surface tension of the liquid condensate
γ_m	Surface tension of mercury
ε	Void fraction in powder bed (powder bed porosity)
ε_d	Dielectric constant of dispersion medium
ζ	Symbol for zeta-potential of dispersed particle
η_L	Liquid viscosity
κ	Thickness of double layer around particle
Θ	Contact angle of mercury at the pore surface
ρ	True particle density
ρ_B	Effective bulk density of powder bed
ρ_e	Envelope particle density (including accessible and in-accessible pores)
ρ_L	Liquid density
ρ_P	Effective particle density
σ	Standard deviation of a distribution
φ_k	Phase angle of the k-th harmonic
ω	Angular velocity of centrifuge ($2\pi N/60$), radians/s

References

1. Allen, T.: Particle Size Measurement. Surface Area and Pore Size Measurement, vol. 2. Chapman & Hall, London (1997)
2. ASTM D281-97: Oil Absorption of Pigments by Spatula Rub-out. American Society for Testing and Materials (2007)
3. ASTM D1483-95: Oil Absorption of Pigments by Gardner-Coleman Method. American Society for Testing and Materials (2007)
4. Beddow, J.K.: Image Analysis Sourcebook. Am. Univ. Sci. Technol. Press, Santa Barbara (1997)
5. Bernhardt, C.: Particle Size Analysis, Classification and Sedimentation Methods. Chapman & Hall, New York (1994)
6. BS 2955: Glossary of Terms Relating to Particle Technology. British Standards Institution (1993)
7. Coppens, P., Deriemaker, L., Finsy, R.: Shape and size determination by laser diffraction: Feasibility of data analysis by neural networks. Part. Part. Syst. Charact. **17**, 117–125 (2000)
8. Cosgrove, T.: Colloid Science; Principles, Methods and Applications. Blackwell, Ames (2005)

9. Delgado, A.V., González-Caballero, F., Hunter, R.J., Koopal, L.K., Lyklema, J.: J. Coll. Interface Sci. **309**, 194–224 (2007). doi:(© IUPAC)
10. Diamond, S.: Mercury porosimetry: An inappropriate method for the measurement of pore size distributions in cement-based materials. Cem. Concr. Res. **30**, 1517–1525 (2000)
11. Dukhin, A.S., Goetz, P.J.: Ultrasound for Characterizing Colloids; Particle Sizing, Zeta Potential, Rheology. Elsevier, Amsterdam (2006).
12. Groen, J.C., Peffer, L.A.A., Pérez-Ramírez, J.: Pore size determination in modified micro- and mesoporous materials. Pitfalls and limitations in gas adsorption data analysis. Micropor. Mesopor. Mater. **60**, 1–17 (2003)
13. Heywood, H. In: The scope of particle size analysis and standardization. Proc. Symp. PSA London 1947, Suppl. to Trans. Inst. Chem. Eng. **25**, 14–24 (1947)
14. Hiemenz, P.C., Rajagopalan, R.: Principles of Colloid and Surface Chemistry. M. Dekker, New York (1997)
15. Hunter, R.J.: Zeta Potential in Colloid Science: Principles and Applications. Academic, New York (1981)
16. ISO 9276-1: Graphical Representation of Particle Size Distributions. **1998** + Corrigendum **2004** International Standards Organisation.
17. ISO 9276-6 (DIS): Particle Shape and Morphology. International Standards Organisation (2009)
18. ISO 9277: Determination of the Specific Surface Area of Solids by Gas Adsorption Using the BET Method (in revision). International Standards Organisation (1995)
19. ISO 13099 (in prep.): Methods for Zeta Potential Determination. International Standards Organisation
20. ISO 14488: Sampling and Sample Splitting for the Determination of Particulate Properties. International Standards Organisation (2007)
21. ISO 14887: Dispersion of Powders in Liquids. International Standards Organisation (2000)
22. ISO 15901-1: Evaluation of Pore Size Distributions of Solid Materials by Mercury Porosimetry. International Standards Organisation (2005)
23. ISO 15901-2: Evaluation of Pore Size Distributions of Solid Materials by Gas Adsorption (and corrigendum). International Standards Organisation (2006/2007).
24. ISO 15901-3: Analysis of Micro-pores by Gas Adsorption. International Standards Organisation (2007)
25. Jillavenkatesa, A., Dapkunas, S.J., Lum, L.-S.H.: Particle Size Characterization, pp. 960–961. NIST Special Publication, Washington, DC (2001)
26. Kaye, B.H.: A random Walk through Fractal Dimensions. VCH, New York (1994)
27. Luerkens, D.W.: Theory and Application of Morphological Analysis: Fine Particles and Surfaces. CRC Press, Boca Raton (1991)
28. Ma, Z., Merkus, H.G., Scarlett, B.: Extending laser diffraction for particle shape characterization: Technical aspects and application. Powder Technol. **118**, 180–187 (2001)
29. Ma, Z.: Measurement of particle size and shape by laser light scattering; Thesis Delft University of Technology (2001)
30. Masuda, H., Higashitani, K., Yoshida, H. (eds.): Powder Technology; Fundamentals of Particles, Powder Beds and Particle Generation. CRC Press, Boca Raton (2007)
31. Merkus, H.G.: Particle Size Measurements; Fundamentals, Practice, Quality. Springer, Dordrecht (2009)
32. Nelson, R.D.: Dispersion of Powders in Liquids. Elsevier, New York (1988)
33. Oshima, H., A simple expansion for Henry's function for the retardation effect in electrophoresis of spherical colloidal particles. J. Coll. Interface Sci. **168**, 269–271 (1994)
34. US Pharmacopoeia. In: Pharmeuropa, chapter 776 on Optical Microscopy, 25th ed., p. 2040
35. Rahaman, M.N.: Ceramic Processing and Sintering. Marcel Dekker, New York (2003)
36. Rouquerol, F., Rouquerol, J., Sing, K.: Adsorption by Powders and Porous Solids. Academic, Boston (1999)
37. Sing, K.S.W., Everett, D.H., Haul, R.A.W., Moscou, L., Pierotti, R.A., Rouquérol, J., Siemieniewska, T.: Physisorption data for gas/solid systems with special reference

to the determination of surface area and porosity. Pure Appl. Chem. **57**, 603–619 (1985). doi:(© 1985 IUPAC)
38. Stroeven, P., Hu, J., Koleva, D.A.: Concrete porosimetry: Aspects of feasibility, reliability and economy. Cem. Concr. Compos. **32**, 291–299 (2010)
39. Vanderhallen, F., Deriemaker, L., Manderick, B., Finsy, R.: Shape and size determination by laser diffraction: Parametric density estimation by neural networks. Part. Part. Syst. Charact. **19**, 65–72 (2002)
40. Wedd, M.W.: The influence of pixel numbers and image centring upon particle size analysis accuracy. Part. Part. Syst. Charact. **27**, 71–75 (2010)
41. Willis, K.L., Abell, A.B.., Lange, D.A.: Image-based characterization of cement pore structure using wood's metal intrusion. Cem. Concr. Res. **28**, 1695–1705 (1998)

Chapter 4
Assessment and Control of Fire and Explosion Hazards and Risks of Particulates

Saul M. Lemkowitz and Hans J. Pasman

Abstract Fires and explosions of particulates have caused, and continue to cause, substantial financial loss and loss of life. This Chapter provides knowledge and insight into preventing or at least mitigating particulate fire and explosion hazards and risks. To this end we distinguish between the concepts hazard and risk, first discussing in some detail the nature of fire and explosion hazards of particulates, the indexes expressing these hazards, the many factors that determine such hazards, and a widely used and internationally accepted system for ranking fire and explosion hazards of particulates. All of these topics relate to a given particulate; i.e., a particulate of a given chemical composition and given physical form (e.g., particle shape and size distribution). We next discuss fire and explosion risks of particulates. Risk is a much more extensive and complex topic than hazard, as risk involves factors extensive to the particulate itself, such as: amounts of particulate; where (region, country) it is made, transported, and stored; the (process) conditions under which it is made, transported, and stored; relevant legislation; and management. Rather than discuss the complex topic of particulate risk as such, we concentrate on a number of fundamental approaches to reduce risk, approaches which are increasingly applied in modern (chemical) engineering design. One general method to reduce risk is to negate the basic factors causing risk. This is the basic logic of Inherently Safer Design (ISD), which, essentially, aims at avoiding or at least greatly reducing hazard by clever design/choice of materials, process, and conditions. Inherently Safer Design forms the starting point of designing layers of defense against mishap. The analysis of the degree of risk reduction by the layers is called Layers of Protection Analysis (LOPA). LOPA recognizes that in practice Inherently Safe Design rarely can eliminate all hazards. Thus, starting from the

S.M. Lemkowitz (✉)
Department of Chemical Engineering, Delft University of Technology, Delft, The Netherlands
e-mail: s.m.lemkowitz@tudelft.nl

H.J. Pasman
Department of Chemical Engineering of Texas A&M University, Mary Kay O'Connor Process Safety Center, College Station, Texas USA

H.G. Merkus and G.M.H. Meesters (eds.), *Particulate Products: Tailoring Properties for Optimal Performance*, Particle Technology Series 19,
DOI 10.1007/978-3-319-00714-4_4,

ideal of Inherently Safer Design, LOPA applies 'layers of protection' around manufacturing the particulate. These 'layers of protection' are applied at various scales, starting from the process itself (e.g., safety devices, such as a device to suppress explosion) to the neighborhood in which the plant manufacturing the particulate is located (e.g., a disaster plan involving local government). The goal of LOPA is to reduce risk to an acceptable level. We discuss Inherently Safer Design and LOPA in some detail, including examples. Relevant legislation and management aspects are also absolutely essential topics in particulate safety. These are mentioned, but because this Chapter focuses more on scientific and technical aspects, these two important topics are discussed in less detail. The Chapter ends with a note on research and also provides an extensive literature list.

4.1 Introduction

Although basics are well-understood, fires and explosions of solid particulates (powders, 'dusts') and liquid particulates ('mists'), especially the former, continue to cause considerable material damage and loss of life.[1] With the goal of reducing particulate fire and explosion hazards and risks to acceptable limits we review scientific and engineering knowledge and insight. Also to this end we provide practical advice, cite relevant legislation and standards, and list general references concerning the design, production, handling, and use of particulate materials. While basics are well-understood, research nevertheless continues in order to more deeply understand fundamentals and to develop new applications; we briefly mention such research. Since this is an introductory chapter to explosion hazards and risks, the reader is advised to look into the recommended literature [1–28] for more detailed information.

4.2 Hazard Versus Risk

Hazard is the potential to create an adverse effect. Typical hazards of particulates are their potential to create adverse effects of fire, explosion, and toxicity. This chapter discusses the first two types of particulate hazards: fire and explosion.

Hazards can be given a quantitative value using certain substances properties, often combined into hazard indexes. We stress, however, that numerical values of hazards are always relative, never absolute. Thus systems quantifying particulate hazards (e.g., National Fire Protection Agency (NFPA) system, discussed in Sect. 4.6), can be used for ranking hazards; i.e., determining which particulate is more, or less, hazardous than another particulate.

Risk is composed of two basic factors: an adverse effect (e.g., death) and the probability of that adverse effect (e.g., death) occurring. Each of the two factors of

[1] Historical reviews of human and material damage caused by dust explosions are given in [1, 2, 12, 28, 68].

risk has a quantitative value (e.g., number of deaths; probability of death (0–100 %)). Unlike hazard, risk can therefore in principle have an absolute value. In practice, however, calculated risks are frequently afflicted by large uncertainties.

While one can speak of the hazard of a given particulate as such (i.e., its potential to cause adverse effects), one can not speak of its risk. Why not? Because the risk associated with that particulate depends on additional factors external to the particulate itself, factors like amounts, conditions (e.g., pressure, temperature), degree and strength of containment, (distance to) possible 'targets', etc. We discuss such factors in Sect. 4.7.

4.3 Fire and Explosion Hazards

Fire is combustion (burning), which consists, overall, of rapid, exothermic (= heat-producing) chemical reactions involving a fuel and producing beside combustion products, such as water and carbon dioxide, heat and, via a flame, also light. Combustion usually occurs with oxygen (in air) and is therefore chemically called oxidation. Fire hazards include intense heat (causing burns/material damage), reduced oxygen concentration (causing asphyxiation), formation of toxic substances (e.g., carbon monoxide), loss of product or product contamination, and loss/damage of equipment and buildings.

Fully oxidized substances, i.e., substances that have completely reacted with oxygen (usually) cannot burn.[2] Examples of fully oxidized materials are oxides (liquids: water (technically: dihydrogen oxide)); solids: sand (silicon dioxide), carbonates (e.g., soda, sodium carbonate), and sulphates (e.g., gypsum, calcium sulphate). Since such substances cannot burn, they cannot explode.

In addition to simply burning, an insidious form of fire, called smoldering, can occur when a large volume of a combustible powder is stored, typically in a pile (e.g., large pile of coal) for an extended time period. Through surface oxidation of the powder, heat is generated, and if the rate of heat generation exceeds the rate of heat loss (as can occur in a large pile), the temperature deep within the pile can reach the ignition temperature. The result is a relatively slow burning (slow because oxygen must diffuse into the pile and reaction gases/vapors must diffuse out). Smoldering is 'insidious' in that it is often not noticed. Aside from generating poisonous gases (e.g., carbon monoxide), smoldering can lead to fires and even explosions [12]. We mention smoldering further in Table 4.2 (under Minimum Ignition Temperature).

More or less all un-oxidized materials, or not fully oxidized materials, can burn. Examples of solid combustible substances are un-oxidized metal powders (e.g., iron, aluminum, titanium) and materials composed chemically of un-oxidized

[2] A very few substances, such as fluorine, are more powerful oxidizers that oxygen. Substances fully oxidized with oxygen, such as water, may even 'burn' (rapidly react exothermically) with fluorine.

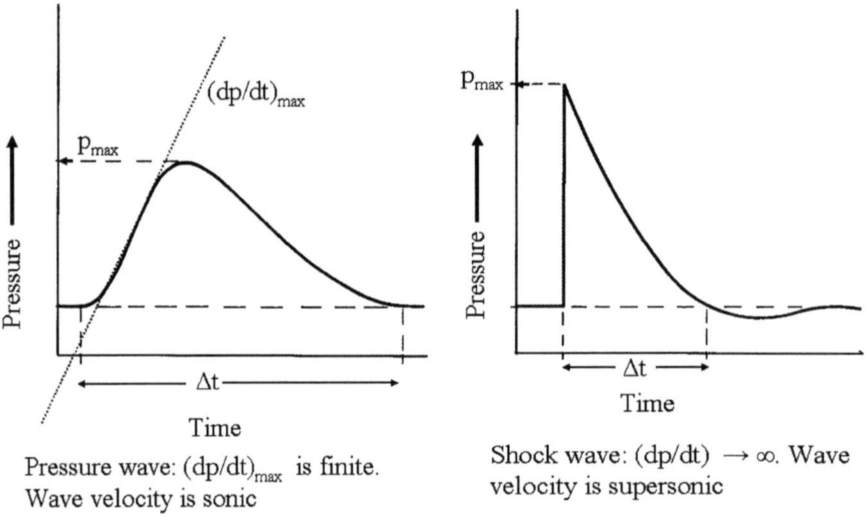

Fig. 4.1 Pressure wave (*left*) and shock wave (*right*) propagating in an open system (i.e., constant pressure system, such as atmospheric)

(or only partially oxidized) carbon, such as (powdered) plastics, aspirin, coal, sugar, flour, grains, even sloughed-off skin cells. Examples of liquid combustible substances are most organic liquids (e.g., hydrocarbons, like gasoline), oils, alcohols, etc., All such non-oxidized/not fully oxidized substances can burn; therefore they can also explode. Most substances industrially produced (e.g., plastics, oils) or processed (e.g., wood, petroleum) are un-oxidized or not fully oxidized and can therefore burn; therefore they can explode. Essential is to be fully aware that:

ALL POWDERS THAT CAN BURN CAN ALSO EXPLODE, WITHOUT EXCEPTION

ALL LIQUIDS THAT CAN BURN CAN – AS PARTICULATES (MISTS) – ALSO EXPLODE, WITHOUT EXCEPTION

Fire is known to all, but what is an explosion? While it is clear that explosion always involves a sudden release of energy, it is perhaps surprising that no definitive, quantitative definition of the term 'explosion' exists. Characteristic of explosion are the following: "much energy" (Joules, J), "in a short time" ((fraction of) seconds (s); thus high power (J/s)), "in a small volume" (m^3, thus high power density (J/(s m^3)). Further: creation of a 'blast' as a pressure wave (Fig. 4.1, left) or as a more destructive shock wave (Fig. 4.1, right), which can wound/kill people, destroy buildings, and create deadly high-velocity fragments.

The energy source of explosions can be physical (e.g., lightning, exploding steam boiler) or chemical (e.g., combustion). Since we are here discussing explosion hazards of particulates, we consider only explosions caused by combustible substances undergoing rapid exothermic (i.e., heat-producing) chemical reactions, most commonly combustion (i.e., burning) in air or oxygen (i.e., oxidation). A reason why no unambiguous scientific definition of explosion exists is that the most common type of chemical explosion, the deflagration (sometimes also called 'explosive burning'), is simply 'a very rapid burning'. How rapid? Rapid enough to create a pressure wave and, when it occurs in the open air and is rapid enough, an audible signal (i.e., a 'bang'). When explosion occurs in a closed vessel and the vessel cannot sustain the pressure, it will produce a blast (i.e., destructive pressure or shock wave) at the moment of bursting.

An important difference between fire and explosion is that in fire the fuel (e.g., burning logs in open hearth) and the oxidizer (air) are clearly separated. Oxygen molecules, necessary to maintain combustion occurring in the fire's flames, reach the flame largely by diffusion. By contrast, explosions are characterized by a 'pre-mixing' of fuel and oxidizer. Examples of increasing intimacy of pre-mixing are: an explosive dust cloud (combustible particles suspended in air); explosive gaseous mixture (e.g., molecular mixture of methane gas and air); solid high explosive e.g., TNT (atoms of fuel (carbon and hydrogen) adjacent to atoms of oxidizer (oxygen) in the same molecule). It is this pre-mixing of fuel and oxidizer, and its degree of 'intimacy' (most 'intimate' at the atomic level, as in TNT), that form the basic reason why explosions occur (much) more rapidly than fires do, thereby resulting in a (much) higher (volumetric) rate of energy release, i.e., power (W or (W/m^3)).

Explosions of powders, commonly called dust explosions, and/or explosions of liquid particulates, called mist explosions, are almost always a type of explosions called deflagrations. The propagation mechanism of deflagration is heat conduction: heat is transferred from the hot reacting zone (the flame) to the much cooler unreacted material ahead of the flame, thus preheating it until it reaches its ignition temperature, at which it ignites and burns (combusts). Behind the flame expanding hot, oxidized gaseous products and (possibly) unburned material are present.

Characteristic of deflagrations is that the deflagrative flame speed is always sub-sonic (i.e., less than the speed of sound, which in air at ambient conditions is ca. 330–340 m/s). In fact, at non-turbulent conditions, deflagrative flame speeds in air at ambient conditions typically range between a few m/s to some tens of m/s.

In a closed vessel filled with a flammable mixture undergoing deflagration, and ignited in the center, the flame expands spherically from the vessel's center until it reaches the vessel's wall. During this process pressure in the vessel continuously rises (Fig. 4.1). Both the pressure and the rate of pressure rise reach a maximum when the flame reaches the vessel's wall. Characteristic of deflagrations in a closed vessel are therefore a maximum explosion pressure, P_{max}, and a finite maximum rate of pressure increase, $(dP/dt)_{max}$. In practice pressure is often expressed as bara (= bar absolute, 1 bara (≈ atmospheric pressure) being 10^5 Pa) and rate of pressure rise (expressed in bar/s.) Maximum pressure, P_{max}, and maximum rate of pressure increase, $(dP/dt)_{max}$, are found at optimal mixture composition. Expansion of the hot gases produced by the combustion process generates pressure waves. In a closed system (e.g., spherical vessel) these pressure waves compress the unburned flammable mixture ahead of the flame, causing pressure rise to a maximum. Maximum explosion pressures of deflagrations of particulates are typically between 5 and 10 times the initial pressure (as explained later via Eqs. 4.1 and 4.2). Typical maximum rates of pressure rise in deflagrations are, for most particulates, less than ca. 300 bar/s, as measured in a sphere of 1 m^3. As noted above, deflagrating flames propagate less rapidly than the speed of sound. However, the pressure waves produced by the expanding flame propagate ahead of the flame with the speed of sound (i.e., 330–340 m/s). Therefore in deflagrations the flame (subsonic velocity) and pressure waves (sonic velocity) are separated from one another, pressure waves preceding the flame. Compared to the speed of sound, the expansion of the hot gases occurs relatively slowly. In a closed system (e.g., sphere) pressure thus has sufficient time to distribute itself more or less equally over the whole volume of the vessel. Therefore the pressure in a deflagration occurring in a spherical vessel is, at any instant of time, roughly equal everywhere in the vessel; i.e., by approximation pressure is independent of position.

In badly managed factories producing flammable particulates (e.g., powdered sugar) housekeeping is often poor. In open parts of such factories, such as a work floor with significant deposits of loose dust, the pressure waves produced by an initial deflagrative dust explosion can whirl the dust up to create a flammable dust-air suspension. The initial local explosion can therefore propagate from one space to another. This phenomenon can cause a small initial explosion to transform itself into secondary and tertiary explosions, which are often more disastrous than the initial explosion.

In closed systems consisting of inter-connected vessels, as an explosion propagates from one vessel to another, the explosion in the secondary vessel is often much more serious (i.e., higher maximum pressure and rate of pressure rise) than the initial explosion in the primary vessel. This occurs because the secondary vessel has been pressurized by the initial explosion before the flame from the initial explosion has arrived. This phenomenon, called 'pressure piling' [11, 12], manifests itself as (often vastly) increased maximum pressures and rate of pressure rise, relative to the initial explosion.

Note an important difference between an explosion occurring in a closed or in an open system. In a *closed system,* the explosion causes a pressure rise that leads to a maximum pressure in the vessel, such as a reactor or silo. For a spherical vessel with ignition at its center, the time scale of the explosion process is largely determined by the radius of the vessel (m) divided by the speed of the deflagrative flame (m/s). Thus the time scale of a deflagration with a flame speed of 10 m/s occurring in a spherical vessel with a radius of 1 m would be approximately 0.1 s (100 ms). (If the maximum explosion pressure was 10 bara and the initial pressure atmospheric, the rate of pressure rise ($= \Delta P/\Delta t$) would be 9 bar/0.1 s, or 90 bar/s.).

A pressure wave propagating in an *open system* (e.g., the atmosphere) is generally a much more transient phenomenon than the process of pressure rise occurring in a closed system, since pressure rise in the former is not limited by enclosing walls. The time scale of the propagating pressure wave is generally much shorter than the time scale of pressure increase caused by explosion in a closed system. Time scales for pressure waves are typically 1 ms or less. Nevertheless, consequences of explosions occurring in open systems may still be severe.

Only under exceptional conditions, briefly mentioned below and in section "Effects of Turbulence" and Sect., can particulates explode via a mechanism other than deflagration: the detonation. A detonation is a chemically reactive shock wave. A shock wave (Fig. 4.1, at right), is a special case of pressure wave. Shock waves have a peak (maximum) pressure, an almost infinite rate of pressure rise, and always proceed faster than the speed of sound. In contrast to deflagrations, in which the deflagrative flame and pressure waves are separate from one another and propagate at different speeds, the detonation flame and the shock wave propagate at the same (supersonic) speed with the combustion occurring in the wave, ignited by the high temperature resulting from the sudden compression. In general, per unit mass of material exploding (e.g., per kg), detonations are much more hazardous and devastating than deflagrations because detonation pressure are considerably higher (ca. 2–8 times greater). Unlike deflagrations, where pressure is usually largely independent of position during the explosion, detonation pressures are highly dependent on position relative to the detonative flame. Because detonations occur so much faster than deflagrations, they are also much more difficult to protect against than deflagrations.

Given the right conditions, slow starting deflagrating flames of particulates can, as they for example propagate through piping connecting various parts of a plant, accelerate to speeds of hundreds of meters per second. This is due to flame turbulence induced by friction with the walls of piping, Under special conditions, even transition to detonation is possible, thus explosions with flame speeds of up to more than two kilometers per second. This feared phenomenon is called a deflagration-to-detonation-transition: DDT [4, 11, 12, 16–19, 69]. We briefly discuss DDTs, and also pressure piling, further in section "Effects of Turbulence" (effects of turbulence) and in Sect. 4.8.5 (Table 4.13). Concerning explosion behavior in general, Abbasi et al. 2009, [29] provide detailed description of types and characteristics of explosions, including deflagrations and detonations.

For deflagrations, by far the most common type of industrial explosion, occurring in a closed volume, the ideal gas law expresses the maximum explosion pressure simply and usually sufficiently accurately (< 10 % error):

$$P_{explosion} \approx \left(n_{explosion}/n_{initial}\right)\cdot\left(T_{explosion}/T_{initial}\right)\cdot P_{initial} \tag{4.1}$$

where:

$P_{explosion}$ = final pressure (bara) of the explosion for given conditions of mixture composition etc.

$(n_{\text{explosion}}/n_{\text{initial}})$ = ratio of number of gaseous moles produced by the explosion relative to the initial number

$T_{\text{explosion}}/T_{\text{initial}}$ = ratio of absolute explosion temperature (K) relative to initial temperature (K)

P_{initial} = initial pressure (bara).

The presence of roughly 80 % inerts (nitrogen + argon) in air causes the factor $(n_{\text{explosion}}/n_{\text{initial}})$ to be approximately equal to 1.0. Therefore Eq. 4.1 simplifies to:

$$P_{explosion} \approx \left(T_{explosion}/T_{initial}\right) \cdot P_{initial} \tag{4.2}$$

At optimum conditions, $P_{\text{explosion}}$ becomes the maximum possible explosion pressure $P_{\max}$, and Eq. 4.2 predicts that the maximum possible explosion pressure of a deflagration is reached when the maximum possible temperature is obtained. This occurs close to optimum combustion conditions, which is near stoichiometric mixture composition. (Actually, the composition is just at the fuel rich side). The maximum combustion temperature is the adiabatic flame temperature, 'adiabatic' referring to an ideal condition in which no heat loss occurs. The adiabatic flame temperature is a so-called chemical thermodynamic property. This means that each substance burning in air at given initial temperature and pressure conditions has a clearly defined maximum combustion temperature. This value is often available in standard reference sources (also on Internet) or, if not thus available, can usually be calculated relatively simply. But does a dust explosion occur adiabatically?

Generally the quicker the rate of the explosion process (and thus the greater the rate of temperature change (dT/dt)), the less time to lose heat (which requires time to occur), and thus the closer the approach to adiabatic conditions, idem adiabatic flame temperature. Maximum adiabatic flame temperatures for common organic powders (i.e., plastics, coal) and/or droplets of combustible liquids exploding in air typically range between 1,500 and 3,000 K. From Eq. 4.2 we see that this means that for typical particulate explosions (deflagrations) occurring in air in closed vessels initially at atmospheric pressure (1 bara) and ambient temperature ($T =$ ca. 300 K), typical maximum explosion pressures vary between ca. 5 and 10 bara (= 4 and 9 baro (baro = bar overpressure)).

What are the hazards of such overpressures? A pressure wave of 0.1 baro (or sometimes even less) typically completely shatters glass windows (forming innumerable jagged glass fragments); 1.0 baro totally destroys buildings (houses; office blocks), causing their collapse and killing those inside; 10 baro kills 100 % of people exposed (people standing perpendicular to the blast wave). Why can seemingly low overpressures, e.g., 0.1–1 baro, cause such devastation? Because pressure is force/area. Overpressures of 0.1 and 1 baro thus correspond to forces of, respectively, 1,000 kg/m^2 (1 t/m^2) and 10,000 kg/m^2 (10 t/m^2). Typical vertical surfaces in buildings (e.g., windows, walls) cannot withstand such huge lateral forces.

As Eq. 4.2 suggests, the main factor causing the high pressures associated with deflagrative dust explosions is a sudden large temperature increase of the hot gas produced. It is a very fast, highly exothermic chemical reaction, i.e., combustion (oxidation), which causes this abrupt high temperature increase, and thus the proportionally large, abrupt pressure increase.

Table 4.1 Explosion severity indexes of particulates dispersed in air (or other oxidative gases)

Explosion severity index	Meaning/significance/dimensions of explosion severity index
Tendency towards detonation	Per kg of material, detonation is the most destructive/most severe explosion. Generally tendency towards detonation increases with increasing mixture reactivity
Maximum explosion pressure	Maximum explosion pressure (follows theoretically from maximum adiabatic flame temperature, but in practice value of P_{max} is usually measured), with initial conditions of 1 bara and room temperature; usually expressed as bara or baro
	Explosion hazard increases with greater maximum explosion pressure
Maximum rate of pressure rise, $(dP/dt)_{max}$	Maximum rate of pressure rise, with initial conditions of 1 bara and room temperature; usually expressed as bar/s. Cannot usually be calculated and must therefore be measured
	Explosion hazard increases with greater maximum rate of pressure rise
Maximum volume-normalized rate of pressure rise, called K_{St}-factor, where: $K_{St} = V^{1/3} \cdot (dP/dt)_{max}$	Maximum rate of pressure rise (bar/s), normalized to a sphere of 1 m^3 volume, with initial conditions of 1 bara and room temperature. (This equation is called the 'cubic law'). The K_{St}-factor must be measured; it is usually expressed as bar m^{-1} s^{-1}
	Explosion hazard increases with increasing value of K_{st}-factor

4.4 Fire and Explosion Indexes

The greater the amount of energy released during combustion and the faster this energy is released, the greater the violence/severity (effects/consequences) of explosion. The smaller the amount of energy needed to initiate combustion, the greater the sensitivity to ignition (to fire and/or explosion), and thus the greater the probability of explosion occurring. Fire and explosion hazards (and also risks) of particulates are therefore expressed in terms of explosion severity indexes and explosion sensitivity indexes. Tables 4.1 and 4.2 list such parameters.

Assessment of fire and explosion hazards of particulates requires accurately knowing the values of the indexes listed in Tables 4.1 and 4.2. We stress most strongly that these values usually cannot be predicted theoretically. Nor is it wise to assume that values found in the literature are applicable to one's own case, as explosion parameters of a given particulate (e.g., grain dust) may vary greatly, depending on particle size, particle size distribution, and even weather conditions (temperature, humidity) under which the powder has been stored, and for how long

Table 4.2 Explosion sensitivity index of particulates dispersed in air (or other oxidative gases)

Explosion sensitivity index	Meaning/significance/dimensions of explosion sensitivity index
Lower Explosion Limit (LEL); also called Lower Flammability Limit (LFL)	Lowest concentration of a substance (commonly expressed as g/m^3) dispersed in air that can explode. Is measured at initial conditions of 1 bara and room temperature. Must be measured Fire and explosion hazard increases as LEL decreases, as even very low concentrations of particulates are still flammable and thus explosive
Upper Explosion Limit (UEL); also called Upper Flammability Limit (UFL)	Highest concentration of a substance (commonly expressed as g/m^3) dispersed in air that can explode. Is measured at initial conditions of 1 bara and room temperature Compared to gases/vapors, which have a more or less definite UEL, the UEL for particulates has a more limited practical utility, as particulate concentrations tend to fall due to particulate deposition. Thus a particulate whose concentration is initially above its UEL, may settle out due to gravity, causing its concentration to decrease to below the UEL and thus enter the explosive range
Explosive range: UEL – LEL	The range of concentration (g/m^3) that is explosive. Generally the larger the explosive range, the greater the probability of ignition, and thus the greater the fire and explosion hazard
Lower Flash point (LFP)	Lowest temperature (T in °C or K) at which a substance reaches a vapor pressure sufficient to form a flammable mixture in air (i.e., the LEL). LFP usually relates to liquids, but is also applicable to solids with high vapor pressure (e.g., naphthalene, benzoic acid). Lower LFP increases fire and explosion hazard
Minimum Ignition Energy (MIE)	Lowest amount of energy (mJ) of an electrical spark that can cause a particulate dispersed in air to explode Fire and explosion hazard usually increase as MIE decreases, as even very weak ignition sources are then capable of causing fire and/or explosion
Minimum Oxygen Concentration (MOC); also called Limiting Oxygen Concentration (LOC)	Lowest concentration (mol- or volume-percent) in gas mixture that can support combustion. Below this oxygen concentration explosion is not possible. The lower the MOC, the more difficult it is to produce a carrier gas in which fire and explosion cannot occur

(continued)

Table 4.2 (continued)

Explosion sensitivity index	Meaning/significance/dimensions of explosion sensitivity index
Maximum Experimental Safe Gap (MESG)	The maximum gap width that allows flame propagation of a given particulate. A smaller gap width quenches the flame, thus preventing fire/explosion propagation. The lower the MESG-value the greater the fire/explosion hazard/risk of a given particulate
Minimum Ignition Temperature (MIT); also sometimes called: Minimum Auto-Ignition Temperature (AIT)	Lowest temperature (T in °C or K) of a surface that can ignite cloud of a particulate dispersed in air to explosion Fire and explosion hazard increases as AIT decreases
For deposited dust layers on a surface: Minimum Ignition Temperature (MIT); also often called Minimum Smoldering Temperature	Lowest temperature (T in °C or K) of a surface that can cause a dust layer of a given depth to catch fire or at least smolder Fire and explosion hazard increases as surface MIT decreases

(this related to possible adsorption/absorption of moisture from air and its effect on particle size).

Since knowing the numerical values of fire and explosion indexes is absolutely critical for particulate safety, and because these indexes generally cannot be calculated/estimated from theory with sufficient accuracy, particulate explosion parameters must be measured. We do not further discuss measurement/testing of particulate explosion indexes, which involves standardized test equipment and test procedures. We do stress further that the measured values of parameters listed in Tables 4.1 and 4.2 and, especially, their application to practice require careful assessment and evaluation, as these values are dependent on specific measurement/testing equipment and procedures, which may differ significantly from practical conditions. For a detailed discussion of test methods and their evaluation the reader is referred to the literature on this topic [7, 12].

Here we concentrate on discussing how particulate factors determine their hazard, as expressed in the values of the fire and explosion indexes listed in Tables 4.1 and 4.2. Reducing particulate hazards requires understanding how such factors determine these parameters and actively applying this understanding to designing and operating equipment handling particulates, so as to reduce hazards and risks to acceptable levels.

4.5 Factors Determining Hazards

4.5.1 Introduction

Particulate hazards depend on chemical and/or physical factors. Section 4.5.2 discusses chemical factors determining particulate hazard. These are relatively simple: if a substance is combustible (i.e., can burn), it can explode; if not combustible, it cannot explode. Combustible substances are divided into categories, like stable and unstable, reactive and non-reactive. These various substance categories and their relation to explosion hazard are also discussed.

Section 4.5.3 discusses physical factors, particle size (distribution) in particular. Fundamental is that, unlike gas/vapor explosions, whose combustion mechanism involves only one phase (gas/vapor), the mechanism of particulate explosions always involve more than one phase. The heterogeneous (i.e., multi-phase) mechanism of particulate combustion is the most basic factor distinguishing particulate explosions from gas/vapor explosions. The mechanism of particulate explosion behavior is complex, requiring detailed explanation. This occurs in successive parts of this Section, which also includes the physical factor of turbulence.

Section 4.5.4 discusses hazards involving both chemical and physical factors, such as hybrid explosions.

4.5.2 Chemical Factors

As noted above, particulates composed of chemically fully oxidized materials (e.g., water, silica, alumina, clay, chalk, rust, gypsum, cement) cannot burn (i.e., are not combustible); thus they cannot explode. Particulates of such substances therefore present no fire or explosion hazard. In fact, as discussed later, such non-combustible substances, being inert, can be added to combustible powders to reduce their fire and explosion hazards.

Particulates composed of un-oxidized or not fully oxidized substances, are combustible and therefore can burn and/or explode. We divide combustible particulates into three substance categories: 1. Reactive substance, 2. Unstable substances, 3. Stable substances.

Reactive substances are of two types. The first type is: *Unstable, reactive* substances: These are substances which ***by themselves*** spontaneously decompose exothermically; they are described in more detail below. The second type of reactive substance is: Substances which can spontaneously burn or explode, but ***only when exposed to air or water***. This second type of reactive substance is therefore not by itself hazardous, but only in combination with air or water. An example of reactive substances of this type is ***pyrophoric*** substances. These can spontaneously burn or explode when exposed to air, examples being finely divided un-oxidized metal powders, such as iron, zirconium, and titanium (see e.g. [21]). A second example of reactive compounds of this type is solid substances that react violently (burn, explode) in contact with water, examples being the alkali metals sodium, potassium, rubidium, and cesium. Many other substances can also react violently with water, such as aluminum bromide and chloride, calcium hydride and carbide, to name a few [21, 49].

The second category of combustible substances is *unstable* substances. These are substances whose decomposition is exothermic. By themselves (e.g., no air present) unstable substances tend to spontaneously decompose, and since decomposition is exothermic, heat is generated. When stored with insufficient cooling, unstable substances therefore tend to heat up, causing temperature increase. Since increased temperature generally exponentially increases chemical reaction rate (of further decomposition, which is exothermic), a vicious cycle may result leading to increasingly fast temperature increase, increasingly fast decomposition reaction, etc., ultimately causing explosion. Examples of unstable substances produced in large quantities are TNT, other explosives, and ammonium nitrate. Such large-volume unstable powders are not commonly used in industry[3]; therefore we do not further discuss them. A few unstable solids are, however, used in small quantities in specialty products, such as medicines; an example is mentioned in Sect. 4.8.3 (Table 4.10). For completeness we also mention that unstable liquids also exist, examples being nitroglycerine, hydrazine, concentrated hydrogen peroxide, and certain organic peroxides [57].

ALL unstable substances (e.g., TNT) are reactive. But not all reactive substances are unstable (e.g., finely divided magnesium powder is reactive, but not unstable).

The third category of substances, *stable* substances, consists of substances that by themselves do not spontaneously decompose and, in contact with air or water, do not spontaneously burn or explode. Stable materials that are non-oxidized (e.g., polyethylene powder, cyclohexane), or incompletely oxidized (e.g., alcohols, grain dust) will only burn/explode when pre-mixed with air, and, additionally, ***subjected to an ignition source*** (e.g., flame, spark) of sufficient energy (i.e., greater than the particulate's MIE). It is such substances – un-oxidized or only partially oxidized substances, and thus combustible - that by far cause the greatest number of fires and explosions in practice. The great majority of combustible materials used in practice in large quantities are stable, and therefore it is these substances, stable substances, which we discuss in most detail.

In general explosion sensitivity and, in particular, explosion severity, tend to increase as the amount of combustion energy (i.e., heat of combustion) per kg increases. Higher heat of combustion usually results in higher adiabatic flame temperature. As Eq. 4.1 predicts, it is the maximum adiabatic flame temperature that largely determines the maximum explosion pressure. Since chemical reaction rates tend to increase exponentially with increasing temperature, substances with high adiabatic flame temperatures tend not only to have high maximum explosion

[3] Powdered ammonium nitrate, mixed with an inert compound such as calcium carbonate to reduce its explosion hazard (i.e., by 'phlegmatizing' it), is widely used as a fertilizer. In spite of the presence of such inerts, serious explosions of large piles of powdered ammonium nitrate have nevertheless occurred.

pressures, but also high rates of pressure increase (i.e., high $(\mathrm{d}P/\mathrm{d}t)_{max}$). Finely divided reactive metals, such as aluminum and magnesium have particularly high heats of combustion, thus high adiabatic flame temperatures, and thus high maximum explosion pressures (typically 12 bara, or even somewhat higher) and high K_{St}-values (well above 300 bar.m/s).

4.5.3 Physical Factors

Similarities of Gas/Vapor and Particulate Explosions

Explosions of particulates (i.e., dust explosions, mist explosions) resemble gas and vapor explosions in that like gases/vapors, particulate explosions occur via chemical combustion; both deflagration and detonation are possible [35]. Additionally, maximum explosion pressures of explosions of particulates are generally similar to maximum explosion pressures of gas/vapor explosions. Thus (for deflagrations) typically 5–10 times the original pressure (as Eq. 4.1 predicts). This similarity is caused by the fact that the maximum adiabatic flame temperatures of common combustible particulates burning in air are usually roughly similar to those of combustible gases/vapors, i.e., usually within 1,500–3,000 K.

Combustion Mechanism of Particulate Explosions

The most fundamental difference between gas/vapor[4] explosions and explosions of particulates is their respective explosion mechanisms. Gas/vapor explosions occur homogenously - within one phase, the gaseous phase. Mechanistically, gas/vapor explosions thus occur in a *pre-mixed mixture of molecules* (e.g., methane molecules and oxygen molecules) which directly react with one another at a molecular level. Compared to this relatively simple mechanism, the mechanism of particulate explosions is much more complex. Particulate combustion occurs heterogeneously (i.e., between at least two distinctly different phases, a solid phase (powder, particles) and/or liquid phase (e.g., mist)) and a gaseous phase (normally air). In some cases, such as explosions of finely divided aluminum, even three phases can be simultaneously present, viz. solid, liquid and gas/vapor. The complexity of the mechanism of particulate explosions is caused by the many sequential steps involved:

[4] Substances are called gases when their boiling point is well below ambient temperature. Substances are in the liquid state (phase) at temperatures below their boiling point (at ambient pressure). If the difference between temperature and boiling point is not too large, part of these liquids generally is in the gaseous phase. This part is called vapor.

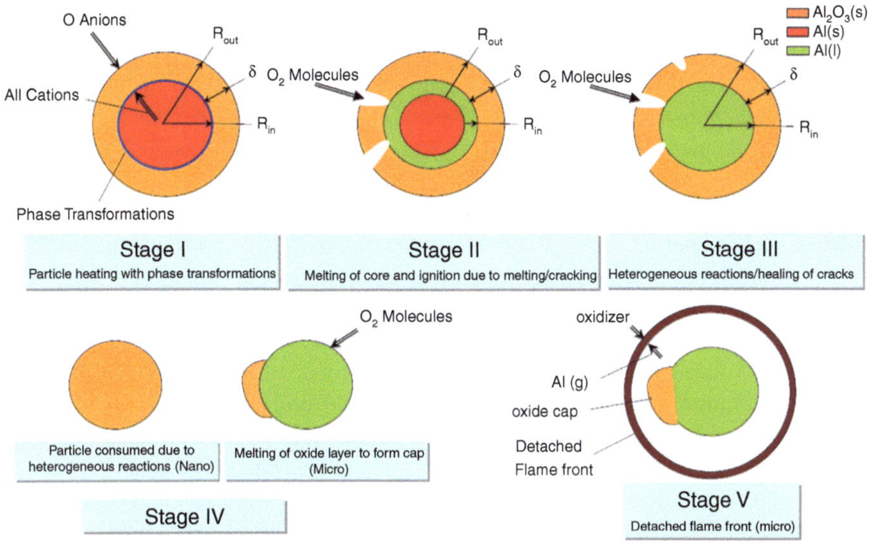

Fig. 4.2 Schematic presentation of various stages of the combustion/explosion mechanism of aluminum particles [67]; courtesy Prof. V. Yang

1. For particles burning on their surface, diffusion of oxygen to the combustion zone and diffusion of combustion products (e.g., carbon dioxide, water vapor) into the gaseous phase.
2. Heating of the particles' surface (due to the heat of combustion occurring on the particles' surface (particle area)) and, via heat conduction, heating of the particles' interior (particle volume).
3. Due to heating, phase transformations: melting, volatilization, boiling. E.g., de-volatilization of flammable gases/vapors from the particles and, via diffusion, transport of these flammable gases/vapors into the gaseous phase.
4. Possibly: combustion exclusively on the surface of the particles (e.g., pure carbon; metals with very high sublimation temperatures, such as titanium). Oxides may be gaseous (e.g., carbon monoxide/carbon dioxide), molten, or solid; the later two types eventually volatilizing (i.e., phase transformation), this depending on melting and boiling points and the adiabatic flame temperature of the relevant particulate [65].
5. Combustion of flammable gases/vapors (e.g., vaporized aluminum) in the gaseous phase.
6. Eventually: further combustion on the surface of the charred remains of out-gassed organic particles (e.g., coal). Reaction rate controlled by diffusion to, from, and within the particle.

Mechanistically, of all these steps, gas/vapor explosions involve only step 5.

Figure 4.2 illustrates the complexity of such a mechanism, here the sequential steps for the ignition and combustion process occurring in the explosion of aluminum particles [67].

Diffusion is the rate determining step of the heterogeneous combustion mechanism process sketched above: oxygen molecules must diffuse from the bulk gaseous phase to the particles' surface, and reaction products from the burning particles must diffuse into the bulk gaseous phase. Diffusion usually occurs much more slowly than the rate of chemical reaction, whose rate tends to increase exponentially with temperature.

The complex mechanism of particulate combustion is influenced by many factors, such as: particle size (distribution); the phase(s) of the particulates involved (solid and/or liquid); for liquids, volatility (vapor pressure as function of temperature (e.g., boiling point)); for solids, idem, including sublimation, melting, and boiling point; surface structure/specific surface area; heat conductivity and heat capacity; for substances forming liquid or solid oxides (e.g., coal; many metals, such as aluminum, magnesium, iron), melting and boiling points of the oxides; turbulence (affects rate of diffusion); flame temperature (K) and thermal intensity (kW/m^2) of the combustion process (affects rate and mechanism of heat transfer). It is obvious that any detailed discussion of the combustion mechanism of particulates is far beyond the scope of this chapter. We therefore sketch only simple, largely qualitative lines. Via literature cited in later sections we refer interested readers to more comprehensive studies.

Perhaps the simplest possible mechanism of explosion of particulates is the rapid combustion (deflagration) of small liquid droplets of a volatile combustible liquid (e.g., a mist of acetone or gasoline droplets) entrained in highly turbulent air. Well before the flame of the explosion reaches such droplets, the explosion's heat evaporates the droplets into the vapor phase, and the turbulence pre-mixes these vapor molecules with the oxygen molecules of air. The ensuing explosion thus occurs homogeneously; i.e., as a gas/vapor explosion [36].

More complicated, and more typical of most particulate explosions, is the case of droplets of a complex organic heavy liquid (e.g., heavy tar oil) or particles of a complex organic solid (e.g., bituminous coal). Also in this case particulates of organic materials undergoing fire or explosion must first be (partially) evaporated and/or (partially) pyrolyzed, i.e., pre-heated and thus brought into the gaseous/vapor phase as molecules, in order to chemically react further [36, 42]. Like the previous case, this evaporation/pyrolysis (pre-heat) is brought about by the heat of the approaching flame. The difference with the previous case, in which the combustion process occurs completely in the gaseous/vapor phase (i.e., homogeneously), is that here the combustion process continues on the surface of the particles i.e., combustion occurs heterogeneously. Because of the existence of two distinct phases, gaseous/vapor and solid (or liquid), the pre-heat zone and the flame zone are much thicker than in gas explosions (cm versus mm). However, once the combustion process is going it may become even more violent than a homogeneous explosion due to micro-mist formation as a result of drag on the particulates by the faster flowing gas [55].

Table 4.3 Effects of particle size (here particle length of a solid cube whose density is assumed to be 1,000 kg/m^3) on various dimensional parameters. Start: a solid cube of $1 \times 1 \times 1$ cm length with a density of 1,000 kg/m^3

Particle length	Particle mass, g	Particle specific surface area, m^2/g	Number of particles per g	Ratio of particle surface area to particle volume, m^{-1}
1 cm	1	$6 \cdot 10^{-4}$	1	$6 \cdot 10^{2}$
1 mm	10^{-3}	$6 \cdot 10^{-3}$	10^{3}	$6 \cdot 10^{3}$
100 μm	10^{-6}	$6 \cdot 10^{-2}$	10^{6}	$6 \cdot 10^{4}$
10 μm	10^{-9}	$6 \cdot 10^{-1}$	10^{9}	$6 \cdot 10^{5}$
1 μm	10^{-12}	6	10^{12}	$6 \cdot 10^{6}$
100 nm	10^{-15}	60	10^{15}	$6 \cdot 10^{7}$
10 nm	10^{-18}	600	10^{18}	$6 \cdot 10^{8}$

Table 4.4 Effects of change of particle size on particulate explosion severity parameters

Explosion severity term	Effect decreasing average particle size
Tendency towards detonation	Greater tendency to detonate
Maximum explosion pressure; P_{max}	Increases
Maximum rate of pressure rise, $(dP/dt)_{max}$	Increases
Volume-normalized maximum rate of pressure rise, called K_{St}-factor, where: $K_{St} = V^{1/3} \cdot (dP/dt)_{max}$	Increases (i.e., higher K_{st}-factor)

Particle Size Effects

General: Smaller Particles = Greater Hazard

Particle size very strongly influences particulate explosion behavior and hazard. For example, the difference between wooden logs quietly burning in an open hearth and a violent explosion of this same material (wood), but present as fine particulates, lies in the huge difference in particle size (and thus specific surface area: m^2/g) between rough wooden logs and wood in the powdered form. Table 4.3 illustrates dimensional parameters (e.g., specific surface area) determined by particle size (for simplicity expressed in terms of solid cubes, starting from a single solid cube of $1 \times 1 \times 1$ cm, and successively dividing this cube into smaller cubes).

To get a sense of dimensional scale: diameter of a human hair is ca. 10–50 μm; diameter of a red blood cell ca. 5 μm; diameter viruses typically ca. 10–100 nm; DNA molecule ca. 2–12 nm, water molecule ca. 0.3 nm.

In general, explosion hazards of particulates increase as particle size decreases, as summarized in Tables 4.4 and 4.5, for explosion severity and explosion sensitivity parameters, respectively. Figures 4.3 and 4.4 show examples of the huge effect of particle size on, respectively, LEL and MIE (explosion sensitivity); Figure 4.5 does the same for maximum pressure and rate of pressure rise (explosion severity). Indeed, while above a certain particle size (typically roughly 300 μm) combustible particulates will still burn, they will, however, no longer explode.

Table 4.5 Effects of change of particle size on particulate explosion sensitivity parameters

Explosion sensitivity term	Effect decreasing particle size
Lower Explosion Limit (LEL); also called Lower Flammability Limit (LFL)	Decreases
Upper Explosion Limit (UEL); also called Upper Flammability Limit (UFL)	Tends to increase. Note that smaller particles tend to remain suspended for a longer time than larger particles, which, due to their weight, tend to settle out more quickly
Explosive range: UEL – LEL	Increases
Lower Flash point (LFP)	Is theoretically unaffected. In practice probably decreases because smaller particles evaporate more quickly than larger particles
Minimum Ignition Energy (MIE)	Decreases
Minimum Oxygen Concentration (MOC)	Decreases
For particulates dispersed in air: Minimum Ignition Temperature (MIT); also sometimes called: Minimum Auto-ignition Temperature (AIT)	Decreases
For particulates that have deposited on a surface forming a surface layer: Minimum Ignition Temperature (MIT); also often called Minimum Smoldering Temperature	Lowest temperature (*T* in °C or K) of a surface that can cause a dust layer of a given depth to catch fire or at least smolder. Must be measured Explosion hazard increases as surface MIT decreases

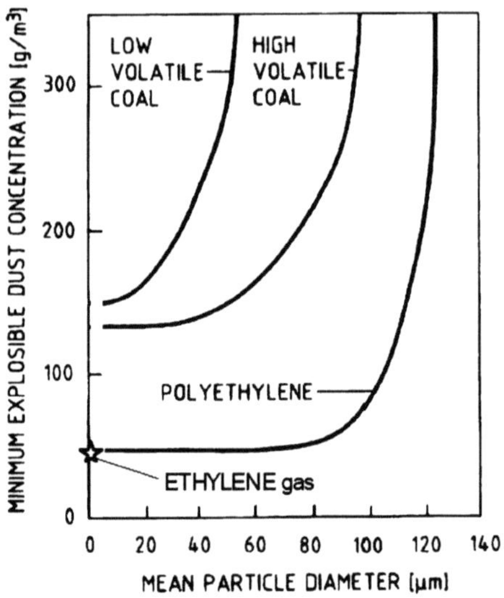

Fig. 4.3 Effect of mean particle diameter on Lower Explosion Limit (LEL; here expressed as 'Minimum Explosive Dust Concentration') of polyethylene powder and high and low volatile coal [44]; courtesy Prof. R. Dobashi

Results in Fig. 4.4 seem to indicate that, within the size range of 20–300 μm, MIE increases with the third power of increasing particle size (i.e. volume). Furthermore, it can be concluded from Figs. 4.3, 4.4, and, 4.5 that above a particle size

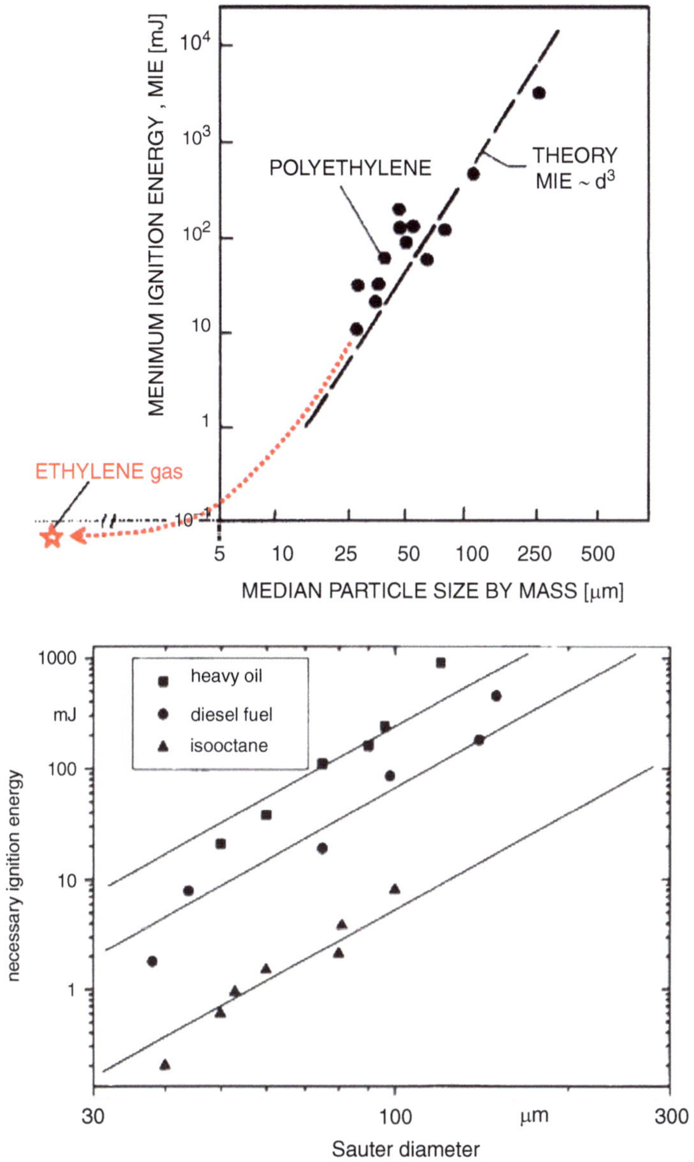

Fig. 4.4 Effect of particle size on Minimum Ignition Energy (MIE). At *above*: Effect of median particle size by mass on Minimum Ignition Energy (MIE) of polyethylene powder [45]; courtesy Prof. R. Dobashi. At *below*: Effect of particle size (Sauter Diameter) on MIE of three mists [50]. Copyright Wiley-VCH Verlag GmbH & Co. KGaA (Reproduced with permission)

of roughly 300 μm combustible particles will still burn, but they will no longer explode.

Given that effects of particle size on particulate explosion hazard can hardly be exaggerated it is surprising that many, especially older (ca. pre-2000), studies

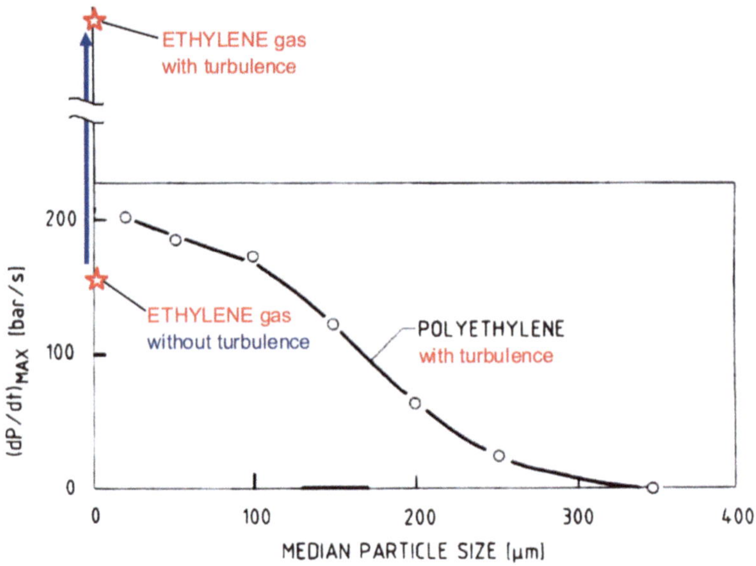

Fig. 4.5 Effect of particle size on maximum rate of pressure rise of explosions of powdered polyethylene [45]; courtesy Prof. R. Dobashi

presenting explosion behavior as a function of particle size are neither clear as to how particle size is defined nor how it is measured. Already in 1980 Ballal [34] noted this hiatus, commenting: "Clearly, from the point of view of fundamental combustion studies, it is the area-weighted Sauter mean diameter (SMD or $D_{3,2}$), particle size distribution, and equivalence ratio that are of great importance." In 2003 Eckhoff wrote [12]: "Particle size analysis is a large field of research and development in itself, and the main purpose of this section is to re-emphasize the major role played by particle size and shape and their distributions in deciding the ignitability and explosibility of a dust of a combustible material." Eckhoff also lists various methods of measuring particle size and size distribution. Nevertheless, like many studies of dust explosion behavior, in his many (highly useful) figures illustrating explosion severity and sensitivity as a function of particle size Eckhoff is not clear as to what he precisely means by 'particle size'. Nowhere in this source, which is presently perhaps the best general source on dust explosions, is the Sauter mean diameter even mentioned. He mentions, however, a related essential point, namely that a measured size distribution may be quite different from that occurring in practice, due to how in practice a powder is dispersed and suspended. Föster does explicitly discuss this point – the importance of how particle size is measured and expressed, including the Sauter mean diameter [50]. Experimental results that do not clearly state how particle size is expressed should be viewed with due care.

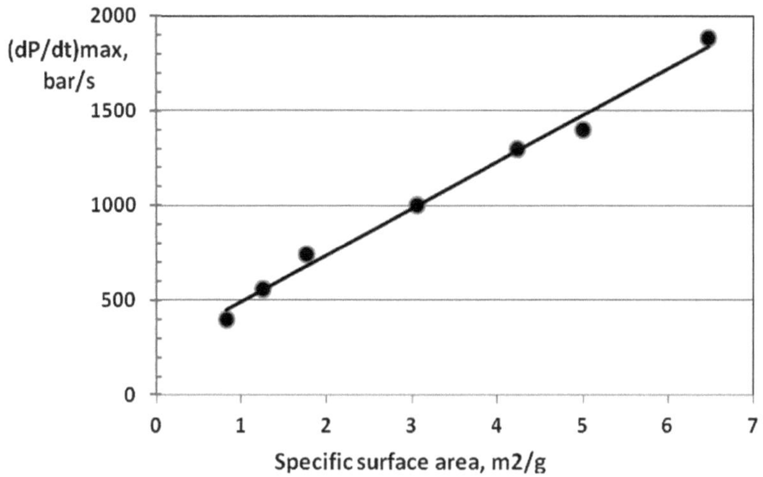

Fig. 4.6 Linear increase of explosion severity [$(dP/dt)_{max}$] with specific surface area for aluminum particles (Adapted from [4])

Particle Size Distribution

Particulates with the same average particle size often have different particle size distributions. This factor can often result in significantly different particulate explosion hazards. For example, a powder having a broad particle size distribution may exhibit high explosion sensitivity and severity, whereas the same powder – with exactly the same chemical composition – but with a narrow particle size distribution around the same median size - exhibits no dust explosion behavior whatsoever. The reason is that the first powder contains a significant fine fraction, which is much more sensitive to ignition than the coarse fraction. These fines start the ignition, and the ensuing combustion is sufficiently energetic to ignite the coarser fraction. Particle size distributions should therefore not be characterized by median size or an undefined mean size, but by a well-chosen mean size, such as the Sauter mean diameter. During size measurements, the size of so-called equivalent spheres is determined, which is dependent on both particle shape and measurement principle [58]. In view of this, measurement results should be treated with due care when comparing results coming from different techniques.

Particles are often assumed to be spheres. However, in general practice particles may well have irregular shapes, resulting in a larger surface area than the corresponding sphere with the same volume. Increasing specific particle surface area (m^2/g) generally increases powder explosion hazards (i.e., greater explosion sensitivity and severity). Figure 4.6 illustrates this for the explosion severity [$(dP/dt)_{max}$] of aluminum powders in the form of flakes [4].

In view of the important role that the particle (external) surface plays in their explosion mechanism, the area-weighted Sauter mean diameter, $D_{3,2}$, should be regarded as the relevant particle size parameter for explosion studies of particulate products. Experimental results that use the median size without stating the PSD width or do not clearly state how particle size is measured should be viewed with due care.

Nanoparticles

Nanoparticles are defined to have at least two dimensions smaller than 100 nm. Previous discussion makes clear that as particle size decreases, both explosion sensitivity and explosion severity increase. Obviously, the following questions arise: as particle size decreases, does a limiting size exist below which further particle size reduction has no effect, and, if so, what is this size limit, and what are the limiting values of explosion sensitivity and explosion severity indexes? It would seem logical to hypothesize that as particulate particle size decreases and approaches molecular size (i.e., into the lower nanometer range and beyond), particulate explosion behavior will increasingly resemble gas/vapor explosion behavior. Thus, the limiting values for explosion sensitivity and explosion severity will approach and eventually become identical to those of gases/vapors,[5] although most often nano-particles may agglomerate to larger clusters, which will limit the effect. We further hypothesize that as the volatility of the particulate increases, the diameter at which the particulate exhibits gas-/vapor-like combustion/explosion behavior will increase; and vice versa. Thus non-volatile particulates, such as powdered pure carbon (graphite or diamond) and tungsten, will exhibit gas/vapor behavior only at the low end of nano-particle diameter range, while volatile substances, such as naphthalene and fuel oil, will already exhibit such behavior well into the micrometer range.

Obviously such behavior is further complicated by the volatility of oxides eventually (e.g., compare gaseous CO and CO_2 with metallic oxides) formed and the intensity (e.g., rate of heat production) of the combustion process.

Figure 4.7 presents indirectly the fundamental reason for such behavior, at least for powders that do not readily volatilize, such as most metals (e.g., Al): as particle size decreases below roughly about 10 μm, the combustion mechanism changes. In the region of larger particles, combustion is diffusion-controlled; in the region of increasingly smaller particles, it becomes controlled by chemical kinetics [47, 66]. In other words: chemical reactions driving gas/vapor explosions are kinetically controlled, and as powder particle size decreases and approaches that of gas molecules, powder explosions progressively become more like gas/vapor explosions. An important exception to this generalization is that it is to be expected that the higher volumetric energy density of dust explosions relative to gas/vapor explosions (section "Upper Explosion Limit, Explosive Range, Volumetric

[5] Note that a 1-nm aluminum sphere contains 32 atoms, a 2-nm sphere 256 atoms. Volume, and thus the number of atoms, increases with the cube of the diameter.

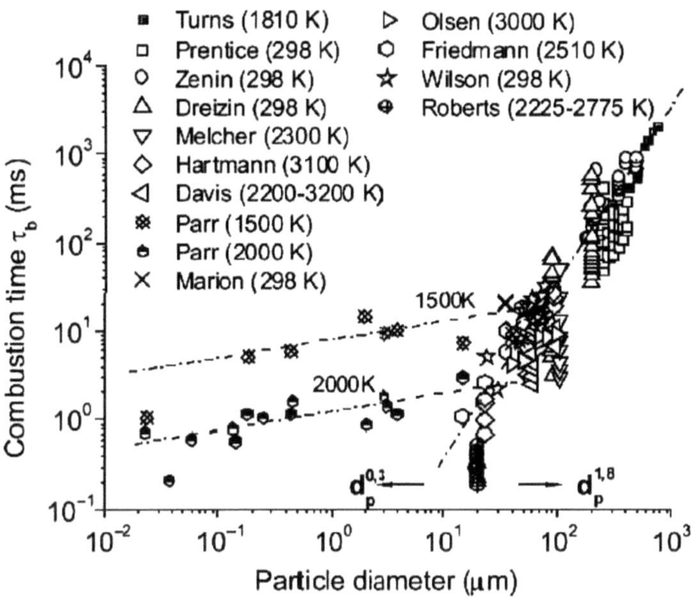

Fig. 4.7 Effect of particle size on combustion time [47]

Explosion Energy") will not be affected by decreasing particle size, even into the nanometer range.

Yang et al. illustrate the transition from diffusion control to chemical kinetics control of reaction rate as particle size decreases from the micrometer range into the nanometer range for the so-called ***laminar burning velocity*** (Fig. 4.8) [72]. This velocity is a fundamental property of the flame of a given mixture of a combustible substance in air at given conditions. It is the speed of the flame in a non-moving (thus laminar) pre-mixed mixture of fuel (e.g., hydrogen, grain powder) and oxidizer (e.g., air). Hydrocarbon fuels exhibit laminar burning velocities of tens of centimeters per second or somewhat higher. Highly reactive combustible gases, such as hydrogen and acetylene, have laminar burning velocities in air typically in the meter per second range. But even for highly reactive combustible particulates in the micrometer range, laminar burning velocities in air are normally typically at least a factor ten smaller. Figure 4.8 illustrates the effect of decreasing particles size on the laminar burning velocity (S_L) of powdered aluminum in air, with extrapolation into the nanometer range and even to the particle size range of molecules. Yang et al. thus predict that as particle size decreases and eventually reaches the molecular range, laminar burning velocities of reactive particulates will be quite similar to those of reactive gases and vapors [72].

This same line of reasoning – nanosized particles in the homogeneous, chemical-kinetically controlled regime, and micrometer-sized particles in the heterogeneous, diffusion controlled regime – is also applicable to ***ignition*** of fine metallic particles (i.e., their MIE and MIT), as is shown in Fig. 4.9 for the Minimum Ignition Energy (MIE) of aluminum particles [47]. As particle size (here expressed as Sauter mean

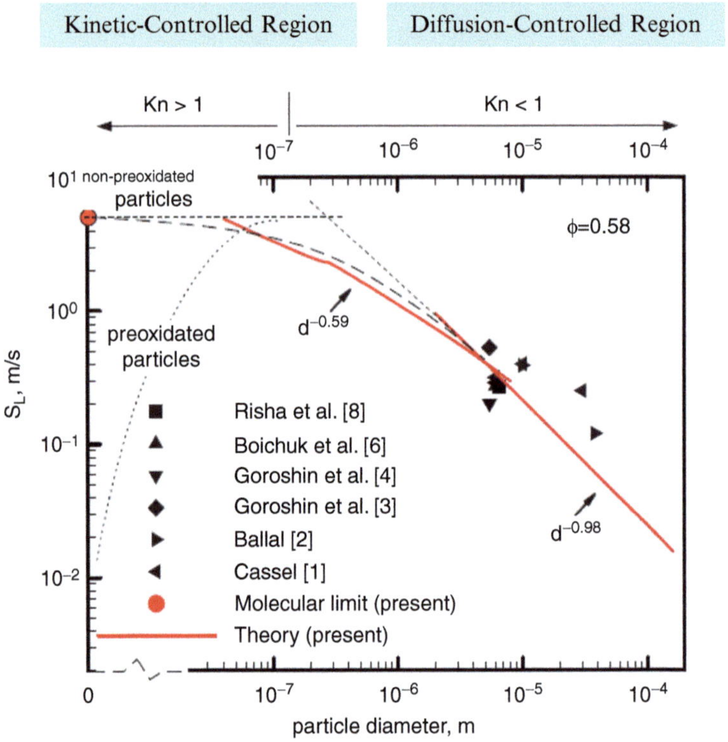

Fig. 4.8 Effect of decreasing particle size on laminar burning velocity (S_L) of reactive (i.e., non-pre-oxidized) aluminum particles. [72]; courtesy Prof. V. Yang

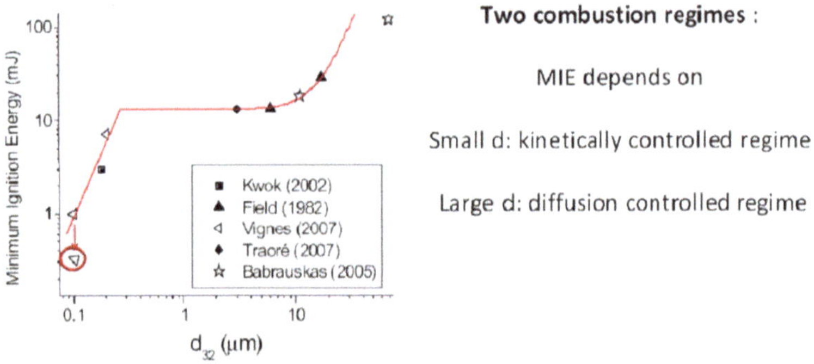

Fig. 4.9 Effect of reduction of particle diameter on Minimum Ignition Energy (MIE) of powdered aluminum (Adapted from [47])

diameter) approaches molecular diameter, MIE-values of particulates approach those of gases/vapors, i.e., for reactive metal powders, MIE below 1 mJ.

Eckhoff nuances some of the above-mentioned conclusions by dividing powders into roughly two types, based on chemical composition: (1) powders of substances that significantly volatilize at relatively low temperatures (e.g., many organic substances, such as polyethylene powder, grain and wood dust, powdered coal) and (2) powders of substances that volatilize only at relatively (very) high temperatures (e.g., many metals (Al, Zi, Ti); pure carbon). Decreasing particle size of powders of the former type will indeed increase explosion severity and sensitivity, but only down to roughly 1–10 μm. The reason for this behavior is that at approximately this particle size, combustion (explosion) already occurs largely in the gaseous phase, and therefore the combustion mechanism is already chemical kinetically determined. For particles of the second type, however, effects of decreasing particle size on increasing explosion sensitivity and explosion severity (the latter expressed as maximum rate of pressure rise (K_{st}-factor)) will continue, even as particle size decreases into the nanometer range [25, 48]. The reason for this is that the diffusion-controlled nature of the combustion process (i.e., slowness of the de-volatilization process) can only be sufficiently negated by particles approaching the size of molecules.

Eckhoff [25, 48] and Dobashi [25, 45] also add the caveat that while individual particles may nominally be in the nanometer range, once distributed into air, intra-particle forces (e.g., van der Waals forces) are, compared to the mass (thus inertia force) of the particles, so large that such nanoparticles rapidly (i.e., within 1 s) agglomerate into much larger clusters in the micrometer range.

Much of the research done on dust explosions of nano-particles relates to reactive metals, such as aluminum. For such reactive substances Dobashi mentions a further possible caveat [25, 45]: Because of their high chemical reactivity, it is difficult to prepare, store, and perform experiments with nanoparticles of reactive metals, such as aluminum, in their pure state, that is, particles that are completely unoxidized. For example, aluminum particles typically have an oxidized surface layer. The presence of such an oxide film obviously reduces the reaction rate by physically interfering with further combustion as well as reduces the heat of combustion. Note that a 4 nm surface layer on a spherical 100 nm particle already takes about 12 % of the total volume.

Concerning effects of reducing particle size from the micrometer into the nanometer range we conclude:

1. As a general rule, the smaller the combustible particles are, the greater is their explosion hazard and risk (as expressed by explosion sensitivity and explosion severity indexes).
2. As particle size increasingly approaches the size range of molecules, limiting values of maximum explosion sensitivity and explosion severity are determined exclusively by chemical properties (as is the case for gases and vapors) and not by physical dimensions.
3. The basic reason for the difference in behavior between explosions of upper-micrometer-sized combustible particles and nanometer-sized particles is that the combustion mechanism of the former is controlled by relatively slow physical processes associated with heterogeneous reaction (e.g., diffusion), while the latter is (largely) controlled by chemical kinetic factors, due to the (more) homogeneous combustion mechanism of such (fine) particulates. Especially at

the high temperatures associated with combustion, chemical kinetic rates are generally (much) faster than physical processes such as diffusion.

4. As particle size decreases from the upper micrometer range to roughly 10 μm, maximum explosion pressure (P_{max}) generally increases. Once particle size falls below this size, however, P_{max} no longer increases. Reason is that maximum explosion pressure is reached at the adiabatic flame temperature, which is a chemical thermodynamic term (i.e., state function), and thus independent of particle size and reaction mechanism.
5. For typical organic products, maximum rate of pressure rise (e.g., K_{St}-value) increases significantly as particle size decreases from the high micrometer range to roughly 10 μm. However, further reduction of particle size seems not to further significantly increase the rate of pressure rise. Possible reason is that as particle sizes fall below roughly 10 μm for such substances, which usually volatilize at relatively low temperatures, combustion occurs largely homogeneously, and is thus controlled by chemical kinetics.
6. Of all the indexes of explosion sensitivity, the Minimum Ignition Energy (MIE) seems to be particularly sensitive (i.e., continues to fall) as particle size decreases from the micrometer into the nanometer range.
7. Investigating and predicting effects on explosion hazard of reducing particle size of particulates into the nanometer range are often complicated by the fact that the smaller the particles, the greater their tendency to agglomerate into larger particles. And, for powders of reactive substances, such as metals: Metal surfaces are very easily oxidized; the oxide film formed reduces both the reactivity (speed of reaction) and the amount of energy available for reaction (e.g., thus reducing adiabatic flame temperature). Both these phenomena tend to reduce, or even negate, effects of particle size reduction on increasing hazard.

Effects of Turbulence

Turbulence[6] is essential for the occurrence of particulate explosions, except when/where gravity and other similarly working forces are absent (e.g., in a non-accelerating vehicle in microgravity). Turbulence is necessary to disperse particulates, mix them with air, and keep them suspended. Turbulent combustion is a complex topic [63]; here we present only a qualitative, highly simplified discussion.

In addition to being essential for creating and maintaining an air-particle mixture, turbulence has a complex effect on dust explosion risk. On the one hand, increased turbulence ***reduces*** risk by reducing *explosion sensitivity* (e.g., increasing MIT, MIE, LEL, etc.). This effect is due to the fact that at the limits of ignition (i.e.,

[6] Turbulence is highly irregular motion of fluids in which local velocities (i.e. speed and direction of fluid motion) exhibit rapid, irregular and apparently random fluctuations (eddies) that are superimposed on the net fluid velocity.

at the conditions of MIT, MIE, LEL, etc.), increased turbulence hampers ignition by dispersing ignition energy. The physical reason for this is that increased turbulence mixes more cold fluid into the just-barely ignitable kernel of air plus particulates, and this quenching effect prevents ignition.

On the other hand, for an already ignited, highly energy-rich particle-air mixture (e.g., an already ignited polyethylene-air mixture in the middle of its explosive range), increased turbulence ***increases*** risk by increasing *explosion severity* (i.e., increasing P_{max}, $(dP/dt)_{max}$, K_{St}-factor). Reasons for this effect are, firstly, that greater turbulence tends to increase the surface area of the flame of a burning mixture. Turbulent eddies cause an initially flat (two dimensional) flame to transform into a (severely) 'wrinkled' three dimensional flame, thus greatly increasing the flame's surface area. Since the flame's surface is the area on which energy is being released (W/m^2), increase in surface area will increase the total rate of energy release. As mentioned above, the greater the rate of energy release, the greater the explosion severity. This effect is also relevant for gaseous explosions occurring in energy-rich mixtures, as Fig. 4.5 illustrates for effect of turbulence on deflagrations of ethylene gas. It forms the mechanism of driving flame acceleration up to hundreds of meters per second leading ultimately to detonation.

Turbulence also tends to decrease the slowing effect of diffusion, diffusion being a factor inherently reducing the rate of combustion of multi-phase (i.e., heterogeneous) mixtures. A simple way to visualize this effect is to consider each particle to be enveloped in a thin laminar 'film', of a given thickness (μm), through which diffusion takes place [32]. Increased turbulence reduces the thickness of this laminar film, thereby reducing the time it takes for diffusion through the film to occur, i.e., diffusion time = (thickness of laminar film (μm))/(average rate of diffusion (m/s)). Since diffusion is the rate-determining step, faster diffusion speeds up the rate of energy release, thus increasing the severity of the explosion. (However, at high turbulence intensity this effect can be countered by flame quenching and extinction due to a high cooling rate.)

On a more practical level turbulence can cause dust explosions to exhibit a unique phenomenon. Through turbulence, an explosion can become self-feeding; i.e., a small explosion starting somewhere in a plant grows due to the settled/deposited dust on floors and machinery being stirred up by the pressure waves and turbulence generated before the flame arrives. Explosions can in such cases propagate from one plant or installation to another – with devastating effects. Many disastrous dust explosions of this kind – in which a primary explosion generates and ignites one or more secondary explosions - are known [70]. The dust deposits are usually caused by neglect due to bad house-keeping [71] (and thus, ultimately, poor management).

Additionally, inside a process installation a small, incipient explosion can, as it propagates through piping connecting vessels within the installation, generate increasing turbulence. It is this self-generated turbulence that can cause flame speeds to dramatically accelerate. As noted earlier, in the extreme case flame acceleration can result in a *deflagration-to-detonation-transition* (DDT) [64]. Another dangerous phenomenon associated with turbulence as an explosion propagates between

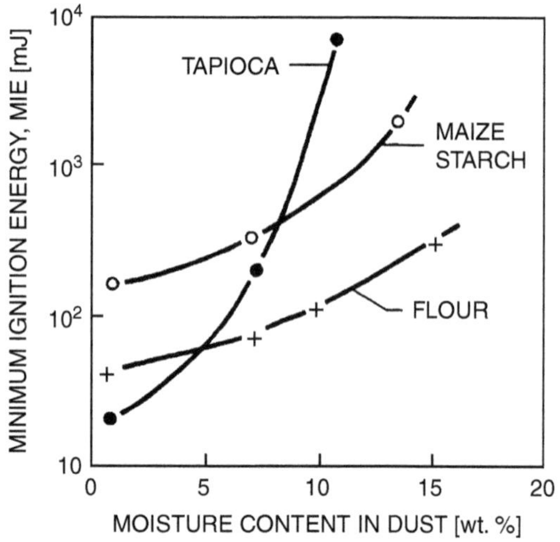

Fig. 4.10 Effect of moisture content on Minimum Ignition Energy (MIE) of three hydrophilic powders [12] (Copyright Elsevier; reproduced with permission)

interconnected vessels is *pressure piling* [11]. Pressure piling results in greatly increased explosion pressures and rates of pressure rise in secondary vessels. It is for these reasons that propagation of incipient explosion must be prevented (e.g., via compartmentalization; see Fig. 4.17).

4.5.4 Combined Chemical and Physical Effects

Effects of Humidity

Chemical composition determines not only combustibility, but also whether and to what degree a particulate is hydrophilic ('water-loving') or hydrophobic ('water-hating'). Hydrophilic powders tend to adsorb/absorb moisture from the air forming a layer of water molecules on the particles' surface. Through surface tension effects, this layer of water causes the particles to stick together (form agglomerates), which can strongly increase virtual particle size and thus reduce external specific surface area. Specific surface area plays an important role in chemical kinetics and adsorption/absorption kinetics and therefore strongly influences explosion hazard. In particular the Minimum Ignition Energy of hydrophilic powders is strongly reduced by increasing concentration of adsorbed/absorbed water, as Fig. 4.10 shows.

Additionally, a layer of adsorbed water tends to increase electrical conductivity of powders, thereby reducing the tendency to generate and retain static electrical charge, thus reducing the hazard of ignition by static electricity [20, 52].

In general, powders of natural origin (e.g., potato and corn starch, wood dust, sugar) are hydrophilic. Many synthetic materials, such as most plastics (e.g., polyethylene) are hydrophobic. Important to note further is that the concentration, and thus the effect, of adsorbed/absorbed water is strongly influenced by the relative humidity and temperature of the air to which a powder is exposed. In general for hydrophilic powders, the higher the relative humidity and the lower the temperature are, the greater the concentration of adsorbed/absorbed moisture; thus the lower the explosion hazard, and vice versa. Obviously exposure time is also important: the longer the time of exposure, the more completely equilibrium is reached between water in the atmosphere (humidity) and water adsorbed/absorbed on the particles (i.e., particle water-content). It is for these reasons that, for example, grain dust explosions in the American mid-west are much more common in the summer, during extended periods of high temperature and low relative humidity, than during the winter.

Upper Explosion Limit, Explosive Range, Volumetric Explosion Energy

Mixtures of combustible gases and vapors with air generally have sharply defined explosion limits (LEL- and UEL-values) and thus sharply defined explosive ranges (UEL – LEL). Additionally, UEL-values of gases and vapors at ambient pressure and temperature are typically roughly two times the stoichiometric concentration.

By contrast, particulates have less sharply delineated explosion limits; in particular, upper explosion limits (UELs) of particulates are poorly defined. Why? For two reasons: Firstly, a particulate-air mixture with a concentration initially above the UEL-value will, through deposition caused by gravity, eventually enter the explosive range. For this reason the UEL-value for a particulate usually has little practical significance and can even be dangerous by causing false confidence.

Secondly, while a gas/vapor explosion takes place between molecules, particulate explosions involves the surface of the particles. As the mass concentration of powder in a powder-air mixture increases to (far) above the theoretical (mass-based) stoichiometric concentration, the powder particles simply burn to a thinner and thinner depth. This is the reason that dust explosions generally show a much ‘flatter’ relationship between explosion severity and concentration compared to gas/vapor explosions. Thus, while gas/vapor explosions generally show a sharp maximum explosion severity at roughly the stoichiometric concentration, and explosion severities fall sharply past this concentration, dust explosion severity is generally reached (far) above the (mass-based) stoichiometric concentration and, additionally, it falls off slowly (often far) above the stoichiometric concentration. Table 4.6 shows these phenomena, by comparing gas explosions (methane in air) with dust explosions (coal dust in air). Note that while methane is hardly explosive at just above about 150 % of the stoichiometric concentration, Pittsburgh coal dust is still explosive at concentrations more than 40 times (!) of its stoichiometric concentration.

Figure 4.11 graphically contrasts the explosion severity of methane in air compared with polyethylene powder in air, as function of concentration [20, 52].

Energy content increases with substance amount. Common gases typically have a density of roughly 1 kg/m^3 (at STP). Solids and liquids generally have a much higher

Table 4.6 Explosion severity (explosion pressure and rate of pressure rise) as a function of concentration (relative to stoichiometric concentration) of a gas explosion (methane in air) and a dust explosion (Pittsburgh coal dust in air). Tests carried out in spherical apparatus of 1 cubic foot (28 l); values estimated from [51]

Fuel Initial concentration percent methane or gram/cubic meter Pit. coal	Initial concentration ratio to stoichiometric	Fraction burned	Maximum pressure (bar)	Maximum rate of pressure rise (bar/sec)
Methane (nonturbulent mixtures)				
6.0	0.53	1.00	4.3	10.1
7.1	0.75	1.00	5.8	40.3
7.9	0.84	1.00	6.4	74
8.9	0.94	1.00	6.9	114
9.4	0.99	1.00	7	107
9.5	1.00	1.00	7.5	141
9.7	1.03	0.98	7.3	148
9.9	1.05	0.96	7	131
10.1	1.07	0.94	7.3	141
11.1	1.16	0.85	7.1	111
12.3	1.30	0.77	6.5	50
13.1	1.38	0.72	5.8	23
14.1	1.49	0.68	4.5	3.4
Pittsburgh coal dust (turbulent mixtures)				
53	0.41	1.00	-	-
106	0.81	0.50	4.7	20.1
212	1.63	0.40	7.1	97
424	3.25	0.30	7.3	128
636	4.88	0.20	7.3	101
848	6.50	0.15	7.1	97
1,060	8.13	0.12	7.1	117
5,300	40.65	0.02	6.7	20

density, typically 1,000 kg/m^3 or more, thus at least about 1,000 times greater than gases. Therefore a typical powder-air mixture can often contain more energy per unit volume (i.e., volumetric energy content, J/m^3) than a gaseous mixture.

The following thought experiment also relates to the much higher density of solids relative is gases/vapors, and the effect of this on the amount of energy liberated by a dust explosion relative to a gas/vapor explosion. Consider a closed spherical volume of 1 m^3 (= 1,000 l) containing a stoichiometric mixture of methane gas and air at room temperature and atmospheric pressure. Since the density of methane at room temperature and atmospheric pressure is ca. 0.68 kg/m^3 = 0.68 g/l, adding 1 g of methane (= ca. 1/0.68 l = ca. 1.5 l of gas), and maintaining pressure at 1 bar, displaces ca. 1.5 l of air from the vessel. This 1.5 l of air is equivalent to ca. 315 ml of oxygen (because air is ca. 21 volume-% oxygen). Consider now the same vessel, but containing a mixture of fine aluminum powder suspended in air (i.e., an explosive powder-air mixture). Aluminum has a density of ca. 2,700 kg/m^3 (= 2.7 kg/l = 2.7 g/cm^3). Adding one extra gram of aluminum to

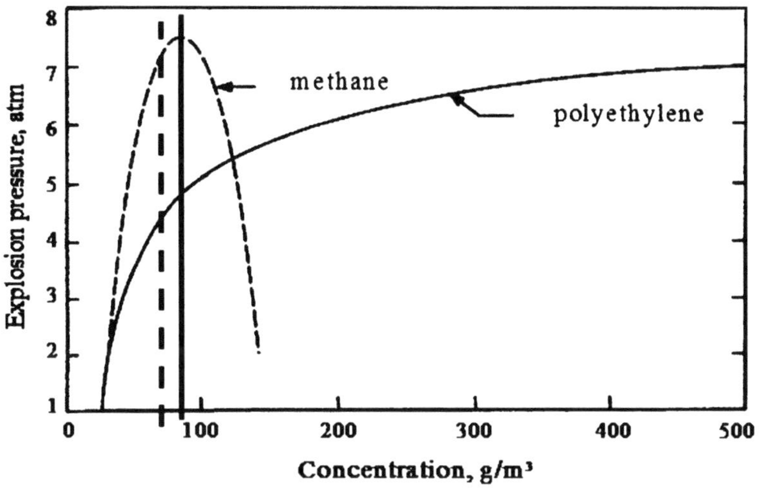

Fig. 4.11 Comparison of explosion limits for a gas explosion (methane) and a dust explosion (polyethylene powder) in air. Stoichiometric concentration of methane = 67.8 g/m^3 (vertical stripped line); stoichiometric concentration of polyethylene in air = 87.1 g/m^3 (solid vertical line) (Adapted from [20])

this mixture of aluminum and air will displace only 1 g/(2.7 g/cm^3) = ca. 0,37 cm^3 of air = ca. 0.07 cm^3 of oxygen, thus only a tiny amount of oxygen, compared to the addition of methane gas.

The point of this discussion is that because solids and liquids are typically a thousand times denser than gases, addition of an additional amount of a combustible particulate to a powder-air mixture will displace ca. 1,000 times ***less*** oxygen than the addition of the same amount of a flammable gas or vapor. Also for this reason flammable (thus explosive) particulate-air mixtures can contain a much higher weight-percentage of fuel than flammable gas/vapor-air mixtures. Since particulate-air mixtures can contain more fuel, with sufficient oxygen, particulate-air mixtures can have a higher volumetric energy density. In other words: the explosion of 1 cubic meter of a typical particulate-air mixture liberates more energy that the explosion of 1 cubic meter of combustible gaseous mixture.

Particulate-air mixtures can therefore contain more energy per unit volume than gaseous mixtures. But do such mixtures explode more violently than gases? Because particulate explosions occur heterogeneously (on particle surface and thus with diffusion limitation), particulate explosions tend to occur more slowly than gas explosions. The slower reaction rate manifests itself in lower rates of pressure rise (dP/dt).

Following from Eq. 4.1, the maximum explosion behavior for deflagrations is determined by the maximum explosion temperature. Gas/vapor and particulate explosions have roughly similar explosion temperatures (i.e., maximum adiabatic flame temperatures between 1,500 and 3,000 K). Therefore, except for dust explosions of highly reactive substances, such as fine powder of metals like aluminum and magnesium, particulate explosion and gas/vapor explosions generally have similar maximum pressures, i.e., between 5 and 10 bara (starting at STP).

If not by higher explosion severity, how then do the higher volumetric energy densities of particulate explosions manifest themselves? By explosions that occur over a longer time period. Explosions lasting longer, and releasing more energy, can be more hazardous than explosions with the same maximum pressure occurring for a shorter time.

Furthermore, since particulate explosions can occur at concentrations far above stoichiometric concentrations, a particulate explosion occurring in and rupturing closed equipment can emit large quantities of hot unburned material to the atmosphere. Upon contact with air this hot unburned material can explode, resulting in a second explosion occurring in the open air.

Hybrid Mixtures

Hybrid mixtures are mixtures of combustible solid particles and air containing a combustible gas and/or vapor.

Minimum ignition energies (MIEs) of typical flammable gas/vapor mixtures (e.g., methane in air, gasoline vapor in air), which are mixtures of individual molecules, are often very low (e.g., MIE-values typically between 0.1 and 1 mJ). Considering the fact that a human body can generate static electrical sparks of more than 10 mJ, such MIE-values are indeed low.

As noted above in section "General: Smaller Particles = Greater Hazard", Minimum Ignition Energies of particulates are very strongly influenced by particle size. Except for some very fine powders of certain substances, such as very fine sulphur powder (and nanoparticles, considered in section "Nanoparticles"), most common non-volatile powders have higher MIE-values than typical flammable gases. Figure 4.12 compares the MIE-values of some common gases with some common powders [12].

In terms of sensitivity to ignition via electrical sparks, most common powders are in general less hazardous than flammable gases.

Figures 4.13 and 4.14 show, respectively, how the presence of a flammable gas and/or vapor can drastically increase explosion sensitivity (MIE and LEL) and explosion severity of various powders, in particular the rate of pressure rise (see also Fig. 4.1) [12].

Important to note is that not only flammable gases can create a hybrid mixture; solvent vapors can too. Powders containing even small concentrations of a flammable solvent (e.g., alcohol, acetone) can, for example during storage, release part of the solvent via evaporation. If such a powder becomes airborne the combination of powder plus flammable vapor (from the evaporated solvent) will create a hybrid mixture, with all the associated increased hazards of explosion sensitivity and severity. There have been cases where unexpectedly explosion occurred with rather coarse particulates not explosive at normal ambient temperature, after they had been heated up and flammable vapors had evolved.

Mixtures of combustible solid particles and air containing a combustible gas and/or vapor (hybrid mixtures) thus combine the worst hazards of dust explosions

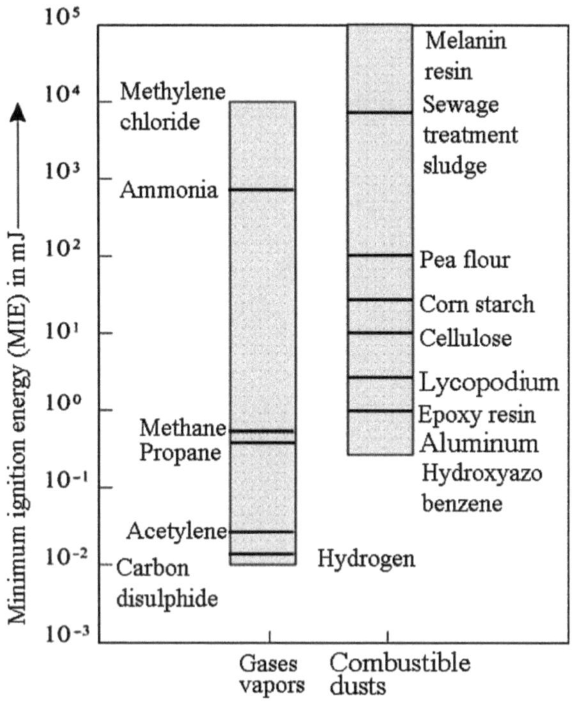

Fig. 4.12 Comparison of Minimum Ignition Energies (MIEs, in mJ) of some common gases and vapors (of carbon disulfide) with some flammable powders. Many common flammable gases (e.g., alkanes, alkenes) and vapors (e.g., alcohol, acetone) have MIEs below 1 mJ, while most common powders, whose particle size is in the micron range, have MIEs (considerably) higher than 1 mJ [12] (Copyright Elsevier; reproduced with permission)

and gas explosions: the high explosion sensitivity of gas explosions, the high energy density of dust explosions, and explosion pressures as high as those of gas/vapor explosions.

All of the above discusses theory. A clearly written industry publication [46] concisely summarizes the most important practical considerations relating to powder ('dust') hazards. In order for a dust to explode:

1. The dust must be combustible.
2. The dust must be capable of becoming airborne (i.e., be below a certain particle size).
3. Sufficient turbulence/flow must exist (have existed) in order to form (have formed) a dust suspension.
4. The dust must have a particle size distribution capable of flame propagation (i.e., be sufficiently fine).
5. The dust concentration must be within the explosion limits (between UEL and LEL).
6. An ignition source of sufficient energy/power/temperature must be present (i.e., above MIE and/or MIT).

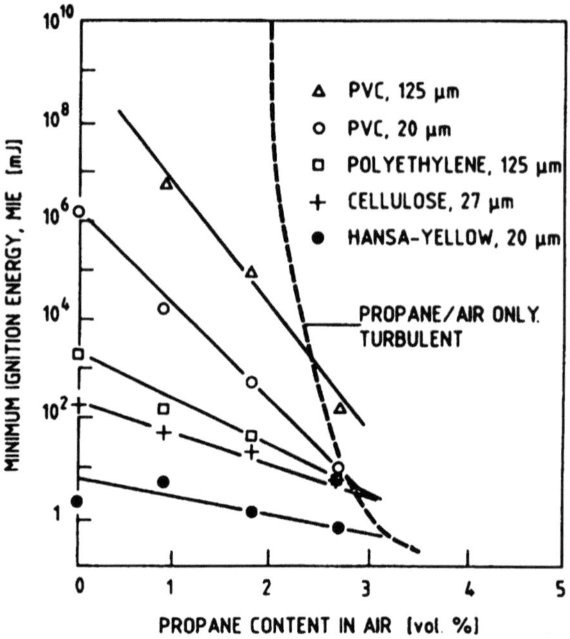

Fig. 4.13 Effect of flammable gas (propane) on drastically reducing the Minimum Ignition Energies (MIEs) of various powders [12] (Copyright Elsevier; reproduced with permission)

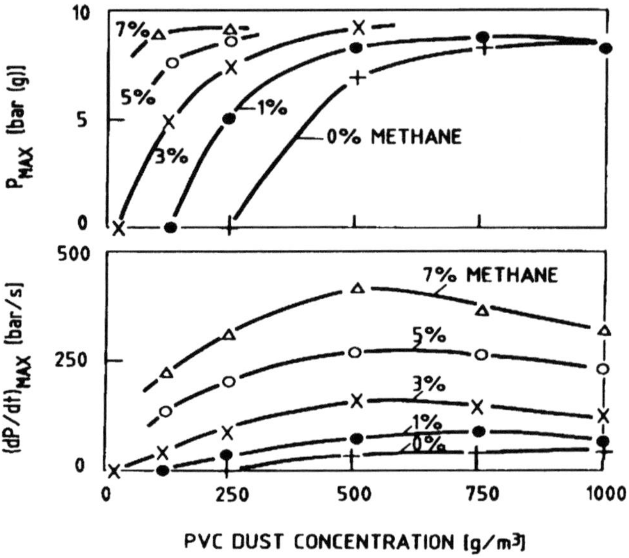

Fig. 4.14 Effect of flammable gas (methane) on reducing the Lower Explosion Limit, increasing the maximum explosion pressure and increasing the maximum rate of pressure rise of PVC-powder [12] (Copyright Elsevier; reproduced with permission)

7. The atmosphere must contain sufficient oxygen to support and sustain combustion (i.e., oxygen concentration above MOC).
8. There must be some degree of confinement of the dust cloud by an enclosure or partial enclosure.

A dust explosion can occur only if ALL of these requirements are met simultaneously. Thus, eliminating just one of these requirements prevents a dust explosion from occurring. Note also that the presence of small concentrations of combustible gas or vapor (i.e. hydbrid mixtures) can enormously increase dust explosion hazards.

4.6 Index Systems Ranking Hazards of Particulates

Hazard index systems combine a number of particulate properties to expresses the hazard of a given particulate. Perhaps the most internationally applied particulate hazard index is that of the National Fire Protection Association (NFPA) of the United States. Figure 4.15 presents the rationale of the NFPA system [10].

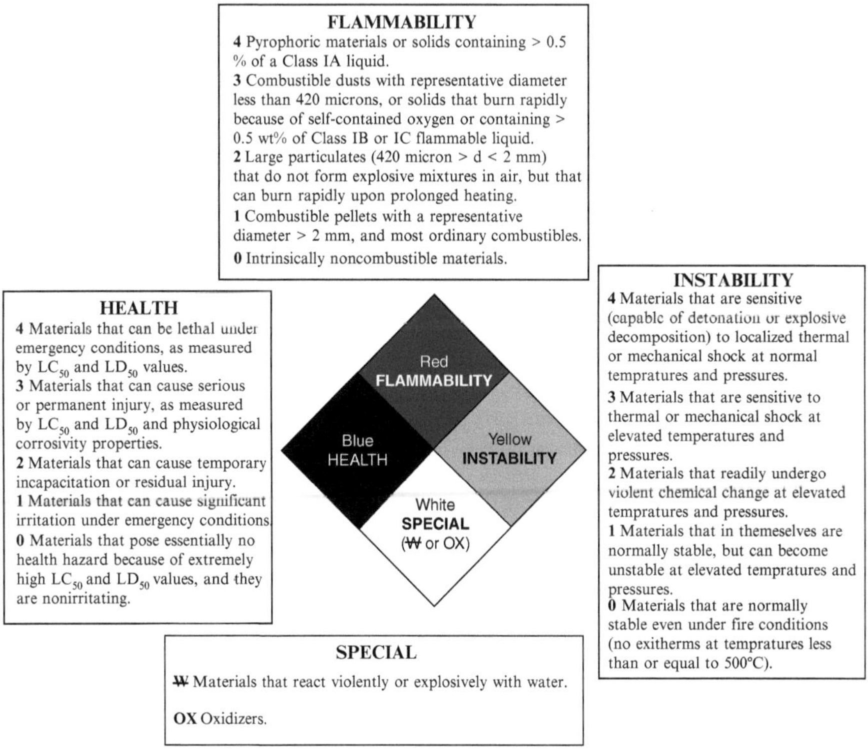

Fig. 4.15 Rationale of NFPA-system for expressing hazard indexes for Health (Toxicity), Flammability (Fire), and Instability (Explosion) for particulates. 0 = No Hazard; 4 = Maximum Hazard [10]

This system ranks particulate hazards in terms of four general parameters: fire ('Flammability'), explosion ('Instability'), toxicity ('Health') and 'Special factors', such as reactivity with water (since water is commonly used to extinguish fires). For the first three parameters a separate rank (0–4) is given (0 = no hazard; 4 = maximum hazard) for a given particulate.

4.7 Fire and Explosion Risks of Particulates

As discussed in Sects. 4.4 and 4.5, fire and explosion ***hazards*** of a given particulate are determined by its physical properties (e.g., particle size) and its chemical properties (e.g., chemical composition). Only one particulate explosion index, namely P_{max}, can be calculated/estimated relatively simply and accurately, as it follows from the adiabatic flame temperature (Eq. 4.2), which is known or can be calculated using standard procedures. The great majority of explosion indexes, however, cannot be calculated, and must therefore be measured. Measurements are done using standardized testing equipment and procedures. We emphasize that particulate explosion indexes refer only to the ***hazard*** of the particulate itself. Questions relevant to the ***risk*** (i.e., magnitudes and probabilities of adverse effects) of a particulate, such as: How much particulate is being used? In which installation? Under what conditions? Where? Who? etc. are not answered by knowing a particulate's explosion indexes.

Calculating/estimating fire and explosion ***risks*** of particulates thus requires taking not only factors intrinsic to a given material itself into consideration (i.e., the particulate's physical properties and explosion indexes), but also many factors external to the material. Table 4.7 lists important factors influencing/determining risk.

Table 4.7 Basic factors determining fire and explosion risks of particulates

1. **Physical properties** (e.g., average particle size and particle size distribution) **and chemical properties** of the particulate (e.g., chemical composition and chemical structure)
2. **Amounts** (in process, transport, storage)
3. **Types of operations** (e.g., for powders: grinding, mixing, transporting pneumatically)
4. **Size, geometry, and inter-connectedness of installation** (e.g., spherically versus cylindrically-shaped equipment; vessels connected to each other via ducts)
5. **Process conditions** to which particulates are subjected (e.g., pressure, temperature, turbulence)
6. **Complexity** of operations
7. **Age of the plant** or equipment (e.g., old plant requiring much maintenance)
8. **Plant layout** (i.e., position of plant units relative to one another; e.g., separation distances between various plant units and offices, control room, etc.)
9. **Plant siting** (i.e., spatial location of plant relative to population centers (towns, cities, hospitals) and distances of separation)
10. **Vulnerability of surroundings** (e.g., office buildings, hospitals, towns/cities, nature preserve)
11. **Preventive and protective measures** taken (e.g., control and safety system/equipment)
12. **Special factors** (e.g., local topological conditions; climatic conditions, regional political stability)
13. Risks of **human error**
14. Design and operation relative to **legislation, legal standards, guidelines,** and **codes of practice**
15. Use/effectiveness of modern **safety and environmental management systems** (e.g., ISO 14,000)

It is obvious that calculating/estimating how (changes in) factors listed in Table 4.7 affect risk is a complex task. Considering the scope of this Chapter, we discuss this topic only briefly, and then not in terms of causative factors, but in terms of controlling risk. Extensive discussions of particulate fire and explosion – causes and control – are found in textbooks, a number of which we list in "A Recommended Literature".

4.8 Control of Fire and Explosion Risks of Particulates

4.8.1 Introduction

We first discuss two modern chemical engineering design methodologies. Using these two methodologies, we, then the design of the particulate itself, the processing operations, the facility requirements, standards, legislation, and finally management issues.

4.8.2 Important General Design Methodologies: Layer of Protection Analysis (LOPA) and Inherently Safer Design (ISD)

Ideally, the starting point of LOPA (Fig. 4.16) is Inherently Safe Design; i.e., the design of a chemical process ('PROCESS' in this figure), including the materials being used and produced, so as to be as 'inherently safe' as possible (i.e., as free of hazards as possible).

The essence of the LOPA-concept is – to the degree that the basic process is itself not completely 'inherently safe' – to design successive rings of 'protective layers' ("Layers of Protection") around the process [38, 53]. Each layer offers protection and/or mitigation in the event of disturbances and failure of a preceding layer. Thus, consecutive layers extend from the basic process through having an increasing scale and they include preventive and mitigative layers around the process and around the complete plant, even extending outwards to the surrounding community. The "Analysis" in LOPA refers to estimating the risk (i.e., consequence and probability) of failure of the process itself and each of the various 'protective layers' surrounding the process. LOPA's goal is to reduce risk to an 'acceptable level' as effectively and efficiently as possible.

Since Table 4.7 lists basic factors and causes of particulate explosion hazard and risk, negating these should – logically – lead to the elimination, or at least the reduction, of such risk. Indeed, negating basic causative factors forms the logic of Inherently Safer Design (ISD). ISD thus converts the descriptive nature of basic factors mentioned in Table 4.7 into active engineering design options [54]. Table 4.8 lists some of the principles of ISD [39].

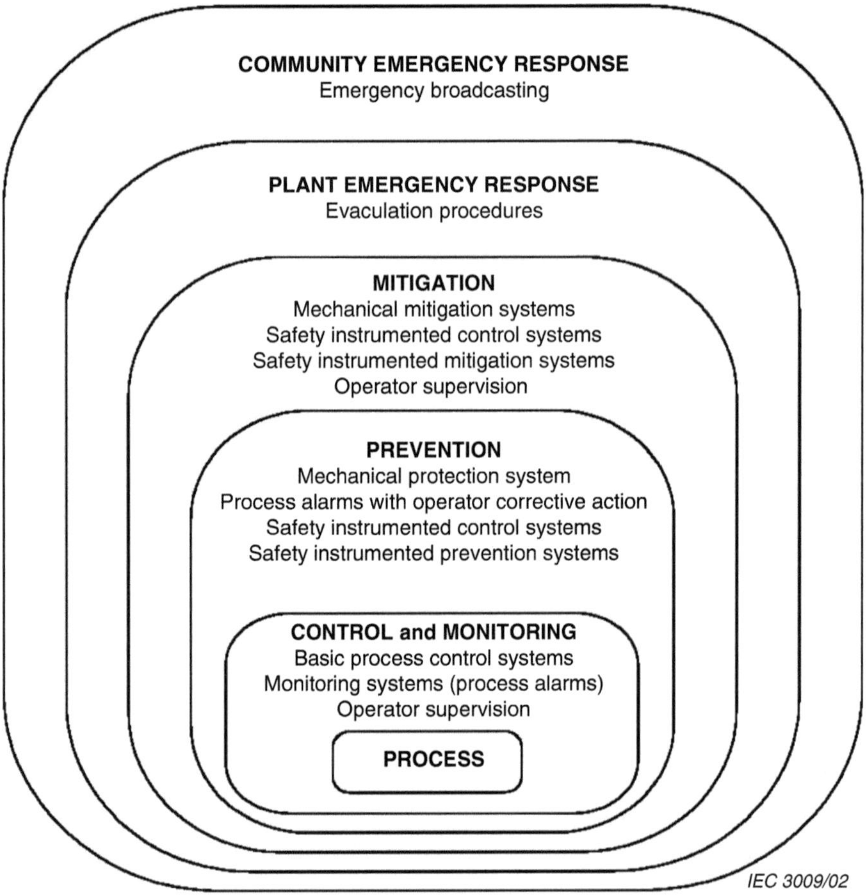

Fig. 4.16 Layer of Protection Analysis (LOPA) [53]. Protective layers at various scales surround the Process (Copyright International Electrotechnical Commission (IEC); reproduced with permission)

Inherently Safer Design also stresses a logical hierarchical approach, namely: preventive methods preferable to protective/mitigative methods; passive safety systems preferable to active systems; active systems preferable to procedural systems. Table 4.9 presents examples of this hierarchical approach.

4.8.3 Designing the Particulate Itself (i.e., Its Physical-Chemical Properties)

For the particulate itself, the ISD-principle of 'Substitution' is most applicable (Table 4.10).

Table 4.8 Some basic principles of Inherently Safer Design (ISD)

Design principle	Factor in Table 4.7	Definition/Description
Substitution	# 1	Use of materials that are non-hazardous or at least less hazardous
Intensification	# 2	Reduction in amount(s) of hazardous materials
Attenuation	# 3 and 5	Running equipment at less hazardous (i.e., safer) conditions
Limitation of effects	# 4, 5, 8, and 11	Changing design and operation, such as separation of units, to create less severe effects should an accident (e.g., explosion) occur

Table 4.9 Examples of hierarchical approach of Inherently Safer Design (ISD)

Design principle	Explanation/mechanism
Preventive design method, passive: Explosion-proof design	Entire system (all components) designed to be sufficiently robust to withstand the most powerful explosion possible. Any explosion will thus be contained within the plant. System is passive in that no component needs to actively function to maintain safety (i.e., to prevent explosion from bursting out of equipment)
Preventive design system, active: Gas inertization	Oxygen-concentration of gas in system (e.g., air) is continuously monitored. Inert gas (e.g., nitrogen) is continuously fed into system to maintain oxygen-concentration below LOC-value. System (measurement, dosing) must actively function in order to be effective
Mitigative design system, passive: Explosion-pressure-proof design	Vulnerable pieces of equipment (i.e., typically those with the largest volumes, such as silos) are designed such that in case of explosion they will deform, but they will not burst open
Mitigative design system, active: explosion venting	The relatively small overpressures developed in early stages of explosion are sufficient to cause large vents in equipment (e.g., a silo) to open, thus preventing high pressures from developing
Procedural approach: operator monitoring	Operator monitors oxygen-concentration in system and adds nitrogen as needed to maintain oxygen concentration below LOC-value

4.8.4 Designing the Process Making the Particulate

Table 4.11 illustrates application of ISD-principles to reduce explosion risk of the process making a particulate.

Amyotte et al. (2003, 2009) discuss in detail the use of ISD in designing powders, including practical examples, to reduce explosion hazard and risk [30, 31].

Table 4.10 Design of the particulate itself, using the ISD-principle of substitution

Substitution design method	Description	Mechanism of reducing hazard	Example
Changing chemical composition 1, eliminating combustibility	Substitute a combustible particulate with a non-combustible particulate	A non-combustible particulate cannot burn. WHAT CAN NOT BURN, CAN NOT EXPLODE	Substituting a red organic pigment (organic, thus combustible, thus explosive) with a non-organic, fully oxidized substance, such as (red) iron oxide
Changing chemical composition 2, reducing chemical instability:	Substitute unstable substance with stable (or less unstable) particulate	The less unstable a substance is (as measured by its energy of dissociation, mJ/kg), the less likely it is to explode and the less powerful the explosion; and vice versa	Substitute mannitol hexanitrate [59] (drug to reduce hypertension; highly explosive!) with isosorbitol dinitrate, which is less unstable but roughly equally effective medically
Changing chemical composition 3, reducing reactivity	Reducing reactivity of pure pyrophoric metal particulates by fully oxidizing their surface	Completely oxidized metal particulates are normally inert due to the presence of an oxide film, which 'seals' the underlying volume from further reaction (oxidation)	Reactive metals, such as finely divided and completely unoxidized iron, aluminum, titanium, magnesium are rendered non-pyrophoric by surface oxidation
Changing chemical composition 4, adding inert solid: particulate inertization	Adding finely divided inert solid (e.g., calcium carbonate) to particulate	By acting as heat-absorbing 'sink', inert material prevents ignition and/or propagation of explosion (process called 'phlegmatization')	Adding calcium carbonate to pills containing isosorbitol dinitrate to completely eliminate explosive hazard [12][a]
Changing the physical nature of particulate by increasing particle size	Increasing particle size by producing larger particles, lowering the amount of very fine particles and/or agglomerating smaller particles into larger ones (e.g., via addition of moisture or binder)	The larger the particle, the greater the amount of energy required to disperse it and to ignite it and the slower its combustion, thus the lower its explosion severity. Above a certain particle size, dust explosions of stable substances are not possible	Finely powdered aspirin is highly explosive[b] [56]. But powdered aspirin agglomerated into pills of 325 or 500 mg is not explosive

[a] In practice, large concentrations of inerts are usually required, often 50–70 % w/w

[b] Measured explosion indexes for aspirin powder are: P_{max} = ca. 7 bara; K_{St}-value = ca. 200 bar.m/s [56]

Table 4.11 Applying ISD-principles to reduce explosion hazard/risk of the particulate-producing process

ISD Methods	Description	Mechanism of reducing hazard	Example
Substitution (of normal air with reduced-oxygen air)	Adding inert gas (e.g., nitrogen; carbon dioxide) to normal air to reduce its oxygen concentration to value below Limiting Oxygen Concentration (LOC; see Table 4.2). This method also called 'inertization' [12]	Reducing oxygen concentration always lowers explosion sensitivity and explosion severity. Each particulate of a given chemical composition and particle size distribution has an oxygen concentration (LOC) below which explosion is not possible	In order to produce particulates of the highest purity (e.g., no oxide film present), preparing reactive metals in the micrometer- and nano-range requires processing and storage in a totally inert gas, such as argon
Intensification	Reducing amounts (of particulate)	Reduced amounts reduce total available energy of combustion and thus result in smaller explosions. And: What one does not have, cannot explode (or burn)	Reducing amounts by developing continuous rather than batch processing; and/or working continuously rather than 8 h per day; reducing inventories via 'just-in-time' logistics
Attenuation	Running equipment at less hazardous (safer) conditions	More hazardous conditions include higher temperature and pressure	Lowering process temperature reduces explosion sensitivity by increasing MIE, reducing generation of static electricity, and increasing average particle size (the latter two by increasing particle humidity)
Limitation of effects	Changing/modifying design and operation to limit effects, should explosion occur. For example, separating units, via *compartmentalization* [12], to prevent a primary explosion from creating secondary explosions, pressure piling, or DDTs	Dust explosions that propagate tend to generate more powerful secondary or even tertiary explosions. Propagation, both inside and outside of equipment, must therefore be prevented. Therefore good housekeeping/preventing settling of dust is essential	For separation or compartmentalization special equipment types exist, such as rotating valves (see Fig. 4.17). Other types include fitting piping connecting various process units with explosion-proof valves that can rapidly close, preventing explosion propagating from one unit to the next

Table 4.12 Relation of preventive measures to explosion sensitivity parameters listed in Table 4.2

Parameter	Theory of operation	Practical working
Explosion limits (LEL, UEL)	Ignition not possible when particulate concentration below LEL or above UEL	In practice usually difficult to achieve and to maintain
Minimum Ignition Temperature (MIT), for dispersions and layers	Ignition not possible when surface temperature less than MIT	Maintain all surface temperatures lower than MIT-values
Minimum Ignition Energy (MIE)	Ignition not possible when (electrical) spark energy less than MIE	Prevent sparks whose energy exceeds MIE-values
Limiting Oxygen Concentration (LOC)	Ignition not possible when oxygen concentration below LOC	Maintain oxygen concentration in gas phase below LOC (e.g., by adding inert gas, such as N_2); process called 'inertization'
Flash points, lower (LFP) and upper (UFP)	At temperatures below lower flash point (LFP) concentration remains below LEL, thus preventing explosion. Analogous reasoning for UFP	Maintain process temperature everywhere below LFP. (Maintaining temperature above UFP is often less practical)

4.8.5 *Designing and Operating the Facility in Which the Particulate Is Made and Stored*

Much of the preceding discussion relates to the starting point of LOPA, Inherently Safer Design (ISD); i.e., designing the particulate itself and the basic process making it to be as inherently safe as possible. Obvious of course is that not all powders and processes can be made completely inherently safe. Indeed, organic particulates are intrinsically flammable and thus potentially explosive. Layer of Protection Analysis takes this reality into account by designing consecutive Preventive and Mitigative/ Protective layers around the basic process (PROCESS in Fig. 4.16).

The basic principal of preventive measures (PREVENTION in Fig. 4.16) is to prevent ignition. Operational principles, presented in Table 4.12. follow logically from explosion sensitivity parameters listed in Table 4.2.

Basic principles of mitigative/protective measures (MITIGATION in Fig. 4.16) are to prevent an incipient explosion from propagating further and thus preventing its development into a serious explosion, or, should serious explosion nevertheless occur, to protect against its effects. Operational principles (Table 4.13) follow from explosion severity parameters listed in Table 4.1.

Extensive literature exists concerning explosion prevention and protection/mitigation. Good reviews exist [1–3, 62]. Specialized textbooks discuss this extensive topic in more detail [7, 12, 23, 69].

Table 4.13 Relation of protective/mitigative measures to explosion severity parameters listed in Table 4.1

Parameter	Theory of operation	Practical working
Detonation	Preventing deflagration from transforming into detonation (preventing DDT). Should detonation nevertheless occur, preventing its further propagation and protecting against its effects	Preventing detonation: Produce particulates as large, humid, and unreactive as possible; reduce pressure and temperature of system; reduce oxygen concentration of gas phase; avoid long inter-connecting piping (L/D < ca. 10); provide inter-connecting piping with valves which can rapidly close (*compartmentalization*) in event of deflagration, thus isolating explosion and thereby preventing flame acceleration (to DDT) and/or pressure piling. Preventing propagation of detonations by means of *detonation arrestors*
		Protecting against detonation: construct all equipment to withstand maximum detonation pressure (usually too expensive)
Maximum explosion pressure (measures listed here also relevant to maximum rate of pressure rise and volume normalized rate of pressure rise: the K_{St}-factor)	Protection again maximum explosion pressure and maximum rate of pressure rise	Provide system that can quench incipient explosion by rapidly injecting heat-absorbing agents (*explosion suppression*)
		Provide equipment with areas which can rapidly open before explosion fully develops (*explosion venting*), thus preventing excessive explosion pressures
		Construct equipment to withstand maximum explosion pressure
		Isolate inter-connected equipment upon occurrence of incipient explosion (*compartmentalization*) via *fast closing valves*, thus prevent against pressure piling

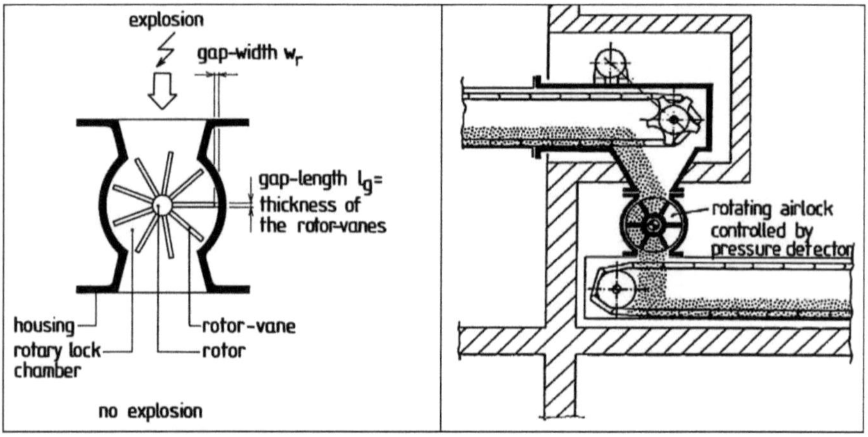

Fig. 4.17 *Left*: Explosion-proof rotating airlock valve blocks/suppresses explosion because its gap-length is less than the Maximum Experimental Safe Gap (MESG; see Table 4.2) of the relevant powder. *Right*: locating rotary airlock to achieve compartmentalization between two parts of a system processing a powder. The rotary valve is actuated to stop rotating by a pressure detector, which detects incipient explosion [12] (Copyright Elsevier; reproduced with permission)

4.8.6 Legislation, Legal Standards, Guidelines, and Codes of Practice

All industrialized countries have extensive legislation, **legal standards, guidelines, and codes of practice** (topic 14 of Table 4.7) concerning industrial safety, including particulate hazards. In the European Union, for example, the two ATEX Directives (from the French: *Appareils destinés à être utilisés en* ***AT****mosphères* ***EX****plosives*) form the most important legislation concerning explosion safety. The ATEX Directives 95 and 137 are meant to protect employees working in an environment with an explosive atmosphere (e.g., containing combustible gases, vapors, mists (liquid droplets) or dusts (solid particulates)) [33].

The basic logic of ATEX is to reduce explosion risk by reducing probability of ignition. A fundamental requirement is to, within a given installation, classify dangerous areas into defined safety zones based on probability of ignition (Zones 2.0, 2.1 and 2.2 in which an explosive mixture exists continuously, occasionally, or rarely, respectively). The zoning puts requirements on the equipment used in the zones with respect to ignition potential. Ignition probability depends, firstly, on the explosion sensitivity indexes of the particulate (e.g., lower values of MIT and MIE; see Table 4.2). Ignition probability depends, secondly, on the type of equipment being used to process a given particulate (e.g., its surface temperature, whether it generates sparks, etc.). As the probability of ignition within a given area where a given particulate substances is used increases (as expressed by a zone number indicating a certain level of ignition probability), equipment allowed by ATEX must have an increasingly lower probability of being able to cause ignition. Reaching a given required safety level (i.e., reduced ignition probability for a given case) is achieved by various ATEX-accepted technical means [12].

It is beyond the scope of this chapter to discuss dust explosion legislation in detail. We provide detailed sources in the bibliography at the end.

Details relating to practical application are described in government publications and, usually in considerably more detail, in standards and norms of trade associations. In the United States, for example, the Occupational Safety and Health Administration (OSHA) of the US Department of Labor is responsible for industrial safety. A part of that responsibility is expressed in publications describing various industrial hazards, such as dust explosions, and their control [61]. Specialized trade associations, such as the National Fire Protection Association (NFPA),[7] publish guidelines providing detailed descriptions of various aspects of industrial safety (e.g., how to recognize, assess, prevent, mitigate, and protect against fire and explosion hazards; how to investigate fires and explosions; how to workers). While written for use in the United States, NFPA guidelines have worldwide application. A similar organization is the German organization DIN (Deutsches Institut für Normung = German Institute for Standardization). Although German in origin, DIN-standards and guidelines also have international authority. Many of DIN's guidelines are published in English as well as in German, which adds to their wide international influence and application [43]. The list of standard issuing organizations is, however, much longer. Mannan presents a succinct discussion of the present tendency to globalize standards and – very importantly – also addresses where to find relevant standards [23]. Similarly, Eckhoff provides a short section on "Standards, Recommendations, and Guidelines" and discusses many of these in his Chaps. 8 and 9 [12].

4.8.7 Managerial Aspects

Experience clearly shows that management is the most critical of all the topics/factors mentioned in Table 4.7. Management must establish and enforce the priority of safety in policy and daily practice and provide the means to achieve it: e.g., clear organization, responsibility, and accountability; competent people; salaries commensurate with responsibility and competence; sufficient investment. It is good management that ensures that the other factors mentioned in Table 4.7 are sufficiently taken into account. Indeed, without good management, the other points mentioned in this Table become irrelevant.

[7] The National Fire Protection Agency was established already in 1897 by a number of American insurance agencies. Nowadays it is an internationally recognized authority on industrial safety relating to fire, explosion and toxicity hazards. An overview of publications is given at [60].

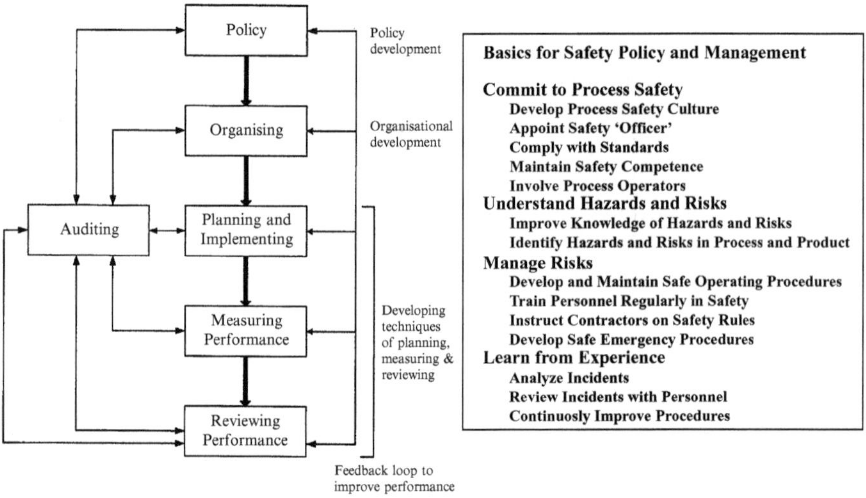

Fig. 4.18 Basic elements for safety policy and management (Adapted from [37, 40])

Systematic Safety and Environment Management Systems have been developed during the last decades, in part due to the huge losses incurred by fires and explosions, (e.g., workers killed/wounded, loss of reputation, plant destroyed (loss of capital), loss of production (loss of cash flow, and customers)). Development and further improvement of such management systems has been, and remains, an important part of the discipline of Loss Prevention.[8] It is beyond our scope to discuss management in any detail. We note that Loss Prevention includes a large, comprehensive, and expanding literature on safety management systems [41].[9] The risk-concept is steadily being worked out in greater comprehensiveness and level of sophistication, and as such is thus increasingly becoming an integral part of modern safety management, especially for establishing priorities [40].

Figures 4.18 and 4.19 schematically illustrate the basic elements of safety policy and management and the main factors that determine safe plant design and operation.

Summarizing, safely producing and using particulate substances requires effectively combining many factors, including policy, scientific knowledge, proper technical measures, responsibility, law and regulation, standards, and good leadership and management. This chapter particularly emphasizes the role of scientific knowledge.

---

[8] Loss Prevention is the systematic and effective approach to designing and operating safe chemical products and chemical plants. See e.g. Chaps. 6 and 27 in [23].

[9] The Center for Chemical Process Safety (CCPS) has been created by the American Institute of Chemical Engineers, to establish and disseminate theory and best practice of safety in the chemical industry. For an overview of literature, see [41].

Fig. 4.19 Major factors determining safe chemical plant design and operation (Adapted from [26])

4.9 A Final Note: Particulate Research

Fire and explosions involving particulates, especially dust explosions, continue to be a significant cause of loss of life and property. Therefore research, particularly concerning dust explosions, continues. Eckhoff [14] gives a good review of current dust explosion research and makes valuable suggestions for the future.

4.10 Summary and Conclusions

Particulate hazards include fire and explosion. Fire (combustion) consists of heat-producing (i.e., exothermic) chemical reactions, usually with oxygen (oxidation), producing a flame, intense heat, and high temperatures. Particulate explosions are also exothermic chemical explosions, usually with oxygen (oxidation). The key difference between fire and explosion is that the rates of physical and chemical processes occurring during explosion are much faster than those occurring during fire. The basic reason for this is that during fire the 'fuel' (i.e., burning substance, such as burning logs) is clearly separated from air (oxygen), while in explosions the fuel is 'pre-mixed' (i.e., intimately interspersed) with air (oxygen), such as occurring in a cloud of combustible dust in air. Particulate explosions, like particulate fires, produce flames, intense heat, and high temperatures, but unlike fire, particulate explosions produce 'blast', a destructive pressure or shock wave, and sometimes also high velocity fragments caused by bursting equipment.

Completely oxidized substances (e.g., sand, calcium carbonate) cannot burn and therefore cannot explode. Conversely, non-oxidized substances (e.g., polyethylene powder; oils) or only partially oxidized substances (e.g., starch powder) are combustible.

DISPERSED POWDERS AND MISTS THAT ARE COMBUSTIBLE (i.e., that can burn) CAN ALSO EXPLODE; WITHOUT EXCEPTION.

Hazards of combustible particulates increase as the chemical energy of their decomposition and/or combustion increases. While some particulates can themselves (i.e., without air) explode (e.g., large masses of powdered TNT), and some particulates may be so chemically reactive that they spontaneously burn or even explode in air (pyrophoric materials, such as un-oxidized iron powder), most particulates will neither spontaneously burn nor explode. They must first be ignited by an external ignition source (e.g., hot surface, electric spark). Additionally, explosion can only occur when the particulate is suspended in and mixed ('pre-mixed') with air; therefore some degree of turbulence is always required (i.e., to suspend the particulate in air). Partial or complete containment (e.g., dust cloud contained in a closed vessel, such as a silo or in rooms of a building) is usually also necessary to create the high pressures of a destructive explosion.

Hazard is a relative term relating to the potential to cause harm (e.g., through fire and/or explosion). Risk is, in principle, an absolute, quantitative term relating to the probability (0–100 %) of a given adverse effect occurring (e.g., death caused by blast wave). Hazards of a given particulate, and also to some degree also its risks, are characterized by two types of fire and explosion indexes. The first, fire and explosion sensitivity indexes, concerns the probability of fire and explosion occurring; these basically relates to the probability of ignition. Examples of such indexes are the Lower and Upper Explosion Limits (LEL and UEL), Minimum Explosion Temperature and Energy (MIT and MIE), and Limiting Oxygen Concentration (LOC). In general particulates are less sensitive to ignition than gases and vapors. The second type of indexes, explosion severity indexes, concerns the magnitude of the adverse effect caused by explosions. Such indexes include the type of explosion occurring (i.e., detonation, rare but enormously destructive; or deflagration, less destructive, but much more common), the maximum explosion pressure (P_{max}), and the (volume) normalized rate of pressure increase, called the K_{St}-factor. In general, particulates produce roughly the same maximum explosion pressures (i.e., 5–10 times initial pressure) as gas explosions, but their rate of pressure rise is usually (much) less than that of gas/vapor explosions. On the other hand, because of the much greater density of particulates relative to gases/vapors, particulate-air mixture contain more energy per unit volume (i.e., have higher volumetric energy density, J/m^3) than mixtures of combustible gas/vapor with air. This greater energy density manifests itself in a longer explosion time duration (s) than gas/vapor explosions, which can result in more damage.

Hazards, and also to some degree risks, of explosions of a given particulate are determined by its chemical and physical properties, as expressed by the values of explosion sensitivity and explosion severity parameters. Increasing chemical reactivity (e.g., higher combustion energy, J/kg) tends to increase all fire and explosion parameters (e.g., lower MIT and MIE; higher P_{max} and K_{St}-factor), thus to increase hazard. The most important physical property affecting fire and explosion parameters is particle size (particle size and particle size distribution). In general the finer the particles are, the greater the hazard, i.e., greater probability of ignition (to fire and/or explosion), higher maximum explosion pressure, and greater explosion speed (higher K_{St}-factor and greater tendency towards detonation). The speed (rate) of flame propagation in an explosion of most particulates, which are in the micrometer range, is controlled by a physical process, diffusion (of oxygen to and reaction products (e.g., carbon dioxide and water) from the particles' surface). Diffusion-control is typical of a heterogeneous (i.e., multi-phase) chemical reaction. The speed (rate) of gas and vapor explosions, which occur homogeneously (i.e., within one phase: gas) is determined not by diffusion, but by chemical kinetics. Compared to rates of chemical reaction, which increase exponentially with temperature, diffusion is a relatively slow process, which is the basic reason why dust explosions generally proceed more slowly than gas/vapor explosions. However, as particle size decreases from the micrometer-range into the nanometer-range, the mechanism of particulate explosions increasingly approaches the mechanism of gas/vapor explosions; i.e., reaction rate is less controlled by diffusion and more by chemical kinetics. In the extreme case, when particulates approach the size of individual molecules, explosion sensitivity and explosion severity increasingly resemble the values of gases and vapors (e.g., smaller and smaller (nano-sized) polyethylene particles increasingly resembles ethylene gas). Thus, nanoparticulates tend generally to be more hazardous than micrometer-sized particulates (not only in terms of fire and explosion, but also in terms of toxicity).

Various chemical, physical and physiological properties of particulates can be combined to create a kind of aggregated hazard index for fire, explosion, and toxicity. For these three general types of hazards, plus water reactivity hazard, the National Fire Protection Agency (NFPA) of the USA has created a system to rank particulates, which is widely used internationally.

While fire and explosion hazards of a given particulate can be determined relatively simply by measuring its sensitivity and severity parameters (via standardized testing procedures), establishing the explosion risks of that particulate in practice is generally much more difficult. Reasons for this is that particulate risk is determined by many factors external to the particulate itself (listed in Table 4.7), such as amount of particulate, how and where (e.g., which country) it is being used, under what conditions, to which legislation (production, storage, transport) it is being subjected, which industrial norms and standards are being applied, etc. Furthermore, experience shows that the most important factor in determining particulate explosion risk in practice is actually quality of management. We stress that good management includes not only policy and strategic goals (e.g., safety priority relative to finance), but most certainly also operational measures, such as 'good housekeeping' (e.g., preventing accumulation of settled dust) and avoiding ignition sources, such as 'hot work', mechanical sparks, hot surfaces, faulty electrical connections and static electricity.

Considering the scope of this chapter and the complexity of particulate risk, rather than discuss each risk factor individually, we approach control of risk from a more general point of view, namely on the basis of addressing major hazard and risk factors and using two modern chemical engineering design principles for safety. These design principles are Layer of Protection Analysis (LOPA) and Inherently Safer Design (ISD). LOPA provides preventive and Mitigative layers at various scales surround the process making the particulate. Ideally, the starting point of LOPA, the process, is based on Inherently Safer Design (ISD). ISD follows logically from the negation of a number of key hazard and risk factors listed in Table 4.7. Safe design principles are discussed concerning designing the particulate itself (Table 4.10), designing the process making the particulate (Table 4.11), and operating the facility in which the particulate is made and stored. Especially at this last scale national legislation and industrial standards and guidelines are vitally important. Examples of legislation and guidelines (ATEX legislation in Europe; DIN guidelines in Germany; NFPA guidelines in US) are mentioned. Safe design, operation, and maintenance is can only be achieved and maintained by effectively applying modern safety and environmental management systems, shown schematically in Figs. 4.18 and 4.19. Good management is absolutely the *sine qua non* of safe design and operation.

4.11 Definitions, Abbreviations and Symbols

Deflagration	Explosion producing a sonic pressure wave
Detonation	Explosion producing a supersonic shock wave
Dust explosion	Explosion of dust (dispersed powder)
Aerosol	Liquid or solid particles dispersed in air (often stable over long periods of time)
Equivalence ratio	Ratio of present fuel-to-oxygen ratio to stoichiometric fuel-to-oxygen ratio, or: $(\text{fuel/oxygen})_{\text{present}}/(\text{fuel/oxygen})_{\text{stoichiometric}}$
Explosion (chemical)	Vigorous reaction of premixed fuel and oxidizer, involving sudden release of energy (heat, pressure/shock wave, 'bang')
Explosion range	UEL – LEL of a substance
Hazard	Potential to create an adverse effect
Risk	An adverse effect, for which an estimated probability is given
Sauter Mean Diameter	Mean value of an area weighted PSD ($D_{3,2}$)
AIT	Minimum Auto-Ignition Temperature
ATEX	Appareils destinés à être utilisés en ATmosphères EXplosives (equipment meant for use in explosive atmospheres)
bara	Bar absolute pressure

(continued)

(continued)

baro	Bar overpressure
CCPS	Center for Chemical Process Safety (USA)
DDT	Deflagration-Detonation Transition
DIN	Deutsches Institut für Normung (German Institute for Standardization)
ISD	Inherently Safer Design
ISO	International Standards Organization
LEL	Lower Explosion Limit
LFL	Lower Flammability Limit
LOC	Limiting Oxygen Concentration (below which explosion is not possible)
LOPA	Layer of Protection Analysis
MESG	Maximum Experimental Safe Gap
MIE	Minimum Ignition Energy
MIT	Minimum Ignition Temperature
NEEM	Nano-Engineered Energetic Materials
NFPA	National Fire Protection Association (USA)
OSHA	Occupational Safety and Health Administration (USA)
PSD	Particle Size Distribution
PVC	PolyVinylChloride
SMD	Sauter Mean Diameter ($D_{3,2}$)
STP	Standard Temperature and Pressure
TNT	TriNitroToluene
UEL	Upper Explosion Limit
UFL	Upper Flammability Limit
D	Particle size (note: for non-spherical particles: measured size depends upon technique)
$D_{3,2}$	Sauter mean diameter, area-weighted mean size, mean value of an area-weighted PSD
$D_{4,3}$	Volume-weighted mean size, mean value of a volume-weighted PSD
K_{st}	Volume-normalized maximum rate of pressure rise in a vessel ($V^{1/3} \cdot (dP/dt)$)
P	Pressure
S_L	Laminar burning velocity
t	Time
T	Temperature
V	Volume of a vessel

References

A Recommended Literature

1. Abbasi, T., Abbasi, S.A.: Dust explosions – cases, causes, consequences, and control. J. Hazard. Mater. **140**, 7–44 (2007)
2. Amyotte, P.R., Eckhoff, R.K.: Dust explosion causation, prevention and mitigation: An overview. J. Chem. Health Saf. Am. Chem. Soc. **17**(1), 15–28 (2010)

3. Amyotte, P.R., Pegg, M.J., Khan, F.I., Nifuku, M., Yingxin, T.: Moderation of dust explosions. J. Loss Prev. Proc. Ind. **20**, 675–87 (2007)
4. Bartknecht, W.: Explosionsschutz – Grundlagen und Anwendung. Springer-Verlag, Heidelberg (1993)
5. Bartknecht, W.: Dust Explosions:_Course, Prevention, Protection. Springer-Verlag, Berlin Heidelberg (1989)
6. Bartknecht, W.: Explosions: Course, Prevention, Protection, Springer-Verlag, Berlin Heidelberg (1981)
7. Barton, J. (ed.): Dust Explosion Prevention and Protection – A Practical Guide. Institution of Chemical Engineers, Rugby (2002)
8. Bjerketvedt, D., Bakke, J.B., van Wingerden, K.: Gas explosion handbook. J. Hazard. Mater. **52**, 1–150 (1997) (In spite of title, much information also relevant to explosions of particulates). (Available as free download from various Internet sources)
9. Cashdollar, K.L.: Overview of dust explosibility characteristics. J. Loss Prev. Proc. Ind. **13**, 183–99 (2000)
10. Center for Chemical Process Safety (CCPS): Guidelines for Safe Handling of Powders and Bulk Solids. American Institute of Chemical Engineers, New York (2005)
11. Crowl, D.A.: Understanding Explosions. Center for Chemical Process Safety (CCPS), American Institute of Chemical Engineers, New York (2003)
12. Eckhoff, R.K.: Dust Explosions in the Process Industries, 3rd edn. Gulf Professional Publishing (an imprint of Elsevier Science), Amsterdam (2003)
13. Eckhoff, R.K.: Explosion Hazards in the Process Industries (Why Explosions Occur and How to Prevent Them, with Case Histories). Gulf Publishing, Houston (2005)
14. Eckhoff, R.K.: Current status and expected future trends in dust explosion research. J. Loss Prev. Proc. Ind. **18**, 225–37 (2005)
15. Eckhoff, R.K.: Understanding dust explosions: The role of powder science and technology. J. Loss Prev. Proc. Ind. **22**, 105–16 (2009)
16. Hattwig, M., Steen, H.: Handbook of Explosion Prevention and Protection. Wiley-VCH, Weinheim (2008)
17. James, H.: 1. Basic Phenomenology of Deflagration, DDT, and Detonation. Lecture available as http://ukelg.ps.ic.ac.uk/41HJ1.pdf (2008)
18. James, H.: 2. Examples of Industrial Incidents Involving Detonations. Lecture available as http://ukelg.ps.ic.ac.uk/41HJ2.pdf (2008)
19. James, H.: 3. *Detonations*; HSE guidance note ref. TD5/039. http://www.hse.gov.uk/foi/internalops/hid/din/539.pdf (2001)
20. Hertzberg, M., Cashdollar, K.L. (eds.).: Industrial Dust Explosions: Symposium on Industrial Dust Explosions. ASTM Special Technical Publication, Philadelphia (1987)
21. Johnson, R.W., Rudy, S.W., Unwin, S.D.: Essential Practices for Managing Chemical Reactivity Hazards (A CCPS Concept Book). Center for Chemical Process Safety (CCPS), American Institute of Chemical Engineer, New York (2003)
22. Lackner, M., Winter, F., Agarwal, A.K. (eds.).: Handbook of Combustion. Wiley-VCH, Weinheim (2010). (Five Volumes Set. See in particular *Vol. 1: Fundamentals and Safety*)
23. Mannan, S. (ed.): Lees' Loss Prevention in the Process Industries, 3rd edn. Elsevier Butterworth Heinemann, Boston (2005) (4 volumes envisaged, of which 2 have been published)
24. Nagy, J.: Development and Control of Dust Explosions. CRC Press, New York (1983)
25. Nanosafe.: *Safe Production and Use of Nanomaterials*. International conferences held biannually in Grenoble, France in 2008, 2010, and 2012. Available on the web. For **2008**: http://www.nanosafe.org/scripts/home/publigen/content/templates/show.asp?P=96&L=EN&ITEMID=42 . For **2010**: http://www.nanosafe.org/scripts/home/publigen/content/templates/show.asp?P=118&L=EN. For **2012**: http://www.nanosafe.org/scripts/home/publigen/content/templates/show.asp?P=124&L=EN

26. Nolan, D.P.: Handbook of Fire and Explosion Protection Engineering Principles for Oil, Gas, Chemical and Related Facilities, 2nd edn. Elsevier, Burlington (2011)
27. Stahl, R.: Basics of explosion protection – introduction to explosion protection for electrical apparatus and installations. http://www.gilsoneng.com/reference/BasicsofEx.pdf
28. UKELG 46th Meeting on *Causes, severity and mitigation of aerosol and particulate explosions*; Department of Chemical Engineering, Imperial College, London, 22nd Sept. (2010). Lectures available at http://ukelg.ps.ic.ac.uk/UKELG46.htm

B Other Literature

29. Abbasi, T., Pasman, H.J., Abbasi, S.A.: A scheme for the classification of explosions in the chemical process industry. J. Hazard. Mat. **174**, 270–80 (2010)
30. Amyotte, P.R., Khan, F.I., Dastidar, A.G.: Reduce dust explosions the inherently safer way. Chem. Eng. Progr. **99**, 36–43, Oct. (2003)
31. Amyotte, P.R., Pegg, M.J., Khan, F.I.: Application of inherent safety principles to dust explosion prevention and mitigation. Proc. Saf. Environ. Prot. **87**, 35–9 (2009)
32. Annamalei, K., et al.: Interactive processes in gasification and combustion-Part III: Coal/Char particle arrays, streams and clouds. Prog. Energy Combust. Sci. **20**, 487–618 (1994)
33. ATEX Guidelines.: Guidelines on the application of directive 94/9/EC of the european parliament and the council of 23 march 1994 on the approximation of the laws of the member states concerning equipment and protective systems intended for use in potentially explosive atmospheres, 3rd edition, June 2009, Update May 2011. http://ec.europa.eu/enterprise/sectors/mechanical/files/atex/guide/atexguidelines-may2011_en.pd
34. Ballal, D.R.: Ignition and flame quenching of quiescent dust clouds of solid fuels. Proc. R. Soc. Lond. A. Math. Phys. Sci. **369**, 479–500 (1980)
35. Borissov, A.A. (ed.): Dynamic Structure of Detonation in Gaseous and Dispersed Media (aerosols). Kluwer Academic Publisher, Dordrecht (1991)
36. Britton, L.G.: Avoiding Static Ignition Hazards in Chemical Operations. Center for Process Safety (CCPS), American Institute of Chemical Engineers, New York (1999)
37. Cameron, I., Raman, R.: Process Systems Risk Management. Elsevier Academic, San Diego (2005)
38. Center for Process Safety (CCPS): Layer of Protection Analysis – Simplified Process Risk Assessment. American Institute of Chemical Engineers, New York (2001)
39. Center for Process Safety (CCPS).: Inherently Safer Processes: A Life Cycle Approach (2nd ed.). American Institute of Chemical Engineers, New York (2009)
40. Center for Process Safety (CCPS): Guidelines for Risk Based Process Safety. Wiley, Hoboken (2007)
41. CCPS-Wiley publications are in general highly recommended. http://eu.wiley.com/WileyCDA/Section/id-291237.html
42. Crowl, D.A., Louvier, J.F.: Chemical Process Safety, Fundamental with Applications, 3rd edn. Pearson Education, Boston (2011)
43. DIN EN 14491:2012-10.: Dust explosion venting protective systems (English translation of the original German version, Schutzsysteme zur Druckentlastung von Staubexplosionen; Deutsche Fassung EN 14491:2012, Publication date: 2012-10). http://www.beuth.de/en/standard/din-en-14491/150360990
44. Dobashi, R.: Risk of dust explosions of combustible nanomaterials. J. Phys.: Confer. Series **170** (2009) 012029. http://iopscience.iop.org/1742-6596/170/1/012029
45. Dobashi, R.: Risk of dust explosions of combustible nanomaterials; lecture at Nanosafe (2008) [25]

46. Dow Technical Bulletin.: Dust Explosions and Solid Epoxy Resins. Dow Chemical Company (2003)
47. Dufaud, O., Vignes, A., Henry, F., Perrin, L., Bouillard, J.: Ignition and explosion of nanopowders: Something new under the dust. Lecture at 2010 [25]. http://www.nanosafe.org/home/liblocal/docs/Nanosafe%202010/2010_oral%20presentations/O11-1_Vignes.pdf (2010)
48. Eckhof, R.K.: Are enhanced dust explosion hazards to be foreseen in production, processing and handling of powders consisting of nano-size particles? Lecture at 2010 [25]. http://www.nanosafe.org/home/liblocal/docs/Nanosafe%202010/2010_oral%20presentations/PL11_Eckhoff.pdf (2010)
49. Environmental Safety Division, University of Georgia: Common water-reactive chemicals. http://www.esd.uga.edu/chem/pub/waterreactivemat.pdf
50. Föster, H.: Properties of flammable mists. Ch. 5 in [16] (2004)
51. Hertzberg, M., Cashdollar, K.L., Zlochhower, I.A.: Flammability limit measurements for dusts and gases. In: Twenty-first International Symposium on Combustion, pp. 303–313. The Combustion Institute (1986)
52. Hertzberg, M., Cashdollar, K.L.: Introduction to dust explosions; Ch. 1 in [20], Philadelphia (USA) (1987)
53. IEC 61511-3.: Functional safety – Safety instrumented systems for the process industry sector, Part 3: Guidance for the determination of the required safety integrity levels, International Electrochemical Commission (IEC), First edition (2003) (http://www.iec.ch). Corrections appearing in 2004 as: CORRIGENDUM 1 (http://webstore.iec.ch/corrigenda/iec61511-3-cor1%7Bed1.0%7Den.pdf) (2004), IEC 61511 is expected to be revised in 2015
54. Kletz, T.: Process Plants: A Handbook for Inherently Safer Design. Taylor & Francis, Philadelphia (1998)
55. Kobiera, A., Szymczyk, J., Wolanski, P., Kuhl, A.: Study of the shock-induced acceleration of hexane droplets. Shock Waves **18**(4), 475–85 (2009)
56. Lunn, G.A.: The venting of dust explosions in a dust collector. J. Hazard. Mat. **12**, 87–107 (1985)
57. Medard, L.A.: Accidental Explosions, vol. 1 and 2. Ellis Horwood, Chichester (1989)
58. Merkus, H.G.: Particle Size Measurements – Fundamentals, Practice, Quality. Springer, Dordrecht (2009)
59. Meyer, R., Köhler, J., Homburg, A.: Explosives, 6th edn. Wiley-VCH, Weinheim (2007)
60. NFPA overview. http://www.nfpa.org/about-nfpa/overview
61. OSHA: Combustible Dust in Industry: Preventing and Mitigating the Effects of Fire and Explosion. OSHA, Washington, DC (2005). https://www.osha.gov/dts/shib/shib073105.html
62. Pekalski, A., Zevenbergen, J.F., Lemkowitz, S.M., Pasman, H.J.: A review of explosion prevention and protection systems suitable as ultimate layer of protection in chemical process installations. Proc. Saf. Environ. Prot. **83**, 1–17 (2005)
63. Peters, N.: Turbulent Combustion. Cambridge University Press, Cambridge (2004)
64. Proust, C.: Dust explosions in pipes: A review. J. Loss Prev. Proc. Ind. **4**(9), 267–77 (1996)
65. Puri, P.: Multiscale modeling of ignition and combustion of micro and nano aluminum particles. Ph.D. Dissertation. Pennsylvania State University, University Park (2008)
66. Puri, P., Lui, T., Ying, H., Yang, V.: Modeling and simulation of supercritical fluid processing of nano energetic materials. Lecture at 2006 MURI NEEM Program Review; Pennsylvania State University. http://www.neem.psu.edu/2006%20Review%20Meeting/Yang%20MURI%20NEEM%20Year%202%20PSU%20Nov%202006.pdf
67. Puri, P., Yang, V.: Multiscale modeling of nano aluminum particle ignition and combustion; lecture at 2005 MURI NEEM Program Review; Aberdeen, Maryland. http://soliton.ae.gatech.edu/people/vigor.yang/projects/Nano%20Aluminum...(Puneesh)/ARO-MURI%20NEEM%20Website-(11-16-05).pdf

68. Santon, R.: Mist explosion incident survey and a new research project; UKELG 46th discussion meeting [28], September 22nd 2010, Imperial College London. http://ukelg.ps.ic.ac.uk/46RCS.pdf (2010)
69. Steen, H.: Measures of Explosion Protection and Prevention (Ch. 6) and Explosion Risks (Ch. 7) in [7]
70. Taveau, J.: Secondary dust explosions: How to prevent them or mitigate their effects? Proc. Saf. Prog. **31**(1), 36–50 (2012)
71. US Chemical Safety and Hazard Investigation Board.: Sugar Dust Explosion and Fire (14 killed, 36 injured); Investigation report no. 2008-05-I-GA, (2009). http://www.csb.gov/assets/1/19/imperial_sugar_report_final_updated.pdf
72. Yang, V., Sundaram, D., Puri, P.: Multiscale modeling of nano aluminum particle ignition and combustion; lecture at 2010 MURI NEEM Program Review; Aberdeen, Maryland. Available at http://www.neem.psu.edu/2010%20Review%20Meeting/GeorgiaTech_Yang_NEEM_MURI_Final_Review_15_March_2010.pdf (2010)

Chapter 5
Particle and Fiber Toxicology

Georgia K. Hinkley and Stephen M. Roberts

Abstract Adverse health effects of inhaled particulates have been noted for centuries, and studies over the last several decades have linked particle and fiber inhalation to several pulmonary diseases, including fibrosis, silicosis, chronic obstructive pulmonary disease, emphysema, asthma, hypersensitivity pneumonitis, and cancer, as well as cardiovascular disease. Inhaled particles and fibers can produce toxicity through several mechanisms, many mediated by the immune system. Particle and fiber toxicity is influenced by composition, size and shape of the material, as well as surface properties, including surface charge, protein corona, and surface contamination. Dissolution rate within the body can also influence toxicity. Although less extensively studied, some particles and fibers can also produce health effects when ingested and as a result of dermal contact. To help protect against adverse effects of particle and fiber exposure, regulatory agencies in numerous countries have developed exposure limits for airborne particulates in general, as well as for specific particles and fibers of special concern.

5.1 Introduction

Particulate matter (PM) was in existence long before dinosaurs walked the earth, with volcanoes and forest fires contributing naturally to atmospheric PM levels. Humans have added to the production of particulates through activities such as the combustion of fossil fuels, production of industrial emissions, refuse incineration, and mechanical generation of dusts from agriculture, mining, and motor vehicle traffic. Humans are exposed to PM in air occupationally, at home, and in fact virtually everywhere they go. Recently, human exposure to PM and airborne fibers has expanded to include not only natural, anthropogenic and occupational sources,

G.K. Hinkley (✉) • S.M. Roberts
Center for Environmental and Human Toxicology, University of Florida, Gainesville, FL, USA
e-mail: ghinkley@ufl.edu; smroberts@ufl.edu

H.G. Merkus and G.M.H. Meesters (eds.), *Particulate Products: Tailoring Properties for Optimal Performance*, Particle Technology Series 19,
DOI 10.1007/978-3-319-00714-4_5, © Springer International Publishing Switzerland 2014

but also engineered particles and tubes on a nanometer scale (i.e., nanoparticles, NPs).

Human health problems associated with the inhalation of particulates such as silica have been reported since the ancient Greeks [83]. In the modern era, the toxicity of inhaled PM and fibers has been a concern in human health, many believe starting with the discovery in the 1930s that asbestos inhalation causes mesothelioma and fibrosis. In addition to asbestos, there were several air pollution incidents across Europe and the United States in the 1940s and 50s which raised awareness about the toxicity of inhaled PM. As one example, in December of 1952, London was trapped under a thick blanket of smog that lasted for 5 days before changes in the weather caused the air to clear. During this time period, it has been estimated that 100,000 people became ill, leading to the premature death of over 10,000 London residents [21]. This incident was important for the development of the 1956 Clean Air Act in the United States, one of the first environmental protection regulations.

This chapter presents an overview of particulate and fiber toxicology, focusing on inhalation exposure as this is generally considered to be the dominant pathway in terms of health risk for PM. Situations in which oral and dermal exposure to particles or fibers have been shown to cause toxicity will also be covered briefly.

5.2 Diseases Associated with Inhalation of Particles and Fibers

As part of their critical role in oxygenating blood, the lungs are responsible for filtering, humidifying and heating the air that we breathe. The role of filtration is very important, preventing systemic exposure to many toxicants in our environment. However, through the act of filtration, the lungs become uniquely susceptible to injury and disease. It is for this reason that inhalation toxicology has long been a key to understanding human health risks when considering environmental and occupational exposures to substances in air. The following is an overview of the diseases that can be caused by the inhalation of PM and fibers, with examples of exposure leading to illness provided from the epidemiological data.

5.2.1 Fibrosis

The flexibility of the lung is functionally a very important aspect of its structure, and is compromised by fibrosis. When the diaphragm contracts, the chest wall expands, pulling the lung with it and creating negative pressure inside the lung, required for air to flow inside. Fibrosis in the lung is the replacement of healthy tissue with inflexible connective tissue. With increased connective tissue, lung flexibility is compromised leading to decreased breathing efficiency. Severe fibrosis from a chronic exposure cannot only limit activity, but can also cause major organ failure and increased mortality rates.

Asbestos fibers have long been recognized to cause pulmonary fibrosis in humans. More recently, PM from several metals and other materials have also been shown to cause fibrosis in humans during occupational exposure, including aluminum, beryllium, chromium (VI), coal dust, cotton dust, wood dust, mineral dust, and livestock associated particulates [16, 35, 97, 105]. Exposure to metal dusts commonly occurs during welding. Epidemiological studies from five populations in the United Kingdom, the United States and Japan all show significant risk for fibrosis associated with exposure to metal dust from welding [97, 98]. After correcting for age and smoking status, two of these populations showed dose-dependent increases in risk, while one study was only able to show increased risk in the sub-population with the longest metal dust exposure period. Like all epidemiological data, these exposures are complicated by several factors involved in the welding process, including: welding materials, electrodes used, operating temperature, and exposure to gases (CO, CO_2, NO_x), as well as variation in confinement and ventilation of the work areas [12]. In addition to welding exposure, it has been shown that chronic/occupational exposure to any form of dust is significantly associated with risk for development of idiopathic pulmonary fibrosis [61].

5.2.2 *Silicosis*

Silicosis is a fibrotic lung disease resulting from inhalation exposure to silica dust. Silica is the most abundant mineral on earth, comprising 67 % of sandstone and up to 40 % of granite. Silica dust is generated during several occupations: quarrying, milling, mining, sand blasting, grinding, excavation, digging, hammering and even ceramics and pottery work. These occupations generate silica dust with particle sizes less than 10 μm, a size sufficiently small enough to be inhaled deep into the lung, leading to disease. The pathology associated with silica inhalation has led to Occupational Safety and Health Administrations (OSHA) to set air quality limits of 0.025–0.35 mg/m^3 (varies by country) for silica dust generation to limit occupational exposure [54].

A tunnel digging project in Hawk's Nest, West Virginia, USA in the 1930s was probably the first major incident to bring attention to the occupational hazards of inhaling silica dust. Of 2,500 underground-exposed workers, 88 % eventually developed silicosis. These events led to 46 of the US states passing occupational safety laws for silica exposure in 1937 [99]. Between 1968 and 2002, the age adjusted mortality rate due to silica exposure has decreased by 93 %; with a mortality rate of 8.9 deaths per million in 1968, dropping to 0.7 deaths per million in 2004 [54, 99]. Despite a decrease in the number of annual deaths due to silicosis, the United States Occupational Safety and Health Administration estimated there were two million industry workers occupationally exposed to silica dust in 2007 [99]. While the annual silica exposure in the United States remains quite high, China leads the way in the number of patients, diagnoses and deaths attributed to silicosis annually. Between 1991 and 1995, there were 500,000 cases of silicosis diagnosed in China, with 6,000 new diagnoses and 24,000 deaths annually [54]. The incidence of silicosis has increased over the last 30 years in some countries; for example, autopsy results from South African gold miners has shown an increase from 3 % to 32 % and 18 % to 22 % for black and white gold miners, respectively, from 1975 to 2007[54]. However, as illustrated with the mortality rate of the United States, the opposite trend is seen in more developed countries due to more stringent occupational safety standards, e.g. use of respirators [99].

5.2.3 Chronic Obstructive Pulmonary Disease and Emphysema

Chronic obstructive pulmonary disease (COPD) is characterized by chronic/progressive irreversible airflow obstruction and is diagnosed by symptoms (e.g., dyspnea, cough, sputum production) and airflow measurement with a spirometer [95]. COPD has been associated with occupational exposures for six centuries and is the fourth leading cause of morbidity and death related to chronic disease [24]. The frequency of COPD has increased over the last 20 years and there are an estimated 16 million people living undiagnosed with COPD in the United States alone [95]. While the leading cause of COPD is cigarette smoking (80–90 % of cases), recognized by "smokers cough", only about 80 % of COPD deaths can be attributed to smoking [7]. However, Blanc et al. also state that there is a synergistic risk relationship between occupational PM exposures and cigarette smoking with an 18-fold increased risk for COPD for these individuals compared to non-exposed smokers. Sources estimate that at least 15 % of COPD deaths are related to occupational exposures, with an estimated 6 % of all COPD cases being among never smokers [7, 11, 95]. Other groups estimate there is a population attributable risk (PAR) for COPD due to occupational exposure of 31 % for never smokers in the rubber, plastics, leather and textile manufacturing industries for US workers [11]. Also summarized by Boschetto et al., a study looking at Swedish workers shows that 52.6 % of COPD cases among never smokers were attributable to occupational exposure to mineral and wood dusts. Heavy metal fumes, mineral dust (coal, hardrock mining and silica) and organic dust (e.g., cotton and hemp) exposures have also been associated with COPD risk [95]. Ambient air exposure and black smoke levels have been associated with COPD deaths in a population of emergency department admissions in Barcelona [48]. Underground coal mining has shown association with emphysema (a significant aspect of COPD) by other groups, with increased risk for emphysema among both never and ever smokers [53, 103].

Animal models have shown that inhalation exposure to cadmium, coal and silica dust have led to the development of emphysema and COPD-like symptoms [11]. However, animal models for emphysema use an α1-antityrpsin deficient mouse model that is not completely representative of the human condition, and is most appropriate for genetically susceptible humans. Another confounding property of analyzing COPD epidemiological data is the lack of consistency in disease description and diagnosis over the last several decades. Some studies make associations between occupational exposures and specific symptoms, while others only use COPD diagnoses for epidemiological analysis [11, 82].

5.2.4 Asthma

Rodent models for asthma are not completely predictive of the human condition. For this reason epidemiological data are often used to investigate how factors (e.g., PM) can influence asthma rates in humans. A problem that plagues all epidemiological studies is the number of confounding variables that are inevitably going to exist. The interpretation of epidemiological data relating to asthma and PM exposure can be greatly influenced by environmental variability. PM is just one of six categories of air pollutants generally investigated: SO_2, O_3, NO_x, CO and lead are also assessed for association with disease [101]. Separating the effects caused by each pollution component is often not possible, and it is unlikely that a location with significant contamination from only one category can be found. Factors such as air humidity must also be considered; high humidity, for example, will significantly increase airway permeability and associated risk for a given exposure [107]. For these reasons, finding significant associations between PM and asthma can be challenging.

While air pollution has long been suspected of being associated with asthma, several studies investigating the possible correlation between PM and asthma have failed to find a significant association [4, 40]. When subjects are separated by age, a significant association between PM pollution and asthma in children was found [107]. Several other investigators have found this same association when separating their study group by age [20, 101, 106]. Another study has found significance in a more diverse age group, but only when analyzing with ethnicity as a factor, showing asthma incidence in African Americans to be associated with PM in the air [30].

5.2.5 Hypersensitivity Pneumonitis

Hypersensitivity pneumonitis (HP) in the lung has been observed after inhalation of several biological dusts and fungi during occupational exposures. These reactions require repeat exposures to PM less than 5 μm, with these sizes being able to penetrate deep enough into the lung to elicit an immune response [39]. In some

Table 5.1 Occupational hypersensitivity pneumonitis-like diseases and the "Trojan Horse" carrier agents (Adapted from [23])

Hypersensitivity pneumonitis disease	Carrier agent	Trojan horse allergen
Farmer's lung	Hay	Bacteria
Fertilizer worker's lung	Soil	Bacteria
Coffee worker's lung	Coffee bean dust	Bacteria
Horseback rider's lung	Hay	Fungus
Papermill worker's lung	Wood chips	Fungus
Wheat weevils disease	Wheat flour	Fungus
Malt worker's lung	Malt and barley	Fungus

cases HP can be directly induced by PM from birds or animals (e.g. fur and feathers) due to a particular protein allergy; however a majority of HP cases are the result of the Trojan horse effect, rather than direct particle induced toxicity. Materials such as wood and grain dusts can act as carriers for fungus and mycobacteria to invade the lung and induce allergic reactions leading to hypersensitivity reactions after repeat exposures [39]. These kinds of reactions can be controlled if exposure to the allergic agent is limited, however with chronic exposures HP can lead to granuloma development and even pulmonary fibrosis and emphysema [72]. A non-comprehensive list of occupational HP-like diseases is provided in Table 5.1, along with the carrier agent that delivers the fungal/bacterial allergen.

5.2.6 Cancer

Inhalation of PM and fibers has been associated with the development of cancer, particularly of the lung. One confounding variable in the assessment of PM exposure and human lung cancer is the typical latency period of decades between the time of exposure and cancer development in humans [105]. When a long latency period exists, it is difficult to assume causation, and instead only associations can be made between lung cancer in humans and exposure to a given particulate. There are several occupational sources of PM exposure, as well as specific elemental exposures that have been identified as occupational carcinogens: welding fumes, wood dust, asphalt fumes, diesel exhaust, coke oven emissions, asphalt fumes, carbon black, asbestos, cadmium, nickel, silica (quartz), and chromium (VI) [14].

Inhalation of asbestos fibers leading to mesothelioma was probably the first instance in which human cancer was shown to be caused by inhalation of fibers or PM. The carcinogenic potential of asbestos and other fibers is related to their length and biopersistence (Fig. 5.1) [22, 65]. Occupational exposure to welding fumes has also been associated with lung cancer development in several cohorts from countries throughout the world. In 1990, the International Agency for Research on Cancer (IARC) stated that there were 30–50 % more lung cancer incidents among welders compared to a control population [93]. This group analyzed standardized incidence ratios (SIR) of lung cancer in Danish welders

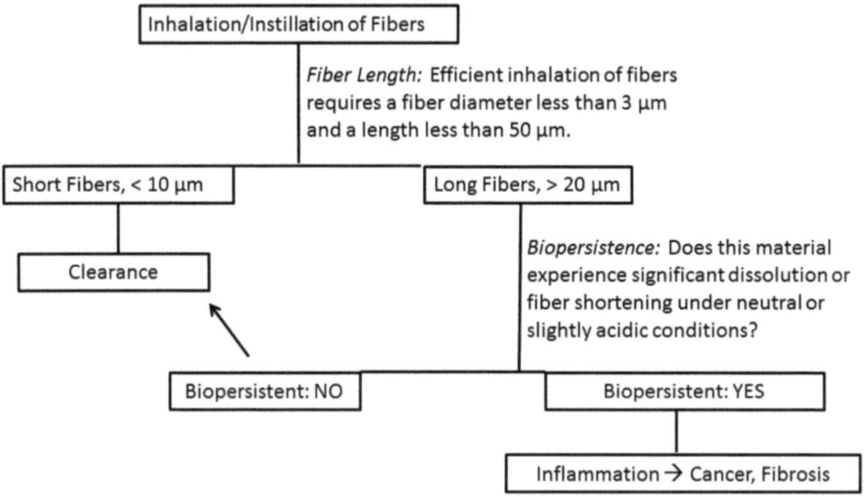

Fig. 5.1 Influence of fiber length and biopersistence on toxicity (Adapted from [22, 38])

working between 1964 and 1984, with more than 21 years of experience, and found a SIR of 1.35 (95 % CI 1.06–1.70). Investigators analyzing cancer risk for welders in Romania, Hungary, Poland, Russia, Slovakia, Czech Republic and the United Kingdom found an odds ratio (OR) of 1.18 (95 % CI 1.01–1.38) when correcting for independent metal exposure to chromium, nickel, cadmium and arsenic [57]. A meta-analysis of risk for specific welding-related jobs was performed for welders in the United States, United Kingdom, Canada, Sweden, New Zealand, Italy, Norway, France, Uruguay, Netherlands, Germany, Argentina, Finland, and Denmark finding no difference in risk between welding jobs, with an overall combined relative risk (CRR) of 1.26 (95 % CI 1.2–1.32) for lung cancer [3]. Investigators analyzing lung cancer risk in Finnish welders born between 1906 and 1945 found significant risk in all exposure groups [90]. For the lowest exposure group, there was an increased RR for all combined cancers (1.09: 95 % CI 1.05–1.14) and for small cell carcinoma (1.15: 95 % CI 1.04–1.27). The medium exposure group was associated with increased risk for all combined cancers (1.16: 95 % CI 1.03–1.31), for squamous cell carcinoma (1.26: 95 % CI 1.04–1.53) and adenocarcinoma (1.42: 95 % CI 1.06–1.91). High exposures were associated with increased risk for squamous cell carcinoma only (1.55: 95 % CI 1.08–2.24) [90]. As mentioned above, specific welding exposure is affected by the materials used, ventilation and operating conditions. Large meta-analyses like these are very important because they include welders from several occupational welding environments, which may help account for these variables.

Unlike most inhalation exposures, breathing leather dust during shoe making or repair has been shown to be associated with risk of sino-nasal (rather than lower respiratory) cancer in five of six analyzed cohorts [10]. Wood dust exposure is also associated with sino-nasal cancers; however, there are several other systemically diffuse cancers associated with wood working as well: digestive, urinary, respiratory, hemopoietic, lymphatic and Hodgkin's disease [63]. Several studies, summarized by [63], have found significant risks for various cancers with hardwood dust exposure: beech, oak and walnut. Another group found epidemiological associations with both hard and soft woods with differential tumor development. Hardwood exposure was associated with adenocarcinoma while softwood dust exposure was associated with squamous cell carcinoma (SCC) and non-Hodgkin's lymphoma [25]. In addition, the type of cancer which developed from softwood exposure seemed to differ based on the woodworkers' occupation; joiners, carpenters and loggers developed SCC, while saw and planing mill workers had an increased risk for non-Hodgkin's lymphoma. The fact that the specific wood working occupation affects the type of cancer that develops may indicate that the particle size of the wood dust generated is important for pathogenesis and deposition location. For wood dust exposure, it is important to consider whether the wood was treated and how either inherent or surface bound (e.g., fungi) properties of the wood may affect the toxicity. Wood that is intended for outdoor use is often treated with metal-based preservatives (chromium, arsenates, copper) to extend its lifetime; however these metals may cause toxicity after wood dust inhalation. Some of the inherent properties of timber that may conceivably be mutagenic were investigated and found to be innocuous using an in vitro mutagenicity test [62].

In addition to occupational exposures, there is a strong correlation between the development of lung cancer and exposure to atmospheric PM, with an 8 % increase in lung cancer mortality for every increase of 10 $\mu g/m^3$ of fine PM [77]. Large anthropogenic contributions to atmospheric PM levels occur through the combustion products of fossil fuels, specifically in the form of diesel exhaust (DE). DE is composed mostly of various gases (e.g., CO_2, CO, N_2, SO_x), hydrocarbons and their aldehyde derivatives, polyaromatic hydrocarbons and PM (elemental carbon) [80]. The particle size distribution of the PM fraction in DE has changed over the last several decades as engine types and operating conditions have evolved. Discussed in more detail later in the chapter, particle size affects the deposition location within the respiratory tract and the body's ability to clear inhaled PM. The surface chemistry of the PM and adsorbed contaminants can also be variable based on post-emission reactions with oxygen and sunlight in the environment [80]. DE was placed on the IARC's category 1 carcinogen list (known human carcinogens) in June 2012 and there have been several epidemiological studies published that show a significant association between inhalation exposure to DE and lung cancer [34]. One group of investigators found an odds ratio of 7.3 (95 % CI 1.46–36.57) for lung cancer among never-smokers in the highest exposure group of United States underground non-metal miners [91]. An increased risk for lung cancer due to DE exposure has also been observed in a population of Chinese industrial workers with 1–19 years of experience: OR of 12.39 with a 95 % C.I. of 2.4–63.94 [102]. This study did not adjust for smoking status; however, the estimated risk did account for age and familial/personal cancer and lung disease history. While these studies focus on chronic exposure to concentrated DE, it is important to consider that carcinogens may operate on a no-threshold model, with even low levels of DE exposure being capable of inducing carcinogenic effects in susceptible individuals.

5.2.7 Cardiovascular Effects

While an inhalation exposure is likely to target the lungs, there have been cardiovascular effects associated with PM in many inhalation studies. In addition, epidemiological data evaluating risk from air pollution show that increased risk in highly polluted areas can be attributed mostly to additional cardiovascular deaths. It is important to consider how particles are able to translocate from the lungs to the vascular system. It has been shown in rats, using an ex vivo lung perfusion, that nanoparticles administered to the lung are able to translocate to the circulation when lung permeability has increased (for example, due to histamine or H_2O_2, both of which may result from immune activation) [2].

The extent of particle translocation into the blood stream is size dependent with a large decrease in bioavailability as particle size increases [47]. Demonstrated with polystyrene particles, 240 nm seems to be the maximum diameter for uptake and translocation of particles into the blood stream following an inhalation exposure [46]. In addition to particle size, protein coating of particles is important for uptake. It has been shown that coating particles (of various materials: chitosan, poly-maleic-oleic acid and phosphatidylcholine) with surfactant protein A increases particle uptake into alveolar macrophages when compared to uncoated particles or particles coated with bovine serum albumin (BSA) [85]. Surfactant coating of particles has also been shown to be important for uptake via epithelial cells [47]. As particle size decreases, coating with surfactant becomes more efficient and may contribute to differences observed in translocation of particles based on size [26].

One of the major possible mechanisms of toxicity leading to cardiovascular effects would be interaction of absorbed particles with clotting factors. It was shown by [84] that TiO_2 nanotubes are capable of decreasing blood clotting time through enhancing the formation of the fibrinogen network. In addition, interference with normal clotting behavior was observed for both positively and negatively charged polystyrene particles, capable of causing platelet activation [67]. These investigators hypothesize that positively charged particles are able to bind to sialic acid groups (negatively charged) on platelets, allowing enhanced platelet-platelet aggregation; the mechanism of interaction for negatively charged particles remains unknown.

Myocardial infarction and stroke are the two major risks associated with plaque formation and rupturing in the arteries. Any aspect of a particle or fiber that affects clotting or plaque stability has the potential to cause devastating cardiovascular problems. Most research done in this area is based on relative risk analysis of epidemiological data, with some conflicting trends. For example, some studies indicate that people exposed to air pollution are most susceptible to cardiovascular effects if they have increased fibrinogen blood levels [74, 78], while another study was unable to show such a correlation [89].

While fibrinogen levels have failed to form a solid link between PM air pollution and cardiovascular risks, there is strong evidence to indicate that levels of C-reactive protein (CRP) may be used to expose high cardiovascular risk groups in relation to air pollution. High levels of CRP have been associated with coronary artery disease and may be indicative of an unstable plaque present in the vasculature; more recently, an association between high PM exposure and high CRP levels has been elucidated [21]. Together, these trends may suggest that exposure to air pollution can induce CRP production, which in turn makes arterial plaques more susceptible to rupture, leading to stroke or heart attack. While these associations have shown to be significant when analyzing epidemiological data, the exact mechanisms underlying cardiovascular risks and air pollution are not well understood.

5.3 Mechanisms of Toxicity for Inhaled Particles and Fibers

5.3.1 Immune Mediated Effects

Dysregulation of Neutrophil Activity and Reactive Oxygen Species Generation

Polymorphonuclear neutrophils (PMNs) are an important mediator for the initiation of an immune response. Neutrophils are generally the first immune cells to be recruited to sites of injury and make up a majority of immune cells in the blood stream. Upon activation, PMNs will release chemokines and proinflammatory cytokines for the recruitment and activation of additional immune cells. This response is tightly controlled to prevent both inappropriate activation and down regulation of the inflammatory pathway. Upregulation of PMN activity can lead to the development of autoimmune diseases such as systemic lupus erythematosus (SLE) or rheumatoid arthritis (RA), while down regulation of PMN responses can lead to increased susceptibility to bacterial infections. Both activation and down regulation of inflammatory responses have been shown by studies investigating the interaction of PM and fibers with neutrophils.

A variety of responses can be observed with TiO_2 particles of various sizes. Small, 1–10 nm, TiO_2 particles were shown to induce IL-8 production, but were also shown to prevent PMN apoptosis. Larger TiO_2 particles, (345 nm–3 μm) were shown to increase ROS production without causing cytotoxicity, while no response was registered with exposure of PMNs to 10 μm TiO_2 particles [31, 52]. In addition to seeing differences in response based on size, material is also an important determinant of inflammatory reaction. Release of lactate dehydrogenase (LDH) from neutrophils was stimulated by 50 nm polymethylmethacrylate (PMMA) particles, while no response was observed for 50 nm polystyrene particles [70]. Differences in response based on material can also be observed with poly (lactic) acid (PLA) particles, in which coating the particle surface with polyethylene glycol (PEG) reduced reactive oxygen species (ROS) production from neutrophils. Neutrophil activation and increased mortality has also been shown to occur with nickel and vanadium particles, leading to LDH and superoxide release, respectively [52].

Despite the fact that many forms of PM and nanoparticles are capable of eliciting an inflammatory response that may mediate their toxicity, interaction with neutrophils can also cause a detoxification reaction. It has been shown that exposure of carbon nanotubes to myeloperoxidase (MPO) released by neutrophils decreases their overall inflammatory nature [45]. Using dynamic light scattering, this group showed that carbon nanotubes exposed to MPO have a decreased length compared to untreated nanotubes. Like asbestos fibers, there is a known positive correlation between nanotube length and toxicity; therefore, [45] hypothesized that the decrease in nanotube size is responsible for the observed decrease in inflammatory response.

T-Helper Cells, Type 1 and 2

The human population has extensive genetic variation, which affects individual responses to environmental factors. When investigating an inflammatory endpoint such as asthma, it is important to consider a person's T helper cell profile. The body has two major types of T-helper cells, types 1 (Th1) and 2 (Th2). Th2 cells produce inflammatory cytokines like interleukins 4 and 5, previously shown to be associated with asthma in humans [33], while Th1 cells produce cytokines that are important for fighting infection. T-helper type 1 cells also help prevent the inflammatory response by releasing interferon-γ which inhibits Th2 cells. All humans have both Th1 and Th2 cells; the ratio between these two cell types can determine whether a person is predisposed to allergic and inflammatory diseases and reactions.

During fetal development we are exposed to maternal Th2 cells and are born with a Th2 dominant profile. Our postnatal environment will determine whether we develop a Th1 or Th2 phenotype as adults. Exposure to *mycobacterium* in soil has been shown to induce a Th1 type phenotype. This is the basis for the hypothesis that letting your children play outside can help prevent allergies. In contrast, inhalation of diesel exhaust particles and residual oil fly ash has been shown to induce a Th2 profile, exacerbating the inflammatory response and inducing fibrosis (See Fig. 5.2) [9].

Inflammatory Response and Fibrosis

Inhaled particles can create a pro-fibrotic environment through several immune-mediated mechanisms (Fig. 5.2). As mentioned above, inhaled particles that lead to a type 2-dominant T helper cell profile will induce an inflammatory response leading to IL-13 release; IL-13 will then induce TGF-β1 release, producing several pro-fibrotic downstream effects. Particles can also create these inflammatory conditions through the induction of TNF-α release, which, like IL-13 will induce TGF-β1. Both collagen deposition and release of connective tissue growth factor are downstream effects of TGF-β1 release, increasing the likelihood of developing fibrosis.

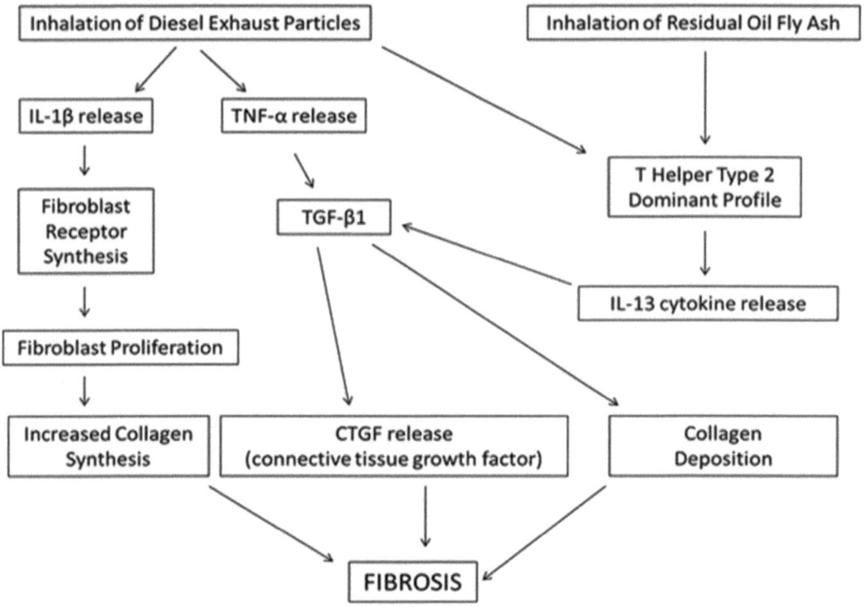

Fig. 5.2 Potential mechanism leading to fibrosis after PM inhalation [9, 33]

5.3.2 Genotoxic Mechanisms

Animal models can be helpful in assessing general carcinogenicity of a given toxicant; however, inter-species differences in cancer rates and target tissues make it difficult to accurately predict human neoplastic responses based entirely on rodent studies. Specific mechanisms of genotoxicity are investigated in vitro to add to the weight of evidence for carcinogenic potential of a given fiber or particle.

All forms of cancer follow a general developmental pathway, starting with initiation (gene mutation), followed by promotion (expansion of mutated population) and then progression (requires additional mutations) leading to cancer development. Particles and fibers contribute to initiation of carcinogenesis through genotoxic mechanisms. For an initiation event to occur, a genome mutation in a critical gene (cell cycle control or repair mechanisms) must avoid DNA repair and be retained through cell division. There are several possible mechanisms, both direct and indirect, for PM and fibers to cause DNA damage.

Oxidative stress is a common indirect method of DNA damage that can lead to tumor initiation. The DNA base pair guanine is susceptible to oxidation, converting to 8-hydroxy guanine. Guanine traditionally base pairs with cytosine; however, 8-hydroxyguanine does not undergo traditional G-C base pairing. If left unrepaired during cell replication, this mutation may ultimately lead to a base transversion from G-C to T-A pairing (illustrated in Fig. 5.3). If this mutation occurs in the coding region of a gene, there will be changes in the amino acid sequence for that protein, potentially leading to large downstream effects. The ability of asbestos

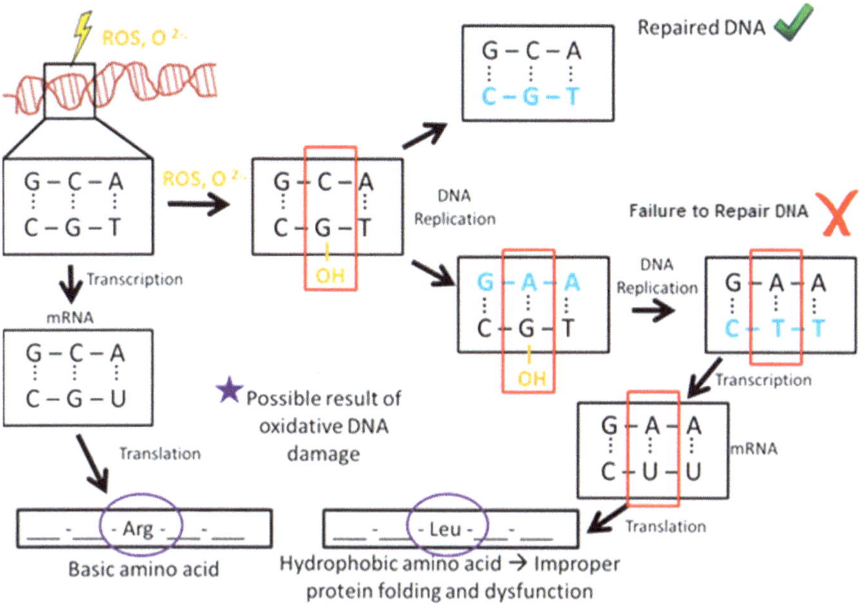

Fig. 5.3 Illustration of possible relationship between unrepaired oxidative DNA damage and cellular dysfunction

fibers to cause this kind of mutation through oxidative stress has been reported by several groups and summarized by [44].

Direct DNA damage is also a common source of genotoxicity, with chromosomal deletions and strand breaks contributing to DNA mutations. Evidence for asbestos fibers causing this kind of genotoxic damage has also been reported [44]. Although genotoxicity of fibers and particles is a likely mechanism of their carcinogenicity, DNA damage is not required for a particle or fiber to be considered carcinogenic. Non-genotoxic mechanisms can contribute to the promotion of a mutation, rather than its initiation. This mechanism of action can be tested by analyzing the ability of a particle or fiber to cause cancer after administration of a known mutagen.

5.4 Particle and Fiber Properties Influencing Toxicity

When characterizing a given particle/fiber exposure or designing an inhalation study it is particularly important to consider dosimetry issues involved in the exposure. As with chemical exposures, the standard for describing exposure to particles and fibers has been on a mass concentration basis. Inhalation toxicology has studied the effects of micrometer-sized inhalants for several decades, but has only in the last 20 years started to focus on smaller, nanometer-sized particles. As these studies have been completed, it has become apparent that lower mass concentrations of nanometer-sized particles are required to elicit the same responses seen with higher concentrations of micrometer-sized particles; and several studies have suggested that expressing exposure response in terms of particle number or surface area can help to better estimate risk from a given exposure [51, 58, 68]. As discussed in more detail later in this chapter, decreases in particle size show an increase in the efficiency of lung deposition and bioavailability, possibly explaining the observed increased toxicity of smaller particles even at lower number concentrations [79]. One of the largest realistic challenges that we are facing with this consideration is occupational exposure monitoring; testing air quality for mass-based contamination is much more straight forward than testing particle number concentrations in a workplace [37, 58]. Another problem with using mass concentration as a measure of exposure is that an emphasis is placed on larger particles under mixed particle size exposure conditions due to the high mass percentage that large particles will account for [37, 51, 58, 75].

5.4.1 Size

The initial investigations into nanoparticle toxicity were stimulated by differences in effects that were observed between particles in the nanometer and micrometer size (based on the diameter of spherical particles). One example of size dependent effects can be demonstrated using TiO_2 and inflammatory response in neutrophils, as described above with three size groups (1–10 nm, 345 nm–3 μm, and 10 μm) [31]. Differences in response based on particle size have also been observed for ROS generation after in vitro exposure to PM collected from the Los Angeles basin area from November 2001 to March 2002 [55]. When analyzing HO-1 expression in BEAS-2B human lung epithelial cells and RAW 264.7 murine macrophages, the smallest particles (ultrafine particles, $<$ 150 nm) showed the most significant induction of HO-1 expression. While larger particles (coarse, 2.5–10 μm and fine, 0.1–2.5 μm) showed a dose dependent reaction for HO-1 induction, even the highest dose of both coarse and fine particles (100 μg/mL) showed a much smaller induction than the ultrafine particles ($<$ 0.1 μm) at a low dose (8 μg/mL) [55]. HO-1 gene expression can be considered an endpoint for the induction of the antioxidant response element (ARE), indicative of ROS production. This group has also shown that exposure to ultrafine and fine particles lead to a 70 % glutathione (GSH) depletion (based on reduced to oxidized glutathione ratio, GSH:GSSG) in RAW 264.7 cells at 12 and 50 μg/mL respectively, compared to no response for coarse particles at 50 μg/mL.

In addition, it is important to consider the in vivo agglomeration state of PM or NPs, as this is a more significant determinant for clearance, deposition and translocation compared to primary particle size [47]. For example, if particle agglomeration is observed to increase with particle concentration, it is possible that a low concentration exposure may be more toxic than a higher concentration because larger, agglomerated particles are cleared more readily in the upper airways. There is an inverse relationship between particle size and retention. Mucociliary action can be considered a major source of particle clearance for large particles, leading to ingestion. However, as mentioned earlier, small particles are more efficient at penetrating the surfactant layer of the lung and it has been hypothesized that this removes the possibility for mucociliary removal. Micrometer sized particles that enter the lung are generally cleared within 24–48 h after exposure, while particles in the nanometer range show more than 75 % retention 48 h after exposure [64].

5.4.2 Shape (Aspect Ratio for Fibers)

There are a few major factors that influence the toxicity of fibers and carbon nanotubes that have been recognized throughout the literature as important aspects for predicting associated risk. First, the in vivo fiber or tube length is very important for determining toxicity, with long particles being most toxic for two reasons. One reason is that long particles are difficult to clear, penetrating very deep within the lung. Macrophages are very efficient at phagocytizing foreign objects in their environment, and will attempt to clear any fibers or nanotubes. This mechanism of clearance is good for small tubes and fibers because the entire fiber can be engulfed by the cell; the problem arises when the fibers are too long to be fully encapsulated by a macrophage. In addition to clearance, long fibers and nanotubes are more toxic than shorter fibers because they seem to have a higher capacity for the generation of reactive species. Another important aspect determining fiber toxicity is the degree to which fibers dissolve or break down in the lung or other biological environment. The most toxic fiber and nanotube exposures have resulted in a high number of long, biopersistent particles being deposited deep in the lung. The relationship between fiber length, biopersistence and toxicity is illustrated in Fig. 5.1.

5.4.3 Surface Charge

Differences in biological response based on particle surface charge have been observed for several endpoints. One example, mentioned briefly earlier, is the effect of polystyrene particles on platelet activation. [67] investigated neutral, positive and negatively charged polystyrene particles with different effects observed for all particle surface chemistries. This group showed increased platelet activation at 50 μg/kg for positively charged particles and at 100 μg/kg for negatively charged particles, while neutral particles were non-toxic, even at doses of 5,000 μg/kg [67].

In general, positively charged particles seem to have more inherent toxicity compared to neutral or negatively charged particles, as observed by several groups [5, 86, 108]. This is hypothesized to be due to particle interaction with negatively charged cell membrane surfaces. Interactions with the plasma membrane can cause a cell to become leaky and undergo apoptosis or necrosis.

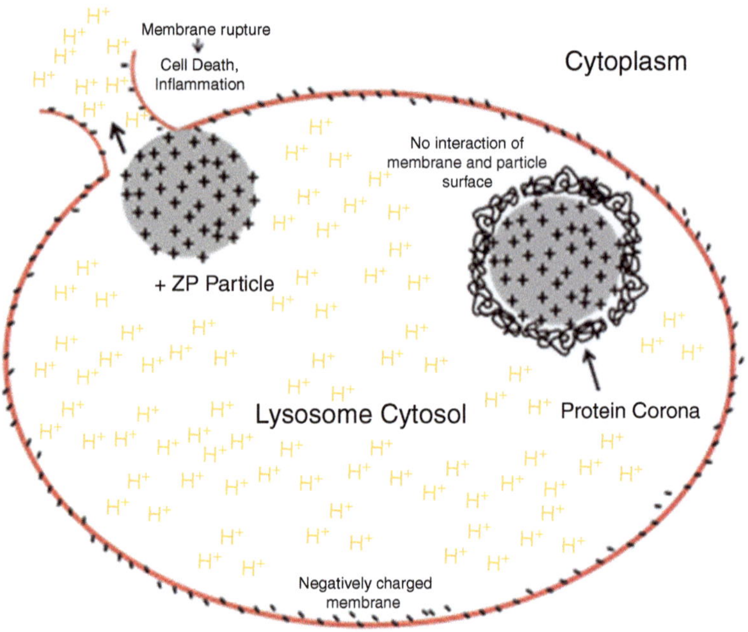

Fig. 5.4 Potential mediation of toxicity by zeta potential and protein corona formation

If a particle is phagocytized and a phagolysosome is formed, the particle has the potential to disrupt lysosomal membranes, leading to intracellular release of cytotoxic lysosomal contents (see Fig. 5.4). This trend was shown in a study by [15] in which particle toxicity was best correlated with ZP for particles with a charge greater than 14 mV. Furthermore, it was shown that particle toxicity was diminished by protein coating, preventing charge-mediated interaction of particles and membranes.

In addition to the direct effects of zeta potential on toxicity, the role that zeta potential plays on agglomeration behavior must also be considered [47]. A particle or fiber with a high absolute value zeta potential (mV > |30|) will be less likely to agglomerate compared to neutral or weakly charged particles. If toxicity occurs as a surface area mediated effect, then particle agglomeration will decrease toxicity, correlating with decreases in surface area. However, it is important to consider the particle/fiber type and the role that zeta potential may play on toxicity, e.g., cationic particles causing lysosomal membrane disruption.

5.4.4 Protein Corona

It has been widely accepted that particles and fibers that enter a biological environment will quickly become coated with proteins [13, 29, 104], with some exceptions (e.g., PEG surface coating prevents protein binding). In many cases, protein coating appears to lessen observed toxicity compared to particle exposures in a protein-free environment [15]. The exact mechanisms by which proteins alleviate toxicity are not completely understood and vary based on the mode of action for a given particle or fiber. One possible protective mechanism is illustrated in Fig. 5.4. As described above, particles with high positive zeta potentials have been shown to be toxic, possibly through interaction with negatively charged membranes. Figure 5.4 illustrates how particles with a protein corona may be unable to interact with the lysosomal membrane because their charge is masked at the surface.

In addition to mediating toxicity, endogenous protein coating diminishes biological activity and immune recognition of foreign particles [92]. However, there are other studies published that provide evidence that proteins bound to the particle surface undergo changes in conformation [17]. Based on this observation, it can be theorized that an immune-mediated reaction may be generated if new, "non-self", epitopes are exposed during particle binding, although this has not yet been reported in the literature.

5.4.5 Particle Surface Contamination

PM, engineered nanoparticles, fibers and nanotubes all have the capacity to induce toxicity through what is called the *trojan horse effect*. This means that rather than the core particulate material itself causing toxicity, contaminants bound to the surface of particles and fibers may be causing toxicity. PM from the combustion of fossil fuels may be contaminated with engine lubricants or unburned gasoline. In support of the trojan horse theory is the knowledge that PM from different sources cause different toxicities [32]. In addition to source, the season in which the particulates are generated has been shown to affect toxicity [36]. These trends suggest that PM toxicity is related not to the core material, but differences in the environment and climate where it was generated.

Engineered nanoparticles, although created in a controlled environment, are not exempt from being contaminated in this fashion. It is important for engineered NPs to be tested for residue surfactant impurities prior to doing any toxicity testing. Surfactants are amphiphilic molecules that are often used for synthesis of nanoparticles, but can disrupt membrane function due to their chemical nature. This form of unintentional toxicity has been reported for CTAB contaminated gold nanorods in cell culture testing [1]. Like engineered NPs, carbon nanotubes (CNT) have several possible sources of contamination from the manufacturing process, including metals and organics [22]. Metals such as cobalt, iron, nickel and molybdenum are used as catalysts for CNT formation. Organics, like carbon dust are used as a raw material for synthesis and may be found bound to nanotubes. In addition to contaminants that originate from synthesis, surfactants are sometimes used to achieve complete dispersion of CNTs before administration to animals or cell culture during toxicity testing [22]. Due to the various sources of particulate and fiber contamination, it is very important to consider how surface bound materials may be confounding your results when conducting or analyzing a toxicity study involving PM or engineered nanomaterials.

5.4.6 Particle Dissolution

Metals are being used extensively in nanomaterials for both electronics and medical supplies to take advantage of several inherent metallic properties, including electromagnetism, conductance, strength and antibacterial function. Metal particles present a unique situation in particle toxicology, in which observed toxicity may be caused directly by the particle or by metal ions that have dissolved from the particle surface. Teasing apart the ion versus particle effects can be challenging, and it is important to consider this dilemma when designing a study involving metal particles.

It has been shown for copper and zinc oxide particles that ion dissolution reaches between 90 % and 100 % after 24 h in a simulated lysosomal environment at pH 5.5 [15]. When analyzing observed toxicity for CuO and ZnO particles it was discovered that ionic dissolution was the particle property which correlated best with the measured inflammatory response. This was not completely surprising, as both copper and zinc ions have been previously shown to be toxic [66, 96]. For metal particles that undergo significant or complete dissolution in a relevant biological environment, toxicity can likely be predicted based on available ion toxicity data.

Recently, it has become apparent that metal particles are generated over time as wear debris in metal based hip replacements. Every year in the United States there are estimated to be 120,000 joint replacement surgeries [76]. The hardware for hip replacement surgeries are made of various materials: ceramic on ceramic (C-O-C), ceramic on plastic (C-O-P), metal on plastic (M-O-P) and metal on metal (M-O-M). Reported metal toxicity from wear debris in hip replacements is rare, with very few reported cases. However, when significant wear has occurred on the metal joint, the effects can be devastating and in some cases permanent. Cobalt, chromium and titanium are typically the metals used for hip replacements [76]. While titanium is biocompatible, there are well known toxicities associated with both cobalt and chromium; cobaltism from non-arthroplasty exposures has exhibited symptoms such as hearing and vision loss [94]. Trivalent chromium (Cr^{3+}) is an essential trace metal, while hexavalent chromium (Cr^{6+}) is a recognized carcinogen and can induce allergic dermal reactions [6]. In at least five of the reported cases of metal toxicity from hip arthroplasty, the toxicity occurred after a revision surgery replacing a fractured ceramic head with either an M-O-M or M-O-P device [41, 69, 73, 81, 94]. Despite careful removal of the ceramic pieces during revision, it is likely that residual ceramic material in the joint is responsible for increased metal wear. It is estimated that under standard post-operative conditions (without residual ceramics in the joint) that 10^{13} particles per year are generated with an average particle size of less than 50 nm [76]. These wear debris particles can become embedded in the surrounding soft tissue or be transported systemically via the lymphatics after phagocytosis by macrophages. Toxicity probably occurs when the particles dissolve, either in the area of the joint or in the lysosomal environment of a macrophage, exhibiting symptoms similar to Co and Cr ionic toxicity. Symptoms reported for these five individuals include hearing and vision loss (some permanent), dermatitis, several cardiac problems (e.g. fibrosis, atrophy), T cell cytotoxicity, weight loss, dysregulation of iron metabolism and paresthesia. Table 5.2 below summarizes the level of Co and Cr in these five patients for various body fluids, with reference values listed where available.

5.5 Oral Exposures to PM

Ingestion can also be a significant route of exposure to environmental and occupational PM. In addition to direct consumption of PM, inhalation exposures can also lead to ingestion of PM. Large particles or agglomerates can be efficiently cleared in the upper airways or by mucociliary action after inhalation, and are subsequently swallowed. The bioavailability of PM and nanoparticles in the gastrointestinal tract has been shown to be very low for polystyrene and gold particles, with less than 0.1 % of the administered dose found in any tissue for very small, 1.4 nm gold particles [42, 88]. Similar to exposure in the lung, systemic uptake of particles in the GI tract is size dependent with smaller particles being taken up more extensively.

Despite low oral bioavailability of particulates, when considering gastric exposure of metal particles, it is likely that dissolution plays a significant role to increase oral bioavailability. As discussed above, both ZnO and CuO particles have been shown to experience significant dissolution at pH 5.5 after 24 h in a simulated lung environment [15]; gastric exposure at pH 2 would likely cause even more rapid dissolution. When the particles dissolve, their bioavailability is likely to increase significantly; this trend is demonstrated by comparing tissue levels of silver between animals treated with particulate and soluble silver [56]. Oral exposure of both ZnO and Ag nanoparticles in rats has been shown to increase liver enzymes compared to vehicle controls, with accompanying histopathological changes observed [50, 71]. Unfortunately, we cannot say whether the observed toxicities were due to particle or ions. Many methods currently used to quantify metal concentrations in tissues, such as inductively

Table 5.2 Cobalt and chromium concentrations in metal-on metal hip replacement patients. Concentrations were measured in the indicated body fluid post revision surgery to replace a fractured ceramic hip with a metal Co-Cr-Ti device. When available, reference values are provided in parentheses

	Co Blood (ppb)	Co Plasma(P) or Serum (S) (ppb)	Co Urine (ppb)	Co CSF (ppb)	Cr Blood (ppb)	Cr Plasma(P) or Serum (S) (ppb)	Cr Urine (ppb)	Cr CSF (ppb)
Oldenburg et al. 2009 7 Months PR*	625		16,500		81		149	
Rizzetti et al. 2009 14 Months PR	549 (0.05–2.7)	90 (P) (0.1–0.6)	1187 (0.1–1.5)	11.4 (0.05–0.15)	54 (0.1–0.5)	210 (P) (0.1–0.5)	510 (0.05–2.2)	4.4 (0.01–0.02)
Pelclova et al. 2012 21 Months PR		506(S) (<0.9)				14.3 (S) (<0.5)		
Steens et al. 2006 24 Months PR		398(S) (0.45)				56 (S) (0.4)		
Ikeda et al. 2010 24 Months PR	400 (0.18)				221 (0.28)			

* Metal-on-plastic rather than metal-on-metal hip replacement

coupled plasma mass spectrometry (ICP-MS) cannot distinguish between particles and ions. Recently, single-particle ICP-MS (SP ICP-MS) has been used to tackle this challenge in aqueous systems [60] and is becoming more common in the analysis of mammalian systems as well.

In addition to particles being absorbed into the blood stream from the GI tract, immune mediated uptake via Peyer's patches in the lower small intestine is a significant source of exposure, especially for large particles. This form of uptake has been investigated mostly with polystyrene particles [43] which do not show any form of toxicity. However, if an immune-active particle was orally administered, both immune suppression and inflammatory activation are possible outcomes, depending on the material.

Direct damage to the intestinal mucosa caused by particles has not been thoroughly investigated in vivo. There is some evidence of ZnO particles causing intestinal inflammation after oral administration [71], although this could be due to the release of toxic Zn^{2+} ions. In addition, there have been a few in vitro studies investigating particle toxicity with an intestinal membrane model, caco-2 cells. Mechanisms of toxicity observed with the caco-2 model include generation of ROS and oxidative DNA damage, GSH depletion and disruption of cellular metabolism [27, 28]. These responses are very similar to observed lung cell culture toxicities discussed earlier.

Another possible mechanism of oral toxicity for PM or fibers is interference with the function of intestinal microbiota. Digestion, absorption and digestive immunity are all affected by our intestinal bacteria. Silver and zinc particles have recently been exploited in the nanoparticle industry because of their antibacterial properties [49, 59]; for example, embedded in food packaging materials. If there is release of those particles into food, oral exposure to antibacterial particles is possible. This source of toxicity has been investigated by one group in quail [87] where differences in the amount of gram-positive bacteria were observed.

5.6 Dermal Exposures to PM

Dermal contact with particulates and fibers can cause skin irritation, and in the case of certain metals such as nickel and cobalt, immune mediated contact dermatitis. Generally, risks of systemic effects from dermal exposure to particulates and fibers are considered to be very low. An interesting exception is beryllium. Occupational beryllium dust exposure can lead to chronic beryllium disease (CBD), a lung disease characterized by allergic beryllium sensitivity and granuloma formation in the lung. Sensitization is required, followed by progressive loss of pulmonary function. There is evidence suggesting that genetics may play a role is susceptibility [19, 100]. CBD is devastating disease, and the importance of worker protection has been recognized since the 1940s. A beryllium processing plant in Lorain, OH opened in 1947, only to close a year later due to the high incidence of lung problems and other health complaints in workers, as well as in residents of the surrounding town [8]. These and

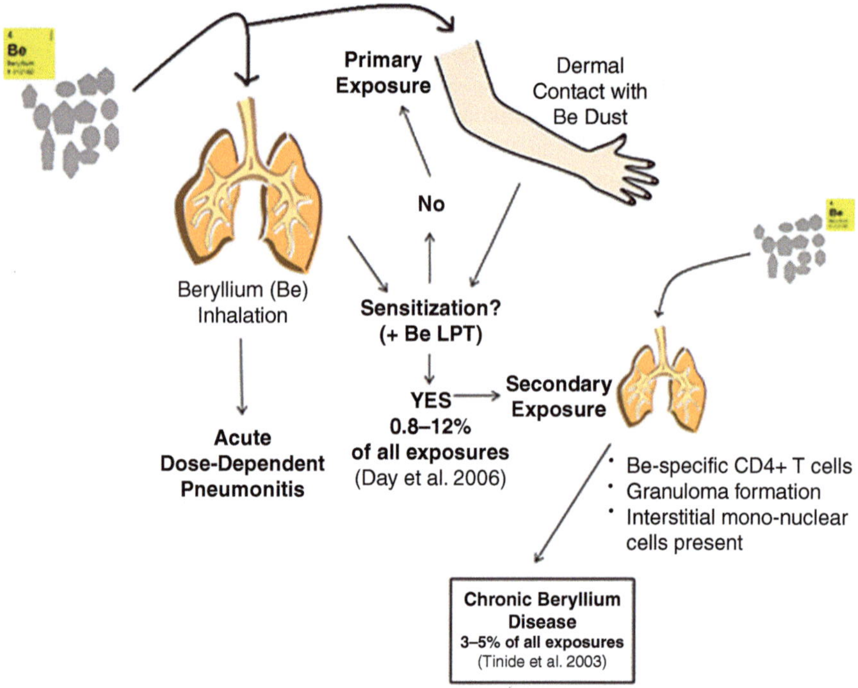

Fig. 5.5 Likely progression for the development of chronic beryllium disease (CBD)

other early occupational exposures to beryllium lead to the regulation and monitoring of beryllium air levels in factories in 1949, limiting beryllium concentrations at 2 μg/m^3 (levels at the Lorain, OH plant were over 43,000 μg/m^3).

Due to presentation as a lung disease, it has long been assumed that inhalation exposure to beryllium dust is responsible for both sensitization and development of CBD. As a result, worker protection has historically focused on limiting beryllium dust inhalation. Despite the use of the most effective respiratory protection available, industry workers were still becoming sensitized to beryllium. It was subsequently found that particles were capable of penetrating human skin during simulated activity and, using a mouse model, that dermal contact with beryllium was sufficient to produce sensitization [100]. The development of CBD and the role that dermal sensitization may play is illustrated in Fig. 5.5. Following dermal sensitization to beryllium, even an inhalation exposure less than the standard 2 μg/m^3, would be sufficient to lead to the development of CBD later in life. This dermal exposure presents a unique situation due to the downstream systemic effects that can occur. There is some evidence from a beryllium oxide ceramics plant that increased dermal protection can help to reduce sensitization rates among exposed workers, with workers from an old regime being 2.1–8.2 times more likely to become sensitized compared to workers under the new protection rules [18].

5.7 Particle and Fiber Exposure Limits

In an effort to protect the public from adverse health effects from particles and fibers, regulatory agencies in many countries have developed exposure limits. In some cases, these limits apply generally to particulate material in ambient air, but most exposure limits are directed to specific airborne particles and fibers of special concern. These are often developed in the context of occupational exposure standards because the workplace is viewed as where particle and fiber exposures are likely to be the highest. Exposure limits can take the form of recommended guidelines or regulatory agency standards. Although both guidelines and standards are useful in evaluating potential health consequences of inhalation of measured concentrations of particles and fibers in air, standards have the additional weight of being legally enforceable.

Air quality guidelines and standards can vary from one organization or agency to another for a variety of reasons. These can include differences in exposure assumptions or in the interpretation of toxicological data upon which the numerical standards are based, or in the level of protection (i.e., acceptable level of risk) sought. Examples of occupational exposure limits for several particles and fibers from the United States (Occupational Safety and Health Administration and the National Institute for Occupational Safety and Health), Great Britain (Health and Safety Executive), and Japan (Japan Society of Occupational Health) are shown in Table 5.3. Exposure limits can also differ for the same material based on exact occupational activity; see for example cotton dust in Table 5.3. Different exposure limits are sometimes provided for short time periods of exposure (< 1 h) versus a full work day of 8 or 10 h exposure; several examples of this can be found in the table.

Until recently, exposure limits were not explicit for nanomaterials; however, new exposure limits are being developed specifically for very small particles. The National Institute for Public Health and the Environment (RIVM) of the Netherlands has developed exposure limits for nanoparticles: e.g., 20,000 particles/cm^3 for TiO_2, ZnO, SiO_2, Al_2O_3 and 40,000 particles/cm^3 for C60, carbon black and carbon nanotubes.

5.8 Conclusions

Despite increased occupational safety regimes, human exposure to PM and fibers via several exposure routes is unavoidable due to the inherent properties of both the natural and modern environments that we live in. We've known for many centuries that inhalation exposure to natural PM can lead to adverse health effects in humans. However, in the last several decades, industrial exposures have broadened our understanding of the vast impact that PM and fiber exposure can have on large numbers of people. Industrial exposures causing disease have led to OSHA guidelines for exposure limits for various materials. Most modern exposures are

Table 5.3 Occupational exposure limits for particles and fibers for the United States (OSHA and NIOSH), Great Britain (HSE) and Japan (JSOH)

	United States		Great Britain	Japan
Table 5.3 Part I	Occupational Safety and Health Administration (OSHA)	National Institute for Occupational Safetyand Health (NIOSH)	Health and Safety Executive (HSE)	Japan Society for Occupational Health (JSOH)
Aluminum				
Total dust †	15 mg/m^3	–	–	2 mg/m^3
Inhalable fraction †	–	–	10 mg/m^3	–
Respirable fraction †	5 mg/m^3	–	4 mg/m^3	0.5 mg/m^3
Aluminum Oxide				
Total dust †	–	–	–	–
Inhalable fraction †	–	–	10 mg/m^3	–
Respirable fraction †	–	–	4 mg/m^3	–
Asbestos				
8-h TWA	0.1 fibers/cm^3	0.1 fibers/cm^3	–	–
30-min TWA	1 fiber/cm^3	–	–	–
Asphalt Fumes				
8-h TWA	–	–	5 mg/m^3	–
15-min TWA	–	5 mg/m^3	10 mg/m^3	–
Beryllium †	–	–	0.002 mg/m^3	0.002 mg/m^3
Cadmium †	–	–	0.025 mg/m^3	0.5 mg/m^3
Cadmium Oxide				
8-h TWA	–	–	0.025 mg/m^3	–
15-min TWA	–	–	0.05 mg/m^3	–
Carbon Black				
Total dust				
8-h TWA	3.5 mg/m^3	–	3.5 mg/m^3	4 mg/m^3
15-min TWA	–	–	7 mg/m^3	–
10-h TWA	–	3.5 mg/m^3	–	–
Respirable fraction †	–	–	–	1 mg/m^3
Cellulose				
Total dust				
8-h TWA	15 mg/m^3	–	–	–
15-min TWA	–	–	20 mg/m^3	–
Inhalable dust †	–	–	10 mg/m^3	–
Respirable dust †	5 mg/m^3	–	4 mg/m^3	–
Chromium metal (Cr) †	1 mg/m^3	0.5 mg/m^3	0.5 mg/m^3	0.5 mg/m^3
Coal dust				
Total dust †	–	–	–	4 mg/m^3
Respirable fraction				
8-h TWA	–	–	–	1 mg/m^3
10-h TWA	–	1 mg/m^3	–	–

(continued)

Table 5.3 (continued)

Table 5.3 Part II	United States		Great Britain	Japan
	Occupational Safety and Health Administration (OSHA)	National Institute for Occupational Safety and Health (NIOSH)	Health and Safety Executive (HSE)	Japan Society for Occupational Health (JSOH)
Cobalt †	0.1 mg/m^3	–	0.1 mg/m^3	0.5 mg/m^3
Cobalt Oxide †	–	–	–	–
Cooper				
Total dust				
8-hour TWA	1 mg/m^3	–	1 mg/m^3	–
15-minuteTWA	–	–	2 mg/m^3	–
Cotton dust				
Total dust (Non-Specific) †	–	< 0.200 mg/m^5	2.5 mg/m^3	4 mg/m^3
Waste Processing †	1 mg/m^3	–	–	–
Yarn Manufacturing †	0.2 mg/m^3	–	–	–
Textile Mills †	0.5 mg/m^3	–	–	–
Textile Slashing/ Weaving †	0.75 mg/m^5	–	–	–
Respirable (Non-Specific) †	–	–	–	1 mg/m^3
Diatomaceous Earth †	–	–	1.2 mg/m^3	–
Grain dust				
Total dust †	10 mg/m^3	–	–	4 mg/m^3
Respirable fraction †	–	–	–	1 mg/m^3
Graphite				
Total dust †	15,000,000 particles/ft^3	–	–	2 mg/m^3
Inhalable fraction †				
Respirable fraction	–	–	10 mg/m^3	–
8-hour TWA				
15-minuteTWA	–	–	–	0.5 mg/m^3
	–	–	4 mg/m^3	–
Limestone				
Total dust †	15 mg/m^3	–	–	8 mg/m^3
Inhalable fraction †	–	–	10 mg/m^3	–
Respirable fraction †	5 mg/m^5	–	4 mg/m^3	2 mg/m^3
Marble				
Total dust †	15 mg/m^5	–	–	–
Respirable fraction †	5 mg/m^5	–	–	–
Nickel †	1 mg/m^3	–	–	1 gm/m^3
Portland Cement				
Total dust †	50,000,000 particle/ft^3	–	–	4 mg/m^3
Inhalable fraction †				
Respirable fraction †	–	–	10 mg/m^3	–
	–	–	4 mg/m^3	1 mg/m^3

(continued)

Table 5.3 (continued)

	United States		Great Britain	Japan
Table 5.3 Part II	Occupational Safety and Health Administration (OSHA)	National Institute for Occupational Safety and Health (NIOSH)	Health and Safety Executive (HSE)	Japan Society for Occupational Health (JSOH)
Amorphous Silica	–	–	–	–
Total dust †	(80 mg/m^3)/ (%SiO$_2$ + 2)	–	–	–
Inhalable fraction †	–	–	6 mg/m^3	–
Respirable fraction †		–	2.3 mg/m^3	–
Crystalline Silica				
Total dust †	(10 mg/m^3)/ (%SiO$_2$ + 2)	–	–	–
Respirable fraction †	–	–	0.1 mg/m^3	–
Talc (w/o asbestos, < 1 % crystalline silica)				
Total dust †	20,000,000 particles/ft^3	–	–	–
Respirable fraction †	–	–	1 mg/m^3	–
Titanium dioxide				
Total dust †	15 mg/m^5	–	–	4 mg/m^3
Inhalable fraction †	–	–	10 mg/m^3	–
Respirable fraction †	–	–	4 mg/m^3	1 mg/m^3
Wood dust				
Softwood †	–	–	5 mg/m^3	–
Hardwood †	–	–	5 mg/m^3	–
Non-Specific				
Total dust †	–	–	–	4 mg/m^3
Respirable fraction †	–	–	–	1 mg/m^3
Total dust (non-specific)				
Total dust †	15 mg/m^3	–	–	–
Respirable fraction †	5 mg/m^5	–	–	–

† 8-hour time weighted average

difficult to fully comprehend due to simultaneous exposure to several compounds. This can be illustrated by the complex analysis of epidemiological data. Estimating risk for particle exposures is also complicated by dosimetry issues: comparing mass, surface area and number concentrations. While analyzing human exposure conditions and outcomes is an intricate process, numerous diseases have been linked to specific PM exposure in several human populations. These associations can in some cases be strengthened with animal exposures; however, human disease progression and presentation is often difficult to simulate in an animal model. Mechanistic investigation on the cellular and molecular levels can be helpful for further elucidation of human disease properties and in many cases involve complex genetic susceptibility. In addition, it is important to consider how intrinsic properties of the exposure material can vastly affect the health impacts observed, as well as how these properties may change over time, altering post-exposure characterization. The prediction and prevention of human disease related to PM exposure will continue to be a challenge, affecting both occupational safety and our everyday lives. As long as technology and human exploration expand, so will our potential exposures to PM and fibers.

5.9 Definitions, Abbreviations and Symbols

ARE	Antioxidant response element
BSA	Bovine serum albumin
CBD	Chronic beryllium disease
CI	Confidence interval
CNT	Carbon nanotube
COC	Ceramic-on-ceramic
COP	Ceramic-on plastic
COPD	Chronic obstructive pulmonary disease
CRP	C-reactive protein
CRR	Combined relative risk
CTAB	Cetyltrimethylammonium bromide: a surfactant used in the synthesis of Au nanorods
DE	Diesel exhaust
GI	Gastro-intestinal
GSH	Glutathione
GSSG	Oxidized glutathione
HP	Hypersensitivity pneumonitis
HSE	Health and safety executive UK
IARC	International agency for research on cancer
ICP	Inductively coupled plasma
JSOH	Japan society of occupational health
MOM	Metal-on-metal
MOP	Metal-on-plastic
MS	Mass spectrometry
NIOSH	National institute for occupational safety and health, USA
NP	nanoparticle
OSHA	Occupational safety and health administration, USA
PAR	Population attributable risk
PEG	Polyethylene glycol
PM	Particulate matter
RIVM	National institute for public health and environment, The Netherlands
ROS	Reactive oxygen species
SCC	Squamous cell carcinoma
SIR	Standardized incidence ratio
SP	Single particle
ZP	Zeta-potential

Bibliography

1. Alkilany, A.M., Nagaria, P.K., Hexel, C.R., Shaw, T.J., Murphy, C.J., Wyatt, M.D.: Cellular uptake and cytotoxicity of gold nanorods: molecular origin of cytotoxicity and surface effects. Small **5**(6), 701–708 (2009)
2. Alfaro-Moreno, E., Nawrot, T.S., Nemmar, A., Nemery, B.: Particulate matter in the environment: pulmonary and cardiovascular effects. Obstruct. Occup. Environ. Dis. **13**, 98–106 (2007)
3. Ambroise, D., Wild, P., Moulin, J.J.: Update of a meta-analysis on lung cancer and welding. Scand. J. Work Environ. Health **32**(1), 22–31 (2006)
4. Anderson, H.R., Butland, B.K., Donkelaar, A., Brauer, M., Strachan, D.P., Clayton, T., Dingenen, R., Amann, M., Brunekreef, B., Cohen, A., Dentener, F., Lai, C., Lamsal, L.N., Martin, R.V.: Satellite-based estimates of ambient air pollution and global variations in childhood asthma prevalence. Environ. Health Perspect. **120**, 1333–1339 (2012)
5. Asati, A., Santra, S., Kaittanis, C., Perez, J.M.: Surface-charge-dependent cell localization and cytotoxicity of cerium oxide nanoparticles. ACS Nano **4**, 5321–5331 (2010)
6. Barceloux, D.G.: Chromium. J. Toxicol. Clin. Toxicol. **37**(2), 173–194 (1999)
7. Blanc, P.D., Eisner, M.D., Earnest, G., Trupin, L., Balmes, J.R., Yelin, E.H., Gregorich, S.E., Kayz, P.P.: Further exporlation of the links between occupational exposure and chronic obstructive pulmonary disease. J. Occup. Environ. Med. **51**(7), 804–810 (2009)
8. Boffetta, P., Fryzek, J.P., Mandel, J.S.: Occupational exposure to beryllium and cancer risk: a review of the epidemiological evidence. Crit. Rev. Toxicol. **42**(2), 107–118 (2012)
9. Bonner, J.C.: Lung fibrotic responses to particle exposure. Toxicol. Pathol. **35**, 148–153 (2007)
10. Bonneterre, V., Deschamps, E., Persoons, R., Bernardet, C., Liaudy, S., Maitre, A., Gaudemaris, R.: Sino-nasal cancer and exposure to leather dust. Occup. Med. **57**, 438–443 (2007)
11. Boschetto, P., Quintavalle, S., Miotto, D., Cascio, N.L., Zeni, E., Mapp, C.E.: Chronic obstructive pulmonary disease (COPD) and occupational exposures. J. Occup. Med. Toxicol. **1**(11), 1–6 (2006)
12. Buerke, U., Schneider, J., Rosler, J., Woitowitz, H.J.: Interstitial pulmonary fibrosis after severe exposure to welding fumes. Am. J. Ind. Med. **41**, 259–268 (2002)
13. Cedervall, T., Lynch, I., Lindman, S., Berggard, T., Thulin, E., Nilsson, H., Dawson, K.A., Linse, S.: Understanding the nanoparticle–protein corona using methods to quantify exchange rates and affinities of proteins for nanoparticles. PNAS **104**(7), 2050–2055 (2007)
14. Centers for Disease Control and Prevention: . Occupational cancer: carcinogen list. *NIOSH and CDC*. Retrieved from http://www.cdc.gov/niosh/topics/cancer/npotocca.htm (2012, May 18). 02 June 2012
15. Cho, W.S., Duffin, R., Thielbeer, F., Bradley, M., Megson, I.L., MacNee, W., Poland, C.A., Tran, C.L., Donaldson, K.: Zeta potential and solubility to toxic ions as mechanisms of lung inflammation caused by metal/metal oxide nanoparticles. Toxicol. Sci. **126**(2), 469–477 (2012)
16. Cohen, R.A.C., Patel, A., Green, F.H.Y.: Lung disease caused by exposure to coal mine and silica dust. Semin. Respir. Crit. Care Med. **29**(6), 651–661 (2008)
17. Cukalevski, R., Lundqvist, M., Oslakovic, C., Dahlbäck, B., Linse, S., Cedervall, T.: Structural changes in apolipoproteins bound to nanoparticles. Langmuir **6**(27), 14360–14369 (2011)
18. Cummings, K.J., Deubner, D.C., Day, G.A., Henneberger, P.K., Kitt, M.M., Kent, M.S., Kreiss, K., Schuler, C.R.: Enhanced preventive programme at a beryllium oxide ceramics facility reduces beryllium sensitization among new workers. Occup. Environ. Med. **64**, 134–140 (2006)
19. Day, G.A., Stefaniak, A.B.., Weston, A., Tinkle, S.S.: Beryllium exposure: dermal and immunological considerations. Int. Arch. Occup. Environ. Health **79**, 161–164 (2006)

20. Delamater, P.L., Finley, A.O., Banerjee, S.: An analysis of asthma hospitalizations, air pollution, and weather conditions in Los Angeles County, California. Sci. Total Environ. **425**, 110–118 (2012)
21. Donaldson, K., Stone, V., Seaton, A., MacNee, W.: Ambient particle inhalation and the cardiovascular system: potential mechanisms. Environ. Health Perspect. **109**, 523–527 (2001)
22. Donaldson, K., Aitken, R., Tran, L., Stone, V., Duffin, R., Forrest, G., Alexander, A.: Carbon nanotubes: a review of their properties in relation to pulmonary toxicology and workplace safety. Toxicol. Sci. **92**(1), 5–22 (2006)
23. Fink, J.N., Lindesmith, L.A., Horvath, E.P.: Hypersensitivity pneumonitis. In: Zenz, C., Dickerson, O.B., Horvath Jr., E.P. (eds.) Occupational Medicine, 3rd edn, pp. 208–209. Mosby, St. Louis (1994)
24. Fishwick, D., Barber, C.M., Darby, A.C.: Review series: occupational and environmental lung disease; chronis obstructive pulmonary disease and the workplace. Chron. Respir. Dis. **7**(2), 113–122 (2010)
25. Flechsig, R., Nedo, G.: Hazardous health effects of occupational exposure to wood dust. Ind. Health **28**, 107–119 (1990)
26. Geiser, M., Kreyling, W.: Deposition and biokinetics of inhaled nanoparticles. Part. Fibre Toxicol. **20**(7), 2 (2010)
27. Gerloff, K., Albretch, C., Boots, A.W., Förster, I., Schins, R.P.F.: Cytotoxicity and oxidative DNA damage by nanoparticles in human intestinal Caco-2 cells. Nanotoxicology **3**(4), 355–364 (2009)
28. Gerloff, K., Fenoglio, I., Carella, E., Kolling, J., Albrecht, C., Boots, A.W., Förster, I., Schins, R.P.: Distinctive toxicity of TiO2 rutile/anatase mixed phase nanoparticles on Caco-2 cells. Chem. Res. Toxicol. **25**(3), 646–655 (2012)
29. Gessner, A., Lieske, A., Paulke, B.R., Muller, R.H.: Influence of surface charge density on protein adsorption on polymeric nanoparticles: analysis by two-dimensional electrophoresis. Eur. J. Pharm. Biopharm. **54**, 165–170 (2002)
30. Glad, J.A., Brink, L.L., Talbott, E.O., Lee, P.C., Xu, X., Saul, M., Rager, J.: The relationship of ambient ozone and PM(2.5) levels and asthma emergency department visits: possible influence of gender and ethnicity. Arch. Environ. Occup Health **67**(2), 103–108 (2012)
31. Goncalves, D.M., Liz, R., Girard, D.: Activation of neutrophils by nanoparticles. Scienitific World J. **11**, 1877–1885 (2011)
32. Gordon, T.: Linking health effects to PM components, size and sources. Inhal. Toxicol. **19**, 3–6 (2007)
33. Graham, L.M.: All I need is the air I breathe: outdoor air quality and asthma. Paediatr. Respir. Rev. **5**, 59–64 (2004)
34. Gulland, A.: Diesel engine exhaust causes lung cancer, says WHO. BMJ **4174**, 1 (2012)
35. Gustafson, T., Dahlman-Höglund, A., Nilsson, K., Ström, K., Tornling, G., Torén, K.: Occupational exposure and severe pulmonary fibrosis. Respir. Med. **101**, 2207–2212 (2007)
36. Ham, W.A., Ruehl, C.R., Kleeman, M.J.: Seasonal variation of airborne particle deposition efficiency in the human respiratory system. Aerosol Sci. Technol. **45**, 795–804 (2011)
37. Heitbrink, W.A., Evans, D.E., Ku, B.K., Maynard, A.D., Slavin, T.J., Peters, T.M.: Relationships among particle number, surface area, and respirable mass concentrations in automotive engine manufacturing. J. Occup. Environ. Hyg. **6**, 19–31 (2009)
38. Hesterberg, T.W., Miller, W.C., McConnell, E.E., Chevalier, J., Hadley, J.G., Bernstein, D.M., Thevenaz, P., Anderson, R.: Chronic inhalation toxicity of size-separated glass fibers in Fischer 344 rats. Fundam. Appl. Toxicol. **20**(4), 464–476 (1993)
39. Hirschmann, J.V., Pipavath, S.N.J., Godwin, J.D.: Hypersensitivity pneumonitis: a historical. Clinical and radiological review. RadioGraphics **29**, 1921–1938 (2009)
40. Hoek, G., Pattenden, S., Willers, S., Antova, T., Fabianova, E., Braun-Fahrländer, C., Forastiere, F., Gehring, U., Luttmann- Gibson, H., Grizes, L., Heinrich, J., Houthuijs, D., Janssen, N., Katsnelson, B., Kosheleva, A., Moshammer, H., Neuberger, M., Privalova, L., Rudnai, P.,

Speizer, F., Slachtova, H., Tomaskova, H., Zlotkowskai, R., Fletcher, T.: PM10 and children's respiratory symptoms and lung function in the PATY study. ERJ Expr. **40**(3), 538–547 (2012)
41. Ikeda, T., Takahashi, K., Kabata, T., Sakagoshi, D., Tomita, K., Yamada, M.: Polyneuropathy caused by cobalt-chromium metallosis after total hip replacement. Muscle Nerve **42**, 140–143 (2010)
42. Jani, P.U., Florence, A., McCarthy, D.E.: Further histological evidence of gastrointestinal absorption of polystyrene nanoparticles in the rat. Int. J. Pharmacol. **84**, 245–252 (1992)
43. Jani, P.U., McCarthy, D.E., Florence, A.T.: Nanosphere and microsphere uptake via Peyer's patches: observation of the rate of uptake in the rat after a single oral dose. Int. J. Pharmacol. **86**, 239–246 (1992)
44. Jaurand, M.C.: Mechanisms of fiber-induced genotoxicity. Environ. Health Perspect. **105**, 1073–1084 (1997)
45. Kagan, V.E., Konduru, N.V., Feng, W., Allen, B.L., Conroy, J., Volkov, Y., Vlasova, I.I., Belikova, N.A., Yanamala, N., Kapralov, A., Tyurina, Y.Y., Shi, J., Kisin, E.R., Murray, A.R., Franks, J., Stolz, D., Gou, P., Klein-Seetharaman, J., Fadeel, B., Star, A., Shvedova, A. A.: Carbon nanotubes degraded by neutrophil myeloperoxidase induce less pulmonary inflammation. Nat. Nanotechnol. **5**(5), 354–359 (2010)
46. Kato, T., Yashiro, T., Murata, Y., Herbert, D.C., Oshikawa, K., Bando, M., Ohno, S., Sugiyama, Y.: Evidence that exogenous substances can be phagocytized by alveolar epithelial cells and transported into blood capillaries. Cell Tissue Res. **311**, 47–51 (2003)
47. Kendall, M., Holgate, S.: Health impacts and toxicological effects of nanomaterials in the lung. Official J. Asian Pacific Soc. Respirology **17**(5), 743–758 (2012)
48. Kennedy, S.M., Chambers, R., Du, W., Dimich-Ward, H.: Environmental and occupational exposures; do they affect chronic obstructive pulmonary disease differently in women and men? Proc. Am. Thorac. Soc. **4**, 692–694 (2007)
49. Kim, J.S., Kuk, E., Yu, K.N., Kim, J.H., Park, S.J., Lee, H.J., Kim, S.H., Park, Y.K., Park, Y. H., Hwang, C.Y., Kim, Y.K., Lee, Y.S., Leong, D.H., Cho, M.H.: Antimicrobial effects of silver nanoparticles. Nanomedicine **3**, 95–101 (2007)
50. Kim, Y.S., Song, M.Y., Park, J.D., Song, K.S., Ryu, H.R., Chung, Y.H., Chang, H.K., Lee, J. H., Oh, K.H., Kelman, B.J., Hwang, I.K., Yu, I.J.: Subchronic oral toxicity of silver nanoparticles. Particle Fibre Toxicol. **7**(20), 1–11 (2010). LOAEL of 125mg/kg, NOAEL 30mg/kg)
51. Kreyling, W.G., Semmler-Behnke, M., Möller, W.: Health implications of nanoparticles. J. Nanoparticle Res. **8**, 543–562 (2006)
52. Kumazawa, R., Watari, F., Takashi, N., Tanimura, Y., Uo, M., Totsuka, Y.: Effects of Ti ions and particles on neutrophil function and morphology. Biomaterials **23**(17), 3757–3764 (2002)
53. Leigh, J., Driscoll, T.R., Cole, B.D., Beck, R.W., Hull, B.P., Yang, J.: Quantitative relation between emphysema and lung mineral content in coal workers. Occup. Environ. Med. **51**, 400–407 (1994)
54. Leung, C.C., Yu, I.T.S., Chen, W.: Silicosis. Lancet **279**, 2008–2018 (2012)
55. Li, N., Sioutas, C., Cho, A., Schmitz, D., Misra, C., Sempf, J., Wang, M., Oberley, T., Froines, J., Nel, A.: Ultrafine particulate pollultants induce oxidative stress and mitochondrial damage. Environ. Health Perspect. **111**(4), 455–460 (2003)
56. Loechner, K., Hadrup, N., Qvortrup, L.A., Gao, X., Vogel, U., Mortensen, A., Lam, H.R., Larsen, E.H.: Distribution of silver in rats following 28 days of repeated oral exposure to silver nanoparticles or silver acetate. Part. Fibre Toxicol. **8**(18), 1–14 (2011)
57. Mannetje, A., Brennan, P., Zaridze, D., Szeszenia-Dabrowska, N., Rudnai, P., Lissowska, J., Fabiánová, E., Cassidy, A., Mates, D., Bencko, V., Foretova, L., Janout, V., Fevotte, J., Fletcher, T., Boffetta, P.: Welding and lung cancer in Centeral and Eastern Europe and the United Kingdom. Am. J. Epidemiol. **175**(7), 706–714 (2012)
58. Maynard, A.D.: Estimating aerosol surface area from number and mass concentration measurements. Ann. Occup. Hyg. **47**(2), 123–144 (2003)

59. Mirzajani, F., Ghassempour, A., Aliahmadi, A., Esmaeili, M.A.: Antibacterial effect of silver nanoparticles on *Staphylococcus aureus*. Res. Microbiol. **162**(5), 542–549 (2011)
60. Mitrano, D.M., Lesher, E.K., Bednar, A., Monserud, J., Higgins, C.P., Ranville, J.F.: Detecting nanoparticulate silver using single-particle inductively coupled plasma–mass spectrometry. Environ. Toxicol. Chem. **31**(1), 115–121 (2011)
61. Miyake, Y., Sasaki, S., Yokoyama, T., Chida, K., Azuma, A., Suda, T., Kudoh, S., Sakamoto, N., Okamoto, K., Kobashi, G., Washio, M., Inaba, Y., Tanaka, H.: Occupational and environmental factors and idiopathic pulmonary fibrosis in Japan. Ann. Occup. Hyg. **49**(3), 259–265 (2005)
62. Mohtashamipur, E., Norpoth, K.: Non-mutagenicity of some wood-related compounds in the bacterial/microsome plate incorporation and microsuspension assays. Int. Arch. Occup. Environ. Health **54**, 83–90 (1984)
63. Mohtashamipur, E., Norpoth, K., Lühmann, F.: Cancer epidemiology of woodworking. J. Cancer Res. Clin. Oncol. **115**, 503–515 (1989)
64. Möller, W., Felten, K., Sommerer, K., Scheuch, G., Meyer, G., Meyer, P., Haussinger, K., Kreyling, W.G.: Deposition, retention, and translocation of ultrafine particles from the central airways and lung periphery. Am. J. Respir. Crit. Care Med. **177**, 426–432 (2008)
65. Mossman, B.T., Lippmann, M., Hesterberg, T.W., Kelsey, K.T., Barchowsky, A., Bonner, J. C.: Pulmonary endpoints (lung carcinomas and asbtestosis) following inhalation exposure to asbestos. J. Toxicol. Environ. Health B Crit. Rev. **14**, 76–121 (2011)
66. Nel, A.E., Madler, L., Velegol, D., Xia, T., Hoek, E.M., Somasundaran, P., Klaessig, F., Castranova, V., Thompson, M.: Understanding biophysiochemical interactions at the nano-bio interface. Nat. Mater. **8**, 543–557 (2009)
67. Nemmar, A., Hoylaerts, M., Hoet, P.H.M., Dinsdale, D., Smith, T., Xu, H., Vermylen, J., Nemery, B.: Ultrafine particles affect experimental thrombosis in an in vivo hamster model. Am. J. Respir. Crit. Care Med. **166**, 998–1004 (2002)
68. Oberdoerster, G., Ferin, J., Lehnert, B.E.: Correlation between particle size, in vivo particle persistence, and lung injury. Environ. Health Perspect. **102**(Suppl 5), 173–179 (1994)
69. Oldenburg, M., Wegner, R., Baur, X.: Sever cobalt intoxication due to prosthesis wear in repeated total hip arthroplasty. J. Arthroplasty **24**(5), 15–20 (2009)
70. Papatheofanis, F.J., Barmada, R.: Polymorphonuclear leukocyte degranulation with exposure to polymethylmethacrylate nanoparticles. J. Biomed. Mater. Res. **25**(6), 761–771 (1991)
71. Pasupuleti, S., Alapati, S., Ganapathy, S., Anumolu, G., Pully, N.R., Prakhya, B.M.: Toxicity of zinc oxide nanoparticles through oral route. Toxicol. Ind. Health. **28**(8), 675–686 (2012)
72. Patel, A.M., Ryu, J.H., Reed, C.E.: Hypersensitivity pneumonitis: current concepts and future questions. Curr. Rev. Allergy Clin. Immunol. **108**, 661–670 (2001)
73. Pelcolva, D., Sklensky, M., Janicek, P., Lach, K.: Severe cobalt intoxication following hip replacement revision: clinical features and outcome. Clin. Toxicol. **50**, 262–265 (2012)
74. Peters, A., Doring, A., Wichmann, H.E., Koenig, W.: Increased plasma viscosity during an air pollution episode: a link to mortality? Lancet **349**, 1582–1587 (1997)
75. Peters, A., Wichmann, H.E., Tuch, T., Heinrich, J., Heyder, J.: Respiratory effects are associated with the number of ultrafine particles. Am. J. Respir. Crit. Care Med. **155**, 1376–1383 (1997)
76. Polyzois, I., Nikolopoulos, D., Michos, I., Patsouris, E., Theocharis, S.: Local and systemic toxicity of nanoscale debris particles in total hip arthroplasty. J. Appl. Toxicol. **32**, 255–269 (2011)
77. Pope III, C.A., Burnett, R.T., Thun, J., Calle, E.E., Krewski, D., Ito, K., Thurston, G.D.: Lung cancer, cardiopulmonary mortality, and long-term exposure to fine particulate air pollution. JAMA **287**(9), 1132–1141 (2002)
78. Prescott, G.J., Lee, R.J., Cohen, G.R., Elton, R.A., Lee, A.J., Fowkes, F.G., Agius, R.M.: Investigation of factors which might indicate susceptibility to particulate air pollution. Occup. Environ. Med. **57**, 53–57 (2000)

79. Ramachandran, G., Paulsen, D., Watts, W., Kittelson, D.: Mass, surface area and number metrics in diesel occupational exposure assessment. J. Environ. Monit. **7**, 728–735 (2005)
80. Ris, C.: U.S. EPA health assessment for diesel engine exhaust: a review. Inhal. Toxicol. **19**, 229–239 (2007)
81. Rizzetti, M.C., Liberini, P., Zarattini, G., Catalani, S., Pazzaglia, U., Apostoli, P., Padovani, A.: Loss of sight and sound. Could it be the hip? Lancet **373**, 1052–1053 (2009)
82. Rodriguez, E., Ferrer, J., Marti, S., Zock, J.P., Plana, E., Morell, F.: Impact of occupational exposure on severity of COPD. Chest **134**(6), 1237–1243 (2008)
83. Rosen, G.: The History of Miners' Diseases: A Medical and Social Interpretation, pp. 450–480. Schuman, New York (1943)
84. Roy, S.C., Oaulosec, M., Grimes, C.A.: The effect of TiO2 nanotubes in the enhancement of blood clotting for the control of hemorrhage. Biomaterials **28**, 4667–4672 (2007)
85. Ruge, C.A., Kirch, J., Cañadas, O., Schneider, M., Perez-Gil, J., Schaefer, U.F., Casals, C., Lehr, C.M.: Uptake of nanoparticles by alveolar macrophages is triggered by surfactant protein A. Nanomedicine **7**, 690–693 (2011)
86. Ruizendaal, L., Bhattacharjee, S., Pournazari, K., Rosso-Vasic, M., de Haan, L.H.J., Alink, G.M., Marcelis, A.T.M., Zuilhof, H.: Synthesis and cytotoxicity of silicon nanoparticles and covalently attached organic monolayers. Nanotoxicology **3**(4), 339–347 (2009)
87. Sawosz, E., Binek, M., Grodzik, M., Ska, M.Z., Sysa, P., Szmidt, M., Niemiec, T., Chwalibog, A.: Influence of hydrocolloidal silver nanoparticles on gastrointestinal microflora and morphology of enterocytes of quails. Arch. Anim. Nutr. **61**(6), 444–451 (2007)
88. Schleh, C., Semmler-Behnke, M., Lipka, J., Wenk, A., Hirn, S., Schäffler, M., Schmid, G., Simon, U., Kreyling, W.G.: Size and surface charge of gold nanoparticles determine absorption across intestinal barriers and accumulation in secondary target organs after oral administration. Nanotoxicology **6**(1), 36–46 (2012)
89. Seaton, A., Soutar, A., Crawford, V., Elton, R., McNerlan, S., Cherrie, J., Watt, M., Agius, R., Stout, R.: Particulate air pollution and the blood. Thorax **54**, 1027–1032 (1999)
90. Siew, S.S., Kauppinen, T., Kyyrönen, P., Heikkilä, P., Pukkala, E.: Exposure to iron and welding fumes and the risk of lung cancer. Scand. J. Work Environ. Health **34**(6), 444–450 (2008)
91. Silverman, D.T., Samanic, C.M., Lubin, J.H., Blair, A.E., Stewart, P.A., Vermeulen, R., Coble, J.B., Rothman, N., Schleiff, P.L., Travis, W.D., Ziegler, R.G., Wachholder, S., Attfield, M.D.: The diesel exhaust in miners study: a nested case–control study of lung cancer and diesel exhaust. J. Natl. Cancer Inst. **104**(11), 855–868 (2012)
92. Smith, M.W., Thomas, N.W., Jenkin, P.G., Miller, N.G.A., Cremaschi, D., Porta, C.: Selective transport of microparticles across Peyer's patch follicle-associated M cells from mice and rats. Exp. Physiol. **80**, 735–743 (1995)
93. Sørensen, A.R., Thulstrup, A.M., Hansen, J., Ramlai-Hansen, C.H., Meersohn, A., Skytthe, A., Bonde, J.P.: Risk of lung cancer according to mild steel and stainless steel welding. Scand. J. Work Environ. Health **33**(5), 379–386 (2007)
94. Steens, W., von Foerster, G., Katxer, A.: Severe cobalt poisoning with loss of sight after ceramic-metal pairing in a hip—a case report. Acta Orthop. **77**(5), 830–832 (2006)
95. Stephens, M.B., Yew, K.S.: Diagnosis of chronic obstructive pulmonary disease. Am. Fam. Physician **78**(1), 87–92 (2008)
96. Studer, A.M., Limbach, L.K., Van Duc, L., Krumeich, F., Athanassiou, E.K., Gerber, L.C., Moch, H., Stark, W.J.: Nanoparticle cytotoxicity depends on intracellular solubility: comparison of stabilized copper metal and degradable copper oxide nanoparticles. Toxicol. Lett. **197**, 169–174 (2010)
97. Taskar, V.S., Coultas, D.B.: Is idiopathic pulmonary fibrosis an environmental disease? Proc. Am. Thorac. Soc. **3**, 293–298 (2006)
98. Taskar, V.S., Coultas, D.B.: Exposures and idiopathic lung disease. Semin. Respir. Crit. Care Med. **29**(6), 670–679 (2008)

99. Thomas, C.R., Kelley, T.R.: A brief review of silicosis in the United States. Environ. Health Insights **4**, 21–26 (2010)
100. Tinkle, S.S., Antonini, J.M., Rich, B.A., Roberts, J.R., Salmen, R., DePree, K., Adkins, E.J.: Skin as a route of exposure and sensitization in chronic beryllium disease. Environ. Health Perspect. **111**(9), 1202–1208 (2003)
101. Trasande, L., Thurston, G.D.: The role of air pollution in asthma and other pediatric morbidities. J. Allergy Clin. Immunol. **115**(4), 689–699 (2005)
102. Tse, L.A., Yu, I.T., Au, J.S.K., Qiu, H., Wang, X.: Silica dust, diesel exhaust, and painting work are the significant occupational risk factors for lung cancer in nonsmoking Chinese men. Br. J. Cancer **10**, 208–213 (2011)
103. Vallyathan, V., Green, F.H.Y., Brower, P., Attfield, M.: The role of coal mine dust exposure in the development of pulmonary emphysema. Ann. Occup. Hyg. **41**(1), 352–357 (1997)
104. Wasdo, S.C., Barber, D.S., Denslow, N.D., Powers, W.P., Palazuelos, M., Jr, S., Moudgil, B., Roberts, S.: Differential binding of serum proteins to nanoparticles. Int. J. Nanotechnol. **5**(1), 92–115 (2008)
105. Witschi, H.R., Pinkerton, K.E., Van Winkle, L.S., Last, J.A.: Toxic responses of the respiratory system. In: Klaassen, C.D. (ed.) Casarett and Doull's Toxicology: The Basic Science of Poisons, 7th edn, pp. 609–630. McGraw Hill Medical, New York (2008)
106. Woodruff, T.J., Parker, J.D., Schoendorf, K.C.: Fine particulate matter (PM2.5) air pollution and selected causes of postneonatal infant mortality in California. Environ. Health Perspect. **114**(5), 786–790 (2006)
107. Yeh, K.W., Chang, C.J., Huang, J.L.: The association of seasonal variations of asthma hospitalization with air pollution among children in Taiwan. Asian Pac. J. Allergy Immunol. **29**, 34–41 (2011)
108. Yin, H., Casey, P.S.: Effects of surface chemistry on cytotoxicity, genotoxicity, and the generation of reactive oxygen species induced by ZnO nanoparticles. Langmuir **26**(19), 15399–15408 (2010)

Chapter 6
Ceramics: Effect of Powder and Slurry Properties on Quality

Makio Naito

Abstract Ceramics are widely used in industry. Especially, advanced ceramics are of growing importance for a large variety of industrial applications in relation to excellent features like high mechanical strength, high stability and high functional properties. Good performance results not only from details of ceramic powders but also from adequate focus on the details of the processing steps for manufacturing the ceramic products. This chapter focuses on the detrimental effect that the presence of few large particles, slurry properties and the structure and mechanical properties of granules in the green body have on the strength of the ceramic product. They have been found to initiate weak spots in ceramic products. Also, some new measurement approaches are explained that allow identification of the fracture origin.

6.1 Introduction

"Ceramics" originates from the Greek word "keramikos" (κεραμίκός). It means 'pottery', which actually has been widely used for human life. The opportunity to utilize ceramics as advanced materials started from the beginning of the twentieth century. Ferrite was invented in the 1930s, the dielectric Barium Titanate ($BaTiO_3$) was discovered in the early 1940s, and then various kinds of ceramics have been developed after the 1950s. This means that microstructure control of ceramics can create specific functional applications including mechanical, thermal, chemical, electrical, magnetic, optical and combinations thereof.

Although ceramics have been used for many applications, still serious problems remain on reliability of mechanical strength and their forming into various shapes. As ceramics are manufactured by using powder materials, their manufacturing

M. Naito (✉)
Joining and Welding Research Institute, Osaka University, Japan
e-mail: m-naito@jwri.osaka-u.ac.jp

H.G. Merkus and G.M.H. Meesters (eds.), *Particulate Products: Tailoring Properties for Optimal Performance*, Particle Technology Series 19,
DOI 10.1007/978-3-319-00714-4_6, © Springer International Publishing Switzerland 2014

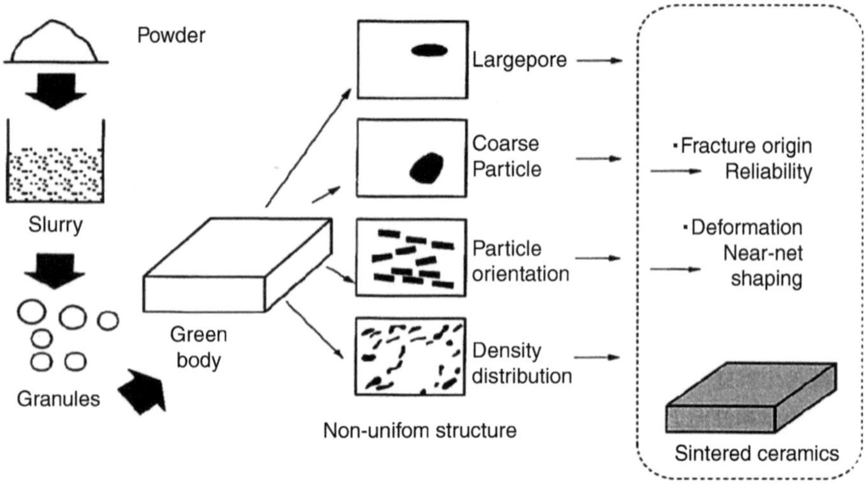

Fig. 6.1 Powder granule compaction process to make sintered ceramics

process has a major influence on the properties of the final product. This chapter addresses various aspects that are essential for manufacture of ceramic products of high quality, viz. details of the particle size distribution (PSD) and powder process conditions. The reason is that a few coarse particles as well as improper conditions in the manufacturing process may lead to significant decrease of ceramic strength. In addition, some new measurement approaches will be described that allow identification of weak spots in ceramic bodies.

Figure 6.1 shows a typical powder granule compaction process used for manufacturing [27] sintered ceramics. The typical process steps are: bringing the dry ceramic powder into an aqueous slurry, spray-drying this slurry into granules, pressing the granules into a green body having the required shape and sintering this shape at about 1,000–1,700 °C into the ceramic product. Note that a non-uniform structure, by presence of a few large pores or coarse particles in the green compact, may become the origin of fracture [3, 4], which affects the reliability of sintered ceramics. On the other hand, optimum particle packing resulting from adequate particle orientation or powder packing density distribution in the green compact, leads to good fracture strength and final shaping. Therefore, the manufacturing process of green compact before firing is crucial for producing high quality ceramics [34].

In this chapter, the manufacturing process of silicon nitride and alumina ceramics is used to elucidate the relationship between the powder processing conditions and the fracture origin in the ceramic products, which is the most basic factor for their quality. The effect of each processing condition of powder, powder slurry, powder granule and green body on the properties of sintered ceramics is discussed. Detailed characterization of powder raw materials, powder slurry, powder granules, green body, and sintered body is found necessary to understand and control ceramic processing conditions and is essential to open the black box of ceramic manufacturing processes using raw powder materials.

6.2 Selection of Raw Powder Materials and Their Grinding Conditions

Raw powder material is very important in ceramics for various aspects. It is one of the most decisive factors for the quality of a product [19]. The use of proper raw material is critical for producing ceramics of high quality at minimal costs. Raw powder materials of high quality are available in the market place. Proper selection of both powders and their processing is crucial for manufacturing high quality of ceramics.

General criteria for good raw powders are shown in Table 6.1 [35]. They may vary, however, with application and the technical level of processing. For example, fine powders are desirable for many applications, at least in principle, for their low sintering temperature, the resultant fine grains after sintering and high quality of products. In practice, however, their processing is often very difficult without high skill, and the quality of ceramics made from them may be lower than that made from coarse powders, if they are inadequately processed. For advanced ceramics, how to process fine powders is critical for high quality.

In the initial phase of the ceramics processing, dry powder is dispersed in water to form a slurry in the so-called grinding or dispersion step, in e.g. ball mills. Conventionally, PSD's of these slurries are characterized by gravitational sedimentation or laser diffraction method. However, these conventional characterization tools are insufficient to detect very small amounts of non-uniform components. For example, Fig. 6.2 shows the particle size distribution of ground silicon nitride powder by wet ball milling [21]. It was measured by X-ray

Table 6.1 Criteria for good raw ceramic powder

Chemistry	Composition	Homogeneous
	Impurities	None
	Stoichiometry	Stoichiometric
	Phase	Stable
Particles	Shape	Uniform and spherical for most applications
	Size range	0.1–30 μm (depending on application)

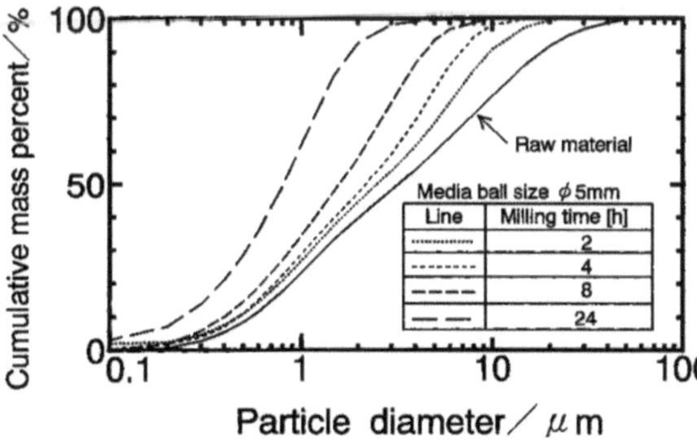

Fig. 6.2 PSD from X-ray sedimentation of Si_3N_4 powder ground with 5 mm media balls, at different milling times

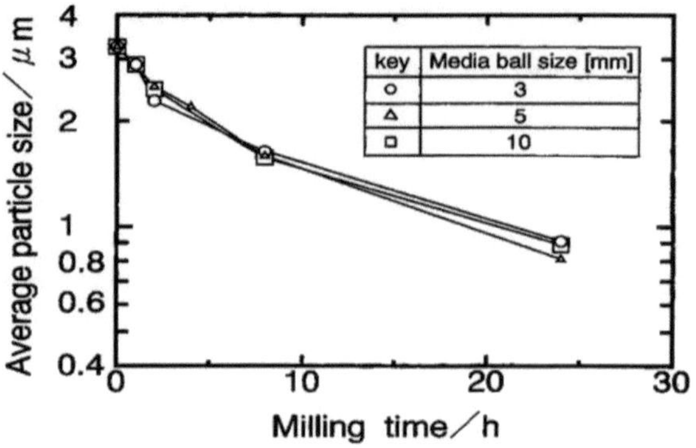

Fig. 6.3 Relationships between average particle size of ground powder and milling time

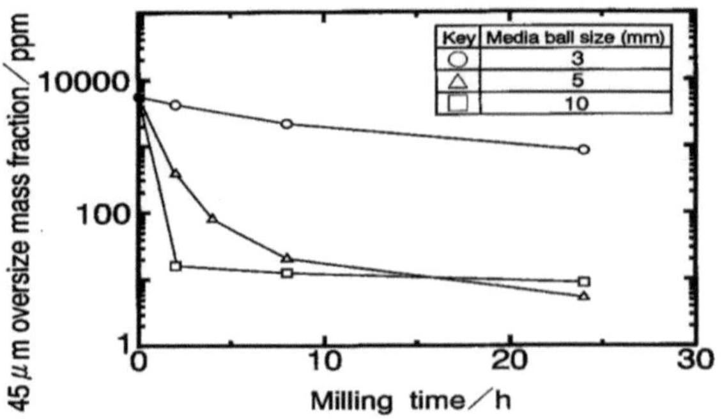

Fig. 6.4 Change of 45 μm oversize mass fraction of ground powder with milling time while using different media ball sizes, measured by wet sieving

sedimentation method [8, 22]. The obtained PSD indicates that the ground powder has no particles coarser than 45 μm. Furthermore, Fig. 6.3 shows the relationships between average particle size of ground powder and ball-milling time [21]. It shows that average particle size usually used for characterizing ceramic powders decreases with the milling time; and, media ball size has no effect in this case. However, the situation was quite different when using wet sieve analysis to examine the 45 μm oversize mass fraction of the ground powder with ball-milling time [21], as can be seen in Fig. 6.4. Different from Fig. 6.2, it shows that coarse particles larger than 45 μm are apparently contained in the ground powder, at a concentration level of about 10–1,000 ppm or about 30–3,000 particles per gram. Media ball size has effect on the change of its mass fraction in the ground powder with milling time. The 10 mm sized media ball is the most effective to grind coarse particles. The results indicate that the wet sieve analysis is a more reliable method to measure a few large particles in the ground powder. Apparently, these small amounts of 45^{+} μm Si_3N_4 particles cannot be detected by the sedimentation method used. There may be two reasons. One is that the sedimentation of these particles in water is very fast (within few seconds), so that they have left the detection zone almost right after starting the analysis. The second reason may be that 10–1,000 ppm of coarse particles is below the lower limit of the concentration measurement of this method.

From Fig. 6.4, we can identify the optimum grinding conditions to reach a minimum amount of coarse particles. As a result, a method to measure coarse particles in advanced ceramic powder was developed [7, 28], which has been filed as an international standard in advanced ceramics [18]. By relating the coarse particle information of raw powder or powder slurry to that of sintered ceramics, we can easily understand how small amounts of coarse particles lower the reliability of sintered ceramics. An example will be explained in Sect. 6.4.

6.3 Effect of Slurry Preparation Conditions on Powder Granule Properties

Slurry preparation conditions can directly affect the powder granule properties made by spray drying and may lead to non-uniform structure of green compact [5]. Figure 6.5 shows the fabrication process of silicon nitride ceramics by powder granule compaction [24].

Commercially available silicon nitride, alumina, and yttrium oxide powders were used as the starting materials. Average particle size of each powder measured by gravitational X-ray sedimentation was 0.44, 0.33, and 0.29 μm, respectively. Silicon nitride powder (270 g) was mixed with alumina (15 g) and yttrium oxide (15 g) by ball-milling with distilled and deionized water (155 g) for 24 h. Dispersant was not added because the pH of the slurry moved to the basic region (up to 10.5) during mixing, at which silicon nitride is deflocculated electrostatically, due to the reaction of silicon nitride and water. After passing through a sieve (32 μm) to remove undesired large agglomerates or coarse particles, the slurry was divided into two parts. One part was kept in the well-dispersed state at pH 10.5. Apparent viscosity of the slurry was measured with a viscometer to be 300 mPa.s at a shear rate of 60 min^{-1}. In the other part, the pH was lowered to 9.4, through slowly adding a dilute aqueous HNO_3 solution while continuous stirring at room temperature. Agglomeration occurred as indicated by the strong increase of the apparent viscosity to 6,500 mPa·s. Both slurries were then spray-dried to form powder granules.

Figure 6.6 shows micrographs of the granules prepared from the well-dispersed slurry and the flocculated slurry [24]. One picture was obtained with SEM, the other two were observed with optical microscope, while using an immersion liquid of about the same refractive index as the particles, which makes the particles transparent [36, 37]. In this latter method, the internal structure of the granules is clearly visible.

Most granules prepared from the well-dispersed slurry have distorted spherical shapes and contain dimples, which are visible in the SEM picture and as light spots in the micrograph taken in the optical transmission mode, as shown in Fig. 6.6a, b. In contrast, the granules prepared from flocculated slurry were essentially spherical and had no dimples, as shown in Fig. 6.6c. It indicates

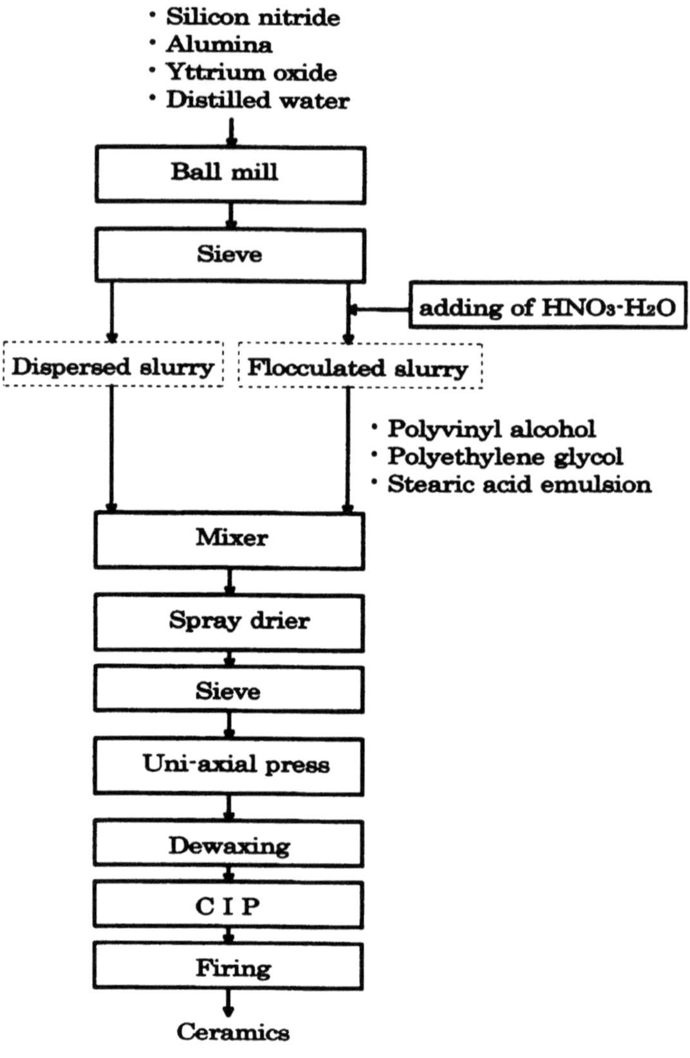

Fig. 6.5 Fabrication process of silicon nitride ceramics

that better characterization tools are necessary to observe possible non-uniform structure in granules.

Figure 6.7 shows the Weibull distribution curves of the flexural strength of the sintered Si_3N_4 bodies. (see Annex 6A) [24]. Average strength[1] of sintered body fabricated from flocculated slurry (717 MPa) was obviously higher than that from well-dispersed slurry (607 MPa), although the calculated Weibull moduli were high for both of them. The fractographical analysis for the fractured specimen

[1] In this chapter, average strength of sintered body means the arithmetic average of measured mechanical strength values.

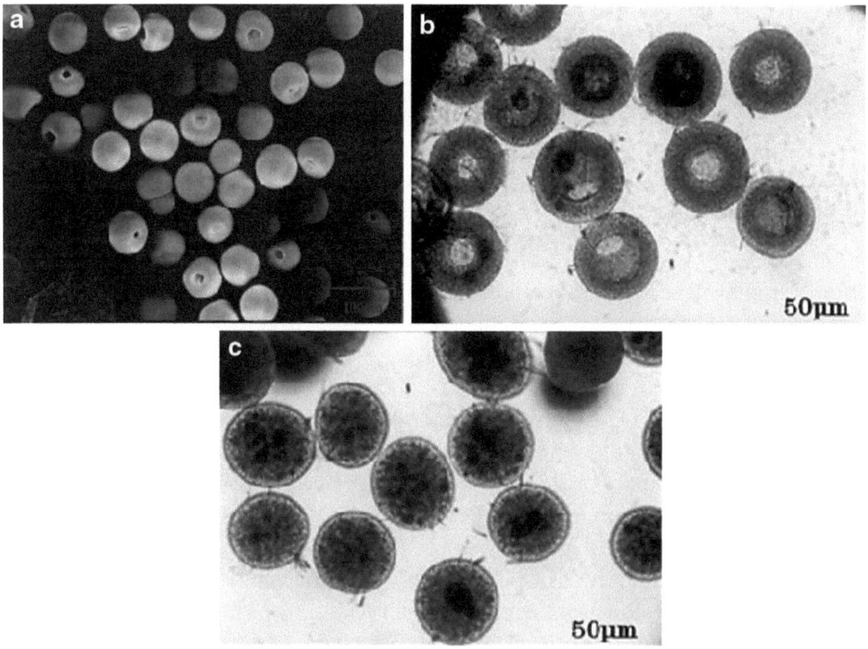

Fig. 6.6 Morphology of Si_3N_4 granules observed by SEM and liquid immersion method; (**a**) SEM picture of granules prepared from well-dispersed slurry, (**b**) liquid immersion method, granules from well-dispersed slurry, (**c**) liquid immersion method, granules from flocculated slurry

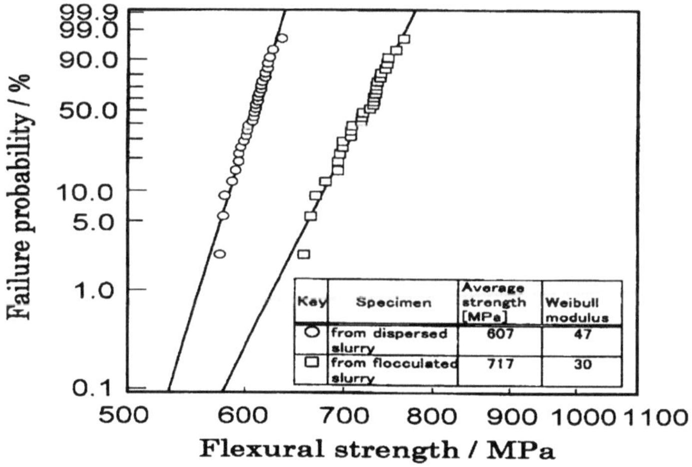

Fig. 6.7 Weibull distributions of the fracture strength measured for sintered Si_3N_4 specimens

clearly indicates that large pore defects were responsible for fracture [24]. Such larger pore defects in the ceramics originated from the shape and the internal structure of particles in the granules. In this case, large pores starting at the centers of granules originated from dimples through powder packing during

Table 6.2 Slurry preparation conditions and properties of the sintered alumina ceramics

No.	pH	Dispersant amount [mass%]	Viscosity [mPa · s]	Density [kg/m^3]	Fracture toughness [MPam$^{1/2}$]
#1	10	0.2	43	3.91×10^3	3.7
#2	9.1	0.5	22	3.94×10^3	3.8
#3	8.1	2.0	54	3.89×10^3	3.8

compaction and those at the boundaries caused by incomplete adhesion between granules affect the fracture strength of ceramics. In the experiment, large pore size was apparently larger in the thinned specimen of the sintered ceramics made by well-dispersed slurry [10, 29].

Although the structure of powder granules is similar, the fracture strength of the sintered bodies may still be affected by other aspects of the slurry preparation conditions [1]. Alumina powder (AL160-SG4, Showadenko, Japan, average particle size: 0.46 μm) was mixed with 0.2 %, 0.5 %, 2.0 % w/w of polymer dispersant (ammonium polyacrylate, CELUNA-D305, Chukyoyushi, Japan) and distilled and deionized water for 24 h by ball milling. The solid concentration of the slurry was 35 % v/v. Table 6.2 indicates the preparation conditions of three slurries and their difference in apparent viscosity [1]. The pH of slurry with 0.2 % w/w dispersant was increased to 10 by adding dilute NH_4OH solution, which decreased its viscosity to 43 mPa·s. The slurries were spray-dried for granulation. The granules were uni-axially pressed at 9.8 MPa, and then isostatically pressed at 176 MPa. The green compacts were sintered at 1550 °C for 2 h in air. As indicated in Table 6.2, the three slurries showed low apparent viscosity. Their granules had distorted spherical shapes and contained dimples, which were clearly visible in the micrograph taken under the optical transmission mode, as shown in Fig. 6.6b. Also, no big difference was observed in size distribution of granules made from the three slurry preparing conditions, and their average sizes were all about 60 μm.

Table 6.2 presents the density and fracture toughness of the specimens. The values are almost the same for these ceramics. Figure 6.8 shows the strength distribution of the alumina ceramics [1]. A significant variation of strength associated with the slurry preparation condition was noted. The average strength is 486, 430 and 363 MPa, respectively, for specimens made from the three slurry preparing conditions.

Figure 6.9 shows a comparison of the compressive strength of the granules prepared from the three different slurries [1]. The strength of granules was measured with a micro compression test machine (MCTM-500, Shimadzu Co., Japan) and the compressive strength was calculated by applying the model proposed for the fracture of spherical rock particle [9, 23].

Figure 6.10 shows the transmission optical micrograph of the thinned ceramics specimens without any immersion liquid [1]. The dark features are large pores in the structures. Clearly, the present ceramics have defects at the center and boundaries of granules. They are developed from the irregularities of packing structure of powder particles in green compact. The pore shape was similar for

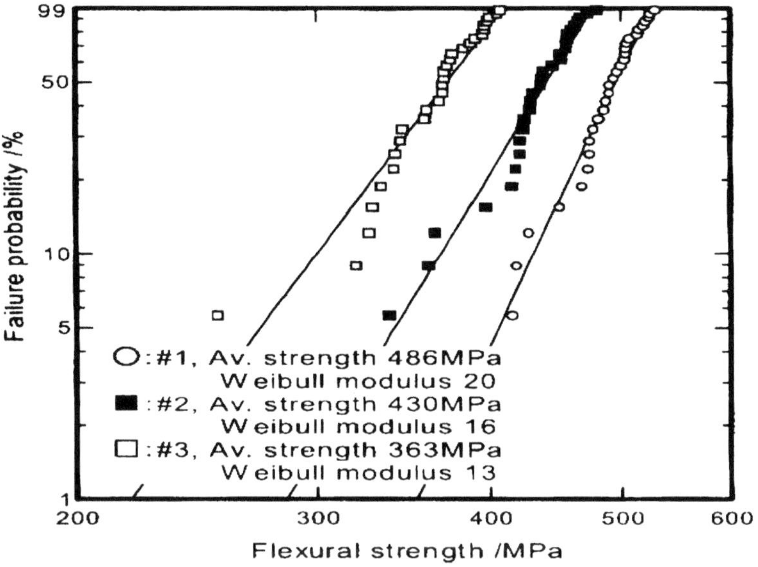

Fig. 6.8 Strength distribution of the alumina ceramics

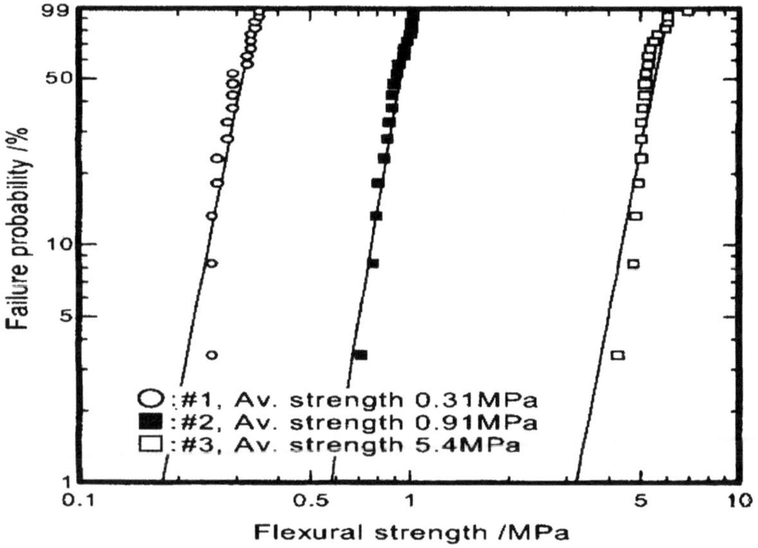

Fig. 6.9 Compressive strength distribution of alumina granules associated with different slurry preparation conditions

the three specimens, except for the size of the pores associated with the slurry preparing condition. The relation between the pore number density versus the pore size of the sintered specimens made from the three slurries was measured [1]. The pore number density was defined as the number of pore per unit volume of

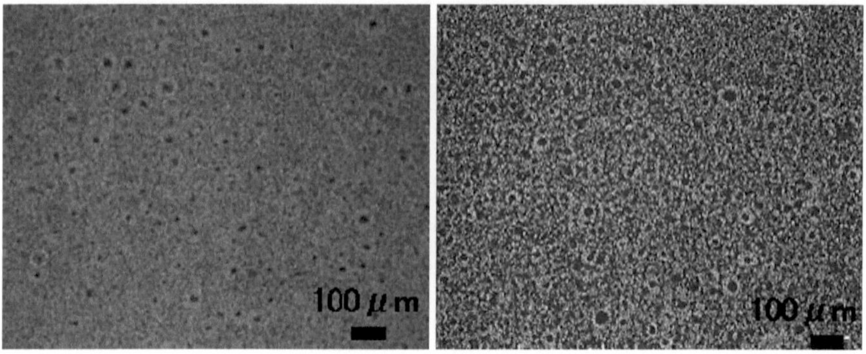

Fig. 6.10 Microstructures of alumina ceramics examined with optical transmission technique; *left*: specimen made from the slurry prepared with #1 condition *right*: specimen made from the slurry prepared with #3 condition

specimen per unit size interval. The effective volume of specimens under the analysis was about 0.5 mm^3. In this study, relatively large pores were analyzed. Pores were assumed to have a spherical shape and their sizes were represented by the equivalent diameter.

According to the fracture mechanics, the strength of ceramics σ, can be related to the size of fracture origin, c, by the following equation:

$$K_{IC} = \sigma Y c^{1/2} \tag{6.1}$$

Where K_{IC} is the fracture toughness and Y is a shape factor.

As expected from the equation, fracture strength of ceramics will depend on the size of fracture origin, provided that the fracture toughness and the shape factor are the same. Focusing on large pores, the size as shown in Fig. 6.10 (right) is 1.29 times larger than that in the specimen shown in Fig. 6.10 (left). Provided the shape factor is the same, the strength of the sintered sample made from the slurry of pH 10 is estimated 1.34 times higher than that made from the slurry of pH 8.1. This estimate is in good agreement with the measured average strength as shown in Fig. 6.8.

The change of the large pore size can be ascribed to the difference in the granule strength as shown in Fig. 6.9. In compaction of harder granules, less deformation can occur at a certain compaction pressure and this leaves larger pores in compacts. The size of large pore increases with the granule strength, which was strongly affected by the amounts of dispersant in slurry as shown in Fig. 6.9.

The granule strength is influenced by various factors, such as the powder packing structure as well as the amount and distribution of dispersant in granules [20]. In this study, the difference of the granule strength can be ascribed to the latter factor dominantly, since the granule structure was almost similar. Note that the amounts of dispersant added, 0.2, 0.5 and 2.0 % w/w, are insufficient, a little excess and excess, resp., for covering the powder surfaces, based on the relation between the amounts of dispersant and the slurry viscosity, respectively. With excessive dispersant, the non-adsorbed dispersant introduces free polymers in the slurry, which increases the apparent viscosity. In spray-drying, the free polymers form solid bridges between powder particles, increasing the granule strength considerably. Clearly, polymeric additives have critical effects on the cohesive force between particles and thus on the powder compaction process.

6.4 Effect of Coarse Particles in Powder Slurry on the Quality of Ceramics

Control of particle size distribution of ceramic powder is also important in slurry preparation. Powder is dispersed in water using mechanical method such as ball milling. A fundamental study was conducted to understand the effect of coarse particles on the fracture strength of ceramics. The specimens were prepared through the procedure as shown in Fig. 6.11 [13].

Low-soda alumina powder (AL-160SG-4, Showa Denko K.K., Japan) was used as raw material. The nominal average particle size was 0.5 μm. The powder was placed in alumina ball mill container (SSA-999, Nikkato, Japan, volume 2000 ml) with 2 kg of alumina balls (SSA-999, Nikkato; diameter 5 mm), and 400 g of aqueous solution (2 % w/w) of dispersant of polyacrylic acid type (CERUNA D-305, Chukyo Yushi, Japan) and mixed for 24 h to make a slurry with 50 % v/v solid content. The slurry was passed through a mesh (2 mm opening) to separate the balls. Weighed slurry was placed in a container and stirred continuously while a small amount of coarse particles (0.01–0.1 % w/w) was added. These coarse particles were prepared from the unground raw material used in the production of the present fine alumina powder. They were classified into three fractions by sieving before being used. A single fraction of coarse particles was added to individual slurry; therefore, three kinds of slurries with coarse particles were prepared. Each slurry was kept stirring for 2 h after the coarse particles were added. Finally each slurry, was cast in gypsum molds (100 × 100 × 9 mm) to prepare green compacts. After drying, the compact was heated at 1,550 °C for 2 h in an electric furnace to sinter the model ceramics [13].

At high magnification by SEM, it was found that the platelet-shape particles formed aggregates of porous structure with the size of 10–20 μm [13]. At lower magnification, these aggregates formed the coarse aggregates in large scale. In this experiment, such coarse aggregates are referred to as coarse particles. Table 6.3 shows the measured densities of green compact and ceramics [13]. The densities were approximately the same for all green bodies and ceramics. Clearly, addition of a small amount of coarse particles has no effect on the densities of ceramics.

Figure 6.12 shows the Weibull plots and the fracture toughness for all specimens [13]. The specimen strength decreased with increasing size of coarse particles added. The Weibull moduli were similar and over 20 for all specimens. All

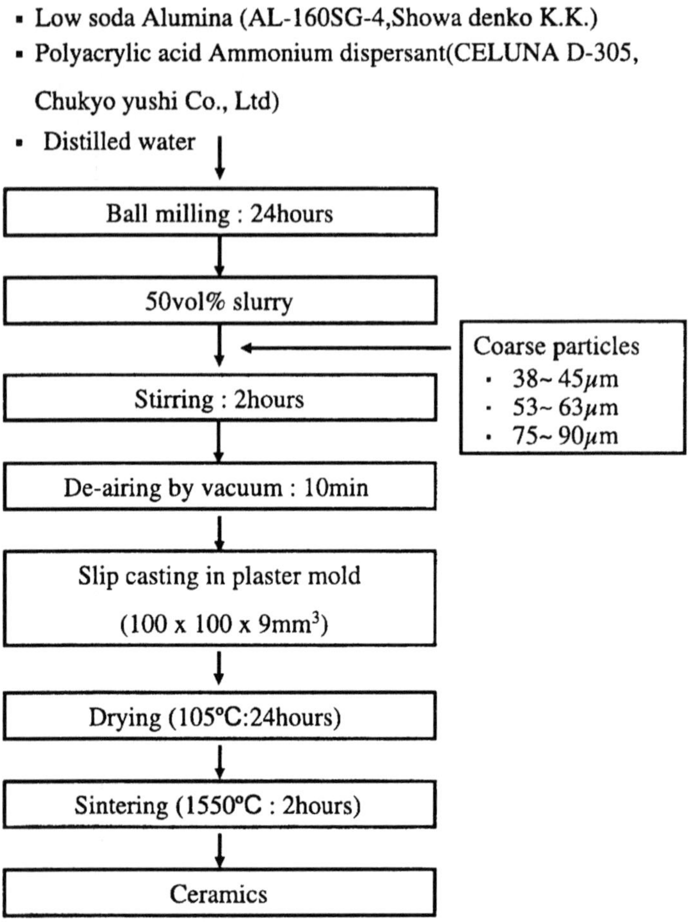

Fig. 6.11 Fabrication process of alumina ceramics

Table 6.3 Densities of green and sintered alumina bodies

	Density/10^3 kg m^{-3}	
Coarse particles size, sieve apertures/μm	Green body	Sintered body
38–45	2.31	3.95
53–63	2.32	3.95
75–90	2.31	3.96

ceramics basically have the same fracture toughness. Figure 6.13 shows SEM micrographs of representative fracture origins found in this study [13]. The specimen contains coarse particles of the size range 75–90 μm. The fracture origin was noted in the specimen of the lowest strength (370 MPa) as seen in Fig. 6.13 (left); Fig. 6.13 (right) shows the origin in the specimen at having a strength of 406 MPa.

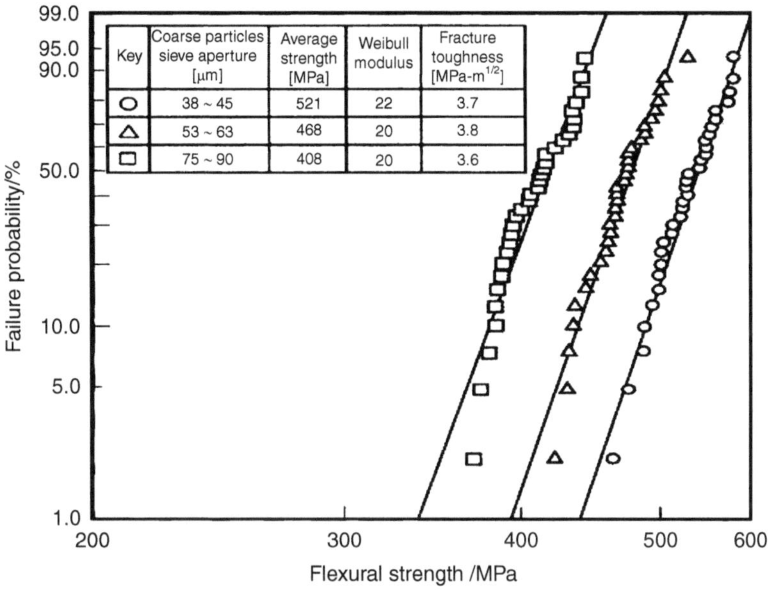

Fig. 6.12 Mechanical properties of sintered alumina bodies

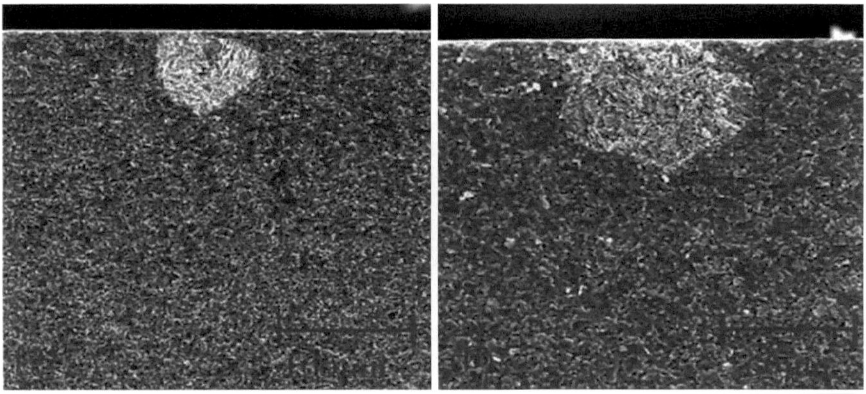

Fig. 6.13 Examples of fracture origins in alumina ceramics containing; 75 ~ 90 μm coarse particles. (*left*) 370 MPa, (*right*) 406 MPa

In both cases, fracture originated at the coarse particles. Lower strength was observed for the specimen containing larger coarse particles.

Figure 6.14 shows IR photomicrographs [38] for the internal structures of ceramics containing coarse particles of various sizes [13]. In the micrographs, the dark spots show coarse particles. The size of coarse particles in the ceramic matrix increased with increasing size of coarse particles added. Again, the sizes of coarse particles are the same as those added in the preparation of specimens [13]. Actually,

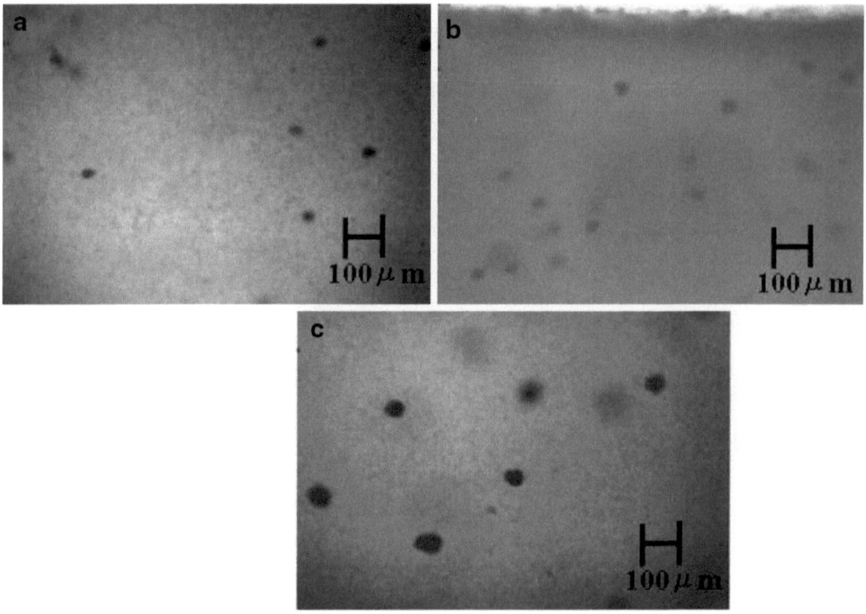

Fig. 6.14 Mid-infrared micrographs of sintered bodies with coarse particles of (**a**) 38 ~ 45 μm, (**b**) 53 ~ 63 μm, (**c**) 75 ~ 90 μm

the estimated values of strength obtained by linear fracture mechanics, assuming that the fracture was always initiated at the coarse particles in the matrix, agrees very well with the measured strength [13]. It means that small amounts of coarse particles decrease the strength of ceramics. Therefore, careful prevention of presence of coarse particles is crucial to maximize the strength of high quality ceramics [14].

6.5 Influence of the Compaction Process on the Quality of Ceramics

Compaction process including uniaxial press and Cold Isostatic Pressing (CIP) affects the properties of green body through powder granule compaction route and thus affects the reliability of sintered ceramics. In this Section, the origin of the strength variation associated with the change of CIP condition in a powder granule compaction processing of silicon nitride ceramics is firstly explained.

The manufacturing process for the experiments is presented in Fig. 6.5. In this experiment, dispersed slurry was applied to prepare powder granule, therefore, the obtained granules have distorted spherical shapes and contain dimples, which are also clearly visible in the micrograph taken in the optical transmission mode, as shown in Fig. 6.6b. All granules were 150 μm-sieved to remove large aggregates and foreign objects. The average granule size was 55 μm. Then, they were

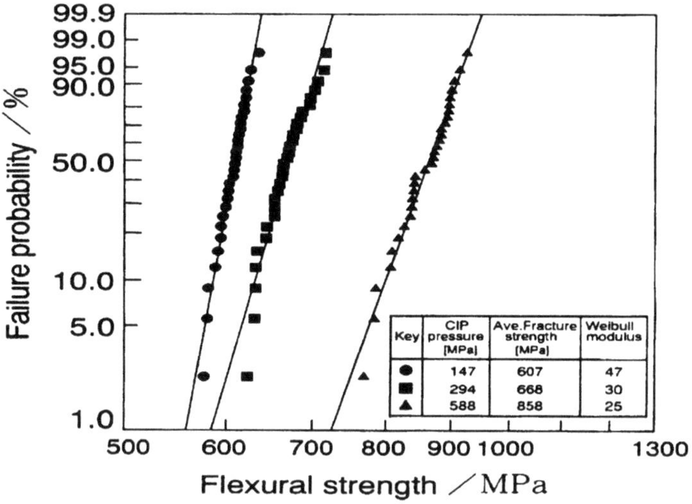

Fig. 6.15 Influence of CIP condition on strength distribution of silicon nitride ceramics

uni-axially pressed in a square die (60 × 50 mm) at an applied pressure of 19.6 MPa. After dewaxing at 530 °C for 15 h in air, the mold-pressed green bodies were cold isostatically pressed (CIP) at 147, 294 and 588 MPa. The green bodies were embedded in boron nitride powders with a silicon nitride container and sintered at 1700 °C for 3 h under nitrogen atmosphere.

Figure 6.15 shows the strength distribution of the ceramics sintered from the three green bodies [2]. The average strength increases with increasing CIP. Their values are 607, 668 and 858 MPa for specimens sintered from green bodies with CIP pressures of 147, 294 and 588 MPa, respectively. The Weibull modulus exceeding 20 was noted for all specimens. The corresponding densities and the values of fracture toughness were essentially the same.

The direct observation methods were successfully applied to examine the origin of strength variation. All relevant structures in the processing, such as the powder granule, their green bodies and sintered bodies were examined by the transmission mode of an optical microscope.

These examinations directly showed the relation between the CIP conditions and the formation of large defects. With increasing CIP pressure, the granules were deformed, packed more tightly and isostatically. The large pores originated from the boundaries of granules even in the green body prepared at high CIP (588 MPa). The variation of fracture strength was quantitatively correlated to the internal large pore size and concentration in sintered bodies [2]. The same tendency was also reported for the sintered ceramics made by flocculated slurry [26] and the manufacturing process of AlN ceramics [6].

Granules often retain their shape in green bodies during consolidation. Intergranular pores and flaws caused by the incomplete deformation and fracture of granules are usually large and often exert a serious influence on the properties of

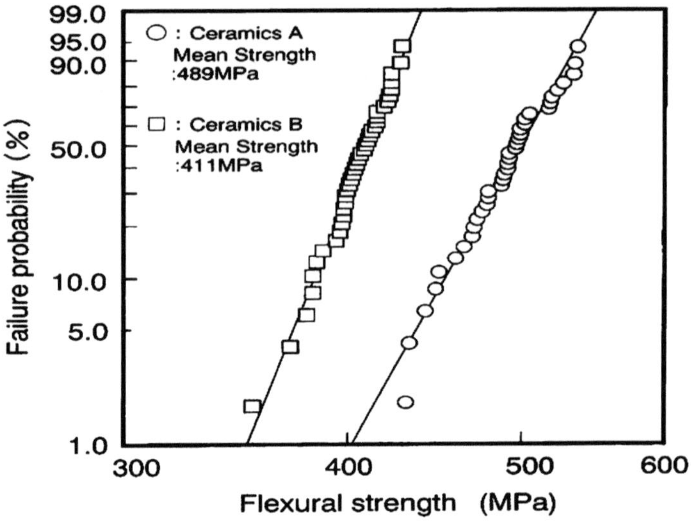

Fig. 6.16 Weibull distribution curves of flexural strength measured for sintered specimens A and B

sintered ceramics. It means that the control of granule properties for high quality green body is key issue for the fabrication of highly reliable ceramics [11, 25, 30, 31]. To examine the effect, the granule strength was changed. In this experiment, alumina granules were prepared by spray-drying, and divided into two groups. One group (granule A) was heat-treated at 500 °C for 5 h to remove added binders. The other group (granule B) was used for testing without prior treatment [33].

Figure 6.16 shows the Weibull distribution curves of flexural strength measured for sintered samples in both groups [33]. The variations shown in fracture strength indicate the marked influence of heat treatment of the granules on the properties of the sintered bodies. The flexural strength of the sintered body A (average 489 MPa) was clearly higher than that of the sintered body B (411 MPa). The Weibull modulus was high for both samples, suggesting that the fracture origins were uniformly large.

The hard and brittle characteristics of the heat-treated granules contributed to achieving a uniform packing structure in the green body, because the pores between granules were efficiently filled with primary powder particles caused by fracture of the granules. On the other hand, the 'as-sprayed' granules preserved more clear interfaces between granules and internal pores in the green compacts, resulting in the development of large pores, and thus, a decrease in fracture strength for the sintered body [12, 33].

The above results suggest that the alteration of CIP and de-waxing procedures affects the pore structure and the fracture strength of sintered ceramics. To examine the effect, the alumina powder granules were uni-axially mold pressed into compacts at 10 MPa with a metal mold. The compacts were then divided into two groups. One was CIP-ed subsequently followed, after mold pressing, and de-waxed

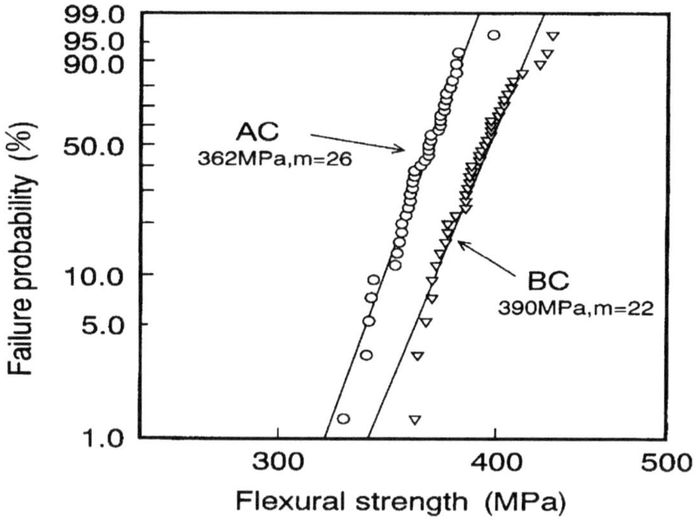

Fig. 6.17 Weibull distributions of the flexural strength measured for the specimens made through AC and BC processes

by heat-treating at 500 °C for 5 h in the electric furnace (referred to as AC). The other was first de-waxed at 500 °C, then CIP-ed for further compaction (referred to as BC). CIP-ing was carried out at 176 MPa for green bodies in both groups and sintered at 1590 °C for 2 h in the electric furnace [32].

Figure 6.17 shows the Weibull distributions of the flexural strength measured for the specimens in both groups [32]. Flexural strength of the BC specimens was clearly higher than that of AC. Weibull modulus was high for both groups, suggesting that the size of fracture origins was rather uniformly large. De-waxing before CIP with the heat-treatment of powder compacts was effective to make the pore defect size in green and sintered bodies small and to improve the fracture strength of sintered bodies. Reduction of pore size in green compact was ascribed to the promotion of rather uniform powder packing for heat-treated and de-waxed granules. This was because the hard and brittle characteristics of the heat-treated granules contributed to achieving a uniform packing structure in the green body [33].

6.6 Conclusions

In this chapter, the effect of powder processing conditions on the quality of ceramics was clarified. Manufacturing processes of silicon nitride ceramics and alumina ceramics were utilized to explain the relationship between ceramics processing conditions and its major fracture origin. Powder slurry preparation conditions were found to potentially create the large pores in powder granules, green body and sintered ceramics. These phenomena can be investigated by using new characterization tools such as the liquid immersion method and the microscopic technique observing thinned ceramics specimen under the transmission mode. Also, very small amounts of coarse particles contained in the powder slurry are weakening the strength of ceramics. This has been found by using wet sieve analysis and the observation technique with thin ceramic specimen. As a result, it is believed that major fracture origin is caused by the large pores and coarse particles, which are generated in the manufacturing steps of ceramics, such as powder slurry preparation, spray drying, and shaping process of green bodies. By making use of these characterization tools, effective processing conditions to eliminate the large pores and coarse particles can be identified for producing high quality ceramics.

6.7 Definitions, Abbreviations and Symbols

Average strength	Arithmetic average of measured mechanical strength values of sintered body.
Flexural strength	Strength according to 4 point bending test method [15]
Fracture toughness	Toughness according to single edge pre-cracked beam (SEPB) method [16]
Weibull distribution	2-parameter model distribution (see Annex 6A)
Weibull modulus	Spread parameter of Weibull function
AlN	Aluminium nitride
CIP	Cold isostatic pressing
HIP	Hot isostatic pressing
IR	Infrared
ISO	International Standards Organization
PSD	Particle size distribution
SEM	Scanning electron microscopy
C	Size of fracture origin (Eq. 6.1)
K_{IC}	Fracture toughness (Eq. 6.1)
Y	Shape factor (Eq. 6.1)

Annex 6A

Weibull Distribution for Ceramic Strength

The Weibull function is a continuous probability distribution function that can be used for many purposes. In materials science, it is often used to present the distribution of life times of objects. For ceramics and other construction materials it is applied to express the results of mechanical strength measurements [17]. Under the name Rosin-Rammler distribution it is one of the 2-parameter model functions used for description of particle size distributions.

In its cumulative form, the function is:

$$F(x) = 1 - e^{-(x/\lambda)^k} \tag{6.2}$$

where:
x = random variable (here mechanical strength, e.g. flexural strength)
λ = scale or location parameter of the distribution
k = shape or spread parameter of the distribution (Weibull modulus in case of strength)

The function can be linearized via:

$$-\ln(1 - F(x)) = (x/\lambda)^k \tag{6.3}$$

to:

$$\ln(-\ln(1 - \mathrm{F}(x))) = k(\ln x) - k(\ln\lambda) \tag{6.4}$$

Thus, by fitting the measured data to the model, or plotting $\ln(-\ln(1{-}\mathrm{F}(x)))$ on the Y-axis against ($\ln x$) on the X-axis of graph paper, the two parameters of the distribution can be easily calculated.

References

1. Abe, H., Hotta, T., Kuroyama, T., Yasutomi, Y., Naito, M., Kamiya, H., Uematsu, K.: Variation of microstrucuture and fracture strength of alumina ceramics made from different slurry preparing condition. Ceram. Process. Sci. Ceram. Trans. **112**, 809–814 (2001); Am. Ceram. Soc
2. Abe, H., Hotta, T., Naito, M., Shinohara, N., Okumiya, M., Kamiya, H., Uematsu, K.: Origin of strength variation of silicon nitride ceramics with CIP condition in a powder compaction process. Powder Technol. **119**, 194–200 (2001)
3. Abe, H., Hotta, T., Naito, M., Shinohara, N., Uematsu, K.: Direct observation of detrimental defects in ceramics. Am. Ceram. Soc. Bull. **81**, 31–34 (2002)
4. Abe, H., Naito, M., Hotta, T., Shinohara, N., Uematsu, K.: Flaw size distribution in high-quality alumina. J. Am. Ceram. Soc. **86**, 1019–1021 (2003)
5. Abe, H., Naito, M., Hotta, T., Kamiya, H., Uematsu, K.: Pore defects related to slurry character and their relevance to strength distribution in alumina cermics. Powder Technnol. **134**, 58–64 (2003)
6. Abe, H., Sato, K., Naito, M., Nogi, K., Hotta, T., Tatami, J., Komeya, K.: Effects of granule compaction procedures on defect structure, fracture strength and thermal conductivity of AlN ceramics. Powder Technol. **159**, 155–160 (2005)
7. Cho, Y.-I., Okumiya, M., Naito, M., Rabe, T., Waesche, R., Morrell, R., Ewsuk, K.G., Hackley, V., Freiman, S., Uematsu, K.: Characterization of coarse particles contained in fine powders at very low concentrations. Ceram. Trans. **133**, 71–76 (2002). Am. Ceram. Soc

8. Hayakawa, O., Nakahira, K., Naito, M., Tsubaki, J.: Experimental analysis on sample preparation conditions for particle size measurement. Powder Technnol. **100**, 61–68 (1998)
9. Hiramatsu, Y., Oka, Y., Kiyama, H.: Rapid determination of tensile strength of rocks with irregular test pieces. Nihon kogyokai-shi (Jap.) **81**, 1024 (1965)
10. Hotta, T., Nakahira, K., Naito, M., Shinohara, N., Okumiya, M., Uematsu, K.: Origin of strength change in ceramics associated with the alteration of spray dryer. J. Mater. Res. **14**, 2974–2979 (1999)
11. Hotta, T., Nakahira, K., Naito, M., Shinohara, N., Okumiya, M., Uematsu, K.: Origin of the strength change of silicon nitride ceramics with the alteration of spray drying conditions. J. Eur. Ceram. Soc. **21**, 603–610 (2001)
12. Hotta, T., Abe, H., Fukui, T., Naito, M., Shinohara, N., Uematsu, K.: Effect of dewaxing procedures of cold isostatically pressed silicon nitiride ceramics on its microstructure and fracture strength. Adv. Powder Technol. **14**, 505–517 (2003)
13. Hotta, T., Abe, H., Naito, M., Takahashi, M., Uematsu, K., Kato, Z.: Effect of coarse particles on the strength of alumina made by slip casting. Powder Technol. **149**, 106–111 (2005)
14. ISO 13383–1.: Fine ceramics – determination of grain size distribution in ceramic powders by image analysis of photomicrographs (2012)
15. ISO 14704.: Fine ceramics – test method for flexural strength (2008)
16. ISO 15732.: Fine ceramics – test method for fracture toughness by single edge pre-cracked beam (SEPB) method (2003)
17. ISO 20501.: Fine ceramics – Weibull statistics for strength data (2003)
18. ISO 24369.: Fine ceramics – determination of coarse particles in ceramic powders by wet sieving method (2005)
19. Lange, L.L.: Powder processing science and technology for increased reliabilty. J. Am. Ceram. Soc. **72**, 3–15 (1989).
20. Naito, M., Fukuda, Y., Yoshikawa, N., Kamiya, H., Tsubaki, J.: Optimization of suspension characteristics for shaping processes. J. Eur. Ceram. Soc. **17**, 251–257 (1997)
21. Naito, M., Hotta, T., Hayakawa, O., Shinohara, N., Uematsu, K.: Ball milling conditions of a very small amount of large particles in silicon nitride powder. J. Ceram. Soc. Jpn (Jap.) **106**, 811–814 (1998)
22. Naito, M., Hayakawa, O., Nakahira, K., Mori, H., Tsubaki, J.: Effect of particle shape on the particle size distribution measured with the commercial equipment. Powder Technol. **100**, 52–60 (1998)
23. Naito, M., Nakahira, K., Hotta, T., Ito, A., Yokoyama, T., Kamiya, H.: Microscopic analysis on the consolidation process of granule beds. Powder Technol. **95**, 214–219 (1998)
24. Naito, M., Hotta, T., Abe, H., Shinohara, N., Okumiya, M., Uematsu, K.: Optical characterization of strength-limiting flaws in silicon nitride ceramics prepared with different slurry flocculation conditions. Proc. Br. Ceram. Soc. **61**, 119–132 (2000)
25. Naito, M., Hotta, T., Abe, H., Shinohara, N., Cho, Y.-I., Okumiya, M., Uematsu, K.: Relevance of the fracture strength to process-related defects in alumina ceramics. Mater. Trans. Jpn Inst. Met. **42**, 114–119 (2001)
26. Naito, M., Abe, H., Hotta, T., Shinohara, N., Uematsu, K.: Fracture strength variability of silicon nitride ceramics made by powder compaction. Ceram. Trans. **133**, 151–157 (2002). Am. Ceram. Soc
27. Naito, M., Abe, H.: The micromeritics. Funsai (Jap.) **48**, 56–62 (2004)
28. Nakahira, K., Hotta, T., Naito, M., Shinohara, N., Cho, Y.-I., Katori, S., Emoto, H., Yamada, T., Takahashi, T., Okumiya, M., Kumagai, C., Uematsu, K.: Characterization of coarse particles in alumina powders by wet sieving method. J. Eur. Ceram. Soc. **23**, 1661–1666 (2003)
29. Shinohara, N., Okumiya, M., Hotta, T., Nakahira, K., Naito, M., Uematsu, K.: Formation mechanisms of processing defects and their relevance to the strength in alumina ceramics made by powder compaction process. J. Mater. Sci. **34**, 4271–4277 (1999)

30. Shinohara, N., Okumiya, M., Hotta, T., Nakahira, K., Naito, M., Uematsu, K.: Effect of seasons on density, strength of alumina. Am. Ceram. Soc. Bull. **78**, 81–84 (1999)
31. Shinohara, N., Okumiya, M., Hotta, T., Nakahira, K., Naito, M., Uematsu, K.: Seasonal variation of microstructure and sintered strength of dry-pressed alumina. J. Am. Ceram. Soc. **82**, 3441–3446 (1999)
32. Shinohara, N., Okumiya, M., Hotta, T., Nakahira, K., Naito, M., Uematsu, K.: Variation of the microstrucuture and fracture strength of cold isostatically pressed alumina ceramics with the alteration of Dewaxing procedures. J. Eur. Ceram. Soc. **20**, 843–849 (2000)
33. Shinohara, N., Katori, S., Okumiya, M., Hotta, T., Nakahira, K., Naito, M., Cho, Y.-I., Uematsu, K.: Effect of heat treatment of alumina granules on the compaction behavior and properties of green and sintered bodies. J. Eur. Ceram. Soc. **22**, 2841–2848 (2002)
34. Shinohara, N., Ohsaka, S., Hotta, T., Abe, H., Naito, M., Uematsu, K.: Relevance of the internal structures to fracture strength of injection molded and sintered silicon nitride ceramics. Adv. Powder Technol. **16**, 425–434 (2005)
35. Somiya, S.,et al.: Handbook of Advanced Ceramics, Vol.1 Materials Science, p. 82. Elsevier/Academic Press, UK (2003)
36. Uematsu, K., Kim, J.-Y., Miyashita, M., Uchida, N., Saito, K.: Direct observation of internal structure in spray-dried alumina granules. J. Am. Ceram. Soc. **73**, 2555–2557 (1990)
37. Uematsu, K., Zhang, Y., Uchida, N., Hotta, T., Naito, M., Shinohara, N., Okumiya, M.: Structural flaw evaluation in green compacts and ceramics by an optical method – for understanding the origin of strength variation in ceramics. Key Eng. Mat. **161–163**, 145–150 (1999)
38. Uematsu, K., Uchida, N., Kato, Z., Tanaka, S., Hotta, T., Naito, M.: Infrared microscopy for examination of strucuture in spray-dried granules and compacts. J. Am. Ceram. Soc. **84**, 254–256 (2001)

Chapter 7
Concrete, from a Centuries-Old Construction Material to Modern Particle-Based Composite Concepts

Piet Stroeven and Huan He

Abstract Chapter 7 is focusing on the material of prime engineering relevance, concrete. The major part of the world's infrastructure is realized in concrete. Moreover, in many high-rise buildings concrete has been employed. The relatively low production costs render possible using high safety factors. As a result, the composition of concrete is governed traditionally by empirical laws. This has been laid down in a large number of practical building codes that confront the structural engineer with selection procedures more than with design requirements on the basis of particulate and hydraulic characteristics of the material. In such procedures, the aggregate's sieve curve is required to fall within defined, relatively wide boundaries. The selection of the water to cement ratio (w/c) is also part of such procedures. This factor significantly influences the material's strength and durability capabilities as well as (partly) provides for proper consistence, involving among other things the capacity to readily fill up the mould during compaction by vibration and enrobe the steel reinforcement completely without a loss of homogeneity. As a consequence, the building code includes requirements for the fresh state in terms of specified slump and flow classes.

Economic and technical demands of relatively recent date have led to material developments that required better insight into the engineering mechanisms of material behavior. This has stimulated extensive experimental and computer simulation research efforts. Concrete is a so-called particulate material on different levels of its microstructure. On meso-level, densely packed aggregate particles build up a skeleton of the material. This skeleton provides concrete with its load-bearing capacity in compression, since aggregate grains are generally relatively

P. Stroeven (✉)
Faculty of Civil Engineering and Geosciences, Delft University of Technology,
Stevinweg 1, 2628 CN, Delft, The Netherlands
e-mail: P.Stroeven@tudelft.nl

H. He
College of Architecture and Civil Engineering, Beijing University of Technology,
Pingleyuan 100, Chaoyang District, 100124, Beijing, P.R. CHINA
e-mail: Huan.He@hotmail.com

H.G. Merkus and G.M.H. Meesters (eds.), *Particulate Products: Tailoring Properties for Optimal Performance*, Particle Technology Series 19,
DOI 10.1007/978-3-319-00714-4_7,

strong and stiff. On micro-level, the aggregate structure is stabilized by the hydraulic cementitious particulate binder that hardens after mixing with water. Tensile stresses are conventionally taken by steel reinforcement, not discussed herein.

The strength and durability of the matured concrete depends upon packing density on both levels of the microstructure and on the complicated details of the (cement) chemistry involved. Concrete performance is at least partly relying on packing phenomena although not explicitly incorporated into the selection procedures of the building code. However, such phenomena play a significant role in research aiming at the development of new categories of concrete with specifically upgraded performance. As an integral part of present day concrete technology, particle packing therefore forms the hard-core of this chapter. Packing on the two levels of the microstructure together with the chemistry involved in the maturation of the binder leads to a process of pore de-percolation. Durability of the material is depending of course on the resulting complex spatial structure of highly tortuous and partly continuous pores so this issue is also discussed in this chapter. Again, traditionally, durability research is based on experimental approaches, whereby short term tests are employed, requiring extrapolation of the results.

7.1 Introduction

Concrete is a *particulate cementitious material*. Mortars and concrete are the fundamental components of much of the construction industry in the world. For daily definition purposes, a concrete is a cementitious material containing aggregates of sizes greater than 5 mm, whereas a mortar is a cementitious material containing aggregates of smaller sizes.[1] The term "mortar" comes from the Latin *mortarium*, the name given to the trough in which the material was mixed, as in "mortar and pestle". In fact, the Romans had a rudimentary but highly effective concrete made of a volcanic ash sourced from Pozzuoli on the slopes of Mt. Vesuvius and slaked lime, hence the term Pozzolanic cement. Cement technology goes back even further, to the Ancient Greeks and Mesopotamian civilizations preceding them, but it was the Romans that developed concretes sufficiently to construct concrete-filled walls, known as "*opus caementicium*", cementitious terracotta flooring, "*opus signinum*" and even used their concretes in building the famous Pantheon in Rome, which still stands today, a testament to their skills. However, it was the patenting of Portland cement by Leeds-based Joseph Aspdin in 1824 that led to the development of the reliable, high-strength concretes and mortars that are used throughout the world today. Named Portland cement because of its similarity in appearance to the highly-admired building stone from Portland Bill in Dorset, it is manufactured by heating slurry of limestone or chalk with clay in a kiln, and grinding the resultant clinker to a fine powder and adding gypsum. Development of building structures in concrete in the Netherlands dates back to around 1900; first standards for regulating all aspects of this industry are from 1912.

[1] BS EN-13139 takes 8 mm as maximum, ASTM takes sieve number 4 (4.76 mm) as maximum.

Regularly, these standards were upgraded, whereby the last decade the standards in the European Union were attuned to each other.

The particulate nature of concrete and mortars extends to different levels of the microstructure [18, 44, 51, 53]. Concrete is made up of a hydraulic matrix and (mostly) hard inclusions. Aggregate occupies at least three-quarters of the volume of concrete, so dominantly governs properties in the fresh and in the hardened state, as well as material costs [15, 64]. The aggregate builds up the load-bearing skeleton in concrete engineering structures subjected to compression; most of the tension is taken up by the reinforcement of steel bars and fibres (see Sect. 7.3.4). However, it also reduces global shrinkage deformations and size of the cracks developed due to shrinkage and differential settlements. Fine cement particles and water, together forming the paste, are added to the aggregate to give the fresh, green concrete mixture a proper workability[2] as well as to allow for the necessary chemical reactions later on for connecting the aggregate particles strongly together (see Sect. 7.2). This fresh concrete mixture is prepared on site or transported to the building side to be poured into the mould. Mostly it is thereupon compacted by vibration to fully fill the mould and enrobe the steel reinforcement bars. Vibration can be avoided when self-compacting concrete is employed (see Sect. 7.6.1).

Concrete is a relatively cheap building material. Hence, design can be crude. High safety factors are applied, leading to material capabilities significantly above expected working stresses. The economic pressure for doing research so that material behavior could be better exploited is as a consequence low. In fact, the conventional design practice is reduced to a selection procedure to fulfill the requirements for the desired construction (see Annex 7A). Empirical relationships form the basis of this procedure, which includes making choices of:

- The relevant exposure class in relation to corrosion risks by the environment;
- The intended life time of the construction;
- Compressive strength class in relation to aforementioned requirements;
- Consistence in terms of slump or flow classes of the fresh concrete in relation to desired workability. Note that higher slump and flow values (see Sect. 7.5) allow easier filling of moulds, though at the expense of concrete strength and durability. Therefore, chemical admixtures often replace part of the water required for adequate consistence.

The result of this procedure leads to a specification that forms the contract between the user and the producer of the concrete. Conformity testing is used to check whether the delivered concrete complies with the specification. Once accepted and used to realize the full-scale structure, testing is still relevant for checking the specification itself. During construction, this can involve testing of workability of the fresh material. Next, specimens are made for checking the grade. Moreover, cores can be drilled from the realized structure to test its ultimate quality.

Conventional concrete technology is based on empiric relationships between composition and behavior. In more recent times it was realized however that

[2] Workability is called *consistence* in the modern European standard for concrete, EN206-1: 2001.

better insight would be fundamental for systematic material improvement. Hence, materials science aspect became more apparent in experimental research efforts. This is a time-consuming, laborious and thus expensive approach, however. As a result, research has been undertaken during the past decades to correlate empirical findings with computer simulation data. Concrete is a particulate material on different levels of its structure. So, packing phenomena and porosity are the main issues in such studies, because packing density is directly related to strength, whereas porosity is related to its underlying durability properties.

Aggregate properties encompass grading, shape, surface texture, stiffness, volumetric density and toughness. For the concrete properties, the amount of aggregate is an important parameter. Aggregate is classified according to size as coarse (river gravel or crushed rock) or fine (sand). The sieve curve represents the volume-weighed size distribution of gravel and sand and of the aggregate mixture (the so called *grading*). The impact of the particle size distribution on packing density has been widely investigated [15]. Optimum and possible ranges of sieve curves have been recommended. Practical concepts have been incorporated in building codes. Considering it from engineering and economic viewpoints, concrete containing a larger *amount* of aggregate that is more densely packed is cheaper, has lower porosity and shrinkage, and will lead to better performance. Therefore, aggregate mixture design plays an important role in concrete technology of *advanced* cementitious composites.

Shape of aggregate has been recognized for a long time as an important factor [15, 17, 23, 37]. It has an enormous influence on workability of fresh concrete. Aggregate grains with angular shape (crushed rock versus aggregate of fluvial origin) reduce workability; higher friction between the paste and aggregate require paste content to be increased for maintaining proper workability conditions, at the expense of increased costs. Rougher aggregate *surface* can impart a better adhesion between aggregate and paste. So, mechanical interlock and increased friction at rougher surfaces can contribute to improving strength, even in flexure. Therefore, high performance concrete also requires careful selection of aggregate [1, 23, 37]. Because of inherent complications in defining shape and in its experimental assessment, an explicit and universal approach is still missing. The influence of particle shape on workability has been implemented in practical construction field by requiring "enough vibration at appropriate workability levels".

Packing also plays a role on micro-level, where cement blending has become a popular way to generate high performance cementitious materials. *Grading* has been demonstrated on this level to be of paramount importance for yielding optimum material performance. As an example, reference can be given to successful blending experiments in which a gap-graded inert mineral admixture was combined with Portland cement (PC) [14]. This similarly holds for concrete with finer fillers such as limestone powder, quartz flour, rice husk ash, or other inert fillers [6].

Particle packing and product porosity are closely related through a complicated, highly tortuous structure of partly connected pores after maturation [11, 42, 49]. Thus, this chapter will deal with these phenomena, particularly because of relatively recent advanced cementitious material developments, exploiting gained insight into aggregate and cement grain packing in the fresh state and its consequences for strength and durability. These advanced cementitious composites will be discussed later on. Of course, cementitious materials derive their capabilities from hydraulic properties. So, the next section will pay attention to elementary notions of cement chemistry.

7.2 Elementary Cement Chemistry

The clinker and the non-hydrated Portland cement are formed from limestone at high temperature (1,450 °C) in the cement kiln. In order to achieve the desired setting qualities in the finished product, a quantity (2–8 %, but typically 5 %) of calcium sulfate (usually gypsum or anhydrite) is added to the clinker and the mixture is finely ground to form the finished cement powder. This is achieved in a cement mill. The grinding process is controlled to obtain a powder with a broad particle size distribution, in which typically 15 % of the particles is smaller than 5 μm and 5 % exceeds 45 μm. Typically, the control is based on size distribution measurements by laser diffraction. Mostly, however, the fineness of the ultimate cement is specified by the specific surface area (Blaine number), which is the total particle surface area per unit mass. The rate of initial reaction (up to 24 h) of cement on addition of water is directly proportional to the specific surface area. Typical values are 320 ~ 380 m^2/kg for general-purpose cements and 450 ~ 650 m^2/kg for "rapid-hardening" cements.

Cement Chemist Notation (CCN) has been introduced to simplify the chemical formulas (Table 7.1). It is a "short-hand" way of writing the formulas of oxides of calcium, silicon and other elements.

The typical composition of Portland cement clinker is presented in Table 7.2. Hence, the four compounds referred to as C_3S, C_2S, C_3A and C_4AF are known as

Table 7.1 Typical oxides in Portland cement in CCN

Oxides in cement	CCN	Mass fraction (%)
Calcium oxide, CaO	C	61 ~ 67
Silicon dioxide, SiO_2	S	19 ~ 23
Aluminum oxide, Al_2O_3	A	2.5 ~ 6
Ferric oxide, Fe_2O_3	F	0 ~ 6
Sulfur trioxide, SO_3	$\overline{S}$	1.5 ~ 4.5
(Crystallization) water, H_2O	H	0.5 ~ 5.5

Table 7.2 Typical constituents of Portland clinker plus gypsum

Clinker, names and composition	CCN	Mass fraction (%)
Tricalcium silicate or Alite, $(CaO)_3 \cdot SiO_2$	C_3S	45 ~ 75
Dicalcium silicate or Belite, $(CaO)_2 \cdot SiO_2$	C_2S	7 ~ 32
Tricalcium aluminate, Aluminate or Celite, $(CaO)_3 \cdot Al_2O_3$	C_3A	0 ~ 13
Tetracalcium aluminoferrite or Ferrite, $(CaO)_4 \cdot Al_2O_3 \cdot Fe_2O_3$	C_4AF	0 ~ 18
Calcium sulfate or Gypsum, $CaSO_4 \cdot 2H_2O$	$C\overline{S}H_2$	2 ~ 10

Table 7.3 Typical composition of basic five Portland cements

Type	C_3S (%)	C_2S (%)	C_3A (%)	C_4AF (%)	MgO (%)	$\overline{S}$(%)	Ignition loss (%)	Free CaO (%)
I	55	19	10	7	2.8	2.9	1.0	1.0
II	51	24	6	11	2.9	2.5	0.8	1.0
III	57	19	10	7	3.0	3.1	0.9	1.3
IV	28	49	4	12	1.8	1.9	0.9	0.8
V	38	43	4	9	1.9	1.8	0.9	0.8

the main phases of Portland cement. The phase composition of particular cements can be quantified through a complex set of calculations known as the Bogue Formula.

The first two phases C_3S and C_2S are crystalline and directly react with water to form calcium silicate hydrate ($C_3S_2H_3$) and calcium hydroxide (CH) in the paste

$$2C_3S + 6H \rightarrow C_3S_2H_3 + 3CH \tag{7.1}$$

$$2C_2S + 4H \rightarrow C_3S_2H_3 + CH \tag{7.2}$$

The cement hydrates are primarily responsible for the strength of the concrete, in addition to the aggregates. CH forms crystals in the pores, thus decreasing porosity, but does not contribute to strength. The first reaction Eq. 7.1 is much faster than Eq. 7.2, so contributes disproportionately to early strength. In fact, hydration products formed in hardened cement pastes are very complicated, because many of these products have nearly the same formula and some are solid-solutions with overlapping formula. Different reactions produce different amounts of heat, so the various cements available on the market have different compositions to fulfill specific requirements, such as low-heat cement for voluminous concrete structures such as a dam.

There are five main types of Portland cements according to ASTM C150-2009 [74] (Table 7.3). The composition may vary within limits.

Type I is the most common used Portland cement, the so called general purpose cement. It is commonly used for general construction especially when making precast and precast-prestressed concrete that is not to be in contact with soils or ground water. C_3A content shall not exceed 15 %.

Type II is intended to have moderate sulfate resistance with or without moderate heat of hydration. This type is meant for general construction that is exposed to moderate sulfate attack, e.g. when the concrete is in contact with soils and ground water with high sulfur content. This type of cement costs about the same as Type I. In view of its properties, Type II Portland cement is the major general purpose cement in Northern America. C_3A content shall not exceed 8 %.

Type III has relatively high early strength. This cement is similar to Type I, but ground finer, typically to a specific surface 50–80 % higher than Type I. The gypsum level may also be increased somewhat. The increased surface area leads to faster hydration reactions. Thus, the concrete obtains a higher compressive strength in a much shorter time that types I and II, but somewhat at the expense of the long-term strength. It is usually used for precast concrete manufacture, where high 1-day strength of concrete allows fast turnover of molds. It may also be used in emergency construction and repairs and construction of machine bases and gate installations.

Type IV Portland cement is generally known for its low heat of hydration. The percentages of C_2S and C_4AF are relatively high and C_3S and C_3A are relatively low. Maximum content of C_3A is 7 % and the maximum content of C_3S 35 %. This causes the heat given off by the hydration reaction to develop at a slower rate. However, as a consequence the strength of the concrete develops slowly. After 1 or 2 years the strength is higher than the other types after full curing. This cement is used for very large concrete structures, such as dams, which have a low surface to volume ratio.

Note: Portland-pozzolan cements and ground granulated blast furnace slag cement additionally offer cheaper and more reliable alternatives and are widely used in Europe.

Type V is used where sulfate resistance is important. This cement has a very low C_3A content with a maximum of 5 %, which accounts for its high sulfate resistance. Also, $C_4AF + 2C_3A$ content cannot exceed 20 %. This type is used in concrete that is to be exposed to alkali soil and ground water sulfates, which react with C_3A causing disruptive expansion. It is unavailable in many places although its use is common in the western United States and Canada. As with Type IV, Type V Portland cement has mainly been supplanted by the use of ordinary cement with added ground granulated blast furnace slag or tertiary blended cements containing slag and fly ash.

Concrete today is no longer a 1:2:4 issue for the ratio of cement: fine aggregate (sand): coarse aggregate (gravel). Concrete has many chemicals added to it besides the basic ingredients, aggregate, cement and water. The chemicals are e.g. retarders to control the setting time, plasticizers for workability, slump retention admixtures for ready mixed concrete and mineral admixtures for durability, like micro silica, fly ash or slag. *Chemical and physical properties of ingredients decide the concrete properties, both in plastic and hardened form.*

7.3 Packing Phenomena

7.3.1 Introduction

Proper particle packing is essential for good concrete strength and durability. Thus, this phenomenon has been given serious attention, particularly in recent decades, both experimentally and analytically (e.g. [44]). Extensive information on packing is also available outside concrete technology. Of course, particle packing on meso- as well as micro-level dominates also the characteristics of the pore network structure formed during maturation of the concrete. Pores will reduce strength;

however, they also render possible the diffusion of water during drying and give access to harmful components from the outside during application in the matured state. The gradual de-percolation phenomenon in the hardening paste between the aggregate grains is a very complicated phenomenon. Hence, analytical and experimental ways to approach the problem can be extremely sophisticated.

With the rapid development of computer facilities, computer simulation of particle packing has become an economic and reliable alternative. In this chapter we make reference to various approaches employing such computer simulation facilities. This requires considering this way of tackling the problems in more detail, because reliability lies in the very details. Concrete technology has witnessed some sort of autonomous development in this respect. Primarily methods of random sequential addition (RSA) of mostly spherical particles have been developed, thereby making use of random generators. In [59], the authors have indicated how in the mid-seventies of the previous century the development of such an RSA system that was based on a sintering mechanism was started at Delft University of Technology. A continuation was based on mutual sliding along the surfaces of interfering spherical particles when compressed in a container of gradual reducing size. A first version was used to predict wall effects in fiber concrete [48]. A new start in the nineties of the previous century, initiated by fast computer developments, led to the SPACE system used in part of this chapter [44]. SPACE, however, is a discrete element method (DEM). Recently, an even more advanced DEM was developed, based on arbitrary-shaped particles. This so-called HADES system is also used in this chapter [15, 54, 60]. These analogue DEMs are in the frontier of research on virtual cementitious systems in the concrete technology field.

Up till now, in concrete technology, predominantly random sequential addition (RSA) systems were in vogue [3, 53, 54]. Outside of concrete technology, only rarely such systems are being used. Herein, particles are placed from large to small on preconceived Poisson field positions [71]. If this leads to physical violations (overlap), the last generated particle is rejected. The number of rejections increases dramatically at moderate volume fractions, making this generation process very time-consuming. However, a fundamental limitation is that particles are more evenly distributed than in practice [54, 66]. Material properties depending on the *dispersion* of the particles, so called structure-sensitive properties [53], will, therefore, be poorly represented by RSA systems. In contrast to RSA systems, DEM are based on concurrent algorithms that reflect the particle interferences when they are poured into a container (either with a cement matrix or not). Static DEM [66] are based on locally shifting the interfering particles. Dynamic solutions have all particles moving inside the container. SPACE [44, 58] and HADES [15] belong to this most advanced category. For a critical discussion on RSA versus DEM, also offering the proper references of the aforementioned publications, see [54]. Dispersion of aggregate or cement particles is far more realistic in DEM than in RSA approaches [15, 66], as confirmed by experiments. Hence, structure-sensitive properties can only reliably be simulated by DEM. This also holds for the pore percolation process in both cement and concrete.

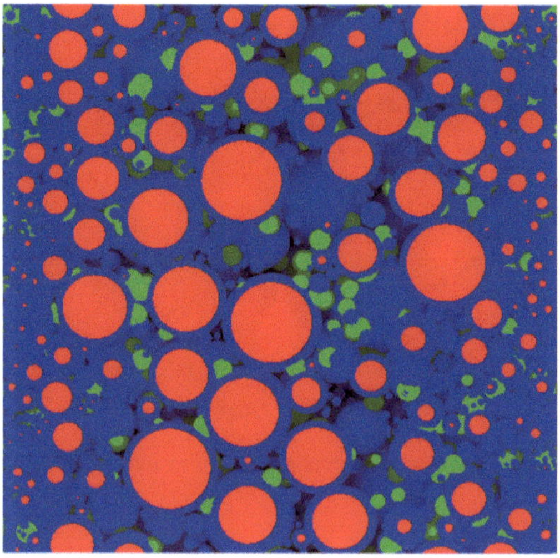

Fig. 7.1 Section of a cube of relative young cement paste with $w/c = 0.25$, in the horizontal direction pocketed between aggregate grain surfaces, simulated by DEM system HADES. Top and bottom surfaces are periodic, simulating a situation of more remote aggregate grain surfaces in the vertical direction. Cement is supposedly only to be composed of C_3S (alite). Un-hydrated cement nuclei are in *red*, gel produced by hydration of C_3S in *blue*, calcium hydroxide CH in *green* and pores in *black*

In the concrete technology field, the simulated fresh cement structure is hydrated, the principles for which are mostly derived from [28]. This is reasonably straight forward for spherical particles [44, 59], but would be quite complicated in systems with arbitrary shaped binder particles, such as pursued by HADES. An example of a cement paste consisting of spherical particles (pocketed between aggregate grains) and hydrated by aforementioned principles (described in [44, 55]) is shown in Fig. 7.1.

7.3.2 *Packing of Aggregate Grains in Concrete*

As stipulated, at the heart of concrete making is the design of particle mixtures. Aggregate of fluvial (or sea) origin or crushed rock has to fulfill a specified series of criteria. The hydrated cement in which the aggregate grains will be dispersed provides an alkali environment with high pH value. Aggregate, the mineralogical composition of which is such that the so called alkali-silica reaction may occur, should be avoided. This reaction leads to local expansion that can cause large-scale crack formation and spalling at concrete surfaces. Extensive damage has occurred in concrete practice before this mechanism was identified.

In the framework of this book, the most characteristic feature of concrete–like materials is grading. Therefore, more explicit attention will be paid to this phenomenon. The basic idea of finding an "ideal" aggregate mixture is illustrated in Fig. 7.2. Nevertheless, in normal practice, the locally available aggregate is employed when the actual grading curve falls within the borders indicated in the building code.

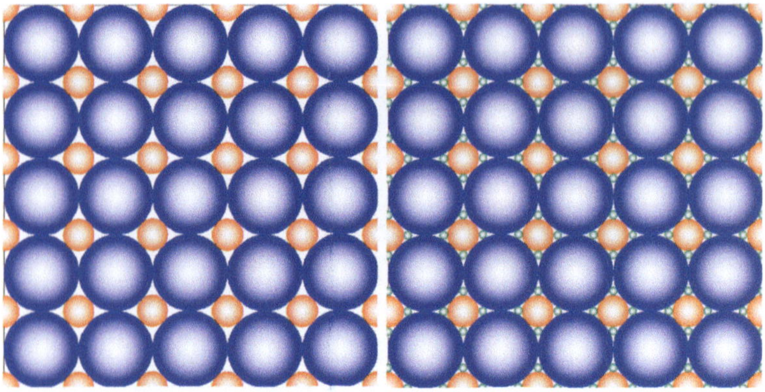

Fig. 7.2 Ordered bimodal (*left*) and tri-modal packed structure of spheres (*right*) [15]

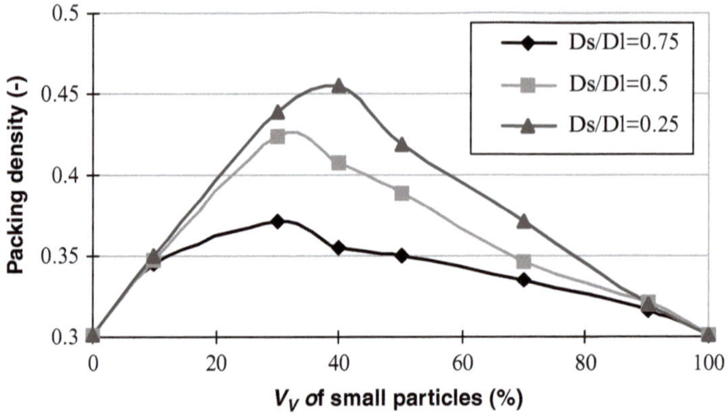

Fig. 7.3 Maximum packing density versus composition of bimodal mixtures for different 3D spherical particle size ratios obtained by a RSA packing system (D_s and D_l: sieve sizes of large and small particles, respectively) [15]

The largest particles leave voids that should ideally be filled up by the fraction of next smaller particles (Fig. 7.2 left bimodal, right tri-modal mixture). Such a concept already renders this possible, demonstrating the effect of (bimodal) grading on packing density (Fig. 7.3).

Thereupon, the next fraction of even smaller particles fills up the voids left in the packing system, and so on. In the literature of the past century, numerous approaches to this packing problem have been developed (see also Sect. 2.2). The "classical" systems are described in [76]. The Fuller curve is renown, leading to near-optimum density in the case of *continuously graded* mixtures (so, encompassing all sieve fractions). It is governed by the parabolic curve $P_d = 100\sqrt{d/D}$, where P_d is the volume percentage of particles passing through the sieve with opening d and D is the maximum grain size in the mixture. In a more generalized

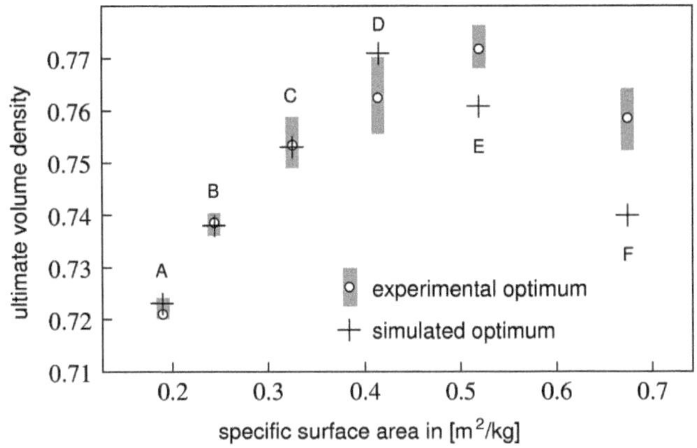

Fig. 7.4 Comparison of the packing density at the jammed states of different mixtures of river gravel aggregate and of SPACE-generated spherical aggregates with similar grading characteristics [44]

version, the square root is replaced by the *i*th root, whereby the value of *i* should be adapted to the field of application and the production method (explicitly involving compaction) [35]. Concrete building codes generally define upper and lower bounds between which the sieve curves of the aggregate should be situated. This is in line with the uncertainty in analytical approaches but is also of economic significance; a local aggregate of which the sieve curve falls inside the delineated area for proper mixtures is cheaper than one composed of separate sieve fractions.

Figure 7.4 shows the packing densities of mixtures of river gravel with different grading, which are determined in a prescribed/standardized way in 8-liters cylinders. Additionally, maximum volume density of mixtures with similar grading has been determined by DEM using the SPACE system that is based on spherical grains. The particle size distributions (PSDs) of mixtures A ~ F cover the 1 ~ 32 mm range. The finenesses of mixtures are increasing from mixture A to mixture F. The PSD of mixture D is corresponding to the Fuller curve. Obviously, correspondence between experiments and simulation is satisfactorily. These tests indeed point out that mixtures close to the so called century-old Fuller PSD (mixtures D and E) have near optimum grading.

Shape is an important parameter for particle packing as well as for properties of concrete. Two typical real aggregates are depicted in Fig. 7.5, revealing aggregate of fluvial origin (river gravel) to have rounded edges, whereas crushed rock aggregate has sharp edges. Crushed rock aggregate tends to reduce consistence and packing density compared to aggregate of fluvial origin. Figure 7.6 shows different density distributions of bimodal packing of different type of aggregate. Particle shape directly influences packing densities of mixture in all proportions. This influence is also illustrated by the mono-size particle packing simulation using HADES in Fig. 7.7, where different types of aggregate are simulated from spherical (of fluvial origin) to tetrahedron-like (crushed rock) shapes. The effect of grain shape on optimum packing density is evident. Packing densities are influenced by

Fig. 7.5 Particles of crushed rock (*left*) and of river gravel aggregate (*right*) [15]

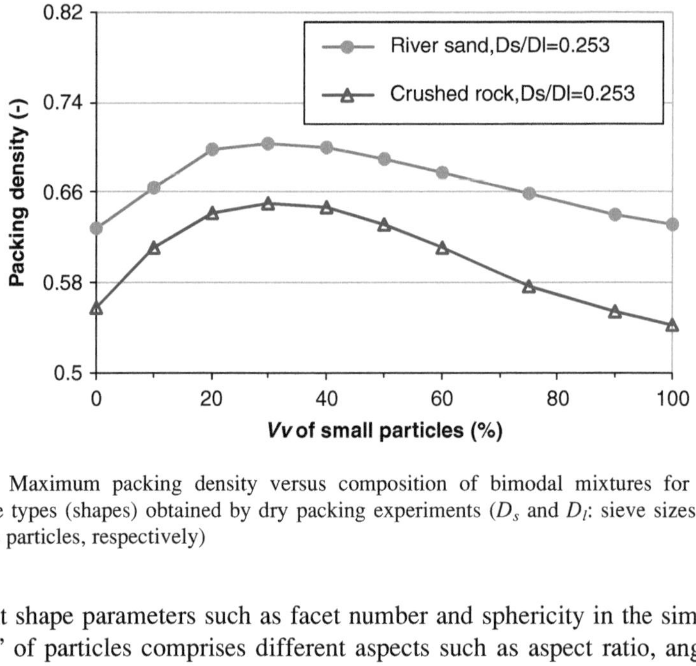

Fig. 7.6 Maximum packing density versus composition of bimodal mixtures for different aggregate types (shapes) obtained by dry packing experiments (D_s and D_l: sieve sizes of small and large particles, respectively)

different shape parameters such as facet number and sphericity in the simulation. "Shape" of particles comprises different aspects such as aspect ratio, angularity, surface roughness and specific surface area and, thus, is hard to define unambiguously. Different approaches have been developed for its assessment [15, 17].

Optimum mixture proportioning is basically focusing on achieving maximum density. This will promote strength and durability of the final product. This can also be achieved with *discontinuous mixtures* (designed on the basis of gap-grading). This could lead to somewhat higher strength than obtained with continuously graded mixtures (containing the same volume fraction of aggregate) [76]. We will see later that in the design of (ultra) high performance concrete (UHPC/HPC) this principle is exploited as to aggregate and binder alike.

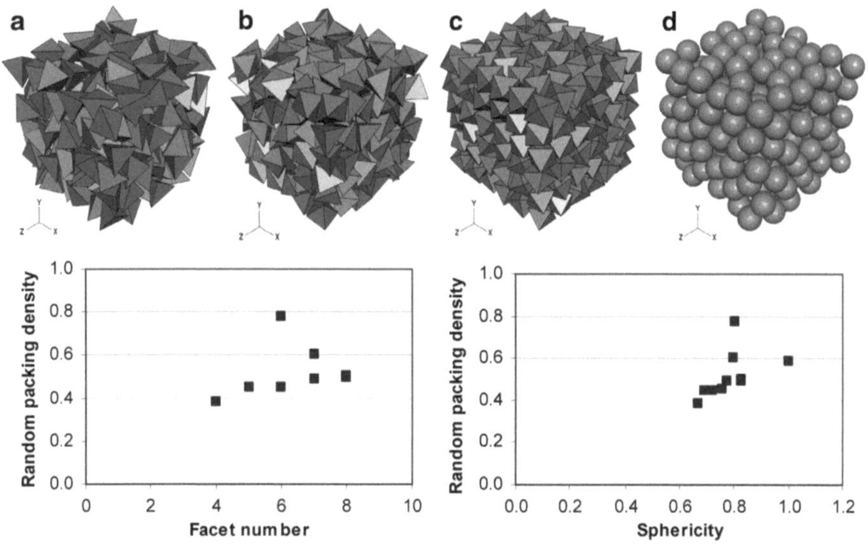

Fig. 7.7 (*top*) Examples of dense random packing of mono-size particles with shape of (**a**) tetrahedron, (**b**) hexahedron, (**c**) octahedron and (**d**) sphere, respectively; (*bottom*) maximum packing density versus (*left*) facet number and (*right*) sphericity of 3D particles shown above [15]

But optimization must as well focus on properties of the fresh state. For, the concrete mixture's consistence should allow proper compaction of the material to fully exploit strength and durability potentialities. Also, fresh mixture's mobility should be large enough to correctly enrobe steel reinforcement. If not, composite behavior will be endangered as well as reinforcement corrosion promoted. On the other hand, mobility should be limited so that large-scale segregation of aggregate particles by vibration during compaction is prevented. As stated earlier, in the design of mixtures use is preferable made of local aggregates that should just fulfill the requirements of the building code. Optimization of grading would increase material costs; this is only relevant for special purpose concretes.

7.3.3 Packing of Binder Particles in Concrete

Hence, proper packing of aggregate is only an obvious design problem for advanced cementitious composites. Particularly in HPC and UHPC also packing of the binder should be carefully designed. It has been experimentally demonstrated that partial replacement of the Portland cement by an inert mineral admixture (carbon black) could lead to higher strength levels, particularly at early age, provided the carbon black is finer than the cement [14]. Then, closer particle packing causes increased (van der Waals) physical forces that more than compensate for the reduced chemical ones. In [6] it was demonstrated experimentally

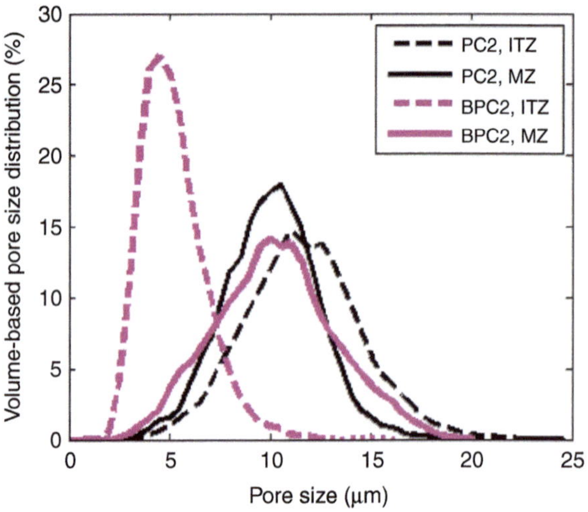

Fig. 7.8 Volume-based pore size distribution functions for plain PC and gap-graded blended PC (BPC) for both the interfacial transition zone (ITZ) and bulk (middle zone: MZ)

and by SPACE simulation that partial replacement (up to 25 %) of Portland cement by fine-grained rice husk ash was only efficient when the size ranges of mineral admixtures and of the cement were *gap-graded*. The significant effect of gap-grading by fine-grained rice husk ash has also favorable effects on pore size distribution (Fig. 7.8) and will therefore also influence durability in a positive way [61]. Also, the use of the pozzolanic mineral admixture silica fume as partial replacement of the Portland cement significantly improves concrete strength and density [2]. So, this gap-graded binder solution combines chemical as well as physical effects (see also Sect. 2.2).

Hydration is influenced by particle shape because of the surface-volume ratio effect [7]. So, aggregate and binder particles should be simulated with proper shape characteristics. Conventional hydration models are mostly based on the assumption of spherical cement grains [28, 44, 70]. But it may introduce serious biases as compared to the surface-volume ratio witnessed with real cement grains, as revealed by X-ray micro-tomography (μCT) [13]. This will also affect the microstructure evolution and properties [7]. After a series of shape analyses, [15] proposed two shape alternatives, i.e. flat ellipsoids and a sort of octahedrons, for a more realistic hydration system. Two example simulations with surface area to volume (*S-V)* curves corresponding to experimental findings are depicted in Fig. 7.9. It proves that these two shapes reflect more realistic *S*/*V* curves and, consequently, could better predict hydrated cement paste properties.

Figure 7.10 may finally demonstrate that a section pattern of a simulated cube of octahedron cement grains shows visual similarity with a section of real cement paste. Hence, for this type of cement this would be the proper shape for simulation of fresh cement grains.

Experiments have revealed inter-particle forces (van der Waals forces, electrostatic forces, etc.) are crucial for packing of cement grains, especially in the drying

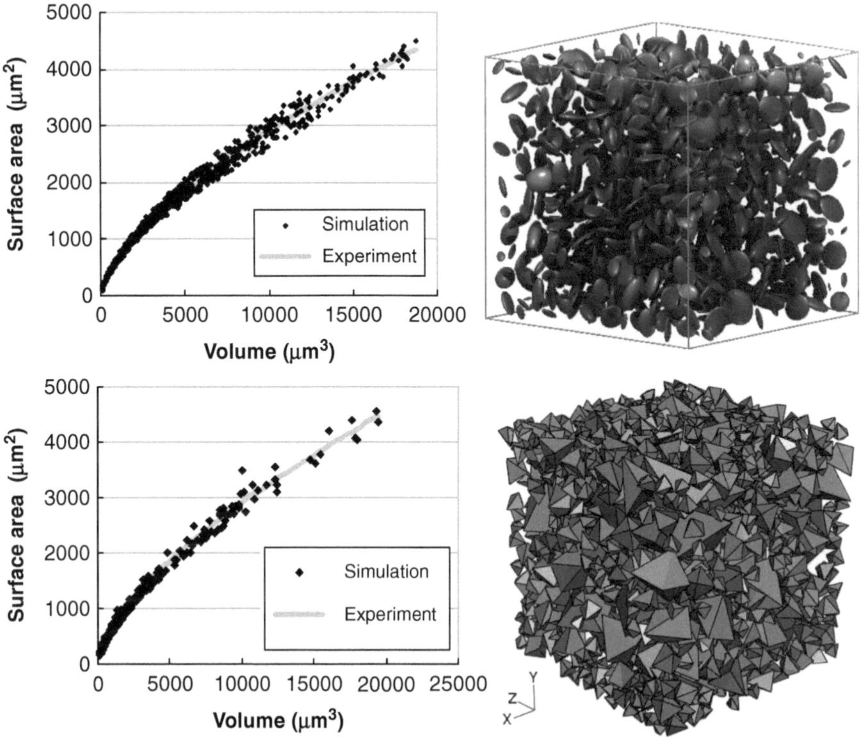

Fig. 7.9 (*left*) *S-V*-relationships of a group of cement grains by flat ellipsoids (*left top*), respectively, polyhedra (*left bottom*). Experimental data are due to [13]. Corresponding simulated structures are displayed at the *right* [15]

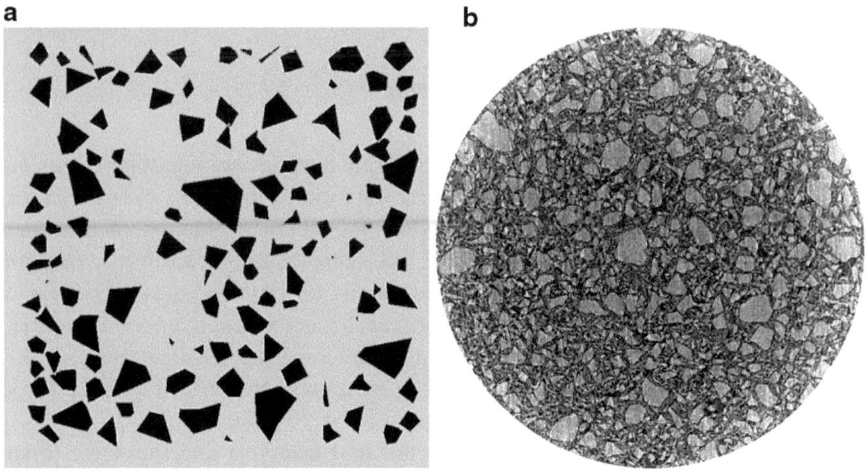

Fig. 7.10 (**a**) a section of the simulated structure (1,000 octahedron grains in 10 ~ 50 μm range); (**b**) a 2D section of a real fresh cement structure (Cement-133) obtained by μCT; $w/c = 0.35$ (Source: http://visiblecement.nist.gov/cement.html. Note that the real particle size range is wider than the one used for the simulation (courtesy NIST))

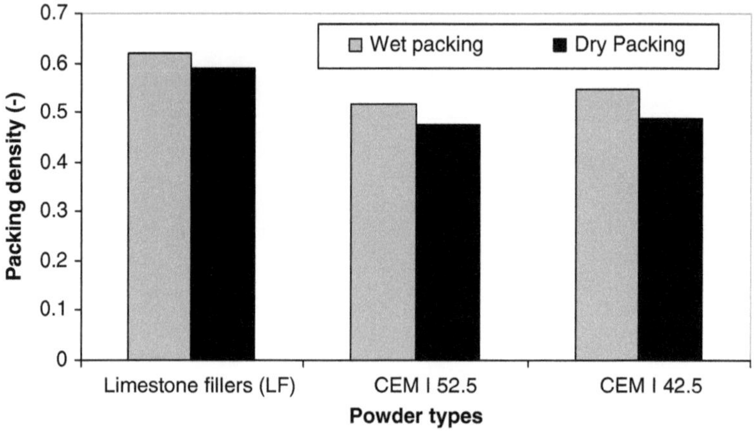

Fig. 7.11 Packing densities of different powder types under different conditions (wet or dry) [69]. For details of the packing procedures, see [16]; LF encompasses a wide particle size distribution; CEM I 52.5 and CEM I 42.5 are Portland cements with higher and lower fineness, respectively

state [16, 69]. Figure 7.11 presents experimental data on the packing of different powders under dry as well as in wet conditions [16]. The wet packing method can reduce the inference of inter-particle forces and thus lead to higher packing density. Particle packing is also affected by particle size distribution. Although inter-particle forces can be dramatically reduced by the wet packing method [69], the internal cement structure may still reveal patch formation despite the addition of a superplasticizer [51]. This can only be simulated by DEM-based systems, of course.

7.3.4 Packing of Fibers in Concrete

Mechanical behavior of fiber reinforced concrete depends among other things on pull-out resistance of the "short" fibers and shearing of the fibers over the crack edge. This has been modeled to determine the efficiency factors of the fiber reinforcement. In design, the fiber packing is assumed isotropic uniformly random (IUR). However, in reality it is influenced locally by large aggregate grains and the mold. Eventually, segregation can occur by compaction under vibration. Hence, the resulting packing can be highly anisometric, as illustrated by the X-ray radiography image of a research-size specimen shown in Fig. 7.12 [56]. See also [19].

A practical assumption introduced by the first author is that the actual fiber dispersion can be considered a summation of a 3D random dispersion (hence, IUR), a planar distribution (referred to as 2D) and a linear one (1D) [45]. This renders possible deriving also the geometric efficiency factors of the fibers in orthogonal

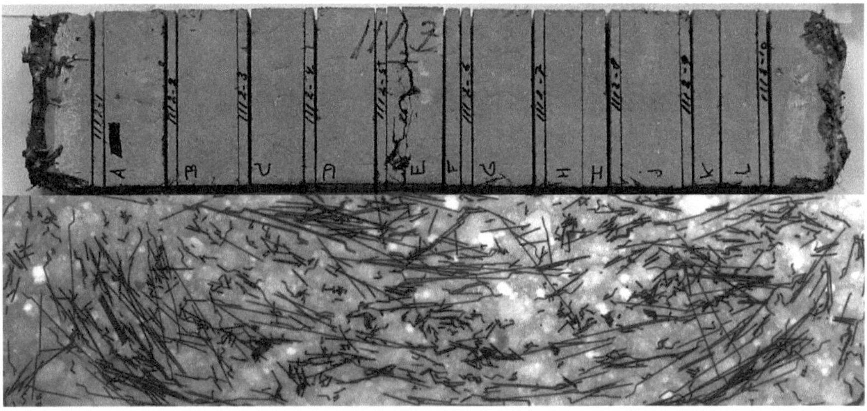

Fig. 7.12 (*top*) Location of vertical slices prepared of prismatic steel fiber reinforced concrete (SFRC) specimen (1,000 × 200 × 50 mm) containing 1.72 % by volume of hooked-end fibers. X-ray radiographs were prepared of all slices of which an example is presented at the *bottom*; this reveals steel fiber dispersion to be highly anisotropic [47] (Courtesy Heron)

directions as a summation of contributions due to 3D, 2D and 1D fiber systems. Estimates of the orthogonal stress transfer capabilities at ultimate $\{\sigma(x), \sigma(y), \sigma(z)\}$ of SFRC specimens subjected successively in the respective coordinate directions to tensile loading are [25, 63]

$$\begin{aligned} \sigma(x) &= \sigma_m(1 - V_f) + \eta(x)a\tau_f V_f \\ \sigma(y) &= \sigma_m(1 - V_f) + \eta(y)a\tau_f V_f \\ \sigma(z) &= \sigma_m(1 - V_f) + \eta(z)a\tau_f V_f \end{aligned} \tag{7.3}$$

in which σ_m stands for the tensile strength of the plain mortar and τ_f is the presumably uniformly distributed shear strength between fiber and cementitious matrix. a and V_f stand for the fiber aspect ratio (slenderness) and volume fraction, respectively. $\eta(x)$, $\eta(y)$ and $\eta(z)$ are the fiber orientation efficiency factors in the respective coordinate directions. The product $aV_f\tau_f$ is generally referred to as the *fiber factor*. Obvious ways to enhance the fiber factor are limited, since too slender fibers (say, $a > 100$) cause "balling" and increased fiber content is undesirable because of economic demands. Shortly after ultimate loading the matrix contribution is exhausted. Thereupon, a steady decline in post-peak load-bearing capacity is found due to pulling-out of the fibers, as shown in Fig. 7.13. Hence, even low volume fractions of fiber reinforcement lead to a more gradual degradation of the overstrained concrete structure.

So far, the discussion implicitly focused on steel fibers; however a variety of fibers is employed in practice (glass, carbon, wood, rice husks, polypropylene, PVC, etc.). Further, *hybrid systems* are applied whereby different types of material are combined or fibers of the same material are employed but of different sizes. In the latter case, the micro-fibers control the early stages of crack formation in the

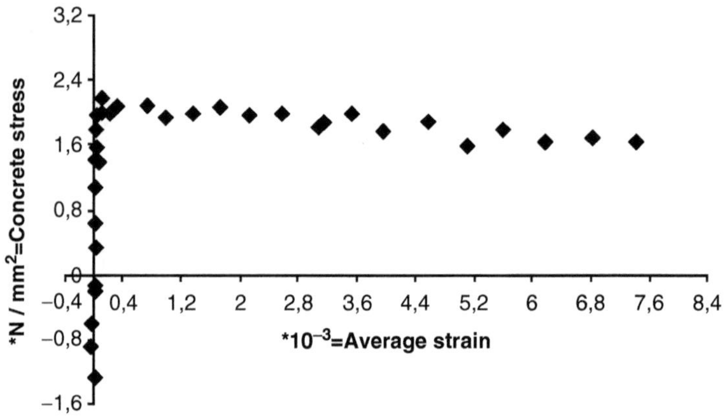

Fig. 7.13 Stress–strain curve in direct tension of SFRC specimens (50 × 200 × 1,000 mm, shown at the top of Fig. 6.11 and containing 1.72 % by volume of 30 × 0.4 mm hooked fibers). Observations are due to 150 mm clip gauges spanning the main crack. For further experimental details, see [47] (Courtesy Heron)

pre-peak range, whereas the macro-fibers do so in the post-peak range. Microfibers require modifications in the mixing procedure to prevent "balling" of the fibers [62]. Fibers of natural origin (wood, rice husks, sisal, coconut fibers) require reduction in the alkalinity of the matrix to extend service life. This is possible because use is limited to sheets without steel reinforcement (that would otherwise easily corrode at reduced alkalinity). Glass (even of high quality) also suffers degradation in the high alkaline environment in concrete.

The effect of compaction by vibration and of the flow of the fresh mix into the mould on actual packing characteristics of the fibers has received major interest, because of economic impact. The external layer of SFRC is different due to this so called wall effect. This can have significantly impact on (surface) crack control capacity. The analytical approach is basically similar to the one for the bulk case. Detailed information can be found in the international literature [52].

The design equation resulting from Eq. 7.1 with $\eta(x) = \eta(y) = \eta(z) = 1/3$ is [46]

$$\sigma_c = \sigma_m\left(1 - V_f\right) + \frac{1}{3}a\tau_f V_f \tag{7.4}$$

This is the tensile strength of the SFRC material, which is governed by a large number of fibers intersecting the macro-crack that resulted from crack coalescence and crack concentration inside the fracture process zone. At a critical crack opening, the contribution by the matrix will be exhausted *and only the load-carrying capacity of the fibers is left*. In the pre-ultimate domain crack openings are small and fibers cannot be assumed fully de-bonded, so a reduction factor has to be added for interfacial stress transfer. Additionally, the small cracks in this domain are to different degrees out of plane with the forthcoming macro-crack. This leads to an additional strength reduction factor, β. So, the resulting stress in the pre-ultimate domain is given by

$$\sigma_c = \sigma_m(1 - V_f) + \frac{1}{3}\beta a\tau_f V_f \tag{7.5}$$

with $\beta < 1$.

This renders possible making the transfer to special concretes (HPFRCC and ECC) discussed in Sect. 7.6. High quality cementitious composites are very brittle and thus require fiber reinforcement. The advanced cementitious materials reveal strain hardening. Hence, post peak strength (after exhausting the matrix contribution) should exceed that of the pre-ultimate material strength. This leads according to Eqs. 7.4 and 7.5 to a critical volume fraction $V_{f\text{crit}}$ given by

$$V_{f\text{crit}} > \frac{\sigma_m}{\frac{1}{3}a\tau_f(1 - \beta)} \tag{7.6}$$

Substitution of practical values of the parameters in Eq. 7.6 ($a = 75$; $\sigma_m = \tau_f$; $\beta = 0.15$) [26] demonstrates that significantly higher volume densities are required to provide concrete with strain hardening capacity. Note that in conventional SFRC, V_f is around 1 % by volume.

The development of SFRC basically started after the Second World War. In the early 1980s a new SFRC material was introduced by Lankard that revealed strain hardening, i.e., "slurry infiltrated fiber concrete" (SIFCON) [27]. Fibers were packed in a mold, whereupon the cementitious slurry was poured into it. Up to 20 % by volume of fibers were employed. Of course, the reinforcement structure was strongly an-isometric. Modern engineered cementitious composites (ECC) is a scientific engineering approach to achieving strain hardening in cementitious composites. For a somewhat more elaborate historical description, see [24], presenting also the proper references to the aforementioned articles.

7.3.5 Wall Effect in Particle Packing

A final phenomenon that should be discussed in this connection is the wall effect. Aggregate near the mould and cement close to the aggregate grain surface will reveal size segregation. Both gradient structures have effects on properties [12, 18, 41, 44]. Volume fraction increases in both cases away from the wall. Near the mould, the mixture is therefore relatively rich of cement paste leading to increased shrinkage and possible surface cracking. Near the aggregate surface, we have a similar phenomenon: more inter-particle space and thus larger porosity. However, as a consequence the degree of hydration is higher (see Fig. 7.14). The wall effects are different in different components of the concrete or in evaluation parameters

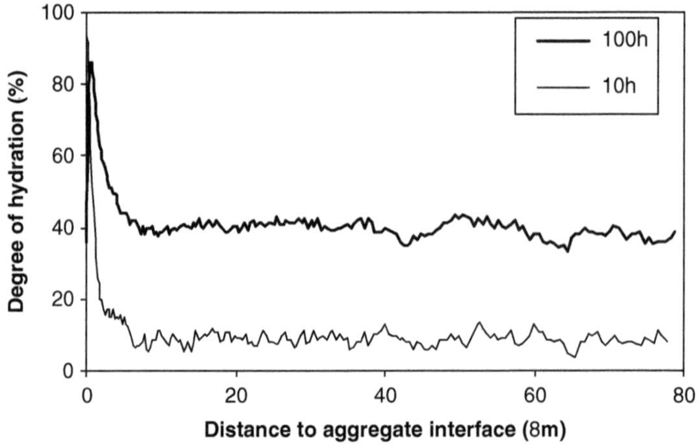

Fig. 7.14 Wall effects in degree of hydration over the ITZ in a model (virtual) concrete after 10-h and 100-h of hydration [18]

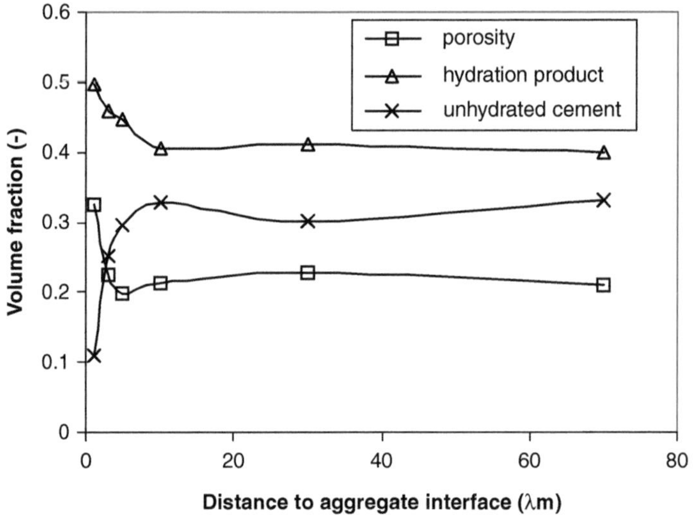

Fig. 7.15 Volume fraction of different phases versus distance to aggregate surface (*at the left*) for model cement at 100-h's hydration. ITZ's extent for unhydrated cement seems exceeding the one for porosity [18]

and they are influenced by design factors (e.g. *w*/*c*) (see Figs. 7.15 and 7.16). Both effects selectively influence the formation of certain chemical compounds (like large calcium hydroxide CH crystals).

This holds for *normal* concrete. When fine mineral admixtures (e.g., silica fume) are used to replace part of the Portland cement and a proper amount of superplasticizer is added for achieving the required consistence level at low *w*/*c*, the fine

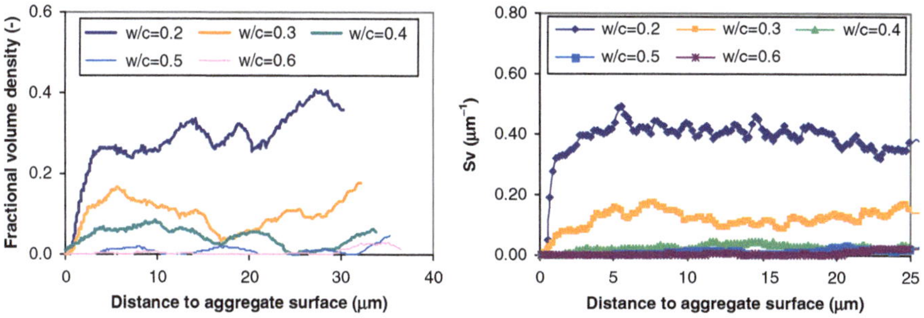

Fig. 7.16 Gradients in volume fraction, V_V (*left*) and specific surface area, S_V (*right*) of unhydrated cement grains in fine-grained (497 m^2/kg) SPACE-generated model cement for five different values for *w*/*c*. It is seen that ITZ's thickness for volume density and specific surface area increase at higher *w*/*c* [15]

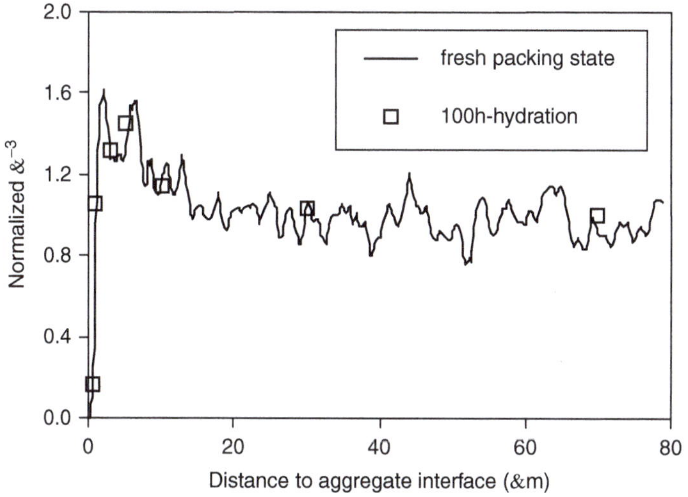

Fig. 7.17 Disproportional bond strength increase in the ITZ remains through hydration process. Quite extensive ITZ can be observed (declining at higher *w*/*c*) [18]

mineral admixture particles will fill up the open spaces between the cement particles, as depicted by Fig. 7.1. Van der Waals forces will be increased to such a degree that the strength level of the interfacial transition zone (ITZ) can exceed the one in bulk (Fig. 7.17). Note that λ is the mean free spacing, defined as the mean value of uninterrupted surface-to-surface distances between all neighbouring particles. λ^{-3} is supposedly associated with local (van der Waals-based) bond capacity [44]. The result of this size segregation phenomenon is that the traditionally high crack initiation capacity in ITZs, which provides concrete in compression with a certain degree of toughness, is eliminated. Under high compressive stresses, the material more-or-less explodes in the brittle mode because of the high crack

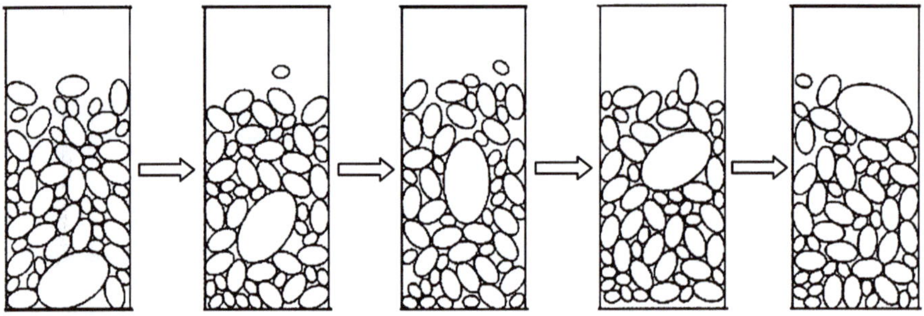

Fig. 7.18 HADES simulation (video snapshots) of size segregation during vibration at container bottom (Brazil Nut Effect) [57]. (Reproduced with permission. Copyright Springer)

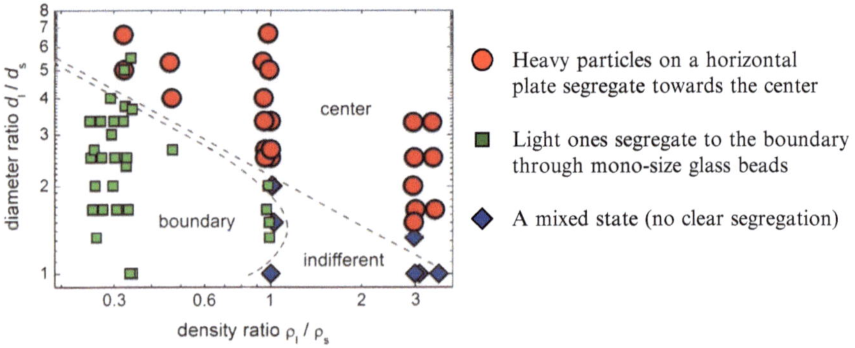

Fig. 7.19 Phase space for particle properties. Each symbol represents one experiment. Shaker amplitude is 1:91 cm and frequency 1:67 Hz. Mechanisms were also depending on the area percentage of filling the horizontal disk with the glass beads (segregation was not observed at lower values) [39] (Courtesy Prof. Rehberg)

energy release rate. This is the basis for the development of HPC and UHPC. To compensate for the loss of toughness, fibres are added to such materials. This may lead as an example to high performance steel fibre reinforced cementitious composites (HPSFRCC).

It is argued in the literature that size segregation might be stimulated by the so called Brazil Nut Effect (BNE) [50] (Fig. 7.18). BNE is a hot item in advanced journals, where the "normal" BNE, i.e., the migration of large particles to the top (so, against the gravity force) of a finer-grained mixture is distinguished from the reversed BNE and the horizontal one; both with respect to the gravitation direction. Research has been conducted showing volumetric density of the migrating particle(s) with respect to that of the rest of the mixture and the size ratio of the two particle systems involved to be leading parameters in whether we can expect dealing with one of these concepts (Fig. 7.19). Some supporting evidence in concrete technology is published in [50], also providing the proper references of the abovementioned contributions to the BNE.

7.4 Porosimetry

7.4.1 Assessment of 3D Pore Geometry

The particulate nature of concrete extends into the micro-level, where binder particles disperse in the watery environment in the fresh state of the material. Packing details of the binder change with reducing *w*/*c*. Such packing details directly influence the structure of the hydrated material and so also the de-percolation process that governs pore space in the matured material. In addition to the aforementioned *w*/*c*, the design of the binder can involve partial replacement of the Portland cement (PC) by a mineral admixture. When the latter is finer than the PC, this can give rise to significant physical contributions (due to van der Waals forces) to the strength of the material. So even inert components have been demonstrated effective [14]. The fineness of the PC (expressed in Blaine number, or specific surface area) is another design parameter influencing the packing details of the binder, and thus of the pores in the matured material [44].

Continuous pores are the way harmful substances can reach the interior of the concrete or the reinforcement steel. So, durability relies on limitation of percolated porosity. Porosity is the space resulting after the hydration products are formed. This is obviously depending on cement particle characteristics, such as size distribution and dispersion. What conventionally is experimentally determined by quantitative image analysis (QIA) is porosity and 2D pore size distribution (PoSD) [53, 67]. Nevertheless, the most popular approach is Mercury Intrusion Porosimetry (MIP). This is a 3D method but offers information orders of magnitude away from more realistic approaches [9, 67], as revealed by Fig. 7.20 (see also section "Mercury Penetration Technique" in Chap. 3).

The 2D pore size distribution obtained in sections cannot easily be transformed into a 3D PoSD function. This can directly be proven by considering sections of a dispersion of mono-size spherical particles with size D. The section reveals circles of different sizes, the diameter of which is defined by x. The probability density function of x is [20, 45]

$$f(x) = \frac{x}{D\sqrt{D^2 - x^2}} \tag{7.7}$$

So, this is different from the Dirac-like probability density function of the spheres. Even average 2D section size of the spheres ($\overline{x} = \pi D/4$) differs from the 3D particle size. So, with first order stereology (classical way of quantitative image analysis), 3D properties of pores cannot be obtained.

Still, 3D information can be obtained on 2D material sections [18, 49, 54]. Of course, sampling is the hot issue in that case. In Fig. 7.20 the star volume measurement technique is illustrated [18]. A random point x on a pore section at the left is provided by a "star" as sketched at the right. A pike in the star represents the unobstructed length, l_i, in the direction of viewing to the nearest perimeter of the pore. A local estimate of 3D pore volume, v^*, is obtained by the given formula in

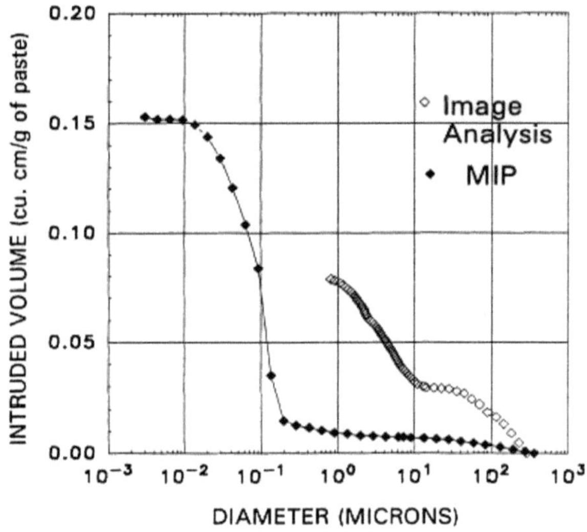

Fig. 7.20 Comparison of MIP and QIA pore size distribution plots for 28 days old cement paste ($w/c = 0.4$) containing air voids [9, 18] (Reproduced with permission; copyright Elsevier)

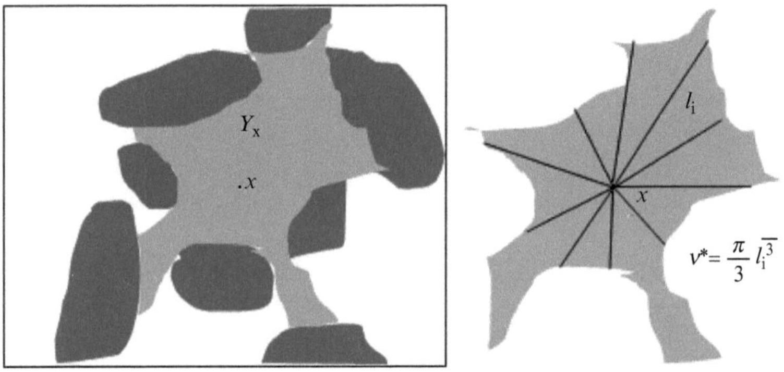

Fig. 7.21 2D schematic drawing of the star of the pores, defined as the volume of the light grey zone v^* averaged for all points x in the pore space. The distribution (histogram) of the star volume as x spans the pore space gives insight into the pore size distribution [18]

Fig. 7.21. A large series of such observations on random points at pore sections can be used for constructing a cumulative PoSD.

Another approach is indicated in Fig. 7.22. The digitized image of a field as part of a section reveals pores in black. In this image the mathematical morphology operator "opening" is applied, meaning subsequent erosion and equivalent dilation within a given opening size, which removes areas smaller than this opening size. By doing this operation at increasing opening sizes, an area-based pore size distribution is obtained, which equals the cumulative 3D volume-based pore size distribution in view of isotropy [18].

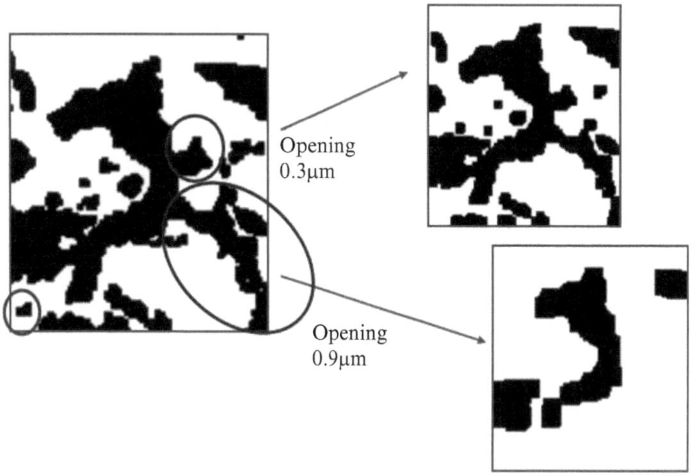

Fig. 7.22 Opening distribution by structuring elements of increasing sizes gives a sort of size classification. By opening with a square 0.9 μm structuring element, the *lower circled region* is removed; hence, this local branch area is contributing to the size range < 0.9 μm [18]

A recently published application is presented in Fig. 7.23 [53]. Two different concretes (N is produced with some sodium chloride) are investigated and opening distribution curves and derivative curve (PoSD) are displayed at the bottom.

7.4.2 Assessment of Pore Topology and Geometry

Analysis of random section is not sufficient to get information on pore percolation, of course. This is however a major feature underlying durability issues of concrete [68]. The most trivial approach is by serial sectioning and 3D reconstruction. This can only be realized economically by computer simulation. Ye [70] developed the necessary algorithms in HYMOSTRUC3D, a sequential random addition system. This has unrealistic cement particle dispersion. Moreover, Ye filled up the pore system by spheres of increasing radii starting from a pre-determined point. This leads to a biased cumulative PoSD [18]. Hence, [8] employed SPACE to more realistically disperse the cement particles in the fresh state, whereupon he implemented the particle system in HYMOSTRUC3D [70]. To reduce computer time, the computer cement was significantly coarser than actual cement with a Blaine value (specific surface area) of 300 m^2/kg. This should be kept in mind when extrapolating his findings, of which Fig. 7.24 presents an example for a relatively low *w*/*c* (so, HPC). It depicts, at the left, the decline in total porosity in the ITZ (aggregate grain surface at the left). At the right, the steeper decline is shown of the percolated fraction of total porosity [8, 50].

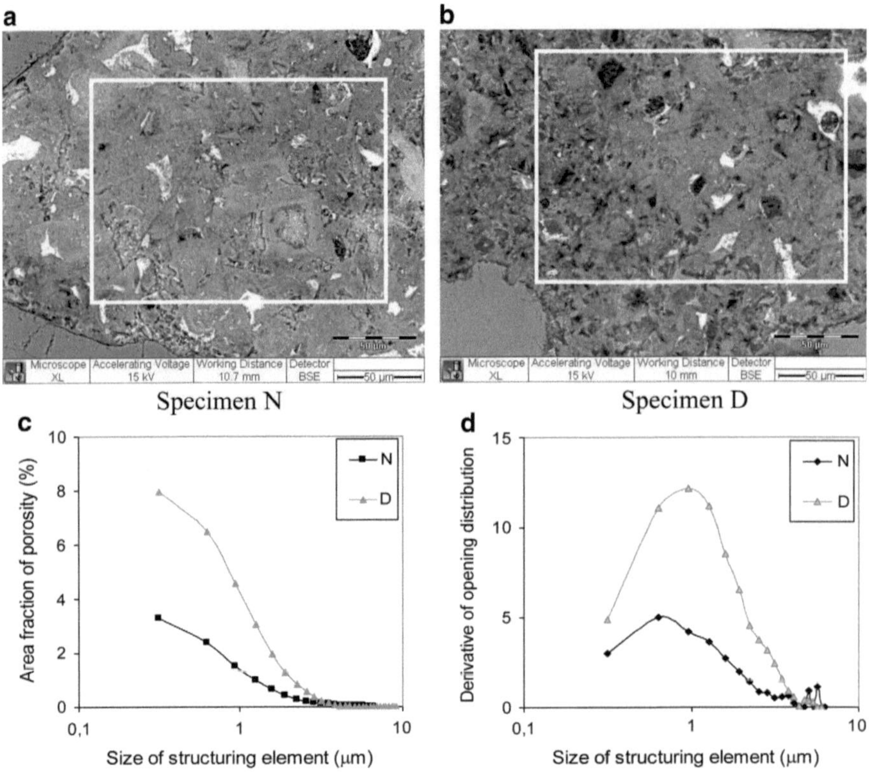

Fig. 7.23 SEM micrographs of concrete specimens N (**a**) and D (produced with some sodium chloride, **b**) and the results due to application of the opening operator on area fraction of porosity (**c**) and on the derivative of the opening distribution (**d**). Lower values of porosity and critical pore size (horizontal value at the top of the curve) are found for N (normal) concrete (**d**) [53] (Reprinted with permission; copyright Elsevier)

Of engineering significance is the conclusion that in this low *w/c* range, so for HPC, percolated porosity *seems* restricted to the ITZs around aggregate particles. These are (partly) connected (percolated) in the dense random packing state of the aggregate. Hence, also for global pore percolation the designed aggregate skeleton would play a major role. Research into the aggregate parameters influencing ITZ percolation [4] revealed the sieve curve and particle shape exerting such influences. Also the ITZ width is obviously an important factor. This is governed by particulate features of the binder (cement fineness, *w/c*, grain distribution), so can also be influenced by design. Figure 7.25 [72] shows influences of volume fraction (V_V) and PSD (Fuller distribution or equal volume fraction (EVF) distribution) on the chloride diffusivity of concrete. Chloride diffusivity is linearly decreased with increasing packing density of aggregate due to the increase of pore tortuosity and reduction in paste content. Concrete with Fuller PSD has a better resistance to chloride diffusion. The newest porosimetry research gives answer to the question

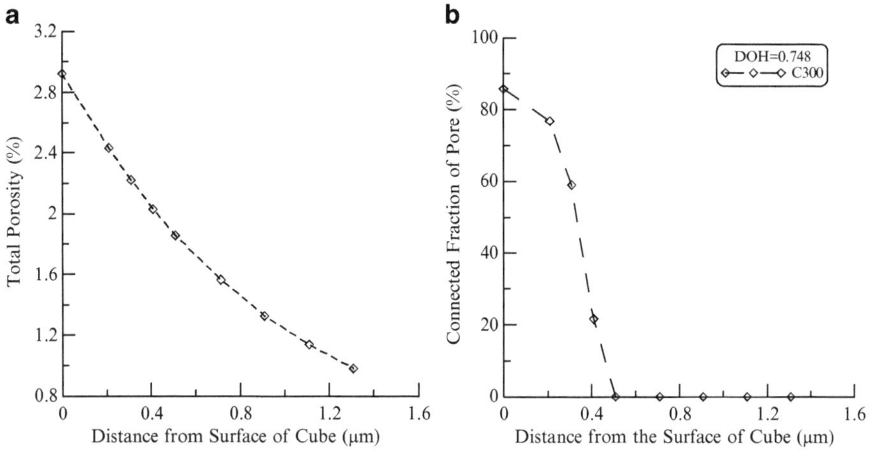

Fig. 7.24 Total porosity (**a**) and percolated fraction of porosity (**b**) inside the ITZ. The involved *w/c* is 0.3, the (ultimate) degree of hydration is 0.748 and cement fineness is 300 m^2/kg [8]. Results obtained by serial sectioning and 3D reconstruction (Reprinted with permission; copyright Elsevier)

Fig. 7.25 Effect of aggregate gradation on chloride diffusivity of concrete [72] (Reproduced with permission; copyright Ice Publ.)

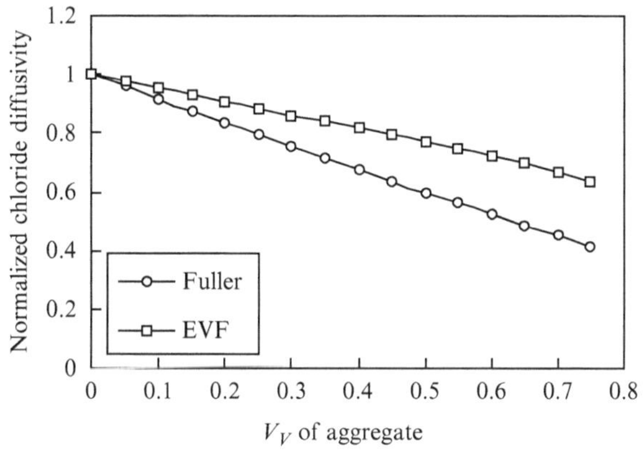

whether ITZ percolation (that occurs increasingly at higher packing densities) would be intimately related to chloride diffusion. This was suggested by Fig. 7.24b that only reveals part of the ITZ liable to pore percolation; the latter is a necessary condition for transport though pores in concrete.

Figure 7.26 is obtained by so called double random multiple-tree structuring (DRMTS), derived from 'Rapidly exploring Random Tree' (RRT) algorithms used in robotics [55, 60]. The two ITZ zones at the left and right cannot easily be distinguished from the central zone! Numerous dead-end pores branch off main pore trunks that directly connect external free surfaces of the specimen (e.g. top and bottom). All these connections modify the pore network almost into one pore tree, offering numerous transport routes. This is revealed by Fig. 7.27, which presents the gradient in continuous porosity between two adjacent aggregate grains

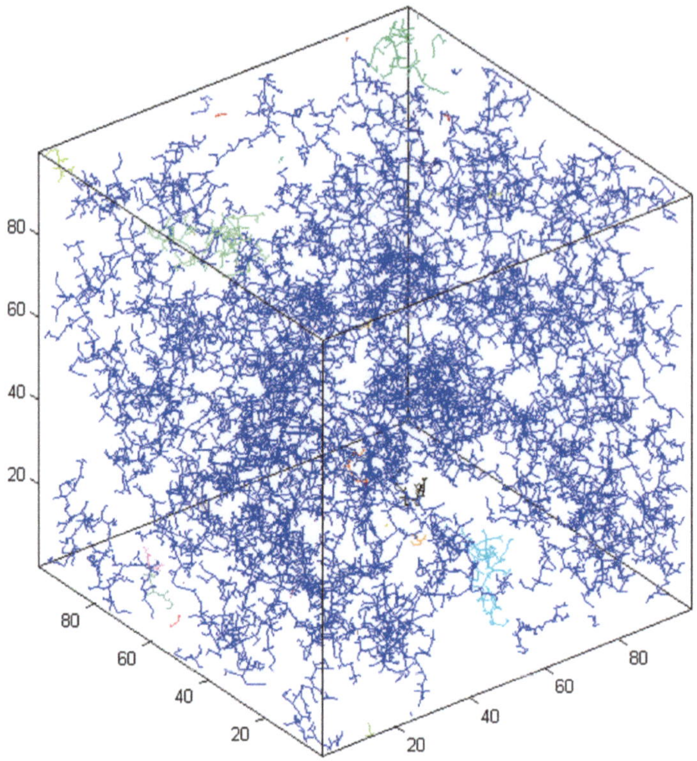

Fig. 7.26 Exploration by DRMTS approach of the pore tree system in virtual hydrated cement paste with $w/c = 0.2$ at hydration time of 1,440 h (2 months); fresh cement paste was simulated by HADES and casted in 100 μm cubes with two rigid (*left and right*) and four periodic surfaces. Finally, it was hydrated by CEMHYD [3] system [55] (Reproduced with permission; courtesy Prof. El-Batahgy)

underlying Fig. 7.26 and the continuous porosity in a specimen with periodic boundaries representing a cement pocket with remote aggregate grain surfaces.

Consequence of Fig. 7.27 as compared to Fig. 7.24b is that outside the narrow zone of direct connections between top and bottom surface of the specimens (parallel to the aggregate grain surfaces at the left and right) along pore "trunks", connections are established between the pores branching far off the main trunks. This will provide a significant contribution to bulk paste and not just to the ITZs. This is supported by only modest effects sorted by ITZ percolation on transport-based phenomena [72].

Generally speaking, this porosimetry research may convincingly demonstrate that particle packing design on both meso-level (aggregate) and on micro-level (cement paste) govern the details of the pore structure in cementitious materials, and, thus, the hydraulic properties of the material. This consequently extends to the durability performance of cementitious composites in engineering structures of the built environment.

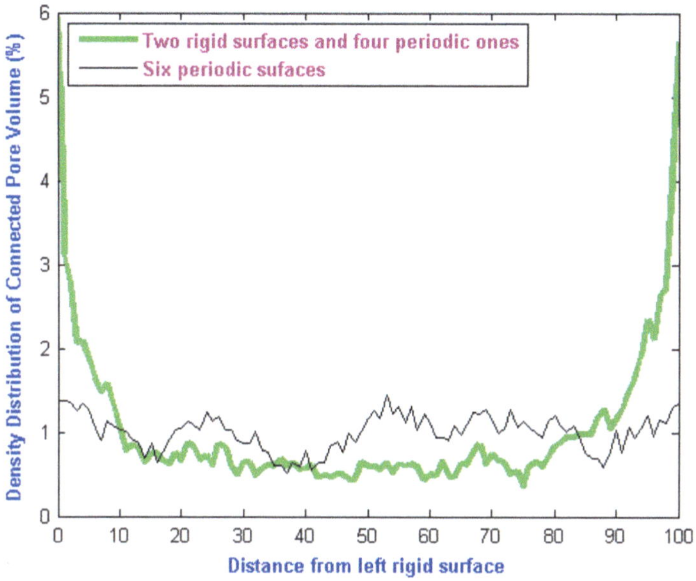

Fig. 7.27 Gradient structures in connected pore volume at two different boundary conditions, of which the *green* one obtained from Fig. 7.18 [55] (Reproduced with permission; courtesy Prof. El-Batahgy)

7.5 Workability or Consistence

Fresh concrete is also called "Green Concrete". It is a stage of concrete in which concrete is in the plastic state that allows the material to be molded. The general notion to describe the state of fresh concrete is workability or consistence; this is the ease with which concrete will flow and can be compacted without loosing its homogeneity, or integrity.

Factors affecting concrete workability are:

- Water cement ratio
- Amount and type of Aggregate
- Amount and type of Cement
- Weather conditions
  - Temperature
  - Wind
- Chemical Admixtures
- Sand to Aggregate ratio

The particulate notions concern the aggregate mixture design, the selection of the type of aggregate and the design of the binder. Firstly, increased internal surface area of the aggregate negatively influences consistence. Secondly, blending the cement by a fine mineral admixture like silica fume dramatically increases internal

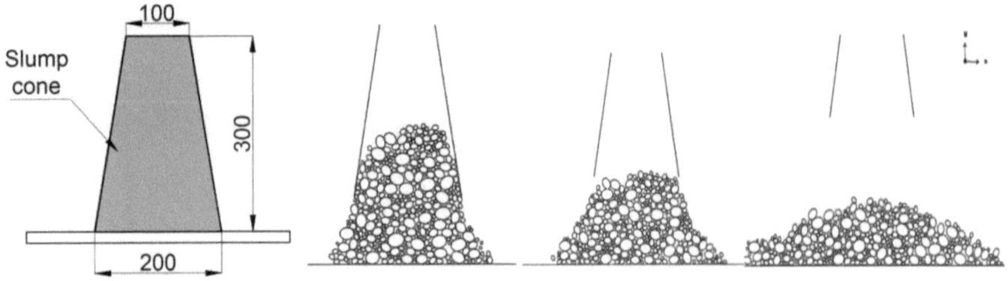

Fig. 7.28 Set up (*left* in mm) and (*right*) stages in the slump test as simulated by a DEM system (HADES) [57] (Reproduced with permission; copyright Springer)

Table 7.4 List of test methods for workability properties of SCC [21]

Test number	Test type
1	Slump-flow by Abrams cone filling ability
2	T 50 cm slump flow filling ability
3	J-ring passing ability
4	V-funnel filling ability
5	V-funnel at T 5 min segregation resistance
6	L-box passing ability
7	U-box passing ability
8	Fill-box passing ability
9	GTM screen stability test segregation resistance
10	Orimet filling ability

surface area, so requires the use of a water reducing agent or superplasticizer. Hence, chemical admixtures are very important in modern concrete technology.

The consistence following from the material design can be checked in a lot of different ways, of which the slump-flow test is a very popular one. In general, the slump flow test is very similar to the standard ASTM C143/C143M (2010) [73] slump test, whereby the sagging of the top surface of the concrete cone is measured as depicted in Fig. 7.28. For that purpose, Abram's cone is placed in the center of the slump flow board, either in the normal orientation (large opening down) or inverted (small opening down). It is filled in one lift (no rodding or other consolidation) with concrete, taking care that the sample is well mixed and not segregated in the sampling process. The cone is then raised in 3 ± 1 s to a height of 230 ± 75 mm (9 ± 3 in.), allowing the fluid concrete to flow onto the slump flow board. The slump flow is the diameter of the resulting concrete "patty" obtained from the average of measuring the greatest diameter and diameter perpendicular to this direction. Large differences between the two diameters indicate a non-level surface, which must be corrected. The result is reported to the nearest 10 mm (half-inch). Self Compacting Concrete (SCC) generally has slump flow of 560–760 mm (22–30 in.). DEM can be employed for the simulation of the slump-flow test as an example shown in Fig. 7.28.

As stipulated, a very wide range of other tests are being employed for similar purposes (Table 7.4), in some cases also involving reinforcement bars as obstruction.

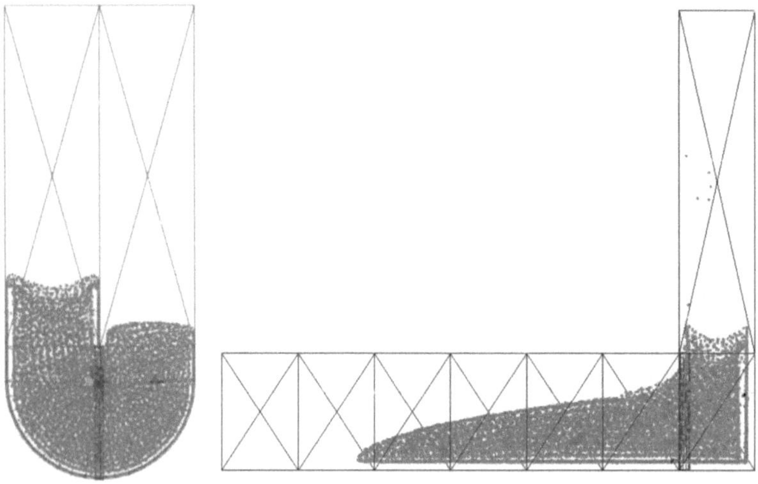

Fig. 7.29 Simulations of SCC in a U-shaped test (*left*) and in an L-Box test (*right*) by a particle-based computational fluid dynamics method

See [21] for a complete survey of such methods all standardized in relevant building codes. Two examples, simulated by computer, are presented in Fig. 7.29 for SCC. Table 7.4 gives a survey of methods standardized for SCC only. The fresh concrete has been modeled using a particle-based computational fluid dynamics (CFD) system. The elements of the simulation use physical properties to control their behavior and can interact with each other and react according to impulses, forces and accelerations. Such approaches make it possible correlating outcomes of the various tests in vogue for concrete.

7.6 Special Concretes

As argued above, the packing design is getting more crucial for modern material developments, such as self-compacting concrete (SCC), HPC and UHPC and engineered cementitious composites (ECC). In the higher performance domain the inclusion of various combinations of short fibers is common. Their packing is also a popular subject of research, because the assumption of isotropic uniform randomness (IUR), as it is generally adopted in design, can be far off reality. Design of the fiber reinforcement thus depends on the effect of such packing details on stress transfer capability over cracks introduced in the material body during maturation and as a result of external loading.

7.6.1 Self Compacting Concrete

Self compacting (also called self-leveling or self-consolidating) concrete is a concrete which compacts itself; there is no further compaction by vibration required. Making concrete structures without vibration has been done in the past, e.g., the placement of concrete under water. Mass concrete, and pile concrete can be successfully placed without vibration. But these concretes are generally of lower strength and it is difficult to obtain consistent quality. Modern application of self-compacting concrete (SCC) is focused on high performance, and thus a better and more reliable and uniform quality [5].

Recognizing the lack of uniformity and complete compaction of concrete by vibration, researchers at the University of Tokyo, Japan, started in late 1980s to develop self compacting concrete, a concrete that did not require vibration to achieve full compaction [29]. The utilization of self compacting concrete started growing rapidly. By the year 2000, SCC had become popular in Japan for prefabricated products and ready mixed concrete [30]. Of course, SCC also spread through Europe and USA.

Self compacting concrete has been described as "the most revolutionary development in concrete construction for several decades". It has proven to be beneficial from the following points:

- Faster construction
- Improved durability
- Reduction in site manpower
- Better surface finish
- Easier placing
- Safer working environment.

SCC is a relative newcomer to the building industry, when one considers the length of time concrete has been in use. The development of a variety of admixtures has allowed the production of a type of concrete that is more cost effective as it reduces construction time significantly, it does not require vibration, compaction and in many cases finishing either. However, the latter is highly dependent on multiple factors, including mix design. In the framework of this chapter it can therefore be stressed that particle packing is also a major phenomenon in SCC [43].

Mix design principles are similar as for normal concrete: successive particle fractions fill up the openings in the skeleton of larger grains. To do so efficiently, generally gap-graded size fractions are employed. Between the fine sand and the Portland cement, a fine powder fraction is added, together indicated as "powder" in Fig. 7.30. Due to deviating significantly from "normal" concrete, special guidelines for designing and testing SCC have been developed [36, 40, 75, 77].

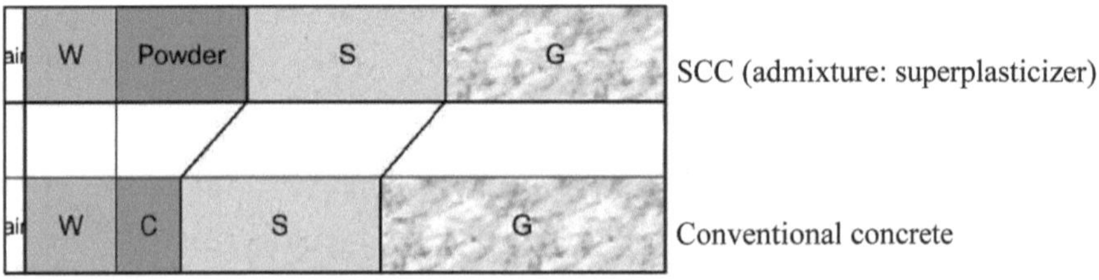

Fig. 7.30 Scheme of compositions of normal concrete and SCC (after [31]); *C* cement, *W* water, *S* sand, *G* gravel (Reproduced with permission; copyright JCI)

As with any field that has practical applications, there are multiple theories relative to mix design and it is quite likely that in the field none of these are applied exactly as written. Practice tends to uncover the impracticalities of theories. However, the mix design methods used in the field and in practice are beyond the scope of this book as it would require discussion with many concrete producers, who would likely be reticent to reveal they do not abide to the letter by the design process. There are three basic variables that have been recognized by all experts as affecting the properties of fresh concrete, namely the properties of the mortar, the ratio of coarse aggregate in the whole mix, and the use of super-plasticizing admixtures to improve fluidity and workability. As with traditional concrete, the mortar will have a serious impact on the final result. The ratio of sand in the mortar will determine the degree of workability/consistence as well as segregation resistance. The super-plasticizer's main role is to improve workability and the quality and quantity used will be the main factors that determine its flowability.

- *The rational mix design method* [32]

This is one of the simplest methods and was developed in 1995 in Japan, being based on the fact that self compacting concrete is affected by the proportions of the various materials as well as their quality. The fundamental principle is that the aggregate content is fixed – both coarse and fine – so that the flowability can be achieved only through the variation of the superplasticizer ratio.

- *Linear optimization mix proportioning* [10]

This type of mix design method for self compacting concrete was developed in the UK as a result of research done to judge whether the materials available were suitable for use in SCC. The method is based on the rational mix design method but also incorporates linear optimization to improve on the initial design method. This may appear as a more complex method but the formulas can be entered into a spreadsheet that allows for the use of linear optimization.

- *Model for SCC* [34]

This is a method that was designed in the field to meet the rigors and specifications of various construction projects. It includes such stages as determining the void content, blocking criteria, mortar proportions and concrete proportions. All these criteria are fulfilled according to the requirements of the project in question, considering strength, durability and so on.

Trial and error methods are often applied in addition to some theory or model on proper mix design, to find optimum basic mixtures for SCC. Further refinement of the SCC composition depends on the project itself, viz. local conditions and requirements. An example of a successful commercial basic mixture for SCC is given in Table 7.5.

Table 7.5 Basic composition of a successful mixture for SCC [www.leyde.com/England/.../BETONAC-SCC.pdf]

Cement (kg/m^3)	Fly ash (kg/m^3)	Water (kg/m^3)	Sand (kg/m^3)	Gravel (kg/m^3)	Superplasticizer (%)
350	200	170–180	650	950	0.8–2.0

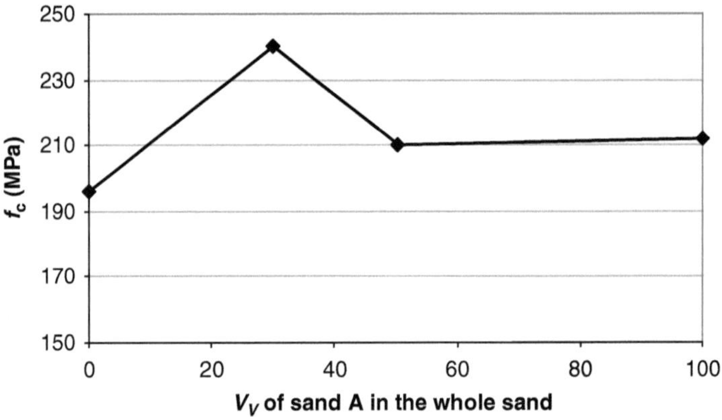

Fig. 7.31 28-days compressive strength of concrete (f_c) containing different proportions of sand A (d_{max} = 0.2 mm) in sand B (d_{max} = 0.8 mm)

7.6.2 High and Ultra High Performance Concrete and Steel Fiber Concrete

Design of the particle mixture is a crucial stage in the development of HPC and UHPC. Large aggregate is excluded (maximum aggregate size is of the order of 1 mm in UHPC), and the finer particle ranges are gap-graded, so packing density is increased, aiming for exclusion of defects in the hardened material. *w/c* is reduced, in UHPC to 0.2 ~ 0.25, accepting that due to water shortage part of the cement cannot contribute to hydration, but will do so to packing density. High-quality cements are blended by silica fume that is considerably finer than the cement. Furthermore, limestone powder or glass powder is added [65]. Short organic (polypropylene, poly-vinyl alcohol (PVA)) and metal (steel or carbon steel) fibers are added to raise the tensile strength [30]. UHPC like the Lafarge-produced Ductal [33] that has been developed in the past decades has very low porosity and displays very low shrinkage and creep. At densities up to 2,600 kg/m^3, compressive strength can be up to 100–250 MPa and a direct tensile strength in the 5–10 MPa range. UHPC is therefore advertised *as a material with properties between concrete and steel.*

Figure 7.31 presents compressive strengths values of UHPC revealing the influence of the proportion of fine sand in the mixes (specified in detail in [65]). The packing parameters have obvious influence on the compressive strength as also demonstrated by Figs. 7.3 and 7.6. The choice of the type of superplasticizer is additionally shown of paramount importance [65]. The amount of superplasticizer in the fine-grained mixtures provided them with self compacting properties, despite the low value of *w/c* (0.2). This shows the gradual transfer from one type/class of special concretes into another one. It is also common to classify such mixtures in the category of HPSFRCC, high performance steel fiber reinforced cementitious composites.

Fiber reinforcement modifies material behavior of normal concrete fundamentally. When normal concrete is subjected to tensile stresses it will crack due to its low tensile strength. Cracking on micro-level generally starts at the surface of the aggregate grains. This may even occur in the so called virgin state, before the material is subjected to any external loading. However, once subjected to such loadings, these interface cracks start coalescing and soon a major crack will cut the specimen in halve. Fibers provide the material with a residual strength after cracking. Residual strength will depend on volume fraction, on fiber geometry, orientation and dispersion (also, on fiber packing!), on the cementitious matrix and on the production conditions (in particular on compaction by vibration). Hence, fiber reinforced concrete is another commodity. Relevant information can be obtained from the vast international literature, encompassing thousands of papers.

7.6.3 Engineered Cementitious Composites

The application of dispersed short fibers has led to the gradual development of HPFRCC and Ultra HPFRCC (UHPFRCC); names adopted by Naaman and Reinhardt since the early 1990s [26]. This also gave rise to the development of engineered cementitious composites (ECC) by Li and co-workers [22], Ductal at Lafarge [33] and by Rossi and co-workers [37]. The latter developed the principle of the multi-scale concept into Multi-Scale Cement Composites (MSCC), which contain three steel fiber geometries totaling more than 10 % by volume aimed for structural use without traditional reinforcement. Specifically, $CEMTEC_{multiscale}$ has been developed for elements in bending. Modulus of Rupture (MOR) values of around 50 MPa have been recorded in experiments [38].

Basically, ECC can be resorted under the category of (U)HPFRCC. Quite obviously, this also involves ECC with self-compacting capabilities. Self-compacting engineered cementitious composite (ECC) was developed by optimizing the micromechanical parameters, which control composite properties in the hardened state, and the processing parameters that govern the rheological properties in the fresh state. In the development concept of self-compacting ECC, micromechanics is adopted to properly select the matrix, fiber, and interface properties to exhibit strain hardening (generally attributed as pseudo-strain hardening) and multiple cracking behavior in the composites. With the selected ingredient materials, the self-compactability of ECC is then realized by the controlled rheological properties of fresh matrix and the uniform dispersion of fibers. The controlled rheological properties of fresh matrix, including deformability, flow rate, and self-compactability is a result of adopting an optimal combination of a superplasticizer and a viscosity agent. According to the measurements of slump flow and the self-placing test result, the ECC developed is proven to be self-compacting. Flexural tests demonstrate that the mechanical performance of self-compacting ECC is insensitive to the externally applied consolidation during placing.

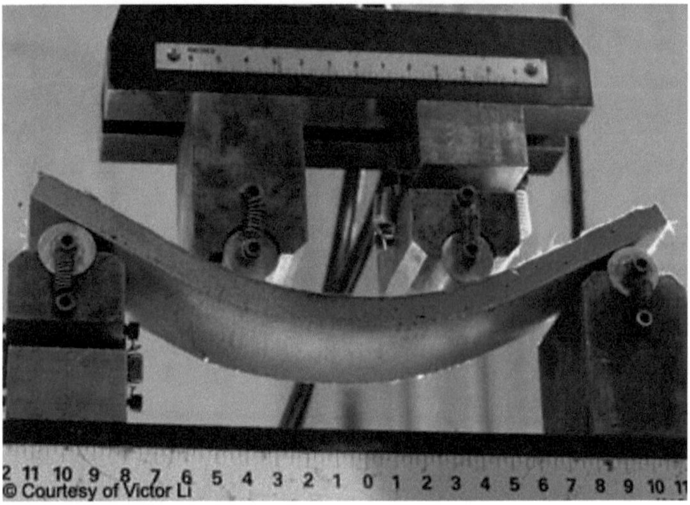

Fig. 7.32 ECC material in pure bending reveals large deformations due to multiple cracking [22] (Courtesy Prof. Li)

The same reasoning holds for ECC in general. It is based on ingredients similar to those of FRC or HPFRC and contains only 2 % by volume or less of discontinuous fibers. Micromechanical optimization or composite constituent tailoring leads to desired performance characteristics of ECC. This explicitly involves pseudo-strain hardening resulting from multiple cracking. Structural elements subjected to bending can reveal behavior similar to that of steel (Fig. 7.32). Therefore, the approach is also denoted as "Performance Driven Design Approach" [22].

Generalizing, the conclusion can be drawn that the optimum packing of particle ranges and of fibers is an integral element of this performance-driven design approach in the field of concrete technology as far as special concretes are at issue.

7.7 Conclusions

Concrete has a particulate nature on different structural levels. Hence, particle packing is a relevant phenomenon in the fresh as well as in mature states. Nevertheless, design was never strictly based on packing principles; it was merely at the background of building code requirements. A vast literature exists in concrete technology that was for a long time predominantly of a trial and error nature. Concrete is a relatively cheap material, dramatically impeding the necessity of designing with safety factors slightly higher than unity. This had similar repercussions for the nature of research. The development of special concretes in competition with other types of material has somewhat changed the scene. Packing has been re-discovered as the fundamental basis for a particulate material such as concrete. Moreover, with the fast development of computer facilities, numerical simulation offers an economic and efficient approach to concrete research. Equally, an extensive literature on packing phenomena allows for a quick start. Unfortunately, during the past decades the RSA method has been developed for that purpose in concrete technology. This hampers the wide-spread employment of superior DEM. Proper DEM is based on a concurrent algorithm that has direct reflections of particle interaction in normal concrete during production. These are therefore more reliable for assessments of structure-sensitive properties that are primarily of engineering relevance, such as those related to strength and durability. An important property is therefore concrete porosity.

DEM used to simulate and study particle packing problems is reflecting modern material developments in concrete technology. Packing principles, which are underlying particulate materials such as concrete, were never explicitly employed in design. However, the development of the advanced concrete composites during the past several decades was only possible on the basis of insight into packing phenomena. This chapter pays therefore special attention to this more recent history of concrete technology. It shows the transformation of this, centuries-old construction material, to far more sophisticated cementitious composites by employing the scientific packing principles also exploited in other material disciplines.

7.7.1 Definitions, Abbreviations and Symbols

Concrete	Cementitious material containing aggregates of sizes larger than 5 mm (or 4.76 mm)
Consistence	Ease of fresh (green) concrete to flow and be compacted without loosing its homogeneity and integrity (note that there are different methods for its determination)
Mortar	Cementitious material containing aggregates of sizes smaller than 5 mm (or 4.76 mm)
Porosity	For concrete: final space left in between the particles after compaction; note that different pore sizes may influence different characteristics, such as strength and diffusion of water and other compounds
Workability	See consistence
A	Aluminum oxide in CCN notation
ASTM	American Society for Testing Materials
BNE	Brazil nut effect
C	Calcium oxide in CCN notation
CCN	Cement chemist notation
CFD	Computational fluid dynamics
CUR	Civieltechnisch Centrum Uitvoering Research en Regelgeving (Dutch Civil-technical Centre for Execution of Research and Regulation)
DEM	Discrete element method
DRMTS	Double random multiple-tree structuring
ECC	Engineered cementitious composites

(continued)

(continued)

EN	European standard
F	Ferric oxide in CCN notation
HADES	HAbanera's Discrete Element Simulator
HPC	High-performance concrete
HPFRCC	High-performance steel fiber reinforced cementitious composites
ITZ	Interfacial transition zone
IUR	Isotropic uniform randomness
MSCC	Multiscale cementitious composite
MOR	Modulus of rupture
PC	Portland cement
PSD	Particle size distribution
PoSD	Pore size distribution
PVA	Polyvinyl alcohol
QIA	Quantitative image analysis
RRT	Rapidly exploring random tree algorithm
S	Silicon dioxide in CCN notation
$\overline{S}$	Sulfur trioxide in CCN notation
SCC	Self-compacting concrete
SEM	Scanning electron microscopy
SFRCC	Steel fiber reinforced cementitious composites
SIFCON	Slurry infiltrated fiber concrete
SPACE	Software Package for the Assessment of Compositional Evolution
UHPC	Ultra-high-performance concrete
UHPFRCC	Ultra-high-performance steel fiber reinforced cementitious composites
UHPP	Ultra-high-performance paste
a	Aspect ratio (=slenderness) of fibers
d	Sieve opening
$d_{\max}$	Maximum sieve opening (=size) of sand
D	Maximum grain size in the mixture
D_l	Sieve opening (=size) of large mono-size particles
D_s	Sieve opening (=size) of small mono-size particles
f_c	28-days compressive strength of concrete
l_{i}	Distance from random point in pore to its perimeter (in section)
P_d	Volume percentage of particles passing through the sieve with opening d
S	Surface area
V	Volume
V_{f}	Volume fraction of fibers
V_{fcrit}	Critical volume fraction of fibers (leading to strain hardening)
V_V	Volume fraction of particles
ν^*	Pore volume
w/c	Water/cement ratio
β	Strength reduction factor for fiber reinforcement
η	Geometric efficiency factor of fiber reinforcement
λ	Mean value of uninterrupted surface-to-surface distances between all neighbouring particles
σ_c	Tensile strength of fiber reinforced concrete
σ_m	Tensile strength of plain concrete
τ_f	Friction resistance between fiber and cement matrix

Annex 7A

Selection Procedure to Fulfil Construction Requirements

In normal practice, the design issue is reduced to finding proper economic solutions in the given situation (using e.g. local aggregates) through the use of building codes (see bibliography). It is therefore more a selection procedure than a scientific design operation. First, the procedure involves selection of the relevant exposure class (Table 7.6).

The next selection step regards the intended life time for the construction (Table 7.7).

The next selection step concerns the required strength of the cover layer on the main reinforcement. With smaller thicknesses (still exceeding the prescribed minimum for the relevant part of the construction), a higher concrete grade would be required to guarantee proper life time of the construction in the relevant environmental conditions (Table 7.8). A higher concrete grade has negative impact on economy of the full structure; a thicker cover layer does not always require increasing thickness of the structural element.

The fresh concrete also sets building code requirements, which specify a range of slump and flow classes for consistence (Table 7.9). Higher slump or flow values as alternatives make filling of the mould easier, but cause a reduction in concrete quality, if done by water addition. Then, not just strength is reduced but also durability is negatively affected. So, once grade is selected, this cannot be the way. Instead of water, chemical admixtures are employed. These, however, increase material costs.

Table 7.6 Main exposure classes

Description	Designation
No risk of corrosion or attack	X0
Corrosion induced by carbonation	XC
Corrosion induced by chlorides other than from seawater	XD
Corrosion induced by chlorides from seawater	XS
Freeze/thaw attack with or without de-icing agents	XF
Chemical attack; Abrasion	ACEC class

Table 7.7 Intended working life time

Temporary structures	10 years
Replaceable structural parts	10–25 years
Agricultural and similar structures	15–30 years
Building structures and other common structures	50 years
Monumental building structures, bridges and other civil engineering structures	100 years

Table 7.8 Compressive strength classes for conventional concrete

Required grade, i.e. minimum cube strength, N/mm^2	Compressive strength class	Required grade, i.e. minimum cube strength, N/mm^2	Compressive strength class
10	C8/10	50	C40/50
15	C12/15	55	C45/55
20	C16/20	60	C50/60
25	C20/25	67	C55/67
30	C25/30	75	C60/75
35	C28/35	85	C70/85
37	C30/37	95	C80/95
40	C32/40	105	C90/105
45	C35/45	115	C100/115

Table 7.9 Target values for consistency of fresh concrete

Target slump value, mm	Slump class	Target flow value, mm	Flow class
20	S1	380	F2
70	S2	450	F3
120	S3	520	F4
180	S4	590	F5

References

1. Alexander, M., Mindess, S.: Aggregates in Concrete. Taylor & Francis, London (2005)
2. Almusallam, A.A., Beshr, H., Maslehuddin, M., Al-Amoudi, O.S.B.: Effect of silica fume on the mechanical properties of low quality coarse aggregate concrete. Cem. Concr. Comp. **26**(7), 891–900 (2004)
3. Bentz, D.P.: Three-dimensional computer simulation of portland cement hydration and microstructure development. J. Am. Ceram. Soc. **80**, 3–21 (1997)
4. Bentz, D.P., Garboczi, E.J.: Percolation of phases in a three-dimensional cement paste microstructure model. Cem. Concr. Res. **21**, 325–344 (1991)
5. Brouwers, H.J.H., Radix, H.J.: Self-compacting concrete: Theoretical and experimental study. Cem. Concr. Res. **35**, 2116–2136 (2005)
6. Bui, D.D., Hu, J., Stroeven, P.: Particle size effect on the strength of rice husk ash blended gap-graded Portland cement concrete. Cem. Concr. Comp. **27**(3), 357–366 (2005)
7. Bullard, J.W., Garboczi, E.J.: A model investigation of the influence of particle shape on Portland cement hydration. Cem. Concr. Res. **36**, 1007–1015 (2006)
8. Chen, H.S., Stroeven, P., Ye, G., Stroeven, M.: Influence of boundary conditions on pore percolation in model cement paste. Key Eng. Mat. **302–303**, 486–492 (2006)
9. Diamond, S.: Mercury porosimetry; an inappropriate method for the measurement of pore size distributions in cement-based materials. Cem. Concr. Res. **30**(10), 1517–1525 (2000)
10. Domone, P.L.J., Xu, Y., Banfill, P.F.G.: Development of the two-point workability test for HPC. Mag. Concr. Res. **51**(3), 171–179 (1999)
11. Elam, W.T., Kersten, A.R., Rehr, J.J.: Critical properties of the void percolation problem for spheres. Phys. Rev. Lett. **52**(17), 1516–1519 (1984)
12. Farran, J.: Contribution minéralogique à l'étude de l'adhérence entre les constituants hydrates des ciments et les matériaux enrobés. Revue des Matériaux de Construction et Travaux Publics **490–491**, 155–172 (1956)

13. Garboczi, E.J., Bullard, J.W.: Shape analysis of a reference cement. Cem. Concr. Res. **34**, 1933–1937 (2004)
14. Goldman, A., Bentur, A.: The influence of microfillers on enhancement of concrete strength. Cem. Concr. Res. **23**(4), 962–972 (1993)
15. He, H.: Computational modelling of particle packing in concrete. Ph.D. Thesis, Delft University of Technology, Ipskamp Drukkers, Delft (2010)
16. He, H., Courard, L., Pirard, E.: Particle packing density and limestone fillers for more sustainable cement. Key Eng. Mater. **517**, 331–337 (2012)
17. Hu, J., Stroeven, P.: Shape characterization of concrete aggregate. Image Anal. Stereology **25**, 43–53 (2006)
18. Hu, J.: Porosity of concrete – Morphological study of model concrete. Ph.D. Thesis, Delft University of Technology, OPTIMA, Delft (2004)
19. Kasperkiewicz, J., Malmberg, B., Skarendahl, A.: Determination of fiber content, distribution and orientation in steel fiber concrete by X-ray technique. In: Swamy, R.N. (ed.) Testing and Test Methods Offiber Cement Composites, pp. 297–314. Constr. Press, Lancaster (1978)
20. Kendall, M.G., Moran, P.A.P.: Geometrical probability, Griffin's Statistical Monographs and Courses (1963)
21. Koehler, E.P., Fowler, D.W.: Summary of concrete workability test methods. Research report ICAR 105–1, International Center for Aggregate Research, The University of Texas at Austin, Austin. (http://www.icar.utexas.edu/publications/105/105_1.pdf) (2003)
22. Li, V.C.: From micromechanics to structural engineering – the design of cementitious composites for civil engineering applications. Struct. Eng./Earthq. Eng. **10**(2), 37–48 (1993)
23. Lees, G.: The measurement of particle shape and its influence in engineering materials. J. Br. Granite Whinstone Fed. **4**(2), 1–22 (1964)
24. Naaman, A.E.: Tensile strain-hardening FRC composites: Historical evolution since the 1960. In: Grosse, C.U. (ed.) Advances in Construction Materials, pp. 181–202. Springer, Berlin/Heidelberg (2007)
25. Naaman, A.E, Moavenzadeh, F., McGarry, J.: Probabilistic analysis of fibre reinforced concrete. J. Eng. Mech. Div. A.S.C.E., EM 2, April, 397–413 (1974)
26. Naaman, A.E., Reinhardt, H.W. (co-eds.).: High Performance Fiber Reinforced Cement Composites. E & FN Spon, London (1992; 1996; 1999; 2003; 2007)
27. Lankard, D.R.: Slurry infiltrated fiber concrete (SIFCON). Concr. Int. **12**, 44–47 (1984)
28. Navi, P., Pignat, C.: Three-dimensional characterization of the pore structure of a simulated cement paste. Cem. Concr. Res. **29**(4), 507–514 (1999)
29. Okamura, H., Ouchi, M.: Development, present use and future. In: Skarendahl, A., Petersson, O. (eds.) Proceedings of the 1st International RILEM Symposium on SCC. RILEM Publ., S.A.R.L., Bagneux, pp. 3–14 (1999)
30. Okamura, H., Ouchi, M.: Development of hybrid polypropylene-steel fibre-reinforced concrete. J. Adv. Concr. Technol. **1**, 5–15 (2003)
31. Okamura, H., Ouchi, M.: Self compacting concrete. J. Adv. Concr. Technol. **1**(1), 5–15 (2003)
32. Okamura, H., Ozawa, K.: Mix design for SCC. Concr. Libr. JSCE **25**, 107–120 (1995)
33. Orange, G., Acker, P., Vernet, C.: A new generation of UHP concrete: Ductal. Damage resistance and micromechanical analysis. In: Naaman, A.E., Reinhardt, H.W. (eds.) High Performance Fiber Reinforced Cement Composites 3. E & FN Spon, London (1999)
34. Petersson, O., Billberg, P., Van, B.K.: A Model for SSS, Proc. Production Methods and Workability of Concrete, pp. 483–492. E & FN Spon, London (1999)
35. Popovics, S.: Formulas on fineness modulus and specific surface area. RILEM Bul. **16**, 19–28 (1962b)
36. Reinhardt, H.W.: DAfStb guidelines on self-compacting concrete. Beton + Fertigteil Technik **67**(12), 54–62 (2001)
37. Rocco, C.G., Elices, M.: Effect of aggregate shape on the mechanical properties of a simple concrete. Eng. Fract. Mech. **76**(2), 286–298 (2009)
38. Rossi, P.: CEMTEC$_{multiscale}$: a new cement composite for building construction. In: Ying-shu Yan, Shah S.P., Heng-lin Lü (eds.) Advances in Concrete and Structures. RILEM, S.A.R.L., Bagneux

39. Schnautz, T., Brito, R., Kruelle, C.A., Rehberg, I.: Horizontal Brazil-nut effect and its reverse. Phys. Rev. Lett. **95**, 028001 (2005). doi:10.1103/PhysRevLett.95. 028001
40. de Schutter, G.: Guidelines for testing self-compacting concrete, Report European Research Project: Measurement of properties of fresh self-compacting concrete (2005)
41. Scrivener, K L.: Characterization of the ITZ and its quantification by test methods. In: Alexander M G., et al. (eds.) Engineering and Transport Properties of the Interfacial Transition Zone in Cementitious Composites, RILEM Report 20, ENS-Cachan: RILEM Publication S.A.R.L., pp. 3–15 (1999)
42. Scrivener, K.L., Nemati, K.M.: The percolation of pore space in the cement paste/aggregate interfacial zone of concrete. Cem. Concr. Res. **26**(1), 35–40 (1996)
43. Sedran, T., de Larrard, F.: Optimization of self-compacting concrete thanks to packing model. In: Skarendahl A., Petersson O., (eds.) Proceedings of the1st International RILEM Symposium on SCC. RILEM Publishers, S.A.R.L., Bagneux, pp. 321–332 (1999)
44. Stroeven, M.: Discrete numerical modeling of composite materials – application to cementitious materials. Ph.D. Thesis, Delft University of Technology, Meinema, Delft, the Netherlands (1999)
45. Stroeven, P.: Some aspects of the micro-mechanics of concrete. Ph.D. Thesis, Delft University of Technology, DUP, Delft, the Netherlands (1973)
46. Stroeven, P.: The analysis of fiber distributions in fiber reinforced materials. J. Microsc. **111**(3), 283–295 (1977)
47. Stroeven, P.: Fibre concrete. Heron **24**(4), 7–40 (1979)
48. Stroeven, P., Guo, W.: Structural modelling and mechanical behaviour of steel fibre reinforced concrete.In: Fibre Cements and Concretes; Recent Developments, pp. 421–434. Elsevier Applied Science, London (1989)
49. Stroeven, P., Guo, Z.: Modern routes to explore concrete's complex pore space. Image Anal. Stereol. **25**(2), 75–85 (2006)
50. Stroeven, P., He, H.: Impact of Brazil nut effect on concrete's structure and its engineering properties. In: Doyle, S.G. (ed.) Construction and Building: Design, Materials and Techniques, pp. 179–194. Nova Science Publishers, Happauge (2011)
51. Stroeven, P., He, H., Guo, Z., Stroeven, M.: Particle packing in a model concrete at different levels of the microstructure: evidence of an intrinsic patchy nature. Mater. Charact. **60**(10), 1088–1092 (2009). doi:10. 1016/j.matchar.2009.02.011
52. Stroeven, P., Hu, J.: Effectiveness near boundaries of fibre reinforcement in concrete. Mater. Struct. **39**, 1001–1013 (2006)
53. Stroeven, P., Hu, J., Koleva, D.A.: Concrete porosimetry: aspects of feasibility, reliability and economy. Cem. Concr Compos. **32**(4), 291–299 (2010). doi:10.1016/j.cemconcomp.2010.01.007
54. Stroeven, P., Hu, J., Stroeven, M.: On the usefulness of discrete element computer modeling of particle packing for material characterization in concrete technology. Comput. Concr. **6**(2), 133–153 (2009)
55. Stroeven, P., Le, L.B.N.: Studying percolated porosity in concrete by DEM. In: El-Batahgy A-M., Waly M. (eds.) Proceedings of the IQCMEA-ICF conference, Processing, Performance and Failure of Engineering Materials, pp. 447–456. Luxor (2011) (available on CD)
56. Stroeven, P., Shah, S.P.: Use of radiography-image analysis for steel fiber reinforced concrete. In: Swamy, R.N. (ed.) Testing and Test Methods of Fiber Cement Composites, pp. 345–353. Construction Press, Lancaster (1978)
57. Stroeven, P., He, H., Stroeven, M.: Discrete element modeling approach to assessment of granular properties in concrete. J. Zejiang Univ. Sci. A (Appl. Phys. Eng.) **12**(5), 335–344 (2001)
58. Stroeven, P., Stroeven, M.: Assessment of packing characteristics by computer simulation. Cem. Concr. Res. **29**, 1201–1206 (1999)
59. Stroeven, P., Stroeven, M.: SPACE approach to concrete's space structure and its mechanical properties. Heron **46**(4), 265–289 (2001)
60. Stroeven, P., Le, L.B.N., Stroeven, M., Sluys, L.J.: Discrete element modeling approach to porosimetry for durability risk estimation of concrete. In: Oñate E., Owen D.R.J., (eds.)

Proceedings PARTICLES 2011, II International Conference Particle-based Methods, Fundamentals and Applications, Barcelona, 26–18 October 2011 (pp. 1–12 on CD)
61. Stroeven, P., Le, L.B.N., He, H.: Methodological approaches to 3D pore structure exploration in cementitious composites. In: Proceedings 13th NOCMAT conference on Novel Construction Materials and Technologies for Sustainability, Changsha, 22–24 Sept 2012, (To be published in Key Engineering Materials)
62. Stroeven, P., Shui, Z., Cheng, Y.: Methods for assessment of uniformity in carbon fibre dispersions in cementitious materials. In: Rossi P., Chanvillard G., (eds.) Proceedings of the Fifth International RILEM Symposium on Fibre-Reinforced Concrete, pp. 159–168. RILEM Publications S.A.R.L., Paris (2000)
63. Swamy, R.N., Mangat, P.S., Rao, C.V.S.K.: The mechanics of fiber reinforcement of cement matrices. ACI Spec. Publ. **SP-44**, 1–28 (1974)
64. Tasdemir, M.A., Karihaloo, B.L.: Effect of aggregate volume fraction on the fracture parameters of concrete: A meso-mechanical approach. Mag. Concr. Res. **53**(6), 405–415 (2001)
65. Wille, K., Naaman, A.E.: Bond stress-slip behaviour of steel fibers embedded in ultra high performance concrete. In: Mechtcherine, V., Kaliske, M. (eds.) Fracture and Damage of Advanced Fibre-reinforced Cement-Based Materials. Aedificatio Publishers, Freiburg (2010)
66. Williams, S.R., Philipse, A.P.: Random packings of spheres and spherocylinders simulated by mechanical contraction. Phys. Rev. E **67**(1), 051301 (2003). doi:10. 1103/PhysRevE.67.051301
67. Willis, K.L., Abell, A.B.., Lange, D.A.: Image-based characterization of cement pore structure using Wood's metal intrusion. Cem. Concr. Res. **28**(12), 1695–1705 (1998)
68. Windslow, D.N., Cohen, M.D., Bentz, D.P., et al.: Percolation and pore structure in mortars and concrete. Cem. Concr. Res. **24**(1), 25–37 (1994)
69. Wong, H.H.C., Kwan, A.K.H.: Packing density of cementitious materials: part 1–measurement using a wet packing method. Mater. Struct. **41**, 689–701 (2008)
70. Ye, G.: Experimental study and numerical simulation of the development of the micro-structure and permeability of cementitious materials. Ph.D. Thesis, Delft University of Technology, Delft (2003)
71. Zheng, J.J., Guo, Z.Q., Deng, D., Stroeven, P., Sluys, L.J.: ITZ volume fraction in concrete with spheroidal aggregate particles and application: part I. numerical algorithm. Mag. Concr. Res. **63**(7), 473–482 (2011)
72. Zheng, J.J., Guo, Z.Q., Huang, X.F., Stroeven, P., Sluys, L.J.: ITZ volume fraction in concrete with spheroidal aggregate particles and application: part II prediction of the chloride diffusivity of concrete. Mag. Concr. Res. **63**(7), 483–491 (2011)
73. ASTM C143/C143M.: Standard Test Method for Slump of Hydraulic-Cement Concrete. ASTM, West Conshohocken (2010)
74. ASTM standard C150/C150M.: Standard Specification for Portland cement. ASTM, West Conshohocken (2009)
75. BIBM, CEMBUREAU, ERMCO, EFCA, EFNARC.: The European guidelines for self-compacting concrete, specification, production and use (Available on Internet) (2005)
76. CUR – Committee for Concrete Research in the Netherland: Optimum composition of concrete, a literature survey, CUR (Concrete Society in the Netherlands), Gouda, Nederland, Report 62 (in Dutch) (1974)
77. CUR – Committee for Concrete Research in the Netherlands.: Self-compacting concrete, CUR (Concrete Society in the Netherlands), Gouda, Nederland, Report 93 (in Dutch) (2002)

Note that [73] to [77] is a small selection of relevant American Standards (ASTM), and reports of the Concrete Society in the Netherlands (CUR). Many more standards exist, e.g. British (BS), European (CEN), German (DIN) and International Standards (ISO). The reader is advised to search the web sites of these organizations for specific standards and for new editions.

Chapter 8
Chocolate

Henk G. Merkus

Abstract Chocolate was introduced as a drink in Central America more than 30 centuries ago. In the sixth century it was brought to Spain and later to other parts of Europe and Northern America, where it became – after sweetening and change of spices – very popular amongst the upper class. In the nineteenth century, Van Houten in the Netherlands invented an improved procedure to separate cocoa powder and butter out of cocoa mass, which led the way to the production of various solid chocolate products. Production proceeds according to a complex process that comprises many stages in order to develop the best aroma, optimum taste and good appearance. The particle size distribution (PSD) is important for both ultimate chocolate sensorial quality and optimum viscosity of the liquid chocolate mass during processing. Optimum particle size for a nice, creamy taste of solid chocolate is in the range 15–30 μm. This is reached during the grinding stages of the process. Viscosity at elevated temperatures shows a complex, non-Newtonian behavior, containing a yield stress and shear-thinning effects. The rheological behavior is conventionally expressed according to the Casson equation. A linear relationship is reported between the percentage of fines and the Casson yield stress. With respect to taste, the presence of few particles larger than 30 μm, usually indicated by the PSD parameter $D_{90;3}$, is used as a quality criterion. In view of the importance of adequate particle size measurement and the fact that the laser diffraction technique seems to have taken a dominant position it is recommended that the refractive index values for chocolate, be agreed and standardized.

H.G. Merkus (✉)
Retired Associate Professor, Delft University of Technology, Park Berkenoord 30, Pijnacker 2641 CZ, The Netherlands
e-mail: henkmerkus@hetnet.nl

H.G. Merkus and G.M.H. Meesters (eds.), *Particulate Products: Tailoring Properties for Optimal Performance*, Particle Technology Series 19,
DOI 10.1007/978-3-319-00714-4_8,

8.1 History

Chocolate products are produced from the seeds of the tropical cacao tree (Theobroma[1] cacao). Excavated pottery, dated around 1150 BC, showed its use as a drink by the Aztecs in Honduras [18, 19, 26]. The analysis results seem to indicate that the early beverage was actually a fruit wine, containing about 6 % alcohol, made by fermentation of cacao pulp, which contains about 12–15 % sugar. Later on, both Mayas and Aztecs used the cacao seeds to make a frothy, bitter, spicy drink through milling, fermentation and mixing the cocoa mass with water, chili pepper, cornmeal, vanilla and some minor ingredients. The Aztecs named this beverage '*xocolatl*' (bitter water). The chocolate drinks were used for various kinds of ceremonial purposes, in addition to everyday life. Chocolate was and still is not only considered as a fancy delicacy but also thought to be healthy and nutritious for people. In contrast, the theobromine ingredient renders chocolate toxic to cats and dogs.

Returning in 1502 from his 4th discovery journey to Central America, Christopher Columbus did not pay much attention to the cacao beans he had received as gifts but brought them to Spain as curiosities. About two decades later, the Spanish conquistador Don Hernándo Cortés saw the commercial value of the new beverage and sent recipes for its preparation back home [5, 33]. The Spaniards sweetened and fattened the drink, replacing the chili pepper by sugar, milk – which were unknown in Central America – and some other spices such as cinnamon. Gradually, optimum conditions for cultivation and preparation were unraveled and eventually trees were grown in Ceylon, Indonesia, the Philippines and West African countries. At present, the main cocoa producing countries are Ivory Coast, Ghana and Indonesia.

Maya painting on chocolate

[1] Theobroma means food of the gods.

Chocolate luxury in Europe

For about a century, the secrets of the chocolate beverage were kept in Spain. Then, its popularity spread to other countries in Europe, although the high price limited its use to the wealthy aristocracy.

Around 1828, Coenraad Van Houten in the Netherlands invented an improved technique for pressing of the cocoa mass to cocoa butter and powder, which ultimately enabled industrial manufacture of molded tablets of dark chocolate. It took until 1847 before Fry developed the first tablet of real eating chocolate. Following this, milk chocolate tablets were developed, in 1875 by the Swiss Daniel Peter, after his compatriot had created milk powder in 1867 [9].

Nowadays, many types of chocolate tablets, bars and figures – with or without fillings – are available, which suit a wide variety of tastes all over the world. The total cocoa production is about 3.5–4 million ton/annum. The price is about 3000 US$/ton [21]. The per capita consumption of chocolate products in the western world is about 2–10 kg/person.

8.2 Manufacturing Process

Varieties of the cacao[2] tree (Theobroma cacao), which only grows under very specific tropical conditions, are at the basis of chocolate manufacture [5, 33, 41]. The flowers on the trunks and thicker branches develop into a relatively

[2] The word cacao is the botanical name and refers to the tree, the pods and the unfermented beans from the pods. The word cocoa refers to all products after fermentation of the beans.

small number of pods, which ripen, not all at the same time but over a period of several months. The ripe pods, which are about 15–30 cm long and 8–10 cm wide, and may show a variety of colors (yellow to orange and purple), do not drop off but have to be harvested. Per annum, harvesting typically takes place during two periods.

After the pods are cut off and broken, the contained seeds and adhering pulp are removed from the pod shells and transferred to heaps or baskets for *natural fermentation*. About 15 % sugar and other carbohydrates is contained within the pulp and seeds. Adequate conditions for even fermentation and good timing are of vital importance for good final taste, since flavor precursors are formed at this stage. Depending on the variety, the fermentation takes between 5 days and 1 week.

Drying starts, from about 60 % moisture to an optimum of about 7 %. This moisture content is critical, since the beans become brittle below about 5 %, and molds start developing above about 8 %. In dry tropical seasons, drying in the sun is preferred since it gives optimum quality. In wet seasons or areas, artificial means for drying are necessary in order to prevent the development of molds. This method of drying takes about 1–2 weeks. The fermented and dried seeds are commercially known as cocoa beans or cocoa. These beans are usually shipped to a chocolate manufacturer for further processing. Main cocoa processing countries are The Netherlands, Ivory Coast, the United States, Germany and Malaysia.

Upon receipt, the beans are *cleaned* to remove fibers, stones, grit, twigs, bean clusters, immature beans and metal in a series of screening, sifting, brushing and magnetic separations.

From pods on a cacao tree to chocolate

The beans are then *roasted* and *sterilized* at temperatures in between 100 °C and 120 °C. The temperature and roasting time depends upon the envisaged product type (chocolate or cocoa powder) and on the type of roaster. The roasting process is essential for developing good flavor and aroma through chemical reactions between the components present. It is also to reduce the moisture content and to render the shell in a loose condition for later removal. Requirements are even heating and good removal of water and volatile components from the beans.

The next step in processing is *winnowing*. Here, the loosened shells are broken, and the contained nibs (cotyledon parts) are separated from the shells by a combination of sieving and air elutriation. In view of the high costs of cocoa, the winnowing equipment has been improved over the years to improve both efficiency and yields.

In the next processing step, the nibs may be treated with an alkali solution (e.g. potassium carbonate) for efficient separation of cocoa powder and butter. This process is named *alkalization* or '*dutching*', after inventor Van Houten's country of origin. It mainly serves to improve color, flavor and dispersability of the ultimate cocoa powder. Additionally, it removes bacterial contamination to virtually zero, which is important in view of potential taste degradation in the further process.

Separation of cocoa powder and (part of the) butter results in a reduction of the fat content from about 55 % (in the nib) to 10–25 %, in relation to the final use of the cocoa powder. This is carried out by *grinding and hydraulic pressing*. The grinding of the cocoa mass determines the fineness of the final cocoa powder. Several mill types are applied, viz. stone mills, disc mills, hammer mills and, more recently, ball mills and pearl mills. In the milling process, the cell walls of the nibs are ruptured and the frictional heat developed keeps the cocoa fat liquefied. After grinding, cocoa powder and butter are separated by hydraulic pressing.

The cocoa powder and butter can now be used as ingredients for further products, viz. cocoa drinks, chocolate drinks, chocolate, ice cream, cosmetics, etc. Each of these products requires the addition of special additional ingredients, such as water, milk, emulsifiers and vanillin, cinnamon and/or other spices. Application of cocoa butter in pharmaceutical and other applications requires deodorization to remove the strong cocoa aroma.

For chocolate production, the essential ingredients are: alkalized cocoa mass or cocoa powder and butter; refined and premilled sugar; lecithin; vanillin; cinnamon and/or other flavoring agents. Milk powder or crumbs are added in the case of milk chocolate. Since cocoa mass and cocoa powder of different origins have different flavor characteristics, some varieties are typically mixed to reach a constant chocolate taste. The ingredients are *mixed* (see Sect. 8.3 for compositions), *refined* to reduce particle size and then *kneaded* for the removal of residual water and undesirable flavors, enabling the development of more pleasant flavors. It also aids the further size reduction and de-agglomeration of the particles. This process is also called *conching*. Temperatures, ranging from 60 °C to 100 °C, together with a content of lecithin (emulsifier) and water are important to optimize the viscosity of the suspension. Good equipment and time control are essential to control maximum particle size in view of taste and structure (see also Sects. 8.4 and 8.5).

Table 8.1 Typical chocolate compositions (% m/m) [12]

Component	Dark chocolate	Milk chocolate	White chocolate
Sugar	50	40	44
Cocoa mass	40	16	–
Cocoa butter	10	19	28
Milk powder	–	25	28
Lecithin	0.4	0.4	0.4
Aroma	0.02	0.02	0.02

The next step in the chocolate production is *tempering* of the suspension. It is a cooling/pre-crystallization step in which an optimum type and structure of fat crystals should be obtained. Note that cocoa butter can crystallize in different forms (polymorphism). The resulting product should melt at about 34 °C, well above room temperature to give the chocolate good solidity and appearance, but just below the human body temperature (37 °C), to provide a good bite and mouth feel properties.

The final process step in chocolate manufacture is *shaping and cooling* of the suspension to the wanted solid form and hereby developing a nice, glossy surface.

8.3 Composition

Chocolates are usually characterized as dark chocolate, milk chocolate and white chocolate. Composition depends on type, brand, grade and country [25]. Typical compositions are given in Table 8.1.

The fat level in chocolate ranges between about 30 and 40 % m/m, leaving a solids concentration of 45–60 % v/v. Note that official food regulations specify lower or upper concentration limits for the components. Customs regulations do not regard white chocolate as chocolate, since it does not contain any cocoa mass.

8.4 Essential Quality Elements During Processing and for Final Product

ISO defines product quality as “the totality of features and characteristics that bear on its ability to satisfy stated or implied needs” [5]. Some quality requirements are described above. In addition, the following elements are essential during processing or for the final product:

- Viscosity of the warm chocolate mass during mixing and grinding.
- Optimum particle size and restriction of maximum size in relation to taste.
- Restriction of amount of small particles in relation to viscosity.

- Mean particle size in relation to release of aroma.
- Correct fat crystal structure in view of adequate melting point.

As can be seen, an adequate control of the different parameters of the size distribution (PSD) of cocoa, milk and sugar particles is essential to both processing and optimum product quality. Literature seems to suggest that the actual PSD of commercial products solely depends on the equipment and conditions for grinding, although application of (mixtures of) specific size fractions, obtained by sieving or air classification, is reported in research studies [35, 39].

In relation to the milling mechanism or to its control, the PSDs are sometimes reported in terms of a modeled distribution, viz. log-normal, Rosin-Rammler (RR) or Gaudin-Schuhmann distribution [17, 34]. PSD parameters are reported to be used in relation to the milling process (e.g. the median size, the $D_{90;3}$ or the RR location parameter), but not to product quality aspects (see below) [6, 17]. Note that all the parameters of a PSD shift in the same direction when changing the grinding conditions in the same equipment, although not necessarily in the same quantitative manner. Therefore, all PSD parameters will show a correlated behavior in this case. However, different grinding equipments may result in different correlations.

8.4.1 *Particle Size Measurements*

The particle size of chocolate was traditionally measured by sedimentation (mostly sediment test) or by dry or wet sieving (mostly residue test on 75 or 45 μm sieve). These techniques yielded only one value (residue) or very few PSD values. Thus, the results may give a correlation with taste (see Sect. 8.4.2) but not with viscosity.

In order to obtain more PSD data, measurements with optical microscopes followed. Here, particle size is derived from the cross-sectional area of magnified particles. From the counting of particles of different size, a number-based PSD is obtained. Early determinations involved inspections of a rough nature or of relatively few particles in view of tedious manual procedures. Nowadays, computers allow automation and, thus analysis of tens of thousands of particles, for size as well as shape. In view of the required low particle concentration to avoid overlapping of particles in the image, many different fields of view have to be analyzed, which requires automated frame selection. Most often, the measurements are performed at a single magnification, which limits the measurement range to about a factor 30 (ratio of maximum to minimum size). The lower size limit of optical microscopy depends upon the magnification and is at minimum about 0.5 μm (while applying a 100 times objective with oil immersion).

The Gravitational sedimentation technique for PSD analysis was upgraded to yield results above about 1 μm [36]. A continuous start method was applied, whilst detection of particulate concentration at the given heights was done by means of a light beam. The size reported in this technique is a Stokes' diameter. The light extinction depends on particle concentration as well as size. The resulting PSD's are only comparable with those coming from other techniques if the extinction is calibrated versus concentration.

The electrical sensing zone technique was also applied, which allows precise measurement of size distributions at high resolution. Here, dispersed particles at high dilution in specified conducting dispersion liquids are classified based upon the resistance change as a particle passes through an orifice where an electric field is maintained. This resistance change is translated into particle size through calibration. Thus, the principal result is a number-based size distribution of particle sizes related to the particle volume. An advantage of this technique is that hundreds of thousands of particles may be analyzed easily. Note that the PSD is limited at both extremes of the size distribution in relation to the size of the orifice (particle size in between about 2 and 60 % of the orifice size). Note also that the electrolyte, used for making the dispersion liquid conducting, may cause agglomeration of particles.

Nowadays, laser diffraction (LD; also named static light scattering) seems to become the preferred technique, as it allows easier application at medium size class resolution. It yields volume-based sizing results and is based on a theoretical model to translate the measured angular light intensity pattern into a PSD. For this translation, a deconvolution procedure is applied to reach the best fitting PSD. The measurement typically takes place in suspensions in isopropyl alcohol or edible oil (to dissolve the fat) after adequate particle dispersion and dilution, and at stated refractive index values for both particles and dispersion medium. For the deconvolution, two optical models are used, viz. the Fraunhofer model and the Mie model. The former is simpler as it only regards the diffraction of light at the particle contours. Mie theory also takes refraction into account and, thus, requires refractive index (RI) parameters for both particles and dispersion medium. The Fraunhofer approximation provides satisfactory results for opaque particles above 10 μm but for transparent and semi-transparent particles Mie theory is preferred [24, 32]. Both models yield the same results above these limits. For chocolate particles, Fig. 8.1 clearly shows that the Fraunhofer interpretation leads to extension of the PSD into the finer particle region. This is due to the Fraunhofer model not accounting for the degree of refraction as compared to the Mie interpretation which, provided an appropriate RI value is employed, provides a more authentic result. Thus, it can be concluded that the Mie model is preferable. Unfortunately, different RI values for use in this Mie model are reported, probably in relationship with the fact that cocoa, milk and sugar particles have different optical properties.[3] This calls for agreed RI values for the different chocolates, established by a knowledgeable organization, like ICA, ISO or ASTM. Note that the laser diffraction results may be influenced by surface roughness of the particles,

[3] Afoakwa reports the use of 1.590 for the real RI part without imaginary part [4, 5]. Malvern Instruments reports 1.500 and 0.001 i respectively [private communication]. Do et al. use a real RI value of 1.45 without imaginary part [13]. This is at either side of the RI of main component sucrose, which is 1.538.

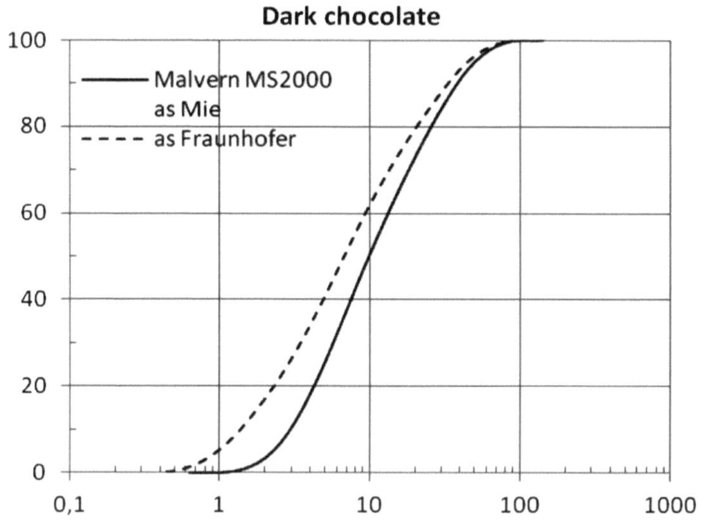

Fig. 8.1 Comparison Mie–Fraunhofer

which leads to more scattered light at larger angles and, thus, to extra, imaginary fine particles. It is often compensated for by applying, for transparent particles, a small imaginary part of the RI[4] (0.01–0.03 i). Pores, if present in the particles, may lead for the same reason to extra, imaginary fine particles. These effects may cause that the fine particles fraction and related size parameters resulting from the LD technique deviate systematically from the real situation.

It is remarked again that all techniques are based on some theoretical model and interpret the size of particles as diameters of spheres (or circles) that behave equivalent according to the principles of that model, together with specified particle properties. This often causes the sizing results for non-spherical particles to vary (somewhat) from technique to technique, since techniques have a different sensitivity to particle shape.

Sampling and sample dispersion are critical for obtaining accurate PSD results in case of wide PSD's. First, it is required that the sample is representative for the product bulk that it represents. A second requirement is that the particles are fully dispersed and de-agglomerated and, for chocolate; that all fat is dissolved in the dispersion medium (see further Chap. 3 and [32]). Ultrasonication of the measurement sample is usually helpful to reach optimum particle dispersion.

As stated above, measured data are sometimes interpreted as model distributions, viz. log-normal or Rosin-Rammler distributions. Figure 8.2 compares the results for a dark chocolate sample. In this case, the log-normal distribution

[4] The imaginary part of the refractive index relates to the degree of light absorption in the particle.

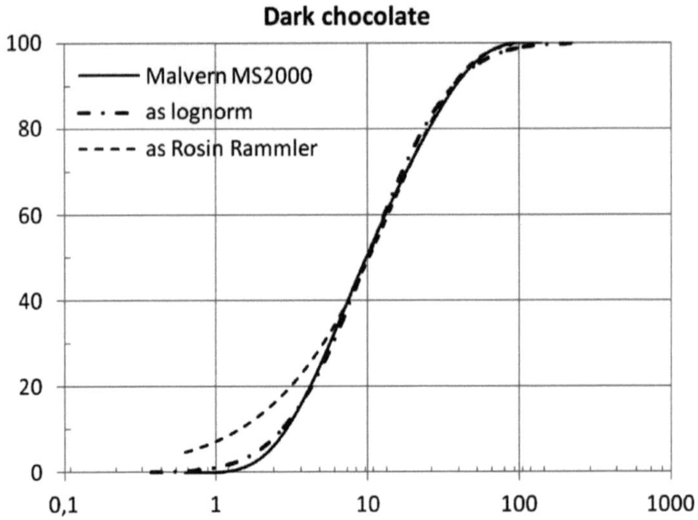

Fig. 8.2 Comparison data with models

agrees reasonably well with the measured data, although it extends the distribution at both ends somewhat; the RR distribution deviates significantly at Q_3's smaller than about 30 %. The imperfect agreement between measured data and models is also indicated by the fact that neither of the plots at the corresponding model graph paper yields a straight line relationship over the full size range, despite the logarithmic scales. These models have the advantage of requiring only two parameters for description of the distribution, which may be attractive for characterization and control of production processes. Product quality and performance, however, should preferably be represented in optimum PSD parameters (see Sect. 1.4).

8.4.2 Particle Size Parameters and Quality Elements

The particle size of cocoa powder, milk powder and sugar has important influences on product quality. Particles larger than about 100 μm show as visible specks in a drink or ice cream and will rapidly settle from suspensions in milk or water. A gritty taste is experienced in chocolate containing particles larger than about 30 μm. Small particles give off their aroma to the suspension or in the mouth faster resulting from their large surface area, but on the other hand, increase the viscosity of chocolate suspensions. Moreover, they may give chocolate a slimy taste [33].

Some typical cumulative particle size distributions (PSDs), analyzed by laser diffraction, are shown in Fig. 8.3. Note that the chocolate PSDs vary from country to country and according to individual taste experiences, although differences seem to decrease with time due to globalization. The above data reveals that the volume-based $D_{10;3}$ is about 2 μm, the median, volume-based size ($D_{50;3}$) about 10 μm, the

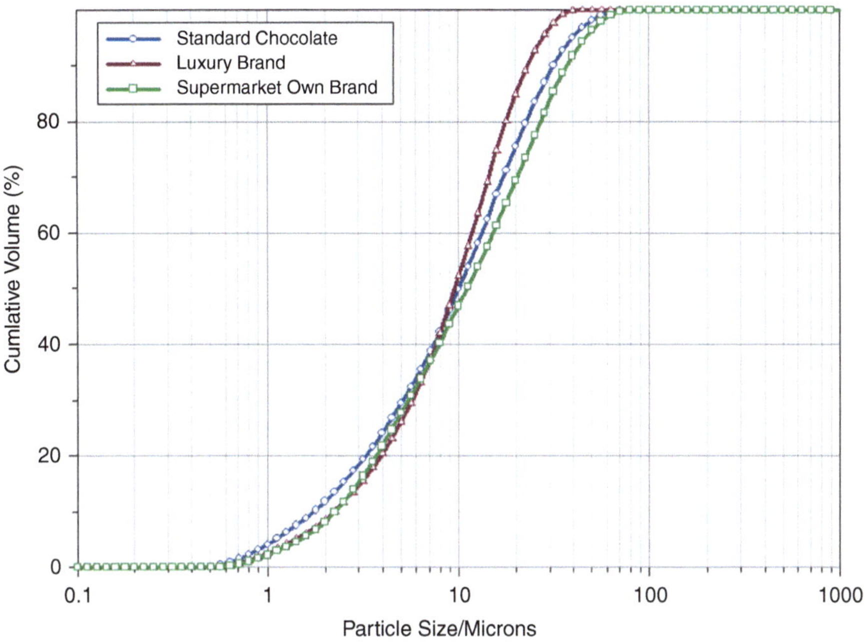

Fig. 8.3 Laser diffraction analysis of PSD of chocolate particles [30] (Mastersizer 2000; courtesy Malvern Instruments)

$D_{90;3}$ (PSD parameter taken for maximum size) about 30 μm, the percentage of particles smaller than 20 μm about 75 % v/v, the volume-weighted mean size $D_{4,3}$ about 15 μm and the calculated area-based Sauter mean diameter $D_{3,2}$ about 7 μm. For the above data, a RI of 1.500 (real) and 0.001 (imaginary) was assumed. These RI values are a compromise as we are dealing with mixtures of sugar and cocoa particles which have different optical properties.

The PSD data from laser diffraction deviate largely from the PSD by optical microscopy, using a 5 times objective and from those coming from the electrical sensing zone technique. Both latter data are similar and show much less fine particles than the LD data (see Figs. 8.4 and 8.5, respectively). These large differences may originate from different sources, viz. different chocolate samples and their pretreatment, and/or limitations of the measurement methods (e.g. size range and optical model used) given in Sect. 8.4.1 and [32]. Note also the large difference between the number-based and the recalculated volume-based PSDs in Fig. 8.5. This illustrates the need to measure a very large number of particles as only at most about 0.2 % of the total number is involved in the upper 10 % by volume of this wide size distribution. On the other hand, more than 50 % n/n of this PSD is involved in the lowest 1 % v/v. Optical microscopy has the advantage that the particles can be seen and that the different RI values of sugar and cocoa particles allow their discrimination.

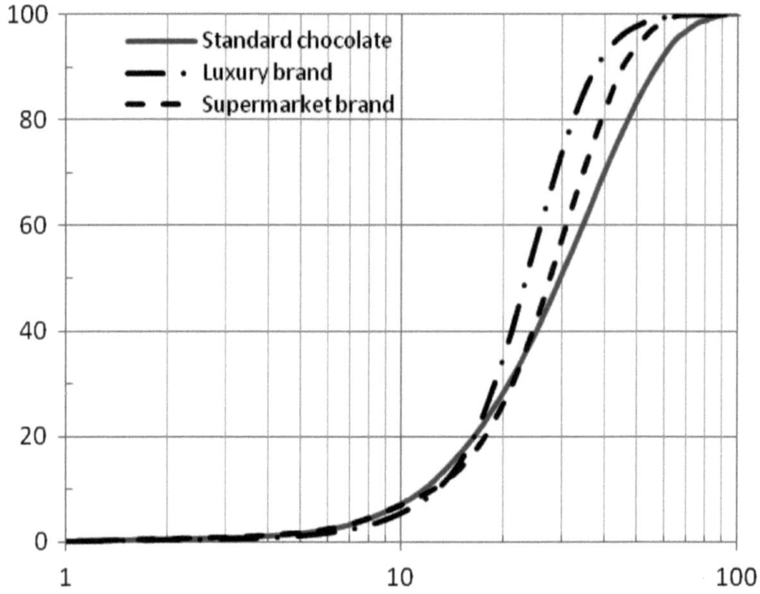

Fig. 8.4 Microscopic PSD analysis of chocolate particles (cumulative v/v data) (Morphologi G3 [31]; courtesy Malvern Instruments)

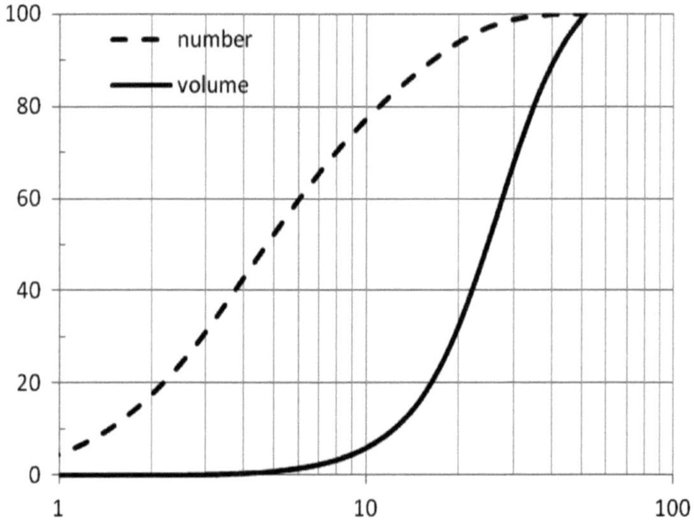

Fig. 8.5 Coulter Counter PSD analysis of chocolate particles (Number data taken from [34])

8.4.3 Sensorial Aspects

Bitter, salt, sour, sweet and umami are presently considered as the five basic tastes [22, 28]. In general, taste results from the combined experience in our mouth, nose and eyes (appearance). For chocolate, the taste experience results primarily from the aroma components present in the cocoa beans, developed during processing, and from added sugar and spices. Particle size also has an important influence on taste, both directly and through its influence on texture. Optimum size for a nice, creamy taste seems to be in the range 15–30 μm [5, 9]. It appears that the presence of a 'few[5]' particles larger than about 30 μm, impart a gritty taste to solid chocolate at the palate, although this experience also depends on e.g. fat content [5, 8, 9, 41, 42]. In view of the better sensitivity of a volume/mass-based characteristic, this is often expressed in a maximum value for the $D_{90;3}$ of 30 μm or for the sieve residue on a sieve of appropriate mesh size. In relation to the data presented in Fig. 8.3 it can be calculated that the number percentage of particles larger than 30 μm in these chocolates is less than about 0.2 % n/n, or about 10^4 of these larger particles per gram chocolate. Note that the mass of a 100 μm particle is about 0.6 μg, so that at minimum 10^5 of such particles are required to yield a reasonable minimum sieve residue of 60 mg. The volume-weighted mean size $D_{4,3}$ is another PSD parameter that has been reported to correlate with taste [35, 43].

A slimy, too creamy taste results when the size of most particles is below about 20 μm. Here, no distinct PSD parameter is reported. It seems that the quantity of such particles is not controlled separately, but results from the operational conditions and grinding techniques employed.

It should be noted that the taste sensations of grittiness not only relate to particle size but also to other aspects, such as particle shape, particulate concentration, flavor, viscosity, texture, presence of gas bubbles and temperature [15, 16, 23].

In the case of cocoa beverages and ice creams, maximum particle size may be greater than stated above, viz. about 100–200 μm. This reflects the influence of the continuous phase, in which the particles are dispersed, upon sensorial aspects such as taste and appearance. The solids concentration and the amount of particles in one bite or swallow are also important. On the other hand, the rapid settling of these large particles in low-viscosity suspensions in milk or water has an undesirable visual effect on the quality of these drinks (through showing sediment).

It is reported that the rate of release of flavor from particles to the surrounding fat or liquid and the mouth is governed by their surface area [4, 5]. This can be related to the Sauter mean diameter $D_{3,2}$, which is area-weighted. Depending upon conditions, diffusion may play a role, which changes from the square relationship with particle size to a linear one.

[5] The term 'few' probably stems from direct microscopic examination where one sees a very small number of large particles in a vast majority of small particles (> 99.8 % n/n). Note that the Hegman gauge is used for similar purpose in other products (see [32] and e.g. Chap. 12).

Note that sugar contributes to the sweetness of chocolate as well as compensates for the bitterness of cocoa solids especially in dark chocolates, although it is often regarded as an inert ingredient with regard to the subtleties of flavor.

For solid chocolates, a melting point of about 34 °C, just below body temperature, causes a pleasant mouth feel experience. It results from a correct cocoa butter crystal structure and network.

A nice, glossy appearance contributes to the (visual) quality of chocolate. This too is mainly a result of an optimum tempering procedure.

Sensorial testing of chocolates involves all sensorial elements and, thus, has a complex nature. Therefore, it is typically done by trained expert panels using standard procedures and terminology [43]. Instrumental analysis (chromatography and mass spectrometry) of aroma components as well as texture and rheology analysis are helpful in providing quantitative basic information [5, 29].

8.4.4 Viscosity of Molten Chocolate

The viscosity of molten chocolate is important in the mixing, grinding and pumping/transportation steps of the production process. Higher viscosities lead to longer processing times for mixing of ingredients and milling, and higher power consumption (costs) for pumping, and vice versa. Also, viscosity has some influence on texture and taste experience [2].

The viscosity of liquid chocolate pastes shows a non-Newtonian flow behavior. To start it to flow, the initial particulate structure has to be destroyed. Here, a minimum force– the yield stress – is required. It mostly depends on the amount of small particles [37]. At increasing forces (shear rates), this structure gradually breaks down, causing a gradual decrease of the apparent viscosity (shear thinning). At still higher forces, viscosity becomes independent of force (as in Newtonian liquids). At this point, no further de-agglomeration takes place. This is the general case of chocolate. However, when the solids concentration comes close to densest particle packing, as in low fat situations, then shear thickening occurs at high shear rates, i.e. a significant viscosity increase with increasing shear rate as the particulate mixture has decreasing possibilities for movement [7]. In the intermediate range, thixotropy may be observed; this means that the viscosity decreases with time under continuous shear (by increased de-agglomeration) but increases again when shear is discontinued (due to some re-agglomeration). This effect can be seen by the different rheological behavior (hysteresis) while increasing the shear rate followed by decreasing it, or by the effect of pre-shearing of the measurement sample before its rheological behavior is measured. (see Chap. 2 for general viscosity information).

Most often, rotation viscosimeters of bob-and-cup or vane types are used for viscosity measurement of molten chocolates, at a temperature of 40 °C. In view of the complex behavior, measurements are conducted, after adequate mixing of the paste and pre-shearing, at both increasing and decreasing shear rates (ramping up and ramping down).

Conventionally, the Casson model is used for expressing viscosity parameters, as agreed in the International Office of Cocoa, Chocolate and Sugar Confectionery (IOCCC) [10, 20, 34]:

$$(\tau)^{1/2} = (\tau_{0,\mathrm{CA}})^{1/2} + (\eta_{\mathrm{CA}})^{1/2} \cdot (\mathrm{d}y/\mathrm{d}t)^{1/2} \quad (8.1)$$

where:

τ = shear stress
$\tau_{0,CA}$ = Casson yield stress
η_{CA} = Casson dynamic (or apparent or plastic) viscosity
$\mathrm{d}\gamma/\mathrm{d}t$ = shear rate.

The Casson yield stress is the minimum force required to cause the chocolate paste to move. A high yield stress may be required e.g. for putting figures on sweets, a low value to give a thin chocolate coating on biscuits [8]. The Casson (apparent) viscosity represents the viscosity at high shear rates, e.g. important for its relation to pumping and the thickness of coatings. With the Casson model, the yield stress is conventionally estimated by graphical extrapolation of the experimental data to zero shear rate, and the Casson viscosity from the linear relationship at higher shear rates. Nowadays, both parameters are most often derived in a computer by regression to find the best fit of measured data to the equation. However, the Casson model was found neither to give a good fit for all products and instruments nor to yield good reproducibility of the calculated yield stress [1]. Therefore, it was later generalized by replacing the square root by a power n, which gives an additional parameter to be fit. Values for n were found to be $0.5 \leq n \leq 1$ [11, 34], for example: $n = 0.67$ for milk chocolate [33]. This equation is very similar to the Herschel-Bulkley equation, which also contains a power n as flow viscosity index and appears to fit the data well [34, 39]:

$$\tau = \tau_0 + \eta_{pl} \cdot (\mathrm{d}y/\mathrm{d}t)^n \quad (8.2)$$

where: η_{pl} is the apparent, plastic viscosity of this model.

Recently, Servais proposed new values for representation of viscosity behavior [38]:

1. A yield stress value derived from the measured yield stress at 5 s^{-1};
2. A new plastic viscosity value from the measured value at 40 s^{-1};
3. A new thixotropy related value from the viscosity data at 40 s^{-1} during ramping up and down.

He states that the reproducibility of these data values is better than of those derived from the Casson model. They also relate better to actual practice.

Cocoa butter is the continuous phase in chocolate, in which cocoa, milk and sugar particles are dispersed. It forms the (Newtonian) basis for the viscosity of the molten mixture (at elevated temperature). On the other hand, the particles increase this viscosity as well as give the molten chocolate a yield stress in a non-Newtonian, shear thinning flow behavior [2, 3, 5].

Lecithin (a phospholipid, typically from soy origin) is the emulsifier that is mainly used to disperse the particles. It also reduces the viscosity and yield value of the molten chocolate mixture during processing. Its emulsifying activity assists in both de-agglomeration of the particles, by which, entrapped butter is freed, and in lowering the surface tension at the sugar-fat interface which allows better dispersion quality. A concentration of 0.3 % of lecithin is equivalent to a replacement of about 5 % m/m of cocoa butter with respect to viscosity [33, 40]. Thus, it allows application of less, expensive cocoa butter in the formulation of chocolates.

The apparent viscosity of chocolates increases with increasing particulate concentrations and decreasing particle size [1, 5, 11]. The increased viscosity relates to the increased inter-particle friction and attractive forces. For the Casson yield stress, a linear relationship is reported with the percentage of fines (< 20 μm) [11]. Note that this size value is still fairly large for fines as about 75 % v/v of the particles is smaller (cf. Fig. 8.3). E.g. Mohos uses a criterion of < 5 μm, which leads to about 25 % v/v smaller particles [34]. Also, the width and the (multi) modality of the PSD influence the viscosity parameters [8, 12, 14, 37, 39, 43]. This is related to how close the actual particle packing comes to maximum packing density. It has been shown that maximum packing density increases in the order monodisperse – polydisperse – multimodal mixtures at the same particulate concentration. Thus, viscosity of dispersions containing the same solids concentration decreases in this order. Consequently, viscosity can be reduced by searching for an optimum PSD for the particulate ingredients, with respect to packing. These effects may be significant as the typical solids concentration in chocolates is 45–60 % v/v (see Sect. 8.3).

The specific surface area (inverse related to particle size; m^2/g), the area-based mean size $D_{3,2}$, the volume-based mean size $D_{4,3}$, the $D_{90;3}$ and the percentage of fines have been reported in relation to chocolate rheology [8–10, 37]. However, no preferential PSD parameter value is stated for use in a quantitative model. Apparently, primary control of the viscosity of the chocolate paste at a given composition is done by direct measurement during processing and, if necessary through adjustment of the composition. The grinding technique and its conditions together with lecithin addition and composition are decisive for the viscosity during grinding and the resulting shear forces, and, thus, for reaching a satisfactory PSD. This is still the basis for the present process.

Other ingredients may influence rheological behavior as well, for example ingredients that replace part of the cocoa butter or sugar in order to reduce calorific value or costs. It seems that they are treated in the process in the same way as the ingredients they replace.

Minute amounts of free moisture greatly increase viscosity of the molten chocolate mixture. Thus, water content is restricted typically to less than 1 %, with higher values for white chocolate compared to dark and milk chocolate (related to the amount of milk powder in the recipe).

8.4.5 *Other Quality Aspects*

In addition to the above mentioned aspects, glossy appearance and constant weight of the bars are considered important for the quality of the final product. These aspects are mainly related to adequate control of tempering, grinding and viscosity. Control charts, e.g. of the Shewhart type, are used to check constant weight [32, 33].

Moreover, absence of spoilage organisms and pathogenic bacteria is essential. Good quality in this respect depends on adequate selection of the cocoa beans and good hygiene during processing.

For powder ingredients, the risk of powder explosions should be taken in consideration and minimized, especially during handling of dry powders like sugar [27] (see further Chap. 4).

8.4.6 *Quality Conclusions with Respect to PSD Parameters*

Although the PSD of cocoa, milk and sugar particles plays an important role both during processing and in the final chocolate, only few PSD parameters are stated in relation to quality. Coarseness in relation to taste is expressed in a $D_{90;3}$ of 30 µm (or 'few' particles larger than 30 µm). Other PSD parameters mentioned are the volume-weighted mean size $D_{4,3}$ in relation to taste, the area-weighted Sauter mean diameter $D_{3,2}$ in relation to aroma release and the percentage fines (< 20 or 5 µm) and the $D_{90;3}$ in relation to viscosity. Maybe, the choice of an indirectly related parameter is not very critical for use in equations when using the same grinding equipment at different conditions, since some correlation between the various PSD parameters is to be expected. For example, an increased grinding time will lead to a smaller $D_{90;3}$, $D_{4,3}$ and $D_{3,2}$, and to a larger amount of fines. However, it is a drawback for good understanding, as different types of mills may cause different relationships between these PSD parameters. It is remarkable that, apart from the $D_{90;3}$ value for grittiness, none of the above mentioned specific parameters nor their values are prescribed in relation to performance aspects. Neither is quantitative mathematical models described for use in product design. Apparently, product properties depend, besides composition, mainly on raw materials and equipment and conditions of the production process. Essential properties are measured directly, where needed, and, if necessary, processing conditions and/or product composition, e.g. fat content, are adapted.

8.5 Definitions, Abbreviations and Symbols

Casson model	Model to describe the relationship between shear stress and shear rate Eq. 8.1 (other models and descriptors are also used)
Cocoa mass	Mixture of cocoa butter and cocoa particles
Equivalent sphere	Sphere that has the same property as the observed particle in relation to a given measurement principle

(continued)

Particle size	Diameter of some defined equivalent sphere
Sauter mean diameter	Average size of a PSD based on surface area
Umami	Fifth basic taste (brothy, savory), imparted by glutamate and ribonucleotides [22, 28]
ASTM	American Society for Testing and Materials
ICA	International Confectionery Association (formerly IOCCC)
ICCO	International Cocoa Organization
IOCCC	International Office of Cocoa, Chocolate and Sugar Confectionery (see also ICA)
ISO	International Standards Organization
LD	Laser diffraction (also named static light scattering)
PSD	Particle size distribution
RR	Rosin-Rammler distribution
Subscripts	0 for number; 1 for length; 2 for area and 3 for volume (in size distributions)
$D_{90;0}$	90th percentile of a cumulative, number-based distribution
$D_{50;3}$	Median, volume-based size; 50th percentile of a cumulative, volume-based distribution
$D_{90;3}$	90th percentile of a cumulative, volume-based distribution
$D_{3,2}$	Sauter mean diameter, area-weighted mean size, mean value of an area-weighted PSD
$D_{4,3}$	Volume-weighted mean size, mean value of a volume-weighted PSD
n	Flow viscosity index in the Herschel-Bulkley model
% m/m	Percentage by mass
% n/n	Percentage by number
% v/v	Percentage by volume
Q_3	Cumulative, undersize volume percentage of a size distribution
$d\gamma/dt$	Shear rate
η_{CA}	Casson dynamic (or apparent or plastic) viscosity
η_{pl}	Apparent, plastic viscosity of the Herschel-Bulkley model
τ	Shear stress
τ_0	Yield stress (in the Herschel-Bulkley model)
c	Casson yield stress

References

1. Aeschlimann, J.M., Becket, S.T.: International inter-laboratory trials to determine the factors affecting the measurement of chocolate viscosity. J. Text. Stud. **31**, 541–576 (2000)
2. Afoakwa, E.O., Paterson, A., Fowler, M., Vieira, J.: Relationship between rheology, texture and melting properties of dark chocolate as influenced by particle size distribution and composition. Eur. Food Res. Technol. **226**, 1215–1223 (2008)
3. Afoakwa, E.O., Paterson, A., Fowler, M.: Effects of particle size distribution and composition on rheological properties of dark chocolate. Eur. Food Res. Technol. **226**, 1259–1268 (2008)
4. Afoakwa, E.O., Paterson, A., Fowler, M., Ryan, A.: Impact of particle size distribution on rheological and textural properties of chocolate models with reduced fat content. Food Chem. **113**, 208–215 (2009)

5. Afoakwa, E.O.: Chocolate Science and Technology. Wiley-Blackwell (2010)
6. Alamprese, C., Datei, L., Semararo, Q.: Optimization of processing parameters of a ball mill refiner for chocolate. J. Food. Eng. **83**, 629–636 (2007)
7. Barnes, H.A.: Shear-thickening ("Dilatancy") in suspensions of non-aggregating solid particles dispersed in Newtonian liquids (Review). J. Rheol. **33**, 329–366 (1989)
8. Beckett, S.T.: The Science of Chocolate. Royal Society London (2000)
9. Beckett, S.T.(ed.).: Industrial Chocolate Manufacture and Use. Blackwell Science (1999)
10. Casson, N.: In: Mill, C.C. (ed.) Rheology of disperse systems, pp. 84–104. Pergamon Press (1959)
11. Chevalley, J.: In: Beckett, S.T. (ed.) Chocolate Flow Properties. Blackwell Science (1999)
12. de Clercq, N.: Changing the functionality of cocoa butter. Ph.D., Thesis Ghent University (2011)
13. Do, T.-A.L., Vieira, J., Hargreaves, J.M., Mitchell, J.R., Wolf, B.: Structural characterization of cocoa particles and their effect on the viscosity of reduced fat chocolate. LWT Food Sci. Technol. **44**, 1207–1211 (2011)
14. Do, T.-A.L., Hargreaves, J.M., Wolf, B., Hort, J., Mitchell, J.R.: Impact of particle size distribution on rheological and textural properties of chocolate models with reduced fat content. J. Food Sci. **72**, E541–E552 (2007)
15. Engelen, L.: A rough guide to texture. Ph.D., Thesis University, Utrecht (2004)
16. Engelen, L., van der Bilt, A., Schipper, M., Bosman, F.: Oral size perception of particles: Effect of size, type, viscosity and method. J. Text. Stud. **36**, 373–386 (2005)
17. Fuerstenau, D.W., Kapur, P.C., Schoenert, K., Marktscheffel, M.: Comparison of energy consumption in the breakage of single particles in a rigidly mounted roll mill with ball mill grinding. Int. J. Min. Proc. **28**, 109–125 (1990)
18. http://archive.fieldmuseum.org/chocolate/about.html
19. http://penn.museum/press-releases/20-the-earliest-chocolate-drink-of-the-world.html
20. ICA 46, Method for determination of Casson viscosity of chocolate (2000); International Confectionery Association
21. ICCO, The World Cocoa Economy; ICCO Report EX/142/6 (2010); International Cocoa Organization
22. Ikeda, K., Tokyo, J.: New Seasonings. Chem. Soc. **30** (1909), 820–836 (Japanese); English translation in: Chem. Senses **27** (2002), 847–849
23. Imai, E., Hatae, K., Shimada, A.: Oral perception of grittiness: Effect of particle size and concentration of the dispersed particles and the dispersion medium. J. Text. Stud. **26**, 561–576 (1995); International Standards Organization
24. ISO 13320, Particle size analysis – Laser diffraction methods. International Standards Organization
25. Jackson, K.: In: Beckett, S.T. (ed.) Chocolate Recipes. Blackwell Science (1999)
26. Kemsley, J.: Archeology, Nov. 14, 2007; http://pubs.acs.org/cen/news/85/i47/8547news3.html
27. Krüger, Ch.: In: Beckett, S.T. (ed.) Sugar and Bulk Sweeteners. Blackwell Science (1999)
28. Lindemann, B., Ogiwara, Y., Ninomiya, Y.: The discovery of umami. Chem. Senses **27**, 843–844 (2002)
29. McFarlane, I.: In: Beckett, S.T. (ed.) Chocolate Instrumentation. Blackwell Science (1999)
30. Malvern Application Note 672–1; Mastersizer 2000; Malvern Instruments Ltd.
31. Malvern Application Note 1223–1; Morphologi G3; Malvern Instruments Ltd.
32. Merkus, H.G.: Particle size measurements: Fundamentals, practice, quality. Springer (2009)
33. Minifie, B.W.: Chocolate, cocoa and confectionery: Science and technology, 2nd edn. AVI Publishing (1980)
34. Mohos, F.A.: Confectionery and chocolate engineering: Principles and applications. Wiley-Blackwell (2010)
35. Mongia, G., Ziegler, G.R.: The role of particle size distribution of suspended solids in defining the flow properties of chocolate. Int. J. Food Prop. **3**, 137–147 (2000)

36. Robinson, A.L.: Myriad ways to measure small particles. Science. J. Food Eng. **212**, 146–152 (10 April 1981)
37. Servais, C., Jones, R., Roberts, I., The influence of particle size distribution on the processing of food. J. Food Eng. **51**, 201–208 (2002)
38. Servais, C., Ranc, H., Roberts, I.D.: Determination of chocolate viscosity. J. Text. Stud. **34**, 467–497 (2004)
39. Sokmen, A., Gunes, G.: Influence of some bulk sweeteners on rheological properties of chocolate. LWT-Food Sci. Technol. **39**, 1053–1058 (2006)
40. Steiner, E.H.: In: Mill, C.C. (ed.) Rheology of Disperse Systems. Pergamon Press, pp. 167–180 (1959)
41. Tannenbaum, G.: Chocolate: A marvelous natural product of chemistry. J. Chem. Educ. **81**, 1131–1135 (2004)
42. Ziegler, G.R., Hogg, R.: In: Beckett, S.T. (ed.) Particle Size Reduction. Blackwell Science (1999)
43. Ziegler, G.R., Mongia, G., Hollender, R.: The role of particle size distribution of suspended solids in defining the sensory properties of milk chocolate. Int. J. Food Prop. **4**, 353–370 (2001)

Chapter 9
Ice Cream

Elke Scholten

Abstract Ice cream is a popular dessert, which owes its sensorial properties (mouth feel) to its complex microstructure. The microstructure is a result of the combination of the ingredients and the production process. Ice cream is produced by simultaneous freezing and shearing of the ice cream mix, which results in the formation of ice crystals, air bubbles and a viscous serum phase. The amount and the size of the ice crystals and air bubbles have significant contributions to the mouth feel, the melting behavior, ease of spooning, etc. It also affects the shelf-life of the products. Therefore, control of the particle size and volume fraction is crucial. The ice crystals have a large influence on the hardness of the ice cream. The particle size, which is normally in the range of 30–50 μm, determines the degree of coldness, which can be controlled by the temperature of the production process. A wide particle distribution will have a negative influence on the shelf-life, as Ostwald ripening and recrystallization will occur more often. Once the recrystallization processes lead to the growth of larger ice crystals above 100 μm, the ice cream will feel icy and gritty. Therefore, to increase mouth feel and shelf-life, a narrow size distribution of small ice crystals is preferred. The air bubbles, in the size range of 20–50 μm, provide the softness and decrease the coldness of the ice cream, and are usually present in high volume fractions. Due to a high volume fraction, the air bubbles are in close proximity, thereby enhancing coalescence. Once coalescence has led to the increase in air bubble size and the formation of channels, the air can escape and the ice cream collapses. Ostwald ripening, due to a wide particle size distribution, will enhance this effect. Once the air has escaped, the ice cream will become harder. As the air is non-conductive, the presence of air slows down the heat transfer, leading to a warmer mouth feel for ice cream with a high volume fraction of air. The collapse of the ice cream will therefore also change the mouth feel from a slightly warmer to a colder ice cream. To increase the shelf-life and mouth feel, the air bubbles should be stabilized.

E. Scholten (✉)
Department of Food Physics, Wageningen University, Bornse Weilanden 9, Wageningen, The Netherlands
e-mail: elke.scholten@wur.nl

H.G. Merkus and G.M.H. Meesters (eds.), *Particulate Products: Tailoring Properties for Optimal Performance*, Particle Technology Series 19,
DOI 10.1007/978-3-319-00714-4_9,

This can be accomplished by coating the air bubbles with fat globules. During the production process, the fat globules will partially coalesce, thereby forming a fat layer around the air bubbles. The process is enhanced by a decrease in fat globule size (to sizes below a micrometer), which is controlled by homogenization of the ice cream mix. To enhance shelf-life and control mouth feel, the particle size and its distribution is important for all elements in the ice cream.

9.1 Introduction

Ice cream is a popular frozen dessert consumed the world over. It is typically enjoyed as a cool warm-weather treat, and a range of varieties can be found across many cultures. Although everyone loves ice cream, Australians and New Zealanders, who eat the frozen treat all year long, seem to have the largest annual per capita consumption with 17 and 16 L, respectively. Ice cream is essentially milk, water, cream and sugar. But change the ingredients slightly and standard ice cream can become gelato (custard-based ice cream), sorbet (nondairy, fruit-based frozen dessert), sherbet (fruit-based frozen dessert with some dairy products), frozen yoghurt, or fruit ice. Ice cream is popular because of its mouth feel, conveying a sensation of decadence on the tongue. Just out of the freezer, it is cold and hard, and as it melts, it turns into a smooth, creamy liquid. The combination of flavor, texture and the cooling sensation in the mouth will determine whether we like it. Although most ice creams have similar ingredients, the mouth feel and the sensations they impart are different. Sorbets, for example, normally feel a bit colder than ice creams, and their texture is completely different. Some ice creams melt very easily and disappear very quickly on the tongue, while others last. Why are there so many differences?, and where do these differences come from? The mouth feel of ice creams and its related sensations are determined by the microstructure of the food product, which is a result of the interplay between the ingredients.

9.1.1 Ingredients

The ingredients of ice cream can be classified into three different groups:

- Components (major) that are present in substantial amounts (about 1–25 %), such as milk proteins, sugar, fat and water.
- Components (minor) that are present in small quantities (less than about 1 %), such as emulsifiers, stabilizers, colors and flavors.
- Extra ingredients, such as chocolate, wafers, fruit pieces, nuts, etc.

The ingredients can be obtained from a variety of raw materials, such as milk, skimmed milk, low- and high-fat cream, yoghurt, buttermilk, milk powder, vegetable fat or butterfat. The total amount of ingredients other than water, such as

Table 9.1 Typical ice cream formulation

Ingredient	Amount (%)
Fat	5–10
Milk protein	4–5
Lactose	5–7
Other sugars	12–16
Stabilizers, emulsifiers and flavors	1
Total solids	**30–40**
water	**60–70**

proteins, fat, and stabilizers, is defined as the total solids. The total amount of solids has a large influence on the quality of the ice cream and can be used to control the mouth feel of the ice cream. In Table 9.1, a standard ice cream formulation is given. A good quality ice cream often contains a total solid content between 30 % and 40 %. Ice cream, with such a high amount of solids, often feel more creamy and smooth. Ice cream that contains lower amounts of solids often feels watery and icy.

9.1.2 Manufacture of Ice Cream

Ice cream is produced by four different steps: (1) mix preparation, including blending, pasteurization and homogenization (2) aging, (3) freezing and (4) hardening. The mixing process is designed to blend, disperse and hydrate the ingredients. Liquid ingredients are dosed first (water, milk, cream) into the mix tank and then heated. The temperature should be kept lower than the denaturation temperature of the proteins to avoid aggregation of the proteins, and should therefore not exceed 65 °C. Solid fats are melted before addition and dry ingredients are added next. The ingredients are mixed in the mix tank. The low shear forces in the tank produce a coarse oil-in-water emulsion with relatively large fat droplets of approximately 2–10 μm. The mix is then pasteurized to reduce the number of micro-organisms. The hot mix is then transferred to a homogenizer to reduce the size of the fat droplets to sizes below 1 μm, which increases the surface area of the fat tremendously. In industry, mostly high pressure homogenizers are used (up to 200 bar), as these apparatus are most effective in decreasing the fat droplet size. However, most small-scale ice-cream producers (craftsman) use high shear mixing devices. During homogenization, the milk proteins adsorb onto the surface of the fat droplets. In the aging step, the mixture is cooled down to 4 °C. During this stage, some of the proteins are replaced by emulsifiers. This allows the fat droplets/particles to come close together, leading to partial coalescence. Since the fat particles are not completely liquid but partially solid, they can only merge to a certain extent and do not result in the formation of a single particle. Instead, this results in large aggregates, in which the fat particles are partially merged creating a strong link between the particles. This process is known as partial coalescence, and is only observed for partially solid particles. Due to their hydrophobic nature, the fat globules are mainly present at the air-water interface. The partial coalescence of the fat globules on the air-water interface leads to the formation of a solid layer around the air bubbles. This takes place during the freezing step, in which the final microstructure of the ice cream is determined. During this step, the mix is aerated and frozen to generate ice crystals, air bubbles and the matrix. Ice cream makers are surface heat exchangers equipped with scrapers. They are designed to extract heat from the viscous solution at the outside of the barrel (the refrigerant is often ammonia or Freon). Inside the barrel, there is a rotating dasher/scraper that scrapes the frozen mix from the wall and redisperses it into the unfrozen ice cream mix. The outlet temperature of the ice cream maker determines the total amount of ice. As the ice cream leaves the freezer, there is a pressure drop of appr. 4 atm. As a result, the air bubbles will expand, which will increase the total volume of air that is incorporated in the system. The temperature of the ice cream is approximately −5 °C when it leaves the ice cream maker, and is transferred to a freezer to harden. In this step, the existing ice crystals will grow until a temperature of −20 °C is reached.

9.2 Structure

Looking closer at ice cream, we can recognize that these frozen products are complex systems whose microstructure can be described as ice crystals and air bubbles embedded in an unfrozen sugar solution, as depicted in Fig. 9.1 [12]. The unfrozen sugar solution is often referred to as the freeze-concentrated matrix, as upon freezing, water is removed from the solution as ice, effectively increasing the

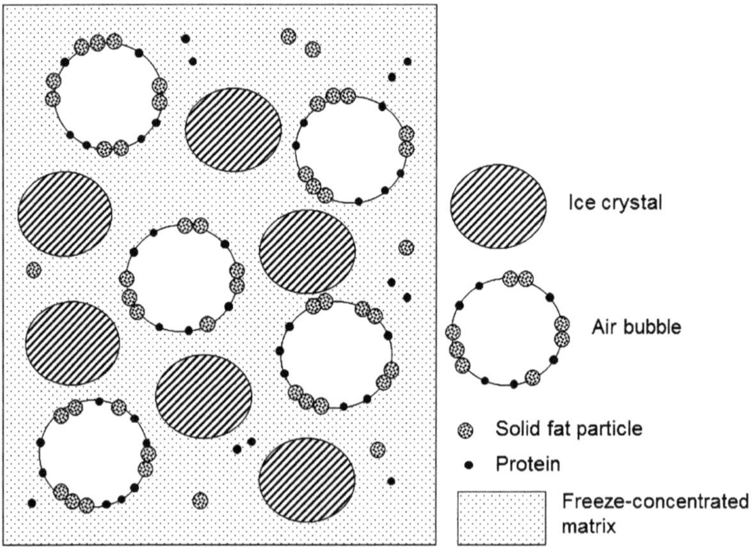

Fig. 9.1 Microstructure of ice cream [12] (Reproduced with permission; copyright Elsevier Ltd.)

sugar concentration in the unfrozen phase. The structure of ice cream – whether it is ice cream, gelato or sorbet – is determined by the proportion of these different specific ingredients and the process by which the ice cream is produced. The microstructure of the ice cream is important because it determines the dessert's sensorial properties, such as hardness, coldness, rate of melting, creaminess, and fluffiness. It thus determines how we perceive ice cream when we eat it, the ease with which the ice cream can be scooped from its container, and how long the ice cream can be stored without losing its characteristic texture. An understanding of how all these elements are interrelated can help us manipulate ice cream's properties to achieve certain sensorial effects.

The volume fraction of ice crystals, air bubbles and serum phase depends on the ingredients and the production of the ice cream. In a conventional ice-cream, the volume fraction of air is roughly 30 %, the volume fraction of ice 40 %, the serum phase 20 % and the fat phase 10 %. However, the composition depends on the ingredients and the production process. For example, industrially-produced ice cream often has an air bubble fraction of over 50 %, whereas craftsman ice-cream usually does not exceed an air bubble fraction of 20 %. This has an effect on the perception of the ice cream and the changes during storage. A very low solid content often has a lower serum phase fraction and a higher ice fraction, leading to a watery and icy feeling.

9.2.1 Ice Crystals

The most important element in ice cream is the ice – solid crystalline water. Usually, 65 % out of the ice cream mix (water + water soluble ingredients) turns into ice. The ice crystal volume fraction is controlled by the ingredients that influence the freezing point depression and the total amount of dry solids that are added to the ice cream mix. A certain amount of sugar is used to control the freezing point depression. This can be done by a combination of different types of sugar, such as sucrose, lactose (from milk), dextrose, saccharose, fructose, etc. Each of these sugars has their own contribution to the freezing point depression and sweetness. (Lactose for example has the same influence on the freezing point depression as sucrose, but has a smaller effect on the sweetness. Fructose has a larger effect on the freezing point depression, but also has a larger effect on the sweetness). In a typical ice cream, the total sweetness is controlled between values of 16–20, where the value is calculated as[1]:

$$\sum \%_{sweetener} \cdot relative\ sweeteness. \tag{9.1}$$

[1] A 20 % sugar solution has a total sweetness of 20; a 20 % solution of lactose has a total sweetness of roughly 5.

During freezing and hardening, the average size of the ice crystals is determined commonly to be roughly 30–50 μm, but values of 1 and 150 μm have also been observed [6, 7, 9, 21]. Since the added sugars do not only play a role in the flavor of the ice-cream but also play a major role in the determination of the total ice fraction, replacement of sugar by artificial sugar is not a straightforward exercise. When regular sugars are replaced by artificial sweeteners, the gained freezing point depression will not be large enough to decrease the ice crystal formation, leading to a very hard and grainy ice cream.

Crystal Shape and Size

The size and the shape of the ice crystals are determined by the production process. The type of freezer and the temperature at which the ice cream is prepared will affect both the ice crystal size and shape. Faster cooling leads to smaller crystal sizes. Crystals appear and grow as a result of a thermodynamic process, in which the energy gain of creating a bulk phase is counteracted by the energy cost of creating a surface. Once the radius of the crystal is larger than a certain critical size, r_c, the growth of the crystal will lead to an energy gain. Crystals, smaller than this critical radius, will melt again and disappear. This critical size will therefore determine to a large extend when crystals will be formed. The smaller the critical size, the easier crystals will be formed, and as a result, more and smaller crystals will appear. The critical size is related to certain parameters, as the interfacial tension, γ, the enthalpy of fusion, $\Delta_F H$, the melting temperature of the ice cream mix, T_m^0, and the difference in temperature (ΔT) between the ice cream mix and the freezer as

$$r_c = \frac{2\gamma}{\Delta u} = \frac{2\gamma T_m^0}{\Delta_F H \Delta T}. \tag{9.2}$$

The larger the difference in temperature between the ice cream mix and the ice cream maker, the smaller the (critical) size of the ice crystals. Therefore, the temperature of the ice cream maker has a large influence on the size, as it becomes smaller for lower temperatures of the ice cream maker. Ice cream prepared with liquid nitrogen (−196 °C) will therefore produce much smaller ice crystals than conventional ice cream makers. However, the very fast freezing in liquid nitrogen is not convenient to include large volume fractions of air. Therefore it is mainly used for the production of water ice, in which air is not incorporated. To incorporate larger volume fractions of air, the mix is cooled down slower, and stirred at the same time. Ice cream makers often exist of surface scrapers that will redistribute the ice crystals back into the ice cream mix, which will partially melt the ice crystals and create new ones on the surface. During the process, air is captured in the ice cream mix. The rate of cooling not only affects the size, but has also shown to affect the crystal shape. The crystal shape of ice crystals has easily been studied by

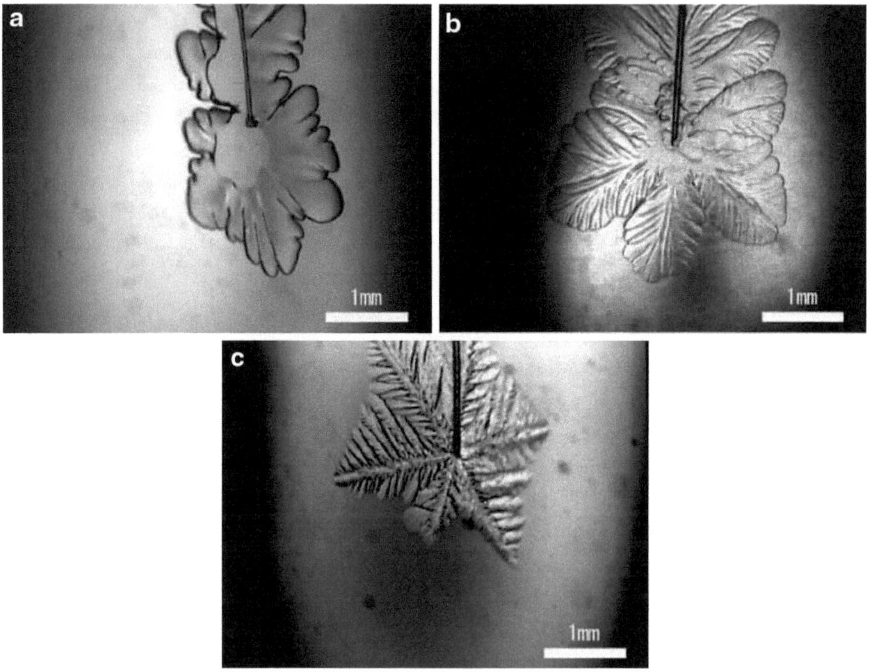

Fig. 9.2 Ice crystal growth at −0.2 °C (**a**), −0.7 °C (**b**) and −2.0 °C (**c**) [27] (Reproduced with permission; copyright Elsevier Ltd.)

investigating the growth of ice crystals placed on a capillary tube and visualized by light microscopy. The growth of the ice crystals reveal their change in shape as a function of the temperature as is depicted in Fig. 9.2 [27]. The shape of the ice crystals in ice cream depend on the production method and the formulation of the ice cream mix.

It has been shown that the rate of cooling has an effect of the final crystal shape. With high cooling rate, dendritic or needle-like shaped ice crystals are often formed [7, 8, 27]. With lower cooling rates, hexagonal (1 h prismatic crystal) or disc-like shaped crystals are more often formed [6, 8]. The shape can also change from one form to another as a result of the production process. Dendritic ice crystals that are created on the freezer surface and redistributed into the warmer ice cream mix by the dasher blades, can partially melt and transform from a dendritic into a disc-like shape [6].

9.2.2 Fat Globules

Since ice cream is prepared with dairy product, it contains fat usually between 5 % and 10 %. Fat plays a large role in the structure of the ice cream as the fat globules stabilize the air bubbles. The stability is increased by the aggregation of the fat globules around the air bubbles, which leads to a certain fraction of a solid layer around the air bubbles. To enhance the effect of the aggregates, also called partially-coalesced fat, the droplets preferably are of small particle size. Therefore, the homogenization step and the aging step in the production process are important. The homogenization step ensures the decrease in the droplet size, which increases the surface area needed for the partial coalescence to take place. However, the dairy proteins present on the fat surface inhibit the coalescence of the fat globules. To increase the coalescence, the proteins have to be removed from the fat surface, which is accomplished in the aging step. During the aging step, the emulsifiers replace the proteins from the surface, which enhances the coalescence, leading to partially-coalesced fat aggregates. The size of the fat globules around the air bubbles and between the air bubbles have an effect on the melting behavior of the ice cream. When the fat aggregates are large enough to form a network between the air bubbles, it will slow down the melting behavior. When the fat aggregates are too small, they will flow out of the matrix. Therefore good control over the size and the coalescence of the fat particles and agglomerates is important. Besides playing a role in the microstructure of the ice cream, fat also plays an important part in the flavor formation of the ice cream. As the fat is of hydrophobic nature, the fat globules hold all the hydrophobic flavor molecules, and provide the dairy taste of the ice cream.

9.2.3 *Air Bubbles*

The air bubbles are created during the production process as a result of the simultaneous freezing and stirring (scraping) of the ice cream mix. The shear, laminar and turbulent forces within the freezer decrease the air bubbles size, [4, 12, 25] which is roughly found to be around 20–50 μm after the freezing and hardening step of the ice cream [5, 6, 12]. The stability of the air bubbles depends on the amount and the type of stabilizers used. The fat globules play a large role in the stability of the air bubbles by forming a solid fat layer around the air bubbles, thereby increasing the stability [14, 29]. The type of fat, the shape and the size of the fat crystals have shown to have an effect on the air bubbles [14, 15]. On average, the volume fraction of air varies from a low 20 % to 30 % to values higher than 50 %. The overrun of the ice cream, i.e. the volume increase of the ice cream mix by inclusion of air, is usually calculated using the density of the ice cream, $\rho_{ice-cream}$, and the density of the ice cream mix, ρ_{mix}, as:

$$\text{overrun} = \frac{(1/\rho_{ice-cream}) - (1/\rho_{mix})}{(1/\rho_{mix})} \times 100 = \left(\frac{\rho_{mix}}{\rho_{ice-cream}} - 1\right) \times 100 \qquad (9.3)$$

The main function of the air in ice cream is to make the ice cream softer and smoother. Since the air bubbles scatter light, they also influence the color of the ice cream. The more air bubbles the ice cream contains, the whiter the ice cream appears. Since air has very poor conductive properties, air delays the transfer of heat; a larger volume fraction of air will inhibit the heat transfer to a larger extent. The heat transfer plays an important role in the rate of melting, and the sensory properties of the ice cream. Therefore, the amount and the stability of the air bubbles have a large influence on the shelf-life, the rate of melting, the smoothness and the coldness of the ice cream.

9.2.4 Serum Phase

As the ice cream mix freezes and the volume fraction of ice crystals becomes larger, the concentration of the sugar in the unfrozen water increases, also known as the unfrozen serum phase. Starting at a conventional total concentration of approximately 25 % of sugar, this concentration increases to nearly 70 % at freezer temperature, as 65 % of water turns into ice. The serum phase therefore consists of a highly concentrated sugar solution, which is close to its saturation concentration and glass transition. However, sugar molecules are rather small and do not have the ability to form a polymer network. A polymer network as a result of high molecular weight ingredients will change the microstructure of the serum phase into an even more viscous serum phase or even a gel-like structure. The presence of polymer networks affects the textural properties (mouth-feel) as each network has its own rheological behavior, such as a brittle, elastic or gummy behavior. These polymer networks are often created by polysaccharides, such as guar gum (GG), locust bean gum (LBG), sodium alginate, pectin, and carboxymethyl cellulose (CMC). The presence of polysaccharides has shown to affect several properties, such as melting behavior, shelf-life, hardness, overrun, etc. [26].

When dairy products (as cream and milk) are used in the production of ice cream, lactose is also present in the unfrozen serum phase next to the other sugars used. Lactose is a natural sugar (a disaccharide of dextrose and galactose) and has a relatively low solubility. Whilst soluble at concentrations in the ice cream mix, its solubility decreases as the concentration increases due to freezing. As a result, the lactose will crystallize out of the solution, creating large arrowhead- or diamond-shaped crystals (see Fig. 9.3). These crystals are larger than 30 μm. Due to their large size, they provide a very unpleasant "sandy" texture. Therefore, care must be taken that not too much lactose from milk or cream is present in the recipe. At high concentrations of sugar, the sucrose can also crystallize out of the solution, forming large aggregates of sugar crystals. These crystals often show up as "white" spots, which are typically seen in water ice. However, since sucrose crystallization is rather slow, it will take a considerable amount of time before these white spots are visible.

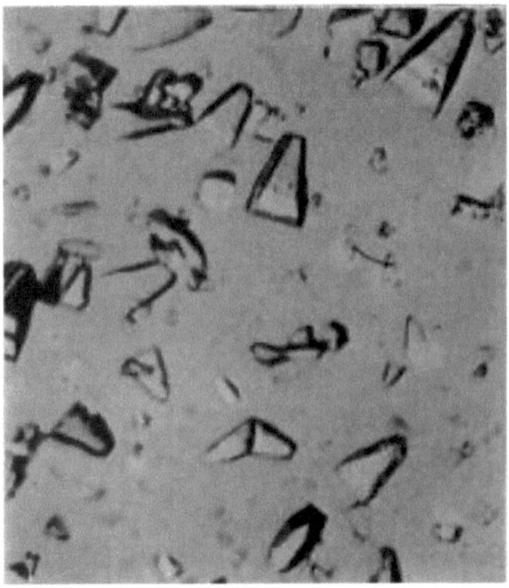

Fig. 9.3 Lactose crystals. The size of the image is 330 μm

9.3 Rheological Properties

The rheological behavior of ice cream is much more complicated than that of simple liquids. The matrix is a solution of small (sugar) and large (polysaccharides) molecules, in which particles of other phases (ice crystals, fat droplets and air bubbles) are suspended. And the volume fraction of these particles even change with temperature. Before the freezing process, the ice cream mix is a solution that contains fat particles, sugar molecules and polysaccharides. The fat particles will increase the viscosity of the mix in a similar way that spherical hard spherical particles do as described in the Krieger-Dougherty equation (see Chap. 2). The large flexible polysaccharides often give the ice cream a shear thinning behavior. With higher shear rates, the polysaccharides line out in the direction of the flow, creating an easier movements, and therefore a lower viscosity. Once the ice cream is frozen, the system will change. Due to the formation of a large volume fraction of ice crystals, the serum phase concentrates, increasing the effective volume fraction of the polysaccharides towards network formation above its overlap concentration. At this moment, the serum phase transforms from a fluid-like to a gel-like system. The increase in polysaccharide concentration will lead to a tremendous increase in its viscosity. The degree of the viscosity increase is very dependent on the type of the polysaccharide used. Some polysaccharides give a very tough and gummy behavior to the ice cream, whereas others make it more brittle.

Ice cream shows a viscoelastic behavior. Its rheological properties can be measured by using large deformation experiments (using a texture analyzer), or small deformation oscillation experiments. In large deformation tests, the ice creams are deformed to large extent to investigate their fracture behavior. From the fracture behavior, their hardness can be deduced. Small deformation tests are often used to measure the

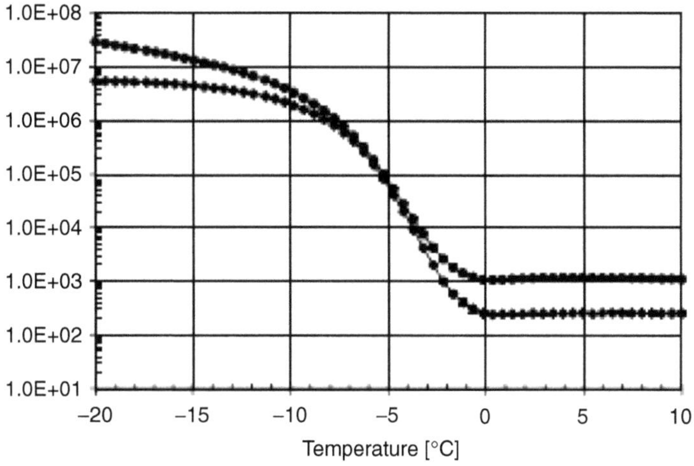

Fig. 9.4 The storage modulus, G', (*upper line*) and the loss modules, G", of ice cream at different temperatures [28] (Reproduced with permission; copyright Elsevier Ltd.)

rheological behavior over a wide range of temperatures. In small deformation test, the ice cream is placed in between two parallel plates, of which one oscillates in a sinusoidal fashion. The rate of the oscillation is expressed as the frequency, and the magnitude of the sinusoidal curve is referred to as the amplitude. When the samples are subjected to an oscillating strain (deformation), the stress on the sample is measured. Depending on the properties of the sample, the sinusoidal stress measured is in or out of phase with the applied sinusoidal curve. This is a result of the energy dissipation within the material. Material with a purely elastic behavior (rubber-like material) will result in a curve which is in phase. Material in which some of the energy is dissipated in order to deform the material (viscous flow) shows a curve which is out of phase. From the lag time, the elastic and viscous properties can be deduced. The elastic properties are expressed as G' (storage modulus), and are mostly related to the solid-like behavior of the material, whereas the viscous properties are expressed as G" (loss modulus), and are related to the viscous (flow) behavior of the material. For ice cream, the solid-like response, G', is believed to be correlated to hardness and stiffness, and the viscous-like response, G", is more representative for the liquid-like response, which is correlated to the scoopability of the sample. Figure 9.4 gives an example of these two parameters as a function of the temperature.

As can be seen, the rheological behavior can be divided into three different zones: [28]

- Zone I: at low temperatures, the volume fraction of the ice crystals is relatively high and thereby dominates the behavior of the ice cream. In this zone, G' is related to the amount of ice crystals and their connectivity. The loss modulus, G", is often related to viscosity of the serum phase and the overrun of the ice cream. A higher volume fraction of air will increase the scoopability, which is measured as a lower G".

- Zone II: as the temperature increases, the volume fraction of the ice crystals decreases, and both moduli decrease. The slope gives an indication of the melting behavior of the ice cream. The steeper the slope, the faster the ice cream melts. The melting behavior is determined by the shape of the crystals and the stability of the air bubbles.
- Zone III: as the temperature reaches values above the freezing point, the ice crystals have completely disappeared. Both moduli will reach their plateau value. The rheological behavior is now determined by the viscosity of the melted sample, which is determined by the network formation of the polysaccharide, the volume fraction of fat particles, and the volume fraction of incorporated air. The higher the overrun, the larger is the loss modulus, G".

9.4 Sensorial Properties

The sensorial properties of ice cream are a result of the microstructure of the ice cream. It is a combination of all phases present and their size distribution: the ice crystals, the air bubbles, the fat particles and the viscous matrix.

9.4.1 Effect of Ice Crystal Size

The main function of the ice crystals is to provide the cooling sensation of ice cream and sorbets. The latent heat (heat needed to melt the ice crystals) determines the heat that will be removed from the mouth and the palate to create the sense of coldness. The amount of heat is determined by the amount of the ice crystals, but also by the size distribution. A similar total volume of ice crystals with a smaller size distribution will contain more surface area. The larger the surface area, the larger is the contact with the body, and the more heat is extracted. Therefore, an ice cream prepared with liquid nitrogen (containing smaller ice crystals due to a smaller critical radius) will feel colder than ice cream prepared at higher temperatures, even though they have the same recipe (and therefore the same total amount of ice) and have been equilibrated at the same temperature. This is particularly noticeable in sorbets that do not contain fat. Small ice spheres prepared with liquid nitrogen (such as Solero shots and Calippo shots) melt very fast in the mouth leading to a very large sensation of coldness. Ice crystals larger than 100 μm are easily detected in the mouth, and are often perceived as a very icy and gritty texture. Recrystallization of ice cream, often due to temperature fluctuations, will therefore lead to a negative influence on the sensory properties.

9.4.2 Effect of Air Bubble Size

The role of the air bubbles is to make the ice cream softer and smoother. The air bubbles interrupt the solid network of the ice crystals and the very viscous serum phase; the more air bubbles are present, the more the solid network is interrupted and the softer and smoother it becomes. The presence of air bubbles will therefore have a large effect on the ease of spooning. They will also have a large effect on the mouth feel. As the air bubbles are not conductive, heat transfer is prevented within the ice cream due to the presence of air. This will decrease the rate of melt, but it will also decrease the heat extraction from the body. An ice cream with more air will therefore feel less cold. The more stable the air bubbles are, the larger the effect will be noticed. As the air bubbles are stabilized primarily by fat globules, the presence of fat will have a large influence on the meltdown rate and the sense of coldness. Ice cream with a network of fat around the air bubbles will increase the stability, thereby decrease the meltdown rate and decrease the sense of coldness. This also explains the difference in coldness between a sorbet and a dairy ice cream. As sorbets do not contain any fat, the air bubbles are less stabilized, leading to a faster heat transfer. A sorbet will therefore always feel colder than a dairy ice cream, even though the total ice crystal content and the size distribution are the same. Sorbets are therefore more preferred on a warm day, while dairy ice creams are more popular on colder days. A change in air bubble size will influence the mouth feel and the ease of spooning. The smaller the air bubbles, the softer and creamier they will feel. Especially when the ice cream melts, smaller air bubbles will last longer in the molten mix and increases the sense of creaminess. Large air bubbles are often more prone to coalescence and escape from the ice cream, thereby decreasing the amount of air and the softness of the ice cream. An increase in air bubble size is a result of pressure fluctuations or time. As soon as the air bubbles form a connecting network, their coalescence can lead to open channels that allow the air to escape and eventually the collapse of the ice cream. This will decrease their spoonability and the smoothness (softness) significantly.

9.4.3 *Effect of Fat*

As discussed, the addition of fat has an effect on the meltdown rate and the sensation of coldness as it is closely related to the stability of air bubbles. Besides these two important properties, the fat also contributes to the creaminess of the ice cream. The fat leaves a coating in the mouth during melting, thereby giving a creamy sensation, even after the ice cream has completely melted. The fat also contains all fat-soluble flavors, and therefore it also greatly contributes to the taste of ice cream.

9.4.4 *Effect of Serum Phase*

The matrix viscosity will have an effect on the gumminess and the thickness of the ice cream. When the ice cream is in its frozen state, the texture is partly dominated by the very concentrated serum phase, besides the presence of the ice crystals and the air bubbles. The properties of the serum phase are determined by the type of network formation, and its rheological behavior. The network is a result of the entanglements of the polysaccharides added and their concentration. This network varies for different polysaccharides used, and the type will therefore influence the mouth feel and the spoonability of the ice cream, e.g., some polysaccharides will make the ice cream more brittle than others. When the ice cream melts, it will transform from a solid-like into a liquid-like material. The amount of sugars and polysaccharides (besides proteins and fats) determines the viscosity of the molten ice cream, and will determine the sensorial properties such as thickness and gumminess.

In the molten state, also the amount of air and the amount of fat in the serum phase plays an important role. The fat usually provides the certain creaminess. When the air bubbles are small enough, they can fulfill the same role as the fat, and creaminess is often enhanced when a large amount of air is still present. Therefore, it is important to keep the air bubbles as small and stable as possible.

9.5 Shelf Life

The shelf life of ice cream is largely determined by the microstructure of the ice cream and the ingredients. For a long shelf-life, the size of the ice crystals and the air bubbles should not change over time.

9.5.1 Effect of Ice Crystal Size

When ice crystals become larger, this often has a negative influence on the texture of the ice cream; it becomes watery and icy. The change in ice crystals is a result of Ostwald ripening, accretion and recrystallization. Ostwald ripening occurs as a result of a wide size distribution of the ice crystals. The solubility of the ice is determined by the size of the ice crystals as:

$$S(r) = S(\infty)\exp\left(\frac{2\gamma V_m}{RTr}\right) \tag{9.4}$$

in which V_m is the molar volume of the solute, γ is the interfacial tension, $S(\infty)$ is the solubility of the solute in the continuous phase for a droplet with infinite curvature and $S(r)$ is the solubility of the solute when contained in a spherical crystal of radius r. For smaller crystals, the solubility is therefore higher than for larger crystals. Since there is always an equilibrium between melting and freezing of water molecules, the melting will occur faster around small molecules, and the freezing faster around large molecules. This will eventually decrease the size of the small ice crystals and increase the size of the large ice crystals. Accretion is the process in which two ice crystals touch, forming a neck in between the ice crystals and the two crystals finally transform into one. Both these processes lead to an increase in the ice crystal size over time, even at low temperatures. Temperature fluctuations will enhance the processes just mentioned. When the temperature is raised (by taking the ice cream from the freezer), the total volume fraction of the ice crystals will decrease, and therefore the surface of the ice crystals will melt. Some of the small crystals will hereby completely disappear, whereas the larger ones only partially melt. When the temperature is decreased again (by putting the ice cream back into the freezer), only the still existing ice crystals can recrystallize, thereby leading to a shift in the crystal size distribution. The small ones disappear, and the larger ones grow. This is therefore an accelerated Ostwald ripening process and will eventually also lead to faster accretion. To increase the shelf life of the ice cream, temperature fluctuations should therefore be minimized, and recrystallization, Ostwald ripening and accretion should be avoided as much as possible. This is usually accomplished by the addition of polysaccharides. The addition of polysaccharides will increase the viscosity of the serum phase, thereby inhibiting the diffusion of water and sucrose. It also acts as a barrier between the ice crystals inhibiting the accretion of the ice crystals. Recently, also other ingredients have been shown to slow down the process dramatically by changing the size and the shape of the ice crystals. These novel ingredients are known as ice structuring proteins (ISP) or anti-freeze proteins, extracted from plants, insects and fish that survive in very cold climates [18, 23, 24]. These ice structuring proteins have the ability to adsorb onto the ice crystal surface, thereby changing the surface properties, and inhibiting further growth of the crystals. As they prevent ice formation to a certain extent, these proteins are referred to as having icephobic properties.

9.5.2 Effect of Air Bubble Size

Similar to the change in ice crystal size, the size distribution of air will also change when stored. The two mechanisms for coarsening of the air bubbles are coalescence (similar to accretion) and disproportionation (similar to Ostwald ripening). Coalescence occurs when two bubbles come into close contact and disproportionation occurs due to a difference in pressure between the two air bubbles. As the pressure of small air bubbles is larger than the pressure of large air bubbles, air will diffuse through the matrix from small air bubbles into large air bubbles. As a result, large air bubbles will grow at the expense of the smaller ones, leading to a growth of the air bubble size. Just as the ice crystals are sensitive to temperature fluctuations, air bubbles are pressure sensitive as a result of a difference in atmospheric pressure at different altitudes. When the pressure is low, the volume will increase, thereby expanding the ice cream. This leads to rupture of the matrix, increasing coalescence between air bubbles, forming a continuous network, and the escape of air through the emerging channels. This eventually leads to the collapse of the ice cream. The change in air bubble size can be prevented by adsorption of proteins or fat on the surface. When the fat partially coalesces, a three dimensional structure stabilizes the air bubbles and will increase its shelf-life.

9.6 Measuring Techniques

9.6.1 Ice Crystals

Particle size of the ice crystals can easily be determined by light microscopy, as the ice crystal size is often in the range of tens of micrometers. The microscope should be mounted with a cold stage to ensure a frozen state of the sample. Both Brightfield and Fluorescence microscopy is used, often in combination with appropriate filters [1, 3, 22]. The cryomicroscopes are often placed in a refrigerated glove box to ensure low temperatures [16]. For a more detailed image of the ice crystals, Low Temperature Scanning Electron Microscopy (LT-SEM) or Cryoscanning Electron Microscopy (Cryo-SEM) are often used [1, 10]. Previously prepared ice cream samples are first placed into liquid nitrogen after which a small piece is fractured and placed into the SEM specimen holder. The samples are then often sputter coated with gold in a cryo-preparation unit to prevent melting of the samples. Figure 9.5 gives an example of such a LT-SEM image from a standard ice cream formulation.

Figure 9.5 shows the microstructure of ice cream, in which the dark smooth regions are the ice crystals. As can be seen, the ice crystals are rather polydisperse in size, and the average size is often found to be in the range between 30 and 50 μm. Similar to SEM, Transmission Electron Microscopy (TEM) can also be used to visualize the samples, from which the crystal size can be deduced. Sample preparation is also done under low temperatures to prevent the samples from changing [21]. The images show similar results.

A disadvantage of these microscopy techniques is that the pictures are often 2D and the preparation of the samples is rather destructive. A lot of the preparation methods (slicing, fracturing, etc.) can lead to artifacts in the system, and therefore the visualization of the microstructure with microscopy is not always accurate. Recently, new methods have been proposed to study the microstructure of the ice cream, and to measure the ice crystal size online. One method involves X-ray micro-tomography, which can be used to visualize the microstructure of the ice cream in 3D. The method uses an image processing protocol, and is described in details by Pinzer and co-workers [20]. With this technique, they investigated the average ice crystal size at different temperatures, going from an ice percentage of 21.9–48.7 %. When the ice crystals would grow homogenously, the size is expected to increase by a factor of 1.3 only. However, they found that the ice crystal size increased by a factor of 4, which can only be explained by a connectivity of the ice crystals of previously unconnected ice

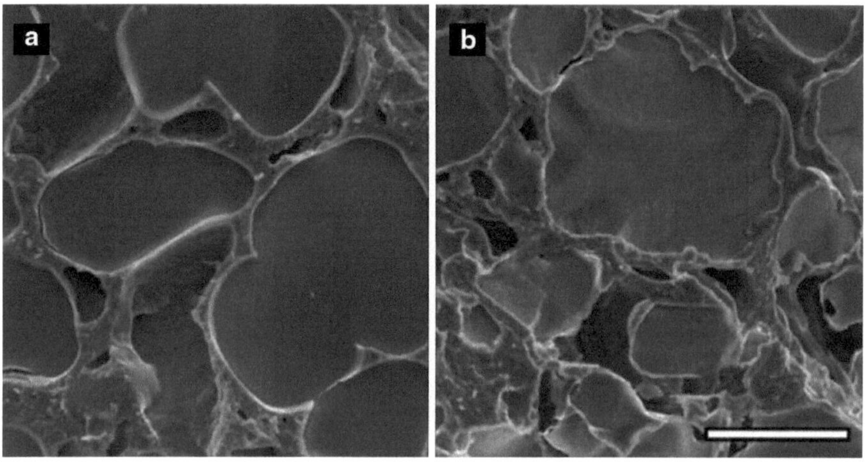

Fig. 9.5 LT-SEM micrographs of a conventional ice cream formulation containing (**a**) no emulsifier and (**b**) 0.3 % emulsifier. The bar is 30 μm. [1] (Reproduced with permission; copyright John Wiley and Sons)

crystals. The opposite is the case when the temperature is increased (thereby lowering the ice volume fraction). The partial melting leads to the decrease of the ice crystals, but also breaking up of the ice crystals. This technique was used to follow the coarsening of the ice crystals as a function of temperature fluctuations, and shows that the mechanism for coarsening is presumably a partial melting refreezing mechanism.

Another method that was recently discussed is the focused beam reflectance measurement (FBRM) technology [2]. This method allows measurement of the size distribution of the ice crystals online. The technique is based on the use of a rotating or vibrating laser beam that is focused just behind a window. A sensor detects the light reflected by particles, registers the time period of reflection, and represents the reflection times as a chord length distribution (CLD) related to particle size distribution (PSD). The advantage of this method is that the method is suitable for in situ measurements of particles at high concentrations of 40 %. However, discarding the volume fraction of air, and taking into account the liquid and solid fraction only, the volume fraction of ice is close to 70 %. This is much larger than the technique allows. Therefore, the technique is more useful for measuring the particle size distribution in situ when the ice cream is transferred from the ice cream maker at draw temperature before the hardening step.

9.6.2 Fat Globules

The particle size of the fat globules is important for the stability and the sensorial properties of ice cream. As a high surface area of fat is preferred to induce partial coalescence and formation of an aggregated network around the air bubbles, fat

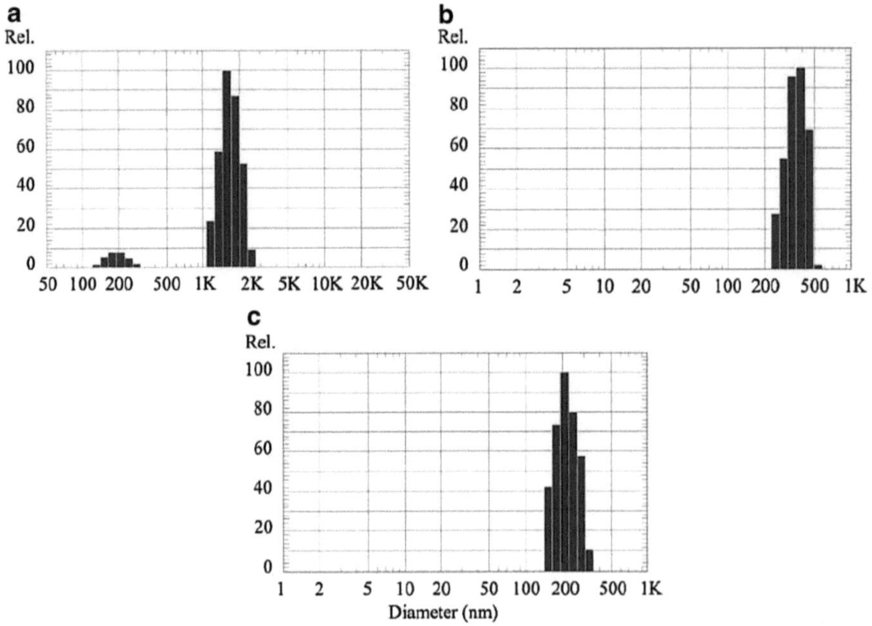

Fig. 9.6 Representative volume-weighted distributions of ice cream mix samples by DLS measurement: (**a**) unhomogenized. (**b**) homogenized at 15 MPa, and (**c**) homogenized at 97 MPa. [17] (Reproduced with permission; copyright Elsevier Ltd.)

globules in dairy products have to be reduced. The size of the fat droplets is controlled by the homogenization of the ice cream mix. High pressure homogenization often leads to a change from a bimodal to a monomodal particle size distribution and a reduction in the mean particle diameter. The particle size of emulsion and fat droplet can be measured with a variety of techniques. The most common ones are dynamic light scattering (DLS) and laser diffraction (LD) (see also Chap. 3 and [19]). DLS techniques measure the size based on the Brownian motion of the particles. The scattered light is detected by one or two detectors, and the intensity is measured as a function of time and interpreted as an auto-correlation or cross-correlation function; through the diffusion coefficient, particles size distribution is calculated. The particle size range of the DLS technique is 0.002–2 μm. LD uses the fact that particles of a certain size scatter light with specific intensities at different angles. Mie theory is used to calculate the particle size distributions. The particle size range of LD is roughly 100 nm–1 mm. Besides light scattering, also other techniques can be used to measure the particle size. One example is the use of nuclear magnetic resonance (NMR) with the use of the Earth's magnetic field (EF) [11]. After applying a pulse field gradient, a NMR signal intensity is measured, which can be related to the diffusion coefficient of the material or the relaxation time. For dairy emulsions (milk, cream), LD or DLS are used most frequently. The particle size is often given as the volume mean, $D_{4,3}$, or the surface mean, $D_{3,2}$. Figure 9.6 gives an example of the fat particle size as a function of the homogenization pressure. The pressure was changed

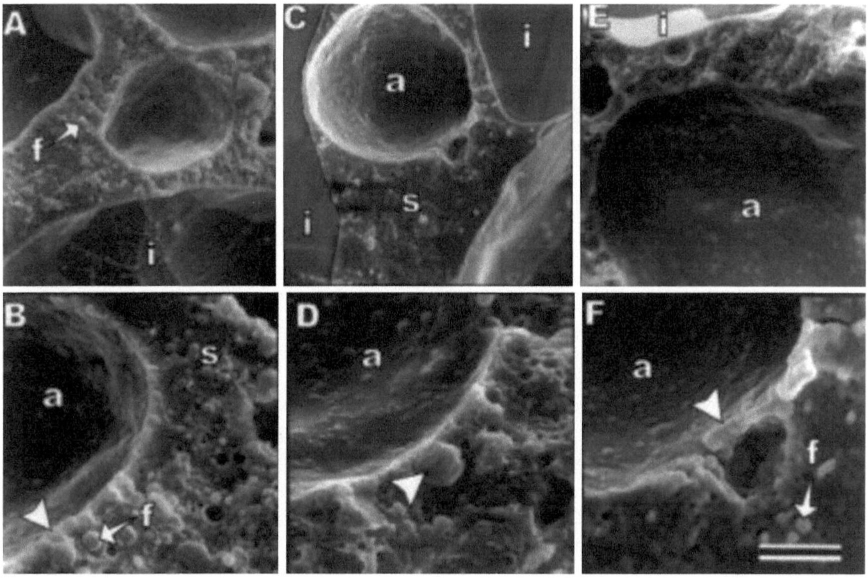

Fig. 9.7 Micrographs of frozen ice mixes at two different magnifications. In the *top* set (**a**, **c**, **e**), the bar is equal to 15 μm, in the *lower* set (**b**, **d**, **f**), the scale bar is equal to 5 μm. *a* refers to air bubbles, *f* to fat globules, *i* to ice crystals and *s* to serum phase. [13] (Reproduced with permission; copyright Elsevier Ltd.)

from 15 MPa (B) to 97 MPa (C). Figure A gives the particle size distribution of unhomogenized cream in which the droplet size shows a bimodal distribution, which is expected for the untreated dairy products. After the homogenization step, the droplet size is considerable lowered and shows a more monomodal distribution with a smaller particle size [17].

The change in the droplet size has a significant effect on the viscosity of the ice cream mix. As the droplet size decreases, the viscosity of the ice cream mix will significantly increase [17].

9.6.3 Air Bubbles

As the air bubble size is usually larger than 20 μm, the air bubbles can be seen with conventional microscopy techniques. For a more detailed visualization, LT-SEM, cryo-SEM or TEM are also used. These techniques are used to provide details of the fat globules around the air bubbles. Figure 9.7 shows micrographs of different ice cream formulations containing fat particles, dairy proteins and emulsifiers. As can be seen, the small fat globules are surrounding the air bubbles. In some of the images they appear to be spherical, whereas in others they appear to be more coalesced, thereby forming larger fat aggregates. Depending on the amount of emulsifiers added, the fat aggregation (partial coalescence) is more pronounced, due to replacement of the dairy proteins. The fat coalescence is clearly seen in Figure F (indicated by the arrow), in which a larger amount of emulsifier was added. The pictures B and D contain less emulsifier, which results in the presence of more globular fat particles [13].

9.7 Definitions

Accretion	Formation of a solid bridge in between adjacent crystals leading to a single particles
Aggregation	Formation of an assemblage of primary particles through strong physical attractive forces and/or chemical bridges
Coalescence	Collision of droplets in an emulsion following by merging to a larger droplet
Partial Coalescence	Collision of fat droplets in an emulsion following by a partial merging of the droplets. The result is the presence of agglomerates with a very strong connection
Ostwald ripening	Growth of larger crystals in suspensions at the expense of smaller ones, due to dissolution and re-crystallization as a result of differences in solubility
Overrun	% increase of ice cream volume over that of the starting mix due to air inclusion
Cryo-SEM	Cryogenic scanning electron microscopy
CLD	Chord length distribution
DLS	Dynamic light scattering
EF	Earth's magnetic field
FBRM	Focused beam reflectance measurement
LD	Laser diffraction
LT-SEM	Low-temperature scanning electron microscopy
NMR	Nuclear magnetic resonance
PSD	Particle size distribution
SEM	Scanning electron microscopy
TEM	Transmission electron microscopy
D	Particle size
$D_{3,2}$	Sauter mean diameter, area-weighted mean size, mean value of an area-based PSD
$D_{4,3}$	Volume-weighted mean size, mean value of a volume-based PSD
$\Delta_F H$	Enthalpy of fusion (J/g)
r	Radius of crystal sphere (m)
R	Gas constant: (8.3 J/molK)
S	Solubility
T	Temperature (K)
T_m	Melting temperature (K)
V_m	Molar volume
γ	Interfacial tension (N/m)
ρ	Density (kg/L)

References

1. Aleong, J.M., Frochot, S., et al.: Ice recrystallization inhibition in ice cream by propylene glycol monostearate. J. Food Sci. **73**(9), E463–E468 (2008)
2. Arellano, M., Benkhelifa, H., et al.: Online ice crystal size measurements during sorbet freezing by means of the focused beam reflectance measurement (FBRM) technology. Influence of operating conditions. J. Food Eng. **113**(2), 351–359 (2012)
3. Bolliger, S., Wildmoser, H., et al.: Relationships between ice cream mix viscoelasticity and ice crystal growth in ice cream. Int. Dairy J. **10**(11), 791–797 (2000)
4. Chang, Y., Hartel, R.W.: Development of air cells in a batch ice cream freezer. J. Food Eng. **55**(1), 71–78 (2002)
5. Chang, Y., Hartel, R.W.: Stability of air cells in ice cream during hardening and storage. J. Food Eng. **55**(1), 59–70 (2002)
6. Cook, K.L.K., Hartel, R.W.: Mechanisms of ice crystallization in ice cream production. Compr. Rev. Food Sci. Food Safety **9**(2), 213–222 (2010)
7. Cook, K.L.K., Hartel, R.W.: Effect of freezing temperature and warming rate on dendrite break-up when freezing ice cream mix. Int. Dairy J. **21**(6), 447–453 (2011)
8. Donhowe, D.P., Hartel, R.W.: Recrystallization of ice in ice cream during controlled accelerated storage. Int. Dairy J. **6**(11–12), 1191–1208 (1996)
9. Flores, A.A., Goff, H.D.: Ice crystal size distributions in dynamically frozen model solutions and ice cream as affected by stabilizers. J. Dairy Sci. **82**(7), 1399–1407 (1999)
10. Flores, A.A., Goff, H.D.: Recrystallization in ice cream after constant and cycling temperature storage conditions as affected by stabilizers. J. Dairy Sci. **82**(7), 1408–1415 (1999)
11. Fridjonsson, E.O., Flux, L.S., et al.: Determination of mean droplet sizes of water-in-oil emulsions using an Earth's field NMR instrument. J. Magn. Reson. **221**, 97–102 (2012)
12. Goff, H.D.: Formation and stabilisation of structure in ice-cream and related products. Curr. Opin. Colloid Interface Sci. **7**(5–6), 432–437 (2002)
13. Goff, H.D., Verespej, E., et al.: A study of fat and air structures in ice cream. Int. Dairy J. **9**(11), 817–829 (1999)
14. Granger, C., Leger, A., et al.: Influence of formulation on the structural networks in ice cream. Int. Dairy J. **15**(3), 255–262 (2005)
15. Granger, C., Schoppe, A., et al.: Influence of formulation on the thermal behavior of ice cream mix and ice cream. J. Am. Oil Chem. Soc. **82**(6), 427–431 (2005)
16. Hagiwara, T., Hartel, R.W., et al.: Relationship between recrystallization rate of ice crystals in sugar solutions and water mobility in freeze-concentrated matrix. Food Biophys. **1**(2), 74–82 (2006)
17. Innocente, N., Biasutti, M., et al.: Effect of high-pressure homogenization on droplet size distribution and rheological properties of ice cream mixes. J. Dairy Sci. **92**(5), 1864–1875 (2009)
18. Kontogiorgos, V., Regand, A., et al.: Isolation and characterization of ice structuring proteins from cold-acclimated winter wheat grass extract for recrystallization inhibition in frozen foods. J. Food Biochem. **31**(2), 139–160 (2007)
19. Merkus, H.G.: Particle size measurements; fundamentals, practice, quality. Springer, New York (2009)
20. Pinzer, B.R., Medebach, A., et al.: 3D-characterization of three-phase systems using X-ray tomography: Tracking the microstructural evolution in ice cream. Soft Matter **8**(17), 4584–4594 (2012)
21. Regand, A., Goff, H.D.: Effect of biopolymers on structure and ice recrystallization in dynamically frozen ice cream model systems. J. Dairy Sci. **85**(11), 2722–2732 (2002)
22. Regand, A., Goff, H.D.: Structure and ice recrystallization in frozen stabilized ice cream model systems. Food Hydrocolloid. **17**(1), 95–102 (2003)
23. Regand, A., Goff, H.D.: Freezing and ice recrystallization properties of sucrose solutions containing ice structuring proteins from cold-acclimated winter wheat grass extract. J. Food Sci. **70**(9), E552–E556 (2005)

24. Regand, A., Goff, H.D.: Ice recrystallization inhibition in ice cream as affected by ice structuring proteins from winter wheat grass. J. Dairy Sci. **89**(1), 49–57 (2006)
25. Sofjan, R.P., Hartel, R.W.: Effects of overrun on structural and physical characteristics of ice cream. Int. Dairy J. **14**(3), 255–262 (2004)
26. Soukoulis, C., Chandrinos, I., et al.: Study of the functionality of selected hydrocolloids and their blends with kappa-carrageenan on storage quality of vanilla ice cream. Lwt-Food Sci. Technol. **41**(10), 1816–1827 (2008)
27. Teraoka, Y., Saito, A., et al.: Ice crystal growth in supercooled solution. Int. J. Refrig. **25**(2), 218–225 (2002)
28. Wildmoser, H., Scheiwiller, J., et al.: Impact of disperse microstructure on rheology and quality aspects of ice cream. Lebensm-Wiss. Technol. **37**(8), 881–891 (2004)
29. Zhang, Z., Goff, H.D.: Protein distribution at air interfaces in dairy foams and ice cream as affected by casein dissociation and emulsifiers. Int. Dairy J. **14**(7), 647–657 (2004)

Chapter 10
Dry Powder Inhalers

Anthony J. Hickey and Zhen Xu

Abstract Successful drug delivery using dry powder inhaler (DPI) technology is based on knowledge of pulmonary deposition, targeting and its relationship to aerodynamic particle size distribution. DPI technologies consist of three notable elements, the formulation, the metering system and the mechanism of dispersion as an aerosol. Each of these is discussed below but emphasis is placed on powder formulation, the forces of interaction between particles that must be overcome to disperse them and the means whereby energy is imparted to achieve this objective. Efficient and reproducible drug delivery with respect to aerosol properties and dose are the objectives of product development and a requirement for regulatory approval for satisfactory disease therapy.

10.1 Introduction

Powdered drugs can be administered to the respiratory system for effective relief of ailments such as asthma and chronic obstructive pulmonary disease. Good design of the particle size distribution is essential for adequate delivery of the drug to the required region. Dry powder inhalers (DPIs) have proven their efficacy in this application.

The purpose of DPIs is to generate aerosols of drug particles in doses sufficient to treat a specific disease when inhaled. Many considerations that have to be addressed in order to develop a DPI for drug administration are presented in the following sections.

A.J. Hickey (✉)
RTI International, 3040 Cornwallis Rd, Research Triangle Park, NC 27709, USA
e-mail: ahickey@rti.org

Z. Xu
School of Pharmacy, University of Maryland, 20 N Pine St, Baltimore, MD 21201, USA
e-mail: zxu@rx.umaryland.edu

H.G. Merkus and G.M.H. Meesters (eds.), *Particulate Products: Tailoring Properties for Optimal Performance*, Particle Technology Series 19,
DOI 10.1007/978-3-319-00714-4_10,

DPI's consist of three important elements: the drug formulation, the metering system and the aerosol dispersion mechanism each of which requires attention for the drug product to achieve efficient and reproducible drug delivery to ensure efficacy and safety.

10.2 Pulmonary Deposition and Targeting: The Clinical Considerations

The efficacy of aerosol dry powder depends upon the deposition of the drug particles in the required dosage to the desired location. An understanding of the physical mechanisms of deposition and the pharmacological mechanisms of action of the drugs are crucial for the successful pulmonary drug delivery.

10.2.1 Deposition

The patient interface with the inhalation platform dictates the path the particles take through the oropharynx to the upper airways and, ultimately, the rest of the respiratory tract. Once in the airways, deposition of aerosol particles in the respiratory tract depends on their aerodynamic particle size distribution (APSD), the width of the respiratory channels, and the mode of inhalation (breathing rate, inhaled volume and pause between inhalation and exhalation etc.).

The main mechanisms of deposition, depending on the particle size, include inertial impaction, gravitational sedimentation, diffusion (Brownian motion), and interception (Fig. 10.1). Either of them dominates in defined parts of the respiratory tract. Inertial impaction occurs mainly in the tracheobronchial airways (i.e. airway generations 1–8 in the Weibel lung model [1]), and it is proportional to the particle size squared and airflow rate [2, 3]. Sedimentation is a time-dependent process. It primarily occurs in the lung periphery including the bronchioles and terminal bronchioles (generations 8–16 in the Weibel model [1]) where the linear velocity of the airflow is low and settling distances are relatively small [2, 3]. Diffusion mainly occurs in the alveoli region. It is inversely proportional to the particle size (<1 μm). Interception refers to elongated particles when they are captured even if they remain aligned in the streamline [3].

The optimal size distribution for regional respiratory deposition differs substantially depending upon a balance of APSD and inter-particulate forces, the mode of inhalation, and the disease state [4]. Generally speaking, it may be concluded that drug particles having an aerodynamic size range of 1–5 μm show optimum deposition in the deep lungs [2]. Those larger than 10 μm are generally deposited in the oropharyngeal region, while smaller than 1 μm are prone to be exhaled [2, 3]. The APSDs of drugs in a DPI are often represented by log-normal distributions because they fit the observed distributions reasonably well and can be expressed in only two parameters [5]. They may be described in terms of mass median aerodynamic diameter and geometric standard deviation [2, 6].

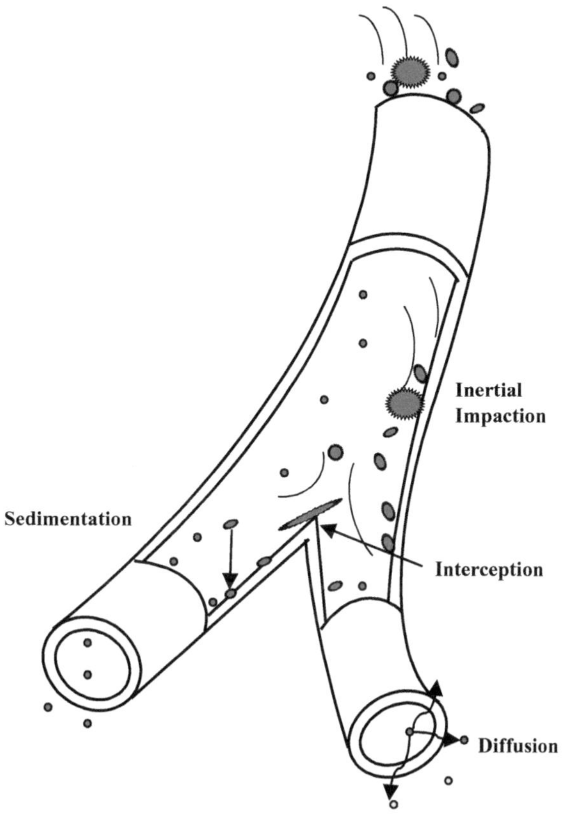

Fig. 10.1 Schematic illustration of particle deposition mechanisms in the lungs

For most pharmaceutical aerosols, the dynamics of particles fall in the Stokes' regime of curvilinear motion and, the aerodynamic diameter is derived from Stokes' law. The same law predicts that low-density (porous or hollow) particles of a particular geometric size behave as if they are smaller and elongated particles perform as if they have a size close to their cross-sectional area size rather than their length [6]. Both physical properties allow particles to penetrate deeper into the lungs. Density can be controlled through particle manufacturing methods. However, elongated particles are more difficult to manufacture reproducibly.

10.2.2 Pulmonary Targeting

Targeted delivery of pharmaceutical aerosols to desired respiratory regions is essential to more effectively exert their therapeutic effect while minimizing side effects of unwanted aerosol exposure. For instance, the location of adrenergic and cholinergic receptors in the lungs responsible for bronchomotor tone through the parasympathetic and sympathetic nervous system are distributed differently.

Adrenergic receptors are more localized in the lung periphery and cholinergic receptors are more centrally located [7]. This should guide the target particle size to achieve the therapeutic effect for beta-adrenergic agonists such as albuterol, or for anticholinergics such as tiotropium. Corticosteroids are delivered to the whole lung to treat inflammation, the underlying cause of asthma and a corollary of COPD. Inflammation occurs throughout the airways in chronic disease [8]. For drugs which are intended for systemic delivery, such as insulin, targeting to the alveolar region is desired for enhanced bioavailability [9]. There is a large supply of blood arriving to the large surface area of alveolar epithelium for oxygenation, which would facilitate rapid drug uptake [10]. However, clearance mechanisms such as mucociliary escalation and phagocytosis by alveolar macrophage should also be considered. The upper airways are supplied from a different arterial system and take the blood back to the vena cava. Consequently, both site of delivery and clearance mechanism play a role in achieving the desired therapeutic effect. The target for delivery of aerosol particles may also vary for the disease state being addressed.

10.2.3 Computational Modeling

Complementary to the effort of clinical lung deposition assessment using gamma scintigraphy technology, advances have been seen in the prediction of total and regional respiratory deposition using more accurate modeling, owing to the high resolution imaging technology, refined physical airway replicas, and the more powerful computational simulation capabilities [11]. Several reviews of these advances have been published recently that cover the lung deposition prediction over a broad range of particle size and breathing pattern using correlated curve fitting, and computational fluid dynamics using simulated respiratory tract [12–15]. These studies represent the progress in the fundamental understanding of targeted pulmonary delivery that may not be observed experimentally. However, further refinement and validation work remains to be done. Uncertainties of the predictions may still exist due to the facts such as the heterogeneous nature of pharmaceutical aerosol, the changes in the oropharyngeal opening due to flow resistance, and the diversity of dynamic breathing regime due to different disease state [14].

10.3 Analysis of Size Distribution of Dry Powder Aerosols

The most important property of a therapeutic aerosol related to clinical effect is particle size distribution. More rapid inspiration through DPIs results in higher flow rate, elevated aerodynamic shear stress, pressure drop, and Reynolds' number, which corresponds to enhanced fine particle dose, but at the same time increases the tendency for throat deposition by inertial impaction. It is clear that the powder dispersion is influenced by flow profiles and more than one flow rate should be adopted for the performance evaluation. Since the aerosol delivered from a passive dry powder inhaler does not exist independently of the inspiratory flow, shifts in the agglomerate state of drug, which can be monitored through light scattering particle size analysis or inertial impaction method at different flow rates.

10.3.1 Inertial Impaction Particle Size Analysis

Current compendia methods for dry powder aerosol testing are based on inertial impaction methods for aerodynamic sizing at fixed airflow rate (30–100 L/min), and these methods are considered the most relevant for the in vitro description of pharmaceutical aerosols [16]. According to the United States Pharmacopoeia standard, any inertial sampling apparatus can be selected for DPI testing, including Marple-Miller impactor (MMI), Andersen cascade impactor (ACI), multi-stage liquid impinger (MSLI), and next generation impactor (NGI). The correct operation of these apparatus in terms of measurement accuracy and reproducibility at established flow profile is important [17].

An induction port standardized for all systems is used. Surface coating of stages to avoid bouncing and re-entrainment is a necessary step to all inertial samplers except MSLI that uses water in each stage. The collected mass must be within 75–125 % of expected, based on delivered dose. The theory of inertial sampling and the apparatus were reviewed before [18–21], so they are not reiterated here.

From a quality control and regulatory perspective, the amount of drug sampled at each stage of the impactor is plotted against the cut-off size for the stage. The particle size distribution represented as cumulative percent undersize can be further analyzed to calculate the mass median aerodynamic diameter and geometric standard deviation [18]. A drawback of impactors is the small number of given size classes. Furthermore, how to interpret the specifications is not yet part of the compendia.

10.3.2 Light Scattering Analysis

Particle de-agglomeration during aerosolization can be seen as changes in particle size distribution. The most commonly used method for volume particle size distribution is laser diffraction (Fig. 10.2). This method is based, depending on particle size, on Mie scattering or Fraunhofer diffraction theory for spherical particles [22]. The fine particle fraction can be obtained by analyzing differential or cumulative distribution data. The volume fine fraction can be converted to aerodynamic fine fraction under the assumption that the sizing results are the same as those

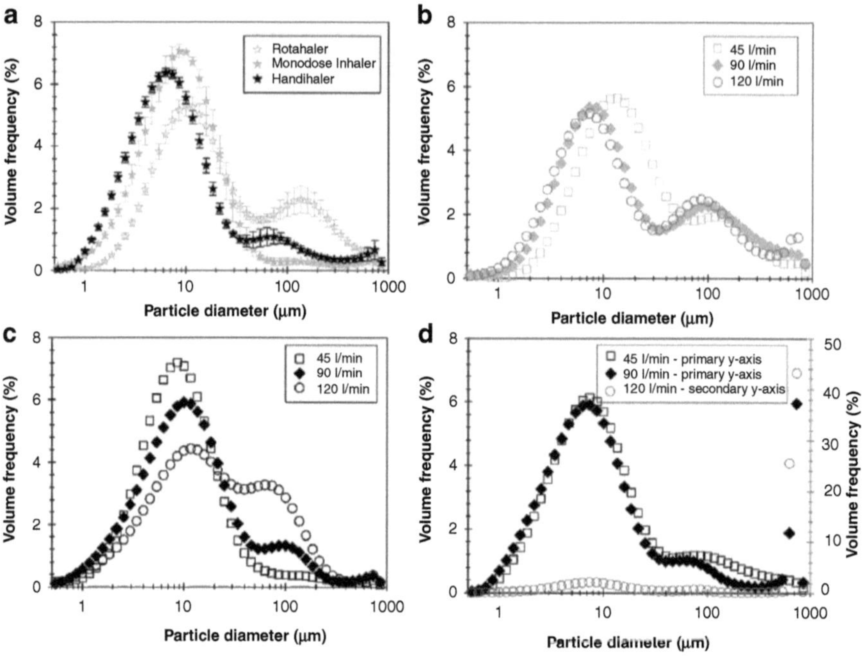

Fig. 10.2 Flow rate dependence of dispersion of dry powder aerosols [141]. Particle size distributions of aerosolised plume of lactohale 300 measured by laser diffraction and aerosolised from Rotahaler, Monodose Inhaler and Handihaler® at: (**a**) 60 l min^{-1} showing typical variability; the mean of five replicates of lactohale 300 aerosolised at 45, 90 and 120 l min^{-1} through (**b**) Rotahaler, (**c**) Monodose Inhaler, and (**d**) Handihaler (120 l min^{-1} on secondary y-axis)

obtained by sedimentation (based on the Stokes' law), or a shape factor is available, and the particle density is known (see also Sect. 10.3.1). The advantage of the laser diffraction method is its efficiency in real time measurement of size distribution at different flow rates, which is valuable for accessing the de-agglomeration [23–25]. This method is mainly used for carrier-free formulations because of the skewed distribution and difficulty in differentiating drug and fine particles. But the measurement of carrier-based formulation can still be achieved by eliminating the inner detector element data which relate to the larger carrier particles [25].

Laser Doppler velocimetry, also known as laser Doppler anemometry, measures the velocity of particles passing through the intersecting point of two laser beams [26]. It is also used to indirectly evaluate the de-agglomeration by measuring the turbulent flow field [27]. Other related laser Doppler methods, such as Phase Doppler analysis (PDA) that measures both particle size and velocity characteristics of irregular shaped particles, have been reviewed previously [18]. PDA optically splits a laser beam, uses the interference of particles with one element of the laser and then converges both portions of the beam to indicate a phase shift proportionate to the size of the particles.

The aerodynamic particle size distribution can be measured by time-of-flight particle sizing systems such as Aerosizer. An aerodisperser generates an aerosol, accelerates the particles which pass two subsequent laser beams. Smaller particles are accelerated faster and reach higher velocities than larger particles due to inertia. The time-of-flight of these particles between the two beams is then converted into corresponding particle size distribution. A shift in the aerodynamic particle size distribution can be achieved by changing the shear force applied for the aerodisperser, and used for evaluation of agglomeration strength of carrier-free powders [28, 29]. The advantage of this method is its in situ generation of aerodynamic data. But it may not be used for formulation blends because it is not possible to differentiate drug and carrier.

10.4 Drug Formulation

10.4.1 Static Powder Properties

Drug particles are ordinarily prepared in micrometer sizes to be respirable, typically an aerodynamic size of less than 5 μm [30]. These small sizes are subject to a variety of forces of interaction that lead to particle clustering. The major fundamental underlying forces of interaction are van der Waals forces, which are inherently important for particles in this size range. They dictate the bulk behavior in dry and uncharged fine powders. The magnitude of the van der Waals forces is a function of the particle separation distance and the molecular properties of the materials involved [31, 32]. They are typically effective within a separation distance less than 100 nm, so they are sensitive to changes in properties including particle size, shape, surface roughness, and contact deformation [33, 34]. From the point of view of thermodynamic work of adhesion, they are also influenced by the surface chemistry and interfacial energy of the powder [35]. It should be noted that van der Waals forces are a manifestation of combining relations of the intermolecular forces and should be operative between molecules (or extended surfaces), nanostructures such as colloids, and microparticles [31].

Electrostatic forces are mainly generated during pharmaceutical powder processing (such as milling, mixing, and filling), when electrostatic charges are accumulated by means of contact, coulombic interaction, and induced charging [36, 37]. They are long range forces compared to van der Waals forces. The magnitude and polarity of electrostatic forces are related to the nature of powder and container wall. Electrostatic charges decay when relative humidity increases [38, 39]. Their effects become negligible at elevated relative humidity (e.g. RH $>$ 65 %), at which the capillary forces become dominant due to capillary condensation of water vapor between neighboring particles. High tensile liquid bridges are formed because of the Laplace pressure difference at the contact interface [31]. The magnitude of capillary forces is a function with respect to the water meniscus and the surface

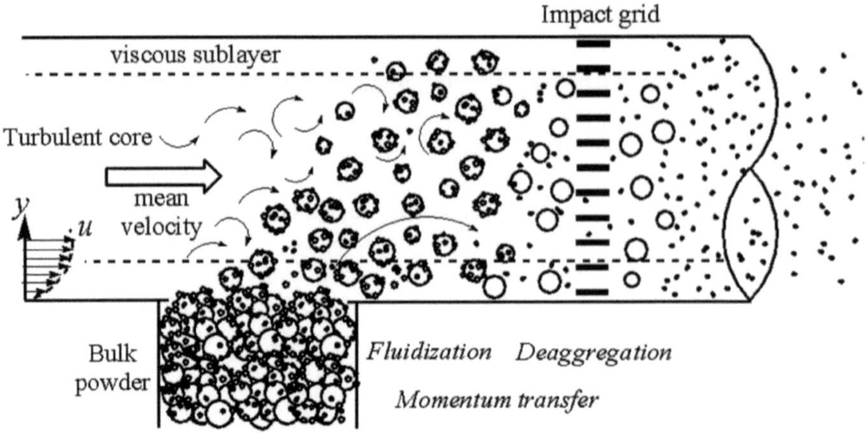

Fig. 10.3 Schematic diagrams of the continuum from powder bed to aerosol

tension between the solid and water interface [40]. The reciprocal but non-linear relationship between capillary and electrostatic forces is a major conundrum in predicting the behavior of dry powders.

In addition, other sources of particle interaction including mechanical interlocking and frictional forces are considered important, though not frequently studied. The magnitude of the mechanical interlocking is dependent on surface roughness and can be enhanced with prolonged mixing [18]. The inter-particulate friction may play an important role at the initiation of powder aerosolization [41]. The special features and implications of dry powder friction forces have been explored previously [32, 42]. Many factors influence and add complexity to the particulate interaction, and they are described in the next section [43].

10.4.2 Powder Dynamics

The aerodynamic fluid phase and the solid phase interacting in the DPI device, play crucial roles in powder dynamics. The powder aerosolization may be considered in terms of static powder, dilation, fluidization, and de-agglomeration (Fig. 10.3). Fluidization is the mobilization of the bulk powder by interaction with air molecules. Depending on the DPI metering system and device design, the fluidization mechanisms include shear force, gas-assisted, and capillary for aerodynamic fluidization, and vibration and impaction for mechanical fluidization [18]. De-agglomeration is the second stage interaction of fluidized powder in air when drug particles are stripped from the carrier surfaces or drug-drug agglomerates and dispersed into primary respirable particles. Turbulent airstreams are generally recognized as the major source for particle de-agglomeration. The primary

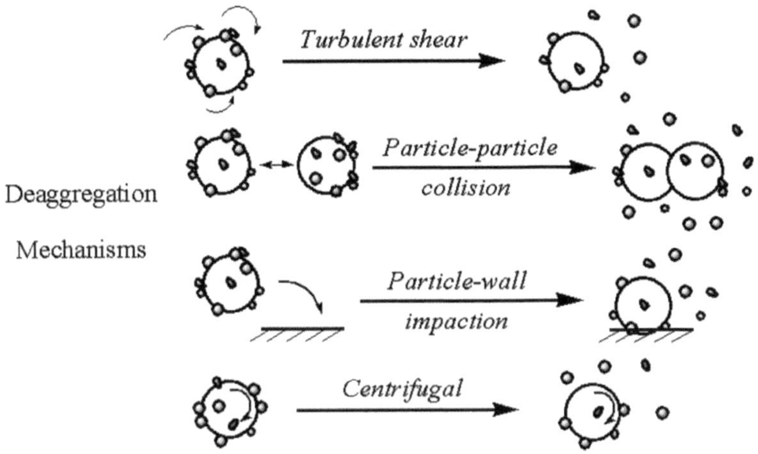

Fig. 10.4 Proposed aerosol de-agglomeration mechanisms of carrier-based powder

mechanisms responsible for de-agglomeration are turbulent shear stress and inertial separation such as that achieved by collision, centrifugal and vibration [18] (Fig. 10.4). Powder entrainment and de-agglomeration within a DPI are dynamic processes with rapid expansion of system volume. However, it is enlightening to consider a related situation when deposition-re-suspension equilibrium is established at a fixed system volume. Two theoretical models, force balance and energy accumulation, were developed for particle re-suspension in turbulent flows [44].

When a turbulent airflow entrains through a DPI device, it exerts aerodynamic forces or moments on the static dry powder to overcome the inter-particulate or surface forces. Once fluid forces reach a critical value to balance adhesive forces, particle detachment occurs. For example, the critical diameter for the separation of two identical spheres can be estimated by equating the viscous shear force to the adhesive van der Waals force [18]. The aerodynamic lift forces could be smaller than adhesion forces by several orders of magnitude because of rolling and sliding [45–47]. Once re-suspended, the particles move/rotate along the downstream airflow or collide with other particles or walls by momentum transfer. The mechanism of turbulent "burst" within the viscous sub-layer has been used to explain particle detachment, but its contribution is controversial [47, 48].

Alternatively, the flow transfers the turbulent energy into the powder bed. Particles can be re-suspended when sufficient energy is accumulated to overcome the adhesion potential resulting in entrainment [49]. There are several advantages in using the energy accumulation model. It accounts for the re-suspension below the critical flow velocity. It also takes into account of the time scale and explains the time dependence of particle re-suspension. Furthermore, it implies that particle re-suspension is analogous to the desorption process that occurs on a molecular level [50].

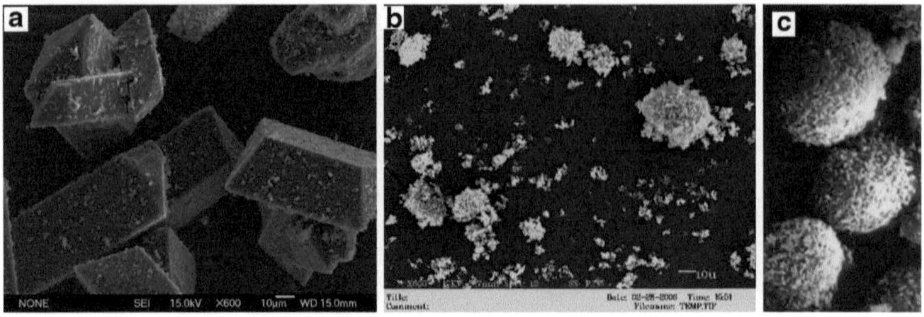

Fig. 10.5 Schematic illustration of DPI powders: (**a**): Carrier based formulation (2 % w/w Albuterol sulfate – Trehalose interactive physical mixture); (**b**): Carrier-free agglomerates (Albuterol sulfate); and (**c**) Low density porous particles generated by spray-drying [6]

10.4.3 Carrier-Based Formulation and Aerosol Performance

The most common DPI formulation consists of interactive physical mixtures of micronized drugs, coarse carriers (and surface fines) maintained by thermodynamically favorable interfacial interactions at the solid-solid interface (Fig. 10.5a). A balance of cohesive and adhesive forces between these components is formed during mixing. The principle involved in this approach is the uniform distribution of micronized drug particles at the surface of a larger carrier (typically 60–90 μm), such as α-lactose monohydrate (abbr. as "lactose" in this chapter) which achieves ease of dose filling, where the lactose is a diluent to facilitate dose metering and a carrier to enhance aerosol dispersion properties, since drug particles can be removed more readily from the surface of lactose than separated from agglomerates with other drug particles. However, this represents a simplified scenario when drug-carrier adhesive forces are much stronger than drug-drug cohesive and drug-fine adhesive forces (so-called ordered mixture [51]). Other scenarios such as the formation of drug-drug or drug-fine agglomerates are common with respect to the heterogeneity and multi-component nature of the powder mixtures. Besides this common carrier-based formulation, topics of specialized forms such as composite micro-particles with coarse carriers [52], or composite carriers [53, 54] are observed but will not be discussed here.

For the purpose of efficient drug particle dispersion, manipulation of the adhesive/cohesive properties and their force balance by a variety of approaches, such as modification of particle size [55], morphology and surface roughness [56, 57], surface coating [58], ternary component inclusion [59], alternative carrier [60, 61], or particle engineering [62, 63], to alter bulk or surface properties of the mixing components has been extensively reviewed [64–71]. It is evident that there are many methods to improve aerosol performance, but the improved performance may be confounded with additional complexities at the heterogeneous solid mixture interface. Previous studies focused on a few featured variables that may or may not be the key influences. For example, increased drug aerosolization was often observed with smaller carrier sizes because they may effectively disrupt the drug-drug cohesion [72]. However, it was also reported that increasing the carrier size had little influence [41] or increased aerosolization [73] in some formulation.

A few recent well-controlled studies on specific formulation variables influencing the aerosolization performance may be highlighted. In one study, a series of model polystyrene spheres used as carriers with different size ranges (volume median size = 82.8; 277.5; 582.9 μm) were studied [55]. Using these carriers, the variables, including carrier shape, morphology, and drug-carrier adhesion, were eliminated. The model drug employed was micronized albuterol sulfate. Selection of these model spheres eliminated other formulation variables while only focusing on the carrier size effect. The aerosol performance increased as carrier size decreased. The decreased carrier size corresponded to increased particle number and surface area, which were presumably attributed to the increased impaction events and associated frictional and rotational collision [55]. A similar study focused on drug loading (varying carrier to drug ratio from 5:1 to 85:1) showed that, as the drug loading increased, the aerosol performance with respect to the fine particle fraction increased after a threshold was reached, at which point, the formulation transitioned from ordered mixture to agglomerated mixture transition [74].

10.4.4 Carrier-Free Formulation and Aerosol Performance

Loose agglomerates of respirable drug (Fig. 10.5b) have also been employed to facilitate aerosol dispersion. These carrier-free formulations avoid possible drug-carrier chemical reaction and are suitable for high drug payload. They are used in commercial DPI formulations such as Pulmicort® [75] and Bricanyl® [23]. Micronized drugs are often very cohesive and have poor aerosolization properties [71]. In order to achieve satisfactory flowability for accurate filling and aerosolization, this type of formulation generally requires well-controlled processes to form loose microparticle agglomerates of different size, a processing technique referred to as pelletization [76]. Special carrier-free formulations such as that inhaled insulin (Exubera®) will not be discussed here.

Increased fluidization and aerosol performance can be achieved by reducing drug-drug cohesive forces. This is typically performed by modification of surface physicochemical properties or particle engineering. For the former method, micronized drugs are often coated with low surface energy additives, known as force control agents (e.g. leucine, lecithin or magnesium stearate), using a variety of coating techniques such as mechanofusion [77–80]. and physical vapor deposition [81, 82]. It has been proposed that the increase in aerosol performance is due to formation of a low surface energy disruption layer to reduce drug-drug cohesion and attrition [58]. Decreasing inter-particulate interactions may also reduce the dependency of aerosolization on the flow rate and inhaler [83]. It should be noted that this method may be limited by the kind of substance that is approved for pulmonary delivery [84]. If this is the case, particle manufacture by spray drying has often been employed to prepare drug particles [62, 85, 86]. Spray dried particles can be prepared having lower bulk density, which decreases cohesive forces. Moreover, they show greater susceptibility to shear in the fluid dynamic environment of the airflow channel of the inhaler because of their larger volumes and lower density [9, 87] (Fig. 10.5c).

10.5 Theories for Dry Powder Aerosol Dispersion

The theories for dispersion of drug from dry powder beds differ for carrier-based versus drug particle alone (carrier-free) formulations (see above). The dispersion of carrier-based formulations will be considered as it has been more extensively studied. The reader interested in theoretical studies of carrier-free formulations is referred to the literature [28, 80, 81, 88–90].

10.5.1 "Active Site" Versus "Agglomeration" Theory

Two complementary theories exist for the improved aerosol performance by introducing ternary fine particle components into carrier-based formulations, namely the "active site" and the "agglomeration" hypotheses [59] (Fig. 10.6).

The former theory proposes that both geometric (asperities larger than drug particles providing shelter) and energetic (raised surface energy) features exist on the coarse carrier surface which are more adhesive than other areas, namely "active" site. The addition of fine particles preferentially binds to these sites, forcing drug particles to the weaker binding site, thus drug particles are more easily liberated from the surface of carrier particles after actuation [91, 92]. The more adhesive sites were shown to exist as was probed by atomic force microscopy [93] and nitrogen adsorption isotherm [94]. The influence of drug payload had little effect on the performance of an albuterol sulfate/lactose formulation until the occupancy of "active" sites was presumably saturated [95]. However, introducing fines of higher energy resulted in an overall increase in surface energy and may have increased the drug-carrier adhesion [96]. The influence of blending sequence studies [91] on aerosol performance have often been used to support "active" site theory, but time-dependent redistribution may occur during mixing [97], and the results may be confounded by other variables such as drug concentrations [98]. Furthermore, the "active" site has a vague definition similar to surface heterogeneity that covers both surface geometric and energetic features. A categorical comparison of "active" vs. "non-active" may be an over-simplified treatment of the complex features on the carrier surface [66].

The latter theory, of particle agglomeration, proposed that drug and fine particles can form loose agglomerates, and can be dispersed more easily than drug particles

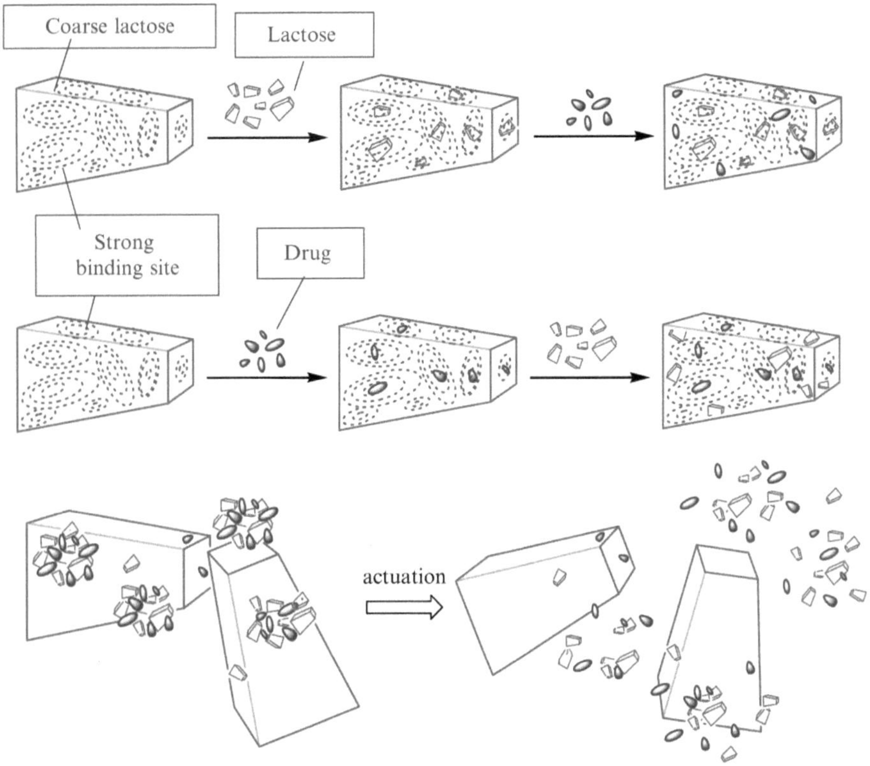

Fig. 10.6 Schematic of 'Active Site' (*above*) and 'Agglomerate' mechanisms (*below*) [66]

alone [99]. An investigation of the mechanism by AFM, employing the cohesive-adhesive balance approach suggested that increasing drug-fine adhesion may give rise to larger agglomerates that experience greater separation forces for de-agglomeration [99]. Surface energy analysis using inverse gas chromatography showed that increasing surface energy interactions between drug and carrier resulted in improved aerosol performance [100]. According to several studies, an optimum drug and lactose fines ratio exists when studying the salmeterol xinafoate blends [101–103]. These findings support the agglomeration theory.

In these theories, it is presumed that carrier lactose helps dilate the powder bed in response to a driving airflow and particles coated with drug are carried into the airstream. In the airstream, impaction with the walls of the inhaler and with other particles in the turbulent environment and surface shear strips drug particles from lactose surfaces, and entrains them towards the patient. The efficiency and reproducibility of the dose delivered is dependent on the forces of interaction between the lactose and drug particles. Theories for this phenomenon have focused on cohesive and adhesive balance of forces. These are thought to underpin the 'active' site theory of binding of drug to lactose and the agglomerate dispersion interpretation both of which explain the behavior of drug in response to airflow parameters.

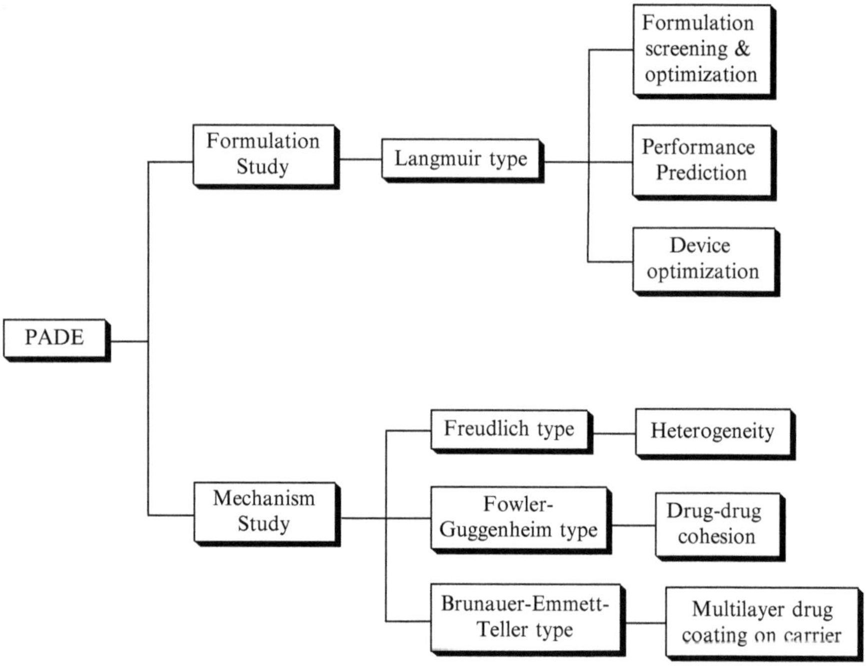

Fig. 10.7 Schematic depicting the application of PADE

10.5.2 Powder Aerosol De-agglomeration Equation

Recently, we have proposed a theory for drug dispersion involving a powder aerosol de-agglomeration equations, or PADE (Fig. 10.7), which directly correlate aerosol performance with an airflow parameter of the standardized entrainment tubes (SETs [104]) (see Sect. 10.6.4) using an algebraic equivalence of Langmuir adsorption equation, namely, the fine particle fraction (*FPF*, %) is correlated with shear stress (τ_s, N/m^2). Increasing τ_s corresponds with increasing *FPF*, until it approaches invariant region (FPF_{max}, %) (Eqs. 10.1 and 10.2) [105]:

$$FPF = \frac{FPF_{\max} k_d \tau_s}{1 + k_d \tau_s} \tag{10.1}$$

$$\frac{\tau_s}{FPF} = \frac{\tau_s}{FPF_{\max}} + \frac{1}{k_d (FPF_{\max})} \tag{10.2}$$

The rationale for the development of the PADE is based on the analogous physical phenomena and meaningful interpretation based on a comprehensive investigation of experimental data [106–108].

Firstly, the behavior of heterogeneous energy distribution and de-agglomeration in a particulate system is considered to be analogous to the surface coverage described in molecular surface adsorption and the number of surface binding sites in colloidal protein binding. During powder mixing, micronized drug particles thermodynamically prefer to adhere onto higher energy sites. When these high energy sites are gradually occupied, the drug particles adhere onto progressively lower energy sites, resulting in a decreased heat of adhesion (or from the guiding analogy, adsorption) with loading. When a certain shear stress is applied, drug particles at lower energy sites are preferentially de-agglomerated while those at higher energy sites remain associated unless higher shear stress is applied [105]. Secondly, it is the notion that the models of molecular surface dissociation described by an adsorption expression may be adapted to fit drug de-agglomeration caused by shear displacement. This is based on the fundamental understanding that inter-molecular and particulate forces that cover these analogous phenomena are essentially the same. Thirdly and most importantly, this theory has shown excellent correlations between experimental data and mathematical approximations using the PADE method.

The adoption of PADE theory has profound potential implications (Fig. 11.7). Expressions describing surface adsorption were intended to allow elucidation of mechanisms including monolayer adsorption heterogeneity (e.g. Freudlich expression [109]), lateral interaction (e.g. Fowler-Guggenheim expression [110]), and multi-layer adsorption (e.g. Brunauer-Emmett-Teller expression [111–113]), the analogy of which may be used to account for heterogeneity, drug-drug cohesion, and multi-layer drug agglomeration, respectively. It should be noted that such an approach may circumvent debate surrounding "active" site and agglomeration theories, and mechanistically evaluates the underlying particle-particle (drug-drug, drug-fines) and particle-surface (drug-carrier, fine-carrier) interactions. In addition, the PADE method can be readily used for fast formulation screening/optimization and robust aerosol performance prediction in a given shear stress range [106].

10.6 Device Technology

10.6.1 Principles of Turbulent Pipe Flow

The development of an efficient DPI device requires fundamental understanding of fluid dynamics. Turbulent airflow is generally recognized as the major source for aerosol dispersion. The airflow paths of DPIs are complex, but it is a useful simplification to consider airflow through a cylindrical pipe, which consists of three regimes: an inviscid turbulent core, a viscous laminar sublayer, and a buffer layer where transition from turbulent to laminar flow occurs [114]. The large-scale motions of airflow are influenced by the boundary conditions, while the small-scale motion is determined by the rate of energy received from the large scales and the fluid viscosity. Eddies carry turbulent kinetic energy distributed over a broad range of scales. The airflow path of the pipe flow can be characterized by several closely related airflow parameters including Reynolds' number (*Re*), shear stress (τ_s), pressure drop (ΔP), and *power*, which may be interpreted as indirect reflections of the energy experienced by the powder during dispersion. Among these parameters, *Re* is the ratio of inertial forces to the viscous forces. The value of *Re* is often used for prediction of turbulence. It is dependent upon device geometry, time scales and initial disturbance [18]. The viscous τ_s is characterized by the energy cascade and Kolmogorov theory [115]. Briefly, the larger scales (eddies) transfer kinetic energy to successively smaller scales until reaching the smallest dissipative scale (Kolmogorov scale), and the energy is dissipated by viscous action into heat. The value of τ_s can be approximated by relating the dynamic and kinematic viscosity of the air, and the energy dissipation rate (ε) which is a function of the nozzle velocity and diameter [116]. The kinetic energy dissipation can also cause ΔP. The value of ΔP measured using manometer up- and downstream of a device yields a combined value of both viscous and inviscid contribution [117], but either contribution can also be measured separately [118]. The *power* is the rate of work done or inspiratory effort during inhalation, so its value is related to the patient condition and inhalation capability [119]. In general these airflow parameters are positively correlated, and they have a close relationship with powder flow and de-agglomeration. In addition, specific resistance (R_D), an airflow independent parameter, is an intrinsic value that is a function of the device internal geometry and dimension. For a given device (determined R_D), inhaled flow rate is proportional to the square root of the ΔP [117].

10.6.2 DPI Device Technologies

The goals of device innovation include design of efficient aerosolization mechanisms, reproducible dose metering, reduced dependence on inspiratory flow rates, and user friendliness. Each type of DPI device has unique airflow paths and internal geometries, which have been designed to help disperse the drug. The airflow path can be characterized in term of the aforementioned four airflow parameters. The DPI device design and innovation features can be classified into aerosolization, dose metering, and actuation methods (Table 10.1). The aerosolization methods include a variety of baffles, impellers, or tortuous channels through which particles will experience collisions of sufficient magnitude to dislodge drug particles and entrain them in the airstream. The dose metering can further be classified into single unit dose (capsule), multiunit dose (capsules, blisters, strips) and multiple doses (reservoir). The actuation can be either passive (breath-actuated) or energy assisted actuation. These features have been extensively reviewed [18, 68, 120–123]. The following discussion focuses on two technologies that can be used to facilitate device design. They include the incorporation of computational fluid dynamics (CFD) for device design and the standardized entrainment tubes (SETs) for the selection of proper airflow parameter for specific formulations.

Table 10.1 Schematic of DPI technologies and mechanisms

DPI Innovation	Design	Mechanisms	DPI examples
Aerosolization	Turbulence, Grid, impeller	Turbulence, Impaction	Rotahaler
	Pressure drop, Shear force	Bernoulli venture, Shear force, Relative motion	Aerolizer, Inhalator
	Spinning impeller	Shear force, Relative motion	Spinhaler
	Helical/spiral wall discharge channel	Turbulence, Shear, Impaction	Turbuhaler, Twisthaler
	Cyclone chamber	Cyclone separation	Novolizer, Twincer
Passive Dose Metering (Breath-actuated)	Single dose capsule	Capillary fluidization Capsule rotation	Aerolizer, Handihaler
	Multi unit-dose capsules, blisters, strips	Bernoulli venture, Shear force	Diskus, Diskhaler
	Multi dose reservoir	Turbulence, Shear, Impaction	Turbuhaler, Twisthaler
Energy Assisted Dose Metering	Transjector, Compressed air	Gas-assisted	Exubera, Powderhale
	Vibration	Piezoelectric transducer	Oriel, MicroDose
	Battery powered impeller	Shear force, Relative motion	Spiros
	Spring-loaded hammer	Mechanical impaction	"Tape based"

10.6.3 Device Development Using CFD Technologies

Despite the many DPI products in the market, CFD has only recently gained its momentum for the study of DPI fluid flow, powder de-agglomeration, and device design, owning to the high speed computer technology. The turbulent flow around each particle can be analyzed using the Reynolds averaged Navier–Stokes equation [124]. The integral scale strain rate (γ_i) of the turbulence flow can be obtained as a function of the rate of energy dissipation (ε) and the rate of turbulence kinetic energy (k) [26]. The magnitude of the γ_i is related to the effectiveness of particle de-agglomeration. By using CFD method, Coates et al. studied the influencing factors to the particle de-agglomeration in the Aerolizer device design by varying grid structure and mouthpiece length [26], capsule size [125], airflow rate [126], mouthpiece geometry [127]. The CFD results were coupled with experimental data using spray dried mannitol agglomerates to gain insight in aerosolization. More recently, a similar computer-aided design using rapid-prototyping 3-D print and CFD have been used for the rational design of devices [128]. This provided a quicker and simpler method for inhaler design than trial and error [128].

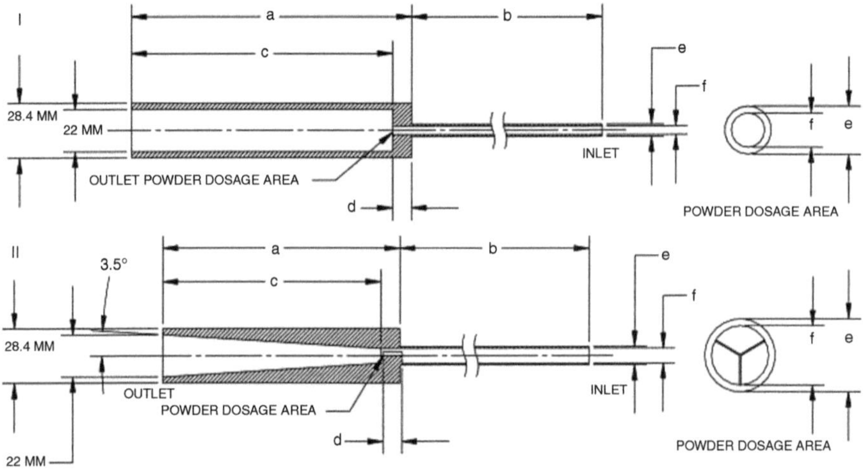

Fig. 10.8 Standardized entrainment tubes (with **I.** peripherally-located dosage table and **II.** centrally-located dosage table [104])

It should be noted that current CFD simulation capability is still limited with regard to the particle number (~10,000), the model formulation (carrier free agglomerates more often studied than carrier-based formulation [129]), and devices (specific de-agglomeration mechanism required for simulation).

10.6.4 Standardized Entrainment Tubes (SETs)

It is clear that mechanistic study of powder aerosolization requires that each contributing factor such as turbulence and impaction be studied separately [130]. This led to the design of specialized dispersion/entrainment tubes for specific mechanism study, such as the standardized entrainment tube that specialized for the study of the particle de-agglomeration influenced by grid structure [27, 131], turbulence [132], and impaction [133], respectively.

We have conducted a number of experiments in which the intention was to control the airflow path and describe it accurately in terms of defined airflow properties to evaluate formulation performance independently of unique device geometries. A series of standardized entrainment tubes with specific resistance range covering commercial DPI devices were built for these purposes [104] (Fig. 10.8). The PADE was first developed when the fine particle fraction of the carrier-based formulation performance was correlated with airflow parameter τ_s [105]. For a specific carrier-based formulation, it can easily estimate by interpolation of the τ_s to the performance invariable region [105]. In addition, two portions of the entrained dry powders: a major portion that involves in particle de-agglomeration and a minor portion remains in the entrainment tubes without

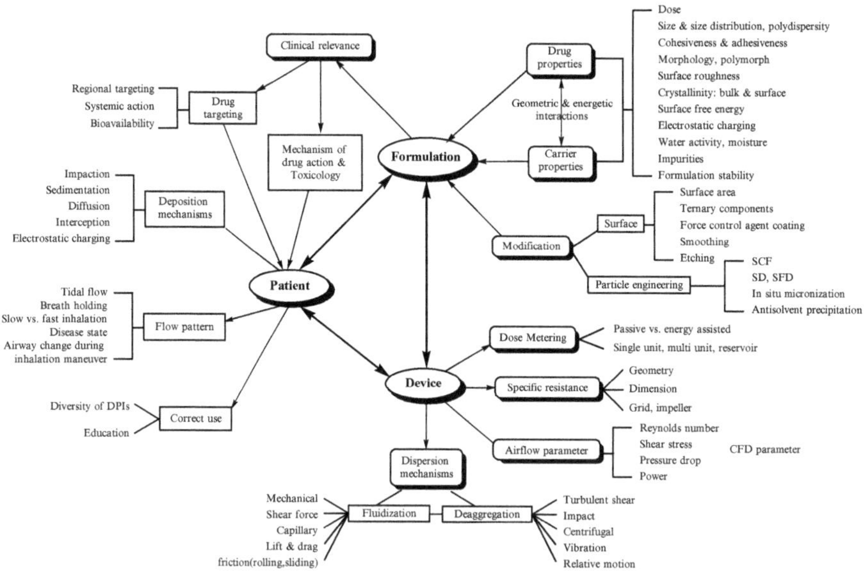

Fig. 10.9 Schematic of the sources of variability in the performance of DPIs

de-agglomeration were differentiated, which is a useful feature of the entrainment tubes since it points exclusively to formulation performance in the absence of specific inhaler design characteristics [106]. This is a novel method of assessing the mass of powder delivered on a portion of the delivered airstream, which is important in assessing where the drug particles would be delivered effectively on the inspiratory flow.

10.7 DPI Product Development

The development of a successful DPI is a challenging task which includes a comprehensive exploration with respect to the formulation/pre-formulation, device innovation, and patient interfaces, let alone the regulatory and marketing aspects. A myriad of factors and considerations for a DPI product are illustrated in Fig. 10.9. However, to be successful and competitive, simple approaches to formulation (minimal processes and excipients) and device development (minimum number of components) are adopted for rapid and efficient product development. For example, the pharmacopeial induction port (throat) was known to be a simplified treatment, but it is recommended for use in product quality control, for the reproducible in vitro measurement [17]. The major variables that influence the product development are often highlighted by the experimental design while any further optimization might not be considered. By minimizing the potential for confounded errors arising from complex strategies the likelihood of achieving manufacturing and performance specifications with sufficient control to achieve limited batch failures is more likely [134, 135]. The recent US FDA interest in quality by design is achieved by adopting engineering practices that allow for control and monitoring that with adequate risk management ensures the product quality. This also serves the commercial needs of the pharmaceutical industry as it ensures a rapid time to market.

10.8 Regulatory Considerations

Characterization of dry powder inhalers involves unique tests that are not employed for active aerosol dispersion systems such as propellant based metered dose inhalers or nebulized solutions/suspensions. For example, the delivered dose uniformity, fine particle mass over patient flow rate range, the airflow resistance are required for the DPI development but not for pMDI or nebulizer [136]. The responsiveness to airflow conditions is an important part of DPI aerosol performance. This is because:

1. The DPI can only be used by a patient capable of inhaling adequately through it;
2. The dose and particle size distribution of the delivered dose determined in vitro may be dependent upon the airflow rate chosen to test the system [137].

There is no clear path to generic equivalence for DPIs [138, 139]. This has resulted in a number of exploratory studies by the Product Quality Research Institute (PQRI) and the International Pharmaceutical Aerosol Consortium on Regulatory Science (IPAC-RS) to address the specific question of appropriate multivariate statistical approaches to the comparison of both particle size distributions and delivered dose uniformity to allow for specifications on in-vitro equivalence [140].

10.9 Conclusion

The foregoing discussion has presented the underlying scientific and engineering principles on which dry powder inhalers are based. There are unique performance variables for these devices, which require assessment and impact on the important measures biological effect delivered dose and particle size distribution. It is important to note that some knowledge of pharmacology and physiology is required to adequately understand the likely outcome of delivering aerosol with defined dose and size properties. Pharmaceutical development requires adequate engineering and process controls to ensure the final product quality. While regulatory guidance exists for the development of DPIs there is currently inadequate direction on statistical methods for comparison of in vitro equivalence.

10.10 Definitions, Abbreviations and Symbols

Aerodynamic particle size	Diameter of a sphere with density 1,000 kg/m^3 having the same aerodynamic property as the particle
Aerosol	Dispersion of liquid or solid particles in a gas (usually air), which is typically stable over long periods of time
Agglomerate	Assemblage of primary particles with intermediate attractive forces (typically by point contacts; sometimes named aggregate)
Aggregate	Assemblage of primary particles with strong attractive forces (typically by face contacts; sometimes named aggregate)
Cluster	Generic name for the total group of agglomerated, aggregated and flocculated particles
De-agglomeration	Formation of primary particles from agglomerates
Fraunhofer diffraction	Diffraction of light at the contour of the particle (assumed to be spherical)
Laser diffraction	Technique for estimating the particle size distribution from the measured scattering pattern by an ensemble of dispersed particles using laser light
Power	The rate of work done or inspiratory effort during inhalation, an airflow parameter
Primary particle	Basic particle, which cannot be separated unless by breakage
Stokes' diameter	Diameter of a sphere that has the same density and settling rate as the particle under conditions of Stokes' law (viscous flow conditions)
ACI	Andersen cascade impactor
AFM	Atomic force microscopy
APSD	Aerodynamic particle size distribution
BET	Brunauer-Emmett-Teller expression
CFD	Computation fluid dynamics
COPD	Chronic obstructive pulmonary disease
ΔP	Pressure drop, an airflow parameter
DPI	Dry powder inhaler
FPF	Fine particle fraction, the proportion of a nominal dose below a certain cutoff size
FPF_{max}	Maximum fine particle fraction, a constant in the powder aerosol deaggregation equation
GSD	Geometric standard deviation
IPAC-RS	International Pharmaceutical Aerosol Consortium on Regulatory Science
MMAD	Mass median aerodynamic diameter
MMI	Marple-Miller impactor
MSLI	Multi-stage liquid impinger
NGI	Next generation impactor
PADE	Powder aerosol deaggregation equation
PDA	Phase Doppler analysis
pMDI	Pressurized metered dose inhaler
PQRI	Product Quality Research Institute
PSD	Particle size distribution
Re	Reynolds' number of a device
SET	Standardized entrainment tube
VMD	Volume median diameter

(continued)

(continued)

k	Turbulence kinetic energy
k_d	Deaggregation constant
R_D	Specific resistance
γ_i	Integral scale strain rate
ε	Energy dissipation rate
τ_s	Shear stress, an airflow parameter

References

1. Weibel, E.R.: Morphometry of the human lung: the state of the art after two decades. Bull. Eur. Physiopath. Res. **15**, 999–1013 (1979)
2. Carvalho, T.C., Peters, J.I., Williams 3rd, R.O.: Influence of particle size on regional lung deposition–what evidence is there? Int. J. Pharm. **406**, 1–10 (2011)
3. Sbirlea-Apiou, G., Katz, I., Caillibotte, G., Martonen, T., Yang, Y.: Deposition mechanics of pharmaceutical particles in human airways. In: Hickey, A.J. (ed.) Inhalation Aerosols: Physical and Biological Basis for Therapy, vol. 221, pp. 1–30. Informa Healthcare USA, New York (2007)
4. Byron, P.R.: Prediction of drug residence times in regions of the human respiratory tract following aerosol inhalation. J. Pharm. Sci. **75**, 433–438 (1986)
5. http://www.mmadcalculator.com/andersen-impactor-mmad.html. Accessed 22 Jan 2012
6. Crowder, T.M., Rosati, J.A., Schroeter, J.D., Hickey, A.J., Martonen, T.B.: Fundamental effects of particle morphology on lung delivery: predictions of Stokes' law and the particular relevance to dry powder inhaler formulation and development. Pharm. Res. **19**, 239–245 (2002)
7. Gardenhire, D.S.: Airway pharmacology. In: Wilkins, R.L., Stoller, J.K., Kacmarek, R.M. (eds.) Egan's Fundamentals of Respiratory Care, pp. 667–692. Mosby/Elsevier Inc., St. Louis (2009)
8. Gardenhire, D.S.: Corticosteroids in respiratory care. In: Gardenhire, D.S. (ed.) Rau's Respiratory Care Pharmacology, pp. 204–225. Mosby/Elsevier Inc., St. Louis (2008)
9. Weers, J.G., Tarara, T.E., Clark, A.R.: Design of fine particles for pulmonary drug delivery. Expert Opin. Drug Deliv. **4**, 297–313 (2007)
10. Raltiere, R.J., Thompson, D.C.: Physiology and pharmacology of the airways. In: Hickey, A.J. (ed.) Inhalation Aerosols: Physical and Biological Basis for Therapy, vol. 221, pp. 83–126. Informa Healthcare USA, Inc., New York (2007)
11. Byron, P.R., Delvadia, R.R., Longest, P.W., Hindle, M.: Stepping into the trachea with realistic physical models: Uncertainties in regional drug deposition from powder inhalers. Respir. Drug Deliv. **1**, 215–224 (2010)
12. Finlay, W.H., Martin, A.R.: Recent advances in predictive understanding of respiratory tract deposition. J. Aerosol Med. Pulm. Drug Deliv. **21**, 189–206 (2008)
13. Rostami, A.A.: Computational modeling of aerosol deposition in respiratory tract: A review. Inhal. Toxicol. **21**, 262–290 (2009)
14. Byron, P.R., Hindle, M., Lange, C.F., Longest, P.W., McRobbie, D., Oldham, M.J., Olsson, B., Thiel, C.G., Wachtel, H., Finlay, W.H.: In vivo-in vitro correlations: Predicting pulmonary drug deposition from pharmaceutical aerosols. J. Aerosol Med. Pulm. Drug Deliv. **23**(Suppl 2), S59–S69 (2010)
15. Longest, P.W., Holbrook, L.T.: In silico models of aerosol delivery to the respiratory tract – development and applications. Adv. Drug Deliv. Rev. **64**, 296–311 (2012)
16. Physical test and determinations, aerosols, nasal sprays, metered-dose inhalers, and dry powder inhalers, USP <601>, pp. 220–240 (2007)
17. Byron, P.R.: Selection and validation of cascade impactor test methods. Respir. Drug Deliv. **IX**, 169–178 (2004)

18. Dunbar, C.A., Hickey, A.J., Holzner, P.: Dispersion and characterization of pharmaceutical dry powder aerosols. KONA **16**, 7–45 (1998)
19. Marple, V.A., Olson, B.A., Santhanakrishnan, K., Mitchell, J.P., Murray, S.C., Hudson-Curtis, B.L.: Next generation pharmaceutical impactor (a new impactor for pharmaceutical inhaler testing). Part II: Archival calibration. J. Aerosol Med. **16**, 301–324 (2003)
20. Marple, V.A., Olson, B.A., Santhanakrishnan, K., Roberts, D.L., Mitchell, J.P., Hudson-Curtis, B.L.: Next generation pharmaceutical impactor: a new impactor for pharmaceutical inhaler testing. Part III. Extension of archival calibration to 15 L/min. J. Aerosol Med. **17**, 335–343 (2004)
21. Marple, V.A., Roberts, D.L., Romay, F.J., Miller, N.C., Truman, K.G., Van Oort, M., Olsson, B., Holroyd, M.J., Mitchell, J.P., Hochrainer, D.: Next generation pharmaceutical impactor (a new impactor for pharmaceutical inhaler testing). Part I: Design. J. Aerosol Med. **16**, 283–299 (2003)
22. Merkus, H.G.: Laser diffraction. In: Particle size measurements: Fundamentals, practice, quality, pp. 259–286. Springer (2009). http://www.amazon.com/Particle-Size-Measurements-Fundamentals-Technology/dp/904818052X
23. Martin, G.P., MacRitchie, H.B., Marriott, C., Zeng, X.M.: Characterisation of a carrier-free dry powder aerosol formulation using inertial impaction and laser diffraction. Pharm. Res. **23**, 2210–2219 (2006)
24. Marriott, C., MacRitchie, H.B., Zeng, X.M., Martin, G.P.: Development of a laser diffraction method for the determination of the particle size of aerosolised powder formulations. Int. J. Pharm. **326**, 39–49 (2006)
25. Zeng, X.M., MacRitchie, H.B., Marriott, C., Martin, G.P.: Correlation between inertial impaction and laser diffraction sizing data for aerosolized carrier-based dry powder formulations. Pharm. Res. **23**, 2200–2209 (2006)
26. Coates, M.S., Fletcher, D.F., Chan, H.K., Raper, J.A.: Effect of design on the performance of a dry powder inhaler using computational fluid dynamics. Part 1: Grid structure and mouthpiece length. J. Pharm. Sci. **93**, 2863–2876 (2004)
27. Voss, A., Finlay, W.H.: Deagglomeration of dry powder pharmaceutical aerosols. Int. J. Pharm. **248**, 39–50 (2002)
28. Begat, P., Morton, D.A., Staniforth, J.N., Price, R.: The cohesive-adhesive balances in dry powder inhaler formulations II: Influence on fine particle delivery characteristics. Pharm. Res. **21**, 1826–1833 (2004)
29. Das, S., Larson, I., Young, P., Stewart, P.: Influence of storage relative humidity on the dispersion of salmeterol xinafoate powders for inhalation. J. Pharm. Sci. **98**, 1015–1027 (2009)
30. Hickey, A.J.: Pharmaceutical Inhalation Aerosol Technology. Marcel Dekker, New York (2004)
31. Israelachvili, J.N.: Intermolecular and Surface Forces. Academic, London (1992)
32. Podczeck, F.: Particle-Particle Adhesion in Pharmaceutical Powder Handling. Imperial College Press, London (1998)
33. Visser, J.: An invited review van der Waals and other other cohesive forces affecting powder fluidization. Powder Technol. **58**, 1–10 (1989)
34. Visser, J.: Particle adhesion and removal: A review. Particul. Sci. Technol. **13**, 169–196 (1995)
35. Derjaguin, B.V.: Friction and adhesion. IV. The theory of adhesion of small particles. Kolloid Z. **69**, 155–164 (1934)
36. Matsusaka, S., Maruyama, H., Matsuyama, T., Ghadiri, M.: Triboelectric charging of powders: A review. Chem. Eng. Sci. **65**, 5781–5807 (2010)
37. Crowder, T.M., Hickey, A.J., Louey, M.D., Orr, N.: A Guide to Pharmaceutical Particulate Science. Interpharm Press/CRC, Boca Raton (2003)
38. Elajnaf, A., Carter, P., Rowley, G.: The effect of relative humidity on electrostatic charge decay of drugs and excipient used in dry powder inhaler formulation. Drug Dev. Ind. Pharm. **33**, 967–974 (2007)

39. Elajnaf, A., Carter, P., Rowley, G.: Electrostatic characterisation of inhaled powders: Effect of contact surface and relative humidity. Eur. J. Pharm. Sci. **29**, 375–384 (2006)
40. Hooton, J.C., German, C.S., Allen, S., Davies, M.C., Roberts, C.J., Tendler, S.J., Williams, P.M.: An atomic force microscopy study of the effect of nanoscale contact geometry and surface chemistry on the adhesion of pharmaceutical particles. Pharm. Res. **21**, 953–961 (2004)
41. Podczeck, F.: The relationship between physical properties of lactose monohydrate and the aerodynamic behaviour of adhered drug particles. Int. J. Pharm. **160**, 119–130 (1998)
42. Podczeck, F., Newton, J.M.: Development of an ultracentrifuge technique to determine the adhesion and friction properties between particles and surfaces. J. Pharm. Sci. **84**, 1067–1071 (1995)
43. Hickey, A.J., Mansour, H.M., Telko, M.J., Xu, Z., Smyth, H.D., Mulder, T., McLean, R., Langridge, J., Papadopoulos, D.: Physical characterization of component particles included in dry powder inhalers. I. Strategy review and static characteristics. J. Pharm. Sci. **96**, 1282–1301 (2007)
44. Ziskind, G., Fichman, M., Gutfinger, C.: Resuspension of particulates from surfaces to turbulent flows – review and analysis. J. Aerosol Sci. **26**, 613–644 (1995)
45. Ziskind, G., Fichman, M., Gutfinger, C.: Adhesion moment model for estimating particle detachment from a surface. J. Aerosol Sci. **28**, 623–634 (1997)
46. Wang, H.C.: Effects of inception motion on particle detachment from surfaces. Aerosol Sci. Technol. **13**, 386–393 (1990)
47. Ibrahim, A.H., Dunn, P.F., Brach, R.M.: Microparticle detachment from surfaces exposed to turbulent air flow: Controlled experiments and modeling. J. Aerosol Sci. **34**, 765–782 (2003)
48. Gradon, L.: Resuspension of particles from surfaces: Technological, environmental and pharmaceutical aspects. Adv. Powder Technol. **20**, 17–28 (2009)
49. Reeks, M.W., Reed, J., Hall, D.: On the resuspension of small particles by a turbulent flow. J. Phys. D Appl. Phys. **21**, 574–589 (1988)
50. Wenand, H.Y., Kasper, G.: On the kinetics of particle reentrainment from surfaces. J. Aerosol Sci. **20**, 483–498 (1989)
51. Hersey, J.A.: Ordered mixing: A new concept in powder mixing practice. Powder Technol. **11**, 41–44 (1975)
52. Corrigan, D.O., Corrigan, O.I., Healy, A.M.: Physicochemical and in vitro deposition properties of salbutamol sulphate/ipratropium bromide and salbutamol sulphate/excipient spray dried mixtures for use in dry powder inhalers. Int. J. Pharm. **322**, 22–30 (2006)
53. Young, P., Roberts, D., Chiou, H., Rae, W., Chan, H.K., Traini, D.: Composite carriers improve the aerosolisation efficiency of drugs for respiratory delivery. J. Aerosol Sci. **39**, 82–93 (2008)
54. Young, P.M., Kwok, P., Adi, H., Chan, H.K., Traini, D.: Lactose composite carriers for respiratory delivery. Pharm. Res. **26**, 802–810 (2009)
55. Ooi, J., Traini, D., Hoe, S., Wong, W., Young, P.M.: Does carrier size matter? A fundamental study of drug aerosolisation from carrier based dry powder inhalation systems. Int. J. Pharm. **413**, 1–9 (2011)
56. Adi, H., Traini, D., Chan, H.K., Young, P.M.: The influence of drug morphology on aerosolisation efficiency of dry powder inhaler formulations. J. Pharm. Sci. **97**, 2780–2788 (2008)
57. Donovan, M.J., Smyth, H.D.: Influence of size and surface roughness of large lactose carrier particles in dry powder inhaler formulations. Int. J. Pharm. **402**, 1–9 (2010)
58. Zhou, Q.T., Morton, D.A.: Drug-lactose binding aspects in adhesive mixtures: Controlling performance in dry powder inhaler formulations by altering lactose carrier surfaces. Adv. Drug Deliv. Rev. (2011)
59. Jones, M.D., Price, R.: The influence of fine excipient particles on the performance of carrier-based dry powder inhalation formulations. Pharm. Res. **23**, 1665–1674 (2006)
60. Hooton, J.C., Jones, M.D., Price, R.: Predicting the behavior of novel sugar carriers for dry powder inhaler formulations via the use of a cohesive-adhesive force balance approach. J. Pharm. Sci. **95**, 1288–1297 (2006)

61. Jones, M.D., Harris, H., Hooton, J.C., Shur, J., King, G.S., Mathoulin, C.A., Nichol, K., Smith, T.L., Dawson, M.L., Ferrie, A.R., Price, R.: An investigation into the relationship between carrier-based dry powder inhalation performance and formulation cohesive-adhesive force balances. Eur. J. Pharm. Biopharm. **69**, 496–507 (2008)
62. Vehring, R.: Pharmaceutical particle engineering via spray drying. Pharm. Res. **25**, 999–1022 (2008)
63. York, P.: Strategies for particle design using supercritical fluid technologies. Pharm. Sci. Technol. Today **2**, 430–440 (1999)
64. Pilcer, G., Wauthoz, N., Amighi, K.: Lactose characteristics and the generation of the aerosol. Adv. Drug Deliv. Rev. **64**, 233–256 (2012)
65. Pilcer, G., Amighi, K.: Formulation strategy and use of excipients in pulmonary drug delivery. Int. J. Pharm. **392**, 1–19 (2010)
66. Xu, Z., Mansour, H.M., Hickey, A.J.: Particle interactions in dry powder inhaler unit processes: A review. J. Adhes. Sci. Technol. **25**, 451–482 (2011)
67. Chan, H.K.: What is the role of particle morphology in pharmaceutical powder aerosols? Expert Opin. Drug Deliv. **5**, 909–914 (2008)
68. Frijlink, H.W., De Boer, A.H.: Dry powder inhalers for pulmonary drug delivery. Expert Opin. Drug Deliv. **1**, 67–86 (2004)
69. Telko, M.J., Hickey, A.J.: Dry powder inhaler formulation. Respir. Care **50**, 1209–1227 (2005)
70. Chan, H.K.: Dry powder aerosol delivery systems: Current and future research directions. J. Aerosol Med. **19**, 21–27 (2006)
71. Smythand, H., Hickey, A.J.: Carriers in dry powder delivery: Implications for inhalation system design. Am. J. Drug Deliv. **3**, 117–132 (2005)
72. Steckeland, H., Muller, B.W.: In vitro evaluation of dry powder inhalers II: Influence of carrier particle size and concentration on in vitro deposition. Int. J. Pharm. **154**, 31–37 (1997)
73. Vanderbist, F., Wery, B., Moyano-Pavon, I., Moes, A.J.: Optimization of a dry powder inhaler formulation of nacystelyn, a new mucoactive agent. J. Pharm. Pharmacol. **51**, 1229–1234 (1999)
74. Young, P.M., Wood, O., Ooi, J., Traini, D.: The influence of drug loading on formulation structure and aerosol performance in carrier based dry powder inhalers. Int. J. Pharm. **416**, 129–135 (2011)
75. Wetterlin, K.: Turbuhaler: A new powder inhaler for administration of drugs to the airways. Pharm. Res. **5**, 506–508 (1988)
76. Edwards, A.M., Chambers, A.: Comparison of a lactose-free formulation of sodium cromoglycate and sodium cromoglycate plus lactose in the treatment of asthma. Curr. Med. Res. Opin. **11**, 283–292 (1989)
77. Zhou, Q.T., Qu, L., Larson, I., Stewart, P.J., Morton, D.A.: Improving aerosolization of drug powders by reducing powder intrinsic cohesion via a mechanical dry coating approach. Int. J. Pharm. **394**, 50–59 (2010)
78. Zhou, Q.T., Armstrong, B., Larson, I., Stewart, P.J., Morton, D.A.: Understanding the influence of powder flowability, fluidization and de-agglomeration characteristics on the aerosolization of pharmaceutical model powders. Eur. J. Pharm. Sci. **40**, 412–421 (2010)
79. Zhou, Q.T., Denman, J.A., Gengenbach, T., Das, S., Qu, L., Zhang, H., Larson, I., Stewart, P. J., Morton, D.A.: Characterization of the surface properties of a model pharmaceutical fine powder modified with a pharmaceutical lubricant to improve flow via a mechanical dry coating approach. J. Pharm. Sci. **100**, 3421–3430 (2011)
80. Begat, P., Morton, D.A., Shur, J., Kippax, P., Staniforth, J.N., Price, R.: The role of force control agents in high-dose dry powder inhaler formulations. J. Pharm. Sci. **98**, 2770–2783 (2009)
81. Raula, J., Lahde, A., Kauppinen, E.I.: Aerosolization behavior of carrier-free L-leucine coated salbutamol sulphate powders. Int. J. Pharm. **365**, 18–25 (2009)
82. Raula, J., Thielmann, F., Naderi, M., Lehto, V.P., Kauppinen, E.I.: Investigations on particle surface characteristics vs. dispersion behaviour of L-leucine coated carrier-free inhalable powders. Int. J. Pharm. **385**, 79–85 (2010)

83. Chew, N.Y., Shekunov, B.Y., Tong, H.H., Chow, A.H., Savage, C., Wu, J., Chan, H.K.: Effect of amino acids on the dispersion of disodium cromoglycate powders. J. Pharm. Sci. **94**, 2289–2300 (2005)
84. Hickey, A.J.: Pulmonary drug delivery: Pharmaceutical chemistry and aerosol technology. In: Wang, B., Siahaan, T., Soltero, R.A. (eds.) Drug Delivery: Principles and Applications, pp. 341–361. Wiley, New Jersey (2005)
85. Van Oort, M., Sacchetti, M.: Spray-drying and supercritical fluid particle generation techniques. In: Hickey, A.J. (ed.) Inhalation Aerosols: Physical and Biological Basis for Therapy, vol. 1, pp. 307–346. Informa Healthcare, New York (2007)
86. Seville, P.C., Li, H.Y., Learoyd, T.P.: Spray-dried powders for pulmonary drug delivery. Crit. Rev. Ther. Drug Carrier Syst. **24**, 307–360 (2007)
87. Edwards, D.A., Hanes, J., Caponetti, G., Hrkach, J., Ben-Jebria, A., Eskew, M.L., Mintzes, J., Deaver, D., Lotan, N., Langer, R.: Large porous particles for pulmonary drug delivery. Science **276**, 1868–1871 (1997). New York, NY
88. Tay, T., Das, S., Stewart, P.: Magnesium stearate increases salbutamol sulphate dispersion: What is the mechanism? Int. J. Pharm. **383**, 62–69 (2010)
89. Young, P.M., Tobyn, M.J., Price, R., Buttrum, M., Dey, F.: The use of colloid probe microscopy to predict aerosolization performance in dry powder inhalers: AFM and in vitro correlation. J. Pharm. Sci. **95**, 1800–1809 (2006)
90. Pilcer, G., Vanderbist, F., Amighi, K.: Spray-dried carrier-free dry powder tobramycin formulations with improved dispersion properties. J. Pharm. Sci. **98**, 1463–1475 (2009)
91. Zeng, X.M., Martin, G.P., Tee, S.K., Ghoush, A.A., Marriott, C.: Effects of particle size and adding sequence of fine lactose on the deposition of salbutamol sulphate from a dry powder formulation. Int. J. Pharm. **182**, 133–144 (1999)
92. Zeng, X.M., Pandhal, K.H., Martin, G.P.: The influence of lactose carrier on the content homogeneity and dispersibility of beclomethasone dipropionate from dry powder aerosols. Int. J. Pharm. **197**, 41–52 (2000)
93. Traini, D., Young, P.M., Jones, M., Edge, S., Price, R.: Comparative study of erythritol and lactose monohydrate as carriers for inhalation: Atomic force microscopy and in vitro correlation. Eur. J. Pharm. Sci. **27**, 243–251 (2006)
94. Sacchetti, M.: The nitrogen adsorption isotherm of alpha-lactose monohydrate. Pharm. Dev. Technol. **11**, 351–358 (2006)
95. Young, P.M., Edge, S., Traini, D., Jones, M.D., Price, R., El-Sabawi, D., Urry, C., Smith, D. C.: The influence of dose on the performance of dry powder inhalation systems. Int. J. Pharm. **296**, 26–33 (2005)
96. Ho, R., Muresan, A.S., Hebbink, G.A., Heng, J.Y.: Influence of fines on the surface energy heterogeneity of lactose for pulmonary drug delivery. Int. J. Pharm. **388**, 88–94 (2010)
97. Louey, M.D., Stewart, P.J.: Particle interactions involved in aerosol dispersion of ternary interactive mixtures. Pharm. Res. **19**, 1524–1531 (2002)
98. Jones, M.D., Santo, J.G., Yakub, B., Dennison, M., Master, H., Buckton, G.: The relationship between drug concentration, mixing time, blending order and ternary dry powder inhalation performance. Int. J. Pharm. **391**, 137–147 (2010)
99. Jones, M.D., Hooton, J.C., Dawson, M.L., Ferrie, A.R., Price, R.: An investigation into the dispersion mechanisms of ternary dry powder inhaler formulations by the quantification of interparticulate forces. Pharm. Res. **25**, 337–348 (2008)
100. Clineand, D., Dalby, R.: Predicting the quality of powders for inhalation from surface energy and area. Pharm. Res. **19**, 1274–1277 (2002)
101. Adi, H., Larson, I., Chiou, H., Young, P., Traini, D., Stewart, P.: Role of agglomeration in the dispersion of salmeterol xinafoate from mixtures for inhalation with differing drug to fine lactose ratios. J. Pharm. Sci. **97**, 3140–3152 (2008)
102. Adi, H., Larson, I., Stewart, P.J.: Adhesion and redistribution of salmeterol xinafoate particles in sugar-based mixtures for inhalation. Int. J. Pharm. **337**, 229–238 (2007)

103. Adi, H., Larson, I., Chiou, H., Young, P., Traini, D., Stewart, P.: Agglomerate strength and dispersion of salmeterol xinafoate from powder mixtures for inhalation. Pharm. Res. **23**, 2556–2565 (2006)
104. Louey, M.D., VanOort, M., Hickey, A.J.: Standardized entrainment tubes for the evaluation of pharmaceutical dry powder dispersion. J. Aerosol Sci. **73**, 1520–1533 (2006)
105. Xu, Z., Mansour, H.M., Mulder, T., McLean, R., Langridge, J., Hickey, A.J.: Heterogeneous particle deaggregation and its implication for therapeutic aerosol performance. J. Pharm. Sci. **99**, 3442–3461 (2010)
106. Xu, Z., Mansour, H.M., Mulder, T., McLean, R., Langridge, J., Hickey, A.J.: Dry powder aerosols generated by standardized entrainment tubes from drug blends with lactose monohydrate: 1. Albuterol sulfate and disodium cromoglycate. J. Pharm. Sci. **99**, 3398–3414 (2010)
107. Xu, Z., Mansour, H.M., Mulder, T., McLean, R., Langridge, J., Hickey, A.J.: Dry powder aerosols generated by standardized entrainment tubes from drug blends with lactose monohydrate: 2. Ipratropium bromide monohydrate and fluticasone propionate. J. Pharm. Sci. **99**, 3415–3429 (2010)
108. Mansour, H.M., Xu, Z., Hickey, A.J.: Dry powder aerosols generated by standardized entrainment tubes from alternative sugar blends: 3. Trehalose dihydrate and D-mannitol carriers. J. Pharm. Sci. **99**, 3430–3441 (2010)
109. Freundlich, H.M.F.: Over the adsorption in solution. J. Phys. Chem. **57**, 385–470 (1906)
110. Do, D.D.: Adsorption Analysis: Equilibria and Kinetics. Imperial College Press, London (1998)
111. Giles, C.H., Smith, D., Huitson, A.: A general treatment and classification of the solute adsorption isotherm I. Theoretical. J. Colloid Interface Sci. **47**, 755–765 (1974)
112. Giles, C.H., Smith, D., Huitson, A.: A general treatment and classification of the solute adsorption isotherm II. Experimental interpretation. J. Colloid Interface Sci. **47**, 766–778 (1974)
113. Brunauer, S., Emmett, P.H., Teller, E.: Adsorption of gases in multimolecular layers. J. Am. Chem. Soc. **60**, 309–319 (1938)
114. Hickey, A.J., Ganderton, D.: Fluid flow. In: Hickey, A.J., Ganderton, D. (eds.) Pharmaceutical Process Engineering, pp. 4–30. Informa Healthcare USA, New York (2010)
115. Pope, S.B.: Wall flows. In: Pope, S.B. (ed.) Turbulent Flows, pp. 264–332. Cambridge University Press, Cambridge, UK (2000)
116. Shekunov, B.Y., Feeley, J.C., Chow, A.H., Tong, H.H., York, P.: Aerosolization behavior of micronized and supercritically-processed powders. J. Aerosol Sci. **34**, 553–568 (2003)
117. Clarkand, A.R., Hollingsworth, A.M.: The relationship between powder inhaler resistance and peak inspiratory conditions in healthy volunteers – implications for in vitro testing. J. Aerosol Med. **6**, 99–110 (1993)
118. Mendes, P.J., Pinto, J.F., Sousa, J.M.M.: A non-dimensional functional relationship for the fine particle fraction produced by dry powder inhalers. J. Aerosol Sci. **38**, 612–624 (2007)
119. de Boer, A.H., Winter, H.M.I., Lerk, C.F.: Inhalation characteristics and their effects on in vitro drug delivery from dry powder inhalers part 1. Inhalation characteristics, work of breathing and volunteers' preference in dependence of the inhaler resistance. Int. J. Pharm. **130**, 231–244 (1996)
120. Islamand, N., Gladki, E.: Dry powder inhalers (DPIs)–a review of device reliability and innovation. Int. J. Pharm. **360**, 1–11 (2008)
121. Newman, S.P., Peart, J.: Dry powder inhalers. In: Newman, S.P. (ed.) Respiratory Drug Delivery: Essential Theory and Practice, pp. 257–307. Respiratory Drug Delivery Online/ VCU, Richmond (2009)
122. Hickey, A.J., Crowder, T.M.: Next generation dry powder inhalation delivery systems. In: Hickey, A.J. (ed.) Inhalation Aerosols: Physical and Biological Basis for Therapy, vol. 221, pp. 445–460. Informa Healthcare USA, New York (2007)

123. Smith, I.J., Parry-Billings, M.: The inhalers of the future?A review of dry powder devices on the market today. Pulm. Pharmacol. Ther. **16**, 79–95 (2003)
124. Menter, F.R.: Two-equation eddy-viscosity models for engineering applications. AIAA J. **32**, 1598–1605 (1994)
125. Coates, M.S., Fletcher, D.F., Chan, H.K., Raper, J.A.: The role of capsule on the performance of a dry powder inhaler using computational and experimental analyses. Pharm. Res. **22**, 923–932 (2005)
126. Coates, M.S., Chan, H.K., Fletcher, D.F., Raper, J.A.: Influence of air flow on the performance of a dry powder inhaler using computational and experimental analyses. Pharm. Res. **22**, 1445–1453 (2005)
127. Coates, M.S., Chan, H.K., Fletcher, D.F., Chiou, H.: Influence of mouthpiece geometry on the aerosol delivery performance of a dry powder inhaler. Pharm. Res. **24**, 1450–1456 (2007)
128. Wong, W., Adi, H., Traini, D., Chan, H.K., Fletcher, D.F., Crapper, J., Young, P.: Use of rapid prototyping and computational fluid dynamics in the design the DPI devices. Respir. Drug Deliv. **3**, 879–882 (2010)
129. Donovan, M.J., Kim, S.H., Raman, V., Smyth, H.D.: Dry powder inhaler device influence on carrier particle performance. J. Pharm. Sci. **101**, 1097–1107 (2012)
130. Calvert, G., Ghadiri, M., Tweedie, R.: Aerodynamic dispersion of cohesive powders: A review of understanding and technology. Adv. Powder Technol. **20**, 4–16 (2009)
131. Wong, W., Fletcher, D.F., Traini, D., Chan, H.K., Crapper, J., Young, P.M.: Particle aerosolisation and break-up in dry powder inhalers: Evaluation and modelling of the influence of grid structures for agglomerated systems. J. Pharm. Sci. **100**, 4710–4721 (2011)
132. Wong, W., Fletcher, D.F., Traini, D., Chan, H.K., Crapper, J., Young, P.M.: Particle aerosolisation and break-up in dry powder inhalers 1: Evaluation and modelling of venturi effects for agglomerated systems. Pharm. Res. **27**, 1367–1376 (2010))
133. Wong, W., Fletcher, D.F., Traini, D., Chan, H.K., Crapper, J., Young, P.M.: Particle aerosolisation and break-up in dry powder inhalers: Evaluation and modelling of impaction effects for agglomerated systems. J. Pharm. Sci. **100**, 2744–2754 (2011)
134. USP: The United States Pharmacopeia. U.S. Pharmacopeial Convention, Rockville (2004)
135. Physical test and determinations, *<601* > Aerosols, nasal sprays, metered-dose inhalers, and dry powder inhalers, USP, pp. 220–240
136. Guidance_for_industry.: Pharmaceutical quality of inhalation and nasal products, pp. 1–32. Canada or EU (2006). http://www.hc-sc.gc.ca/dhp-mps/alt_formats/hpfb-dgpsa/pdf/prodpharma/inhalationnas-eng.pdf
137. Committee_for_proprietary_medicinal_products.: Note for guidance on dry powder inhalers, pp. 1–6. London (1998)
138. Persson, G., Ankerst, J., Gillen, M., Bengtsson, T., Thorsson, L.: Relative systemic availability of budesonide in patients with asthma after inhalation from two dry powder inhalers. Curr. Med. Res. Opin. **24**, 1511–1517 (2008)
139. Taki, M., Ahmed, S., Marriott, C., Zeng, X.M., Martin, G.P.: The 'stage-by-stage' deposition of drugs from commercial single-active and combination dry powder inhaler formulations. Eur. J. Pharm. Sci. **43**, 225–235 (2011)
140. Shiand, S., Hickey, A.J.: Multivariate data analysis as a semi-quantitative tool for interpretive evaluation of comparability or equivalence of aerodynamic particle size distribution profiles. AAPS Pharm. Sci. Technol. **10**, 1113–1120 (2009)
141. Behara, S.R., Larson, I., Kippax, P., Morton, D.A., Stewart, P.: The kinetics of cohesive powder de-agglomeration from three inhaler devices. Int. J. Pharm. **421**, 72–81 (2011)

Chapter 11
The Role of Particle Size in Drug Release and Absorption

Giuseppina Sandri, M. Cristina Bonferoni, Franca Ferrari, Silvia Rossi, and Carla M. Caramella

Abstract Solid drug delivery systems are crucial formulations via the oral route. In such drug systems, particle size has a strong impact on drug dissolution and on drug absorption. Its role in dissolution rate is described starting from the Noyes–Whitney equation, the modified form by Nerst–Brunner and the cube root equation. According to these equations diffusion of solute through a boundary layer around the particles is the rate limiting step in both drug dissolution and absorption and, thus, depends upon the specific (external) surface area of the particles, the diffusion coefficient of the solute, the thickness of the boundary layer and the solute solubility. In relation to this, good wetting of the particle surface by the surrounding liquid and adequate particle dispersion play an essential role. Information from dissolution rates suggests that the thickness of the boundary layer is constant for larger particle sizes but dependent upon size for smaller particles. Given the larger surface area of smaller particles, the attention is directed to nanosystems and on their relevance to the bioavailability of poorly soluble drugs. A second advantage of such drug systems is that the solubility increases with decreasing particle size, according to the Freundlich–Ostwald equation. Since dissolution and absorption are closely related, the impact of particle size on drug absorption is described. Moreover, regulatory implications of particle size are reviewed.

11.1 Introduction

It has been recognized that the availability of a drug for gastro-intestinal absorption from solid dosage forms is often reflected by in vitro dissolution rates. It has also been recognized that among the new chemical entities the poorly soluble drugs are more and more common. For these drugs the rate-determining step in the absorption

---

G. Sandri (✉) • M.C. Bonferoni • F. Ferrari • S. Rossi • C.M. Caramella
School of Pharmacy, University of Pavia, Pavia, Italy
e-mail: giuseppina.sandri@unipv.it

H.G. Merkus and G.M.H. Meesters (eds.), *Particulate Products: Tailoring Properties for Optimal Performance*, Particle Technology Series 19,
DOI 10.1007/978-3-319-00714-4_11,

of drugs is the dissolution rate of drugs in the gastro-intestinal fluids rather than the rapidity of their diffusion across the intestine wall. The granulometric properties of pharmaceutical powders (both active ingredients and excipients) play an important role on their behavior in technological as well as biopharmaceutically relevant processes. Furthermore they are fundamental in the design and development of drug delivery systems where the drug is loaded as a solid. The solubility and dissolution behavior of drug particles is fundamental to allow drug absorption. Developing a drug delivery system taking into account the granulometric parameters means not only to characterize raw materials (mainly active ingredients and excipients) but also to consider the influence of particle size on technological processes and on biopharmaceutical properties of formulations. In particular as for biopharmaceutical properties dissolution rate has a great impact on drug absorption. Furthermore, the bioavailability of poorly soluble drugs, in particular hydrophobic drugs with extremely low aqueous solubility, is dramatically influenced by drug dissolution rate. The investigation of drug dissolution rate enhancement is a crucial point; in this perspective particle size plays a key role since dissolution rate is mainly dependent on interfacial surface area. Adequate dispersion of powdered solid in a dissolution medium is a prerequisite to start the dissolution process: if particles tend to form clumps in the dissolution medium, then the dissolution surface area available for dissolution is reduced. This effect may be overcome by the addition of a wetting agent (surfactant) to improve the dispersion of clumps into primary powder particles. Therefore, if the particles can be easily wetted by dissolution medium, an increased surface area (a decreased particle size) results and, in turn, an increased dissolution rate. The present chapter is focused on the role of particle size in solubility and dissolution properties. In addition, the effect of nanonization technology on bioavailability of poorly soluble drugs is reviewed.

11.2 Particle Size and Dissolution Properties

The relevance of particle size to dissolution of poorly soluble drugs is well known to scientists involved in oral dosage forms development.

The relationship between particle size and dissolution rate can be derived from the Noyes–Whitney Eq. 11.1 that was introduced in 1897 to explain the dissolution phenomenon [40]. This equation explains the correlation between particle size and dissolution rate and is based on the assumption that in the dissolution process the solubilization is fast, whereas the subsequent diffusion of drug molecules through a diffusion boundary layer surrounding the particle is the rate limiting step.

$$\frac{dm}{dt} = k \cdot (C_s - C_t) \tag{11.1}$$

where:

dm/dt = dissolution rate

C_s = equilibrium solubility of the substance, i.e. the concentration of its saturated solution

C_t = concentration of the substance in the bulk medium at time t

k = constant

t = time.

In 1904 the equation was modified by Nerst and Brunner [6, 36]. They explained the above constant k in terms of surface area, diffusion coefficient and diffusion layer thickness. In this form, it became one of the most used equations to describe drug dissolution:

$$\frac{dm}{dt} = \frac{A \cdot \mathbb{D}}{\delta}(C_s - C_t) \tag{11.2}$$

where:

A = surface area of the undissolved solid drug in contact with the solvent

$\mathbb{D}$ = diffusion coefficient of the solute

δ = thickness of the diffusion boundary layer.

The medium viscosity influences the diffusion coefficient $\mathbb{D}$, which decreases on increasing the viscosity. The thickness of the diffusion layer δ is influenced by both medium viscosity and stirring rate. At a given viscosity δ decreases with increase of stirring rate, while at the same level of stirring rate it increases with increasing viscosity.

When C_t does not exceed 10 % of C_s, the so called sink conditions are satisfied, and C_t can be considered negligible in comparison to C_s. Under these conditions, the dissolution rate can be considered directly proportional to drug solubility and to the interfacial surface area, which is in turn dependent on drug particle size and shape.

The key role played by the initial radius of the particles that undergo dissolution is evident in the Hixson–Crowell equation [16], also known as Cube root equation, derived from the Noyes–Whitney equation:

$$\sqrt[3]{m_t} = \sqrt[3]{m_0} - Kt \tag{11.3}$$

where:

K = dissolution constant "cube root"

m_t = drug amount undissolved at time t

m_0 = initial drug amount

t = time.

The amount of drug undissolved, m_t, is related to the remaining volume of solid powder and, thus, to the cube of the volume-based mean particle size, $D_{4,3}$. In fact, the Hixson–Crowell equation is rigorously applicable to systems made of spherical particles, but is also quite commonly used to describe the dissolution behavior of populations of non-spherical particles in an approximateve way.

Kaneniwa et al. [22] used the Cube root equation to correlate the dissolution rate to the particle size of sulfamethizole and demonstrated that the increase in dimensions from 60 to 700 μm causes the corresponding decrease in dissolution rate.

As illustrated in Eq. 11.2, the dissolution rate directly relates to the specific surface area of the solid. Dissolution rates are, therefore, measured either by maintaining the surface area constant (in intrinsic dissolution rate measurements), or by taking into account the variation of the particle surface area during dissolution. Nystrom et al. [41] proposed to measure of particle size by means of a Coulter Counter to calculate the surface area of a solid undergoing dissolution. This calculation involved the measurement of the weight of particle population dissolved as a function of time. Under the assumption that the shape of the particles remains constant [8], the authors introduced a new parameter, the surface specific dissolution rate (G):

$$G = \frac{m_0 - m_t}{t((A_0 - A_t)/2)} \tag{11.4}$$

where:

G = surface specific dissolution rate
m_0 = initial weight of solid particles
m_t = weight of solid particles at time t
A_0, A_t = the surface area of solid particles at time zero and time t, respectively
t = time.

The proposed parameter G relates to the loss of weight divided by the loss of surface and, thus, to the inverse of the specific surface area (the surface area per unit weight).

In fact, among the different mean diameters used in particle size characterization, the area-weighted Sauter mean diameter, $D_{3,2}$, gives information that directly relates to specific surface area and, thus, it is profitably used in the pharmaceutical field.

It was however noticed that the increase in dissolution rate by a decrease in particle size can be higher than expected on the basis of the surface area increase. This finding can be explained by taking into account that a decrease in particle size is accompanied by a decreased thickness of the boundary layer δ [5]. This effect is especially pronounced for materials with mean particle size less than 5 μm [3, 5] and was confirmed by Mosharraf and Nystrom [32] in a study involving sparingly soluble materials such as griseofulvin, glibenclamide, barium sulphate and oxazepam.

These topics became of particular interest in pharmaceutical technology after the introduction of biopharmaceutical concerns in the evaluation of the performance of drug delivery systems. One of the paradigms at the basis of biopharmaceutics is that drug absorption requires that the drug is in solution at the absorption site [1, 9]. This focused particular attention on dissolution rate because, as dissolution is the limiting step for absorption, all the factors that affect dissolution are potentially determinant for drug bioavailability.

Among the first results dealing with the use of particle size to explain or predict dissolution rate and therefore absorption, are some findings involving sulfamidics such as sulfadiazine, sulfaisossazole, sulfathiazole and sulfamethizole [22, 23, 48].

The relationship between particle size and dissolution rate appeared evident especially for poorly soluble drugs, while no influence could be found for freely soluble drugs. A quite dramatic influence of particle size on rabbit gastro-intestinal absorption was found for sulfadimetossine: the increase in particle size corresponded to a decrease in maximum plasma concentration, C_{max}, and an increase in the time to reach the maximum concentration, t_{max}, due to slower gastro-intestinal absorption [23].

Similar results were found in man for other drugs, a.o. phenytoin [37]. Neuvonen put clearly in evidence that various biopharmaceutical factors, among which drug particle size, has an important effect on the oral absorption of this drug, and in turn, due to the dose dependent metabolism and narrow therapeutic range, can result in serious clinical consequences [37]. The positive effect of particle size reduction on bioavailability was observed also for digoxin [20, 45] and, although with some discordant results, for griseofulvin [4, 12].

Simoes et al. [47] evaluated the influence of particle size and related properties in particular specific surface area and solubility on the dissolution rate of a sparingly soluble drug indomethacin. As expected a strong influence of the fraction size on the dissolution rate was found. A correlation was established between the mean dissolution time (MDT) and the mean particle size of the various indomethacin fractions. As expected, the dissolution rate increased with a reduction in particle size. The authors confirmed a correlation between the MDT and the mean particle size.

Micronization by means of dry milling (jet milling, ball milling, pin milling) to 2–5 μm sizes was therefore proposed as a useful method to improve bioavailability of this category of drugs. This strategy was proven successful also in more recent studies, such as that performed by Ning et al. [39], where micronization gave results positive to solid dispersion in improving bioavailability, as evidenced by AUC (area under blood plasma concentration-time curve), C_{max} and t_{max} values of glimepiride administered to beagle dogs.

Oh et al. [42] reported that the dose fraction absorbed from suspensions is a function of four dimensionless parameters: absorption number, dissolution number, dose number and initial saturation. The absorption number (*An*) is the ratio of radial absorption rate to axial convection rate; the dissolution number (*Dn*) is the ratio of residence time in intestine to the dissolution time. The dissolution number depends on drug properties such as solubility, diffusivity, density and, in particular, *Dn* is inversely proportional to the square of the initial particle size. Thus, a way to increase *Dn* is to decrease particle size [43]. The dose number (*Do*) is the ratio of dose concentration to solubility. The initial saturation (*Is*) is the ratio of drug luminal concentration at the beginning of the intestine to drug solubility.

After drug suspension administration via oral route, particle size decreases down the tube so that the dissolution rate from particles increases. Drugs with low *Do* and low *Dn* can be completely absorbed by reducing their particle size, while the absorption of drugs with high *Do* and low *Dn* is solubility (*Dn*) limited and requires a higher solubility, in addition to micronization, to enhance the fraction absorbed.

A good knowledge of bioavailability of a drug is also important for regulatory reasons, in view of assuring a constant level of quality and reproducibility and, consequently, adequate predictability of therapeutical effects. This has led during

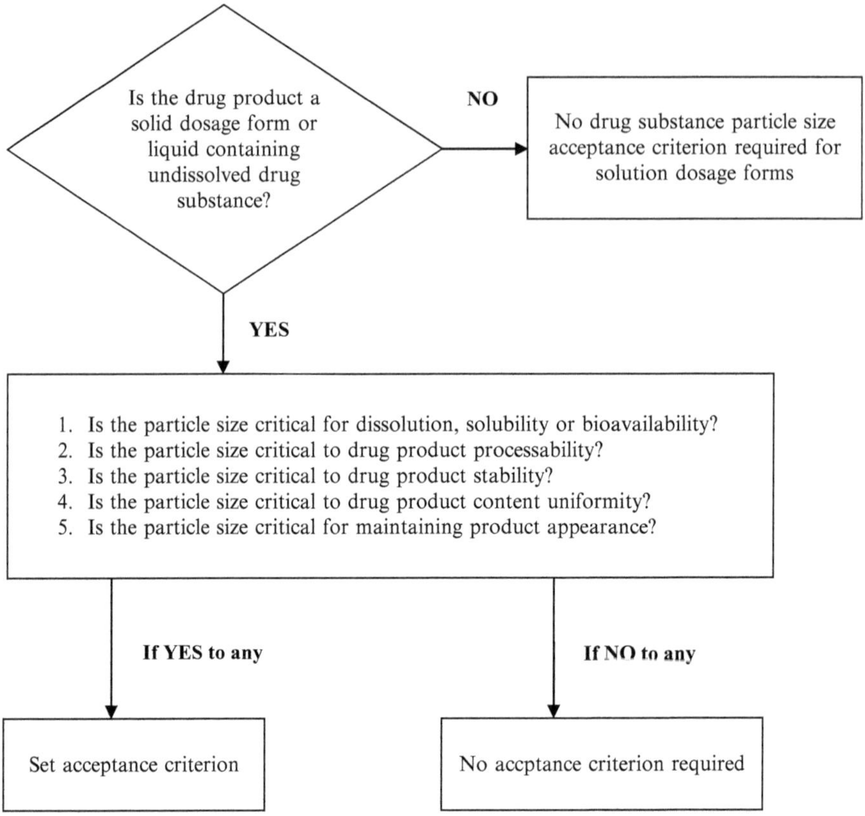

Fig. 11.1 Decision tree on particle size reported in ICH Guideline Q6A [17]; courtesy ICH

the last decennia to the identification of all the possible critical parameters of any formulation. For solid oral dosage forms and liquids containing undissolved drug substances (suspensions), particle size is considered as a parameter to be necessarily taken into account in the definition and control of quality specifications. In fact, the ICH Guideline Q6A [17] on product specifications under the paragraph 3.3.1, New Drug Substances, states that "testing for particle size distribution should be carried out, and acceptance criteria should be decided if particle size affects dissolution rate, bioavailability or stability of a new drug substance in solid or suspension drug product". Indications relevant to the possible role of particle size on drug product properties, in particular on bioavailability, are detailed in decision tree # 3 of the same guideline (Fig. 11.1). In this document the role of particle size in processability, content uniformity and stability of a drug product is clearly recognized. The first question, however, regards the possible effect of particle size on solubility, dissolution and bioavailability, which are strictly related to each other. In case of possible effects of particle size on product performance, definition of specifications will be required (ICH GL).

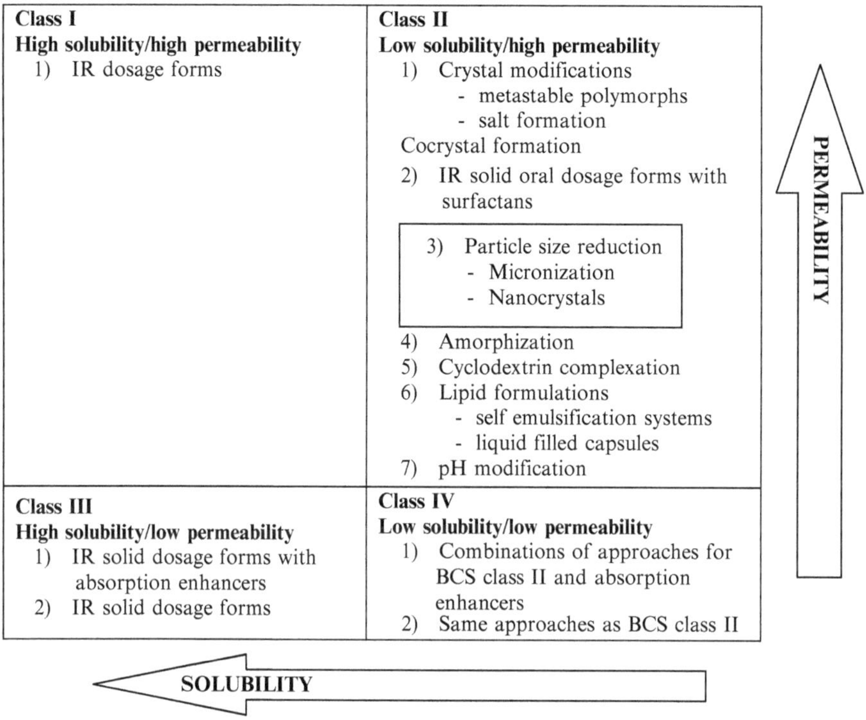

Fig. 11.2 Formulation design for poorly water soluble drugs based on BCS: basic approaches and practical applications [24] (Copyright Elsevier Ltd; reproduced with permission)

The observations reported in early studies [7] on the relevance of particle size on dissolution rate, that is particularly evident when poorly soluble drugs are concerned, have been more recently revised and reconsidered, in a regulatory perspective, in the frame of the Biopharmaceutical Classification System (BCS) [2].

BCS was developed to avoid in vivo bioequivalence studies for drug products characterized by rapid dissolution in vitro, being based on drugs that are very soluble and have good permeability (class I). In these cases, in vivo studies can be substituted by in vitro studies based on dissolution tests. BCS based bio-waiver approach can be extended also to class III drugs (with good solubility and poor permeability). Drugs with low solubility and good permeation (class II) and with low solubility and poor permeation (class IV) pose much more concerns from a regulatory point of view, since small changes in physical-chemical properties can significantly affect dissolution and absorption and require more extensive in vivo characterization. The increasing attention for this topic depends on the fact that it is estimated that about 40 % and 60 % of the new drug entities exhibit poor solubility, with related uncertainty on bioavailability and clinical use [28].

BCS represents therefore a reference also for the formulative approach, since poorly soluble drugs are a major challenge for formulation process. As evidenced by Kawabata et al. [24], the physical-chemical properties of a drug, and particle size among them, are key factors for reproducibility, bioavailability and therapeutic effect of a given formulation (Fig. 11.2). Particle size reduction was therefore suggested as technological strategy for class II and IV drug formulation, both as micronization and as reduction to nano-crystals.

The regulatory concerns and the recognition of the particle size relevance for bioavailability led in more recent years to the development of an intensive research about possible prediction of bioavailability in physiologically based models simulating gastrointestinal transit and absorption. In some studies, samples containing drugs of different formulas, characterized by different dissolution profiles according to Noyes–Whitney equation, have been evaluated through this modeling approach. Among the commercial software available for pharmacokinetic simulation are GastroPlus™, PK-Sim® and Stella®.

11.3 Nanonization

Although micronization represents a simple and attractive technology to enhance dissolution rate by increasing surface area, it has often experienced limited success. In particular, poorly soluble drugs even though micronized, tend to be eliminated from the gastro-intestinal tract before there are fully absorbed due to an insufficient solubility in the gastro-intestinal fluids. Although the micronization improves the dissolution rate in comparison to conventional milling, this is, sometimes, not enough to significantly and effectively increase drug solubility/dissolution. This results in highly intra – and inter- individual variability of plasma profiles and lack the of dose response, especially for substances belonging to BCS class II.

The nanonization of a drug substance is the logical subsequent step to micronization for surface area increase and dissolution rate enhancement. Nanonization refers to the reduction of drug particle size to the sub-micrometer range. There are two main approaches for nanonization: "top down" and "bottom up" technologies. The "top down" approach is by far the more popular and it basically relies on mechanical attrition to render large crystalline particles into nanoparticles [25]. Examples of the "top down" approach include Elan's NanoCrystal® wet-milling technology [31] and SkyePharma's Dissocubes® high-pressure homogenization technology [25, 34]. The "bottom up" approach is based on controlled precipitation/crystallization [44]. This process involves dissolving the drug in a solvent and precipitating it in a controlled manner to nanoparticles through addition of an anti-solvent (usually water). This technology is available from DowPharma (Midland, MI, USA) and BASF Pharma Solutions (Florham Park, NJ, USA). A hybrid approach is also feasible. Baxter's NANOEDGE® technology employs both "bottom up" and "top down" approaches through microprecipitation and homogenization [25].

The sub-micrometer particles are stabilized with surfactants or polymers in nanosuspensions to avoid agglomeration. These nanosuspensions increase dissolution rates of drug compounds and complement other technologies developed to enhance bioavailability of insoluble drugs (BCS Class II and IV) such as the use of solubility enhancers (i.e. surfactants), liquid-filled capsules or solid dispersions of drugs in their amorphous state. The success of the nanonization approach has been confirmed by several commercial available products containing nanosized drugs (Table 11.1).

The underlying basis for dissolution-limited bioavailability and its improvement by nanonization is illustrated in Fig. 11.3 (modified from Gao et al. [11]).

According to the Nernst–Brunner equation, the dissolution of particles is not only dependent on particle surface area, but also on the thickness of the diffusion boundary layer δ [21]. There are different theories on the relation between δ and particle size. In particular, Hixson assumes δ to be constant and consequently the increase in surface area is the unique driving force for dissolution rate enhancement. However, other models do not assume that the thickness of the boundary layer remains constant on changing particle size.

For instance, Higuchi et al. (1963) assume that the diffusion boundary layer thickness decreases proportionally with the particle size. According to this assumption a 12.5 fold increase in surface area would cause a theoretical increase in dissolution rate by 156.25 factor compared with the initial dissolution rate of micronized material [14].

Other relationships between the thickness of the diffusion boundary layer, δ, and particle size have also been considered. For instance, Niebergall et al. [38] proposed that δ correlates with the square root of the particle radius; Hintz and Johnson proposed the concept of a transitional particle size of 30 μm, implying that δ is constantly 30 μm for particles larger than 30 μm, whereas the diffusion layer is equal to particle radius for particles smaller than 30 μm [15]. Sheng et al. [46] found a dependency of δ on medium agitation. From these results it can be concluded that calculations of the dissolution rate are strongly dependent on the choice of the theoretical model.

Most modern software packages for plasma level simulation (e.g. GastroPlus®) use a constant δ value for simulating the dissolution of a drug. This causes only an approximate predictability of the mathematical models and suggests that the dissolution rate should be experimentally determined to obtain reliable results.

Besides the dissolution rate, nanonization increases saturation solubility, according to the Freundlich–Ostwald equation:

$$C_S = C_{\infty} \exp\left(\frac{2\lambda M}{r\rho RT}\right) \tag{11.5}$$

where:

C_s = saturation solubility of the nanonized drug
C_{∞} = saturation solubility of an infinitely large drug crystal
λ = the crystal/medium interfacial tension
M = molecular weight of drug
r = particle radius

Table 11.1 Marketed products containing drug nanocrystals [11] (Copyright Elsevier Ltd.; reproduced with permission)

Trade name	INN name	FDA approval	Company	Adminstration route	Indications	Therapeutic benefit
Casamet®	Nabilone	2005	Lilly	Oral	Antiemetic	High bioavailability
Rapamune®	Sirolimus	2000	Pfizer (Wyeth)	Oral	Innumosuppresant	High patient compliance, tablet formulation instead of solution
Emend®	Aprepitant	2003	Merck	Oral	Antiemetic	High bioavailability, no food effects
Tricor® Lyphantyl®	Fenofibrate	2004	Fournier Pharma, Abbott Laboratories	Oral	Hypocolesterolemic	No food effects
Triglide®	Fenofibrate	2005	Sciele, Shionogi Pharma Inc.	Oral	Hypocolesterolemic	No food effects
Megace® ES	Megestrole acetate	2005	PAR Pharmaceuticals	Oral	Appetite stimulant	No food effects, high patient compliance
Invega® Sustenna® Xeplion®	Paliperidone palmitate	2009	Janssen	Parenteral, Intramuscular	Antidepressant	High bioavailability

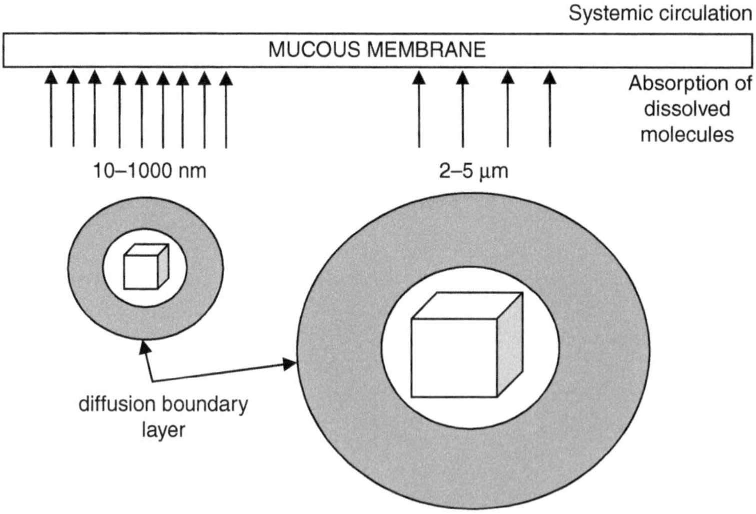

Fig. 11.3 Thickness of boundary layers of nanocrystals (10–1000 nm) and micronized particles (2–5 μm) (Modified from [11]; copyright Elsevier Ltd.; reproduced with permission)

ρ = particle density
R = gas constant
T = temperature.

Equation 11.5 shows that solubility of a given substance at given conditions depends upon particle size and also upon drug molecular weight. During dissolution, Ostwald ripening may occur: large particles can grow at the expense of dissolving small particles.

According to Eq. 11.5, 10–15 % increase in solubility could be expected for a drug having M = 500, ρ =1 g/ml, particle size of 100 nm and crystal-intestinal fluid interfacial tension λ = 15–20 mN m^{-1} [26]. However, a more pronounced increase in solubility was experimentally determined; for instance, Muller and Peters [35] reported a 50 % increase in solubility of an insoluble antimicrobial compound when the particle size was reduced from 2.4 μm to 800 or 300 nm. Such an increase in solubility causes an increase in dissolution rate proportional to solubility and, thus, a significantly higher bioavailability. The presence of surfactants as stabilizers of nanosuspensions, further enhanced dissolution rate of the insoluble antimicrobial compound in comparison with micronized suspensions due to increases in surface wetting.

Drug nanocrystals are generally reported as safe and well tolerated systems for many administration routes when compared with conventional products. In particular, nanocrystals can be profitably administered via conventional routes, suitable as oral and parenteral, as well as employed for ophthalmic and pulmonary administration [11]. A fine particle size helps in improving safety of oral administered poorly soluble drugs, by increasing the distribution uniformity in the gastrointestinal fluid and avoiding a high and prolonged local concentration [29]. Nanosized particles are also beneficial for a better tolerability in mucosal administration.

Table 11.2 Changes of pharmacokinetic properties of oral drug nanocrystal formulations compared with the conventional formulations

	Dosage form		Ref
Fenofibrate	Aqueous nanosuspension (194–356 nm)	1.8–12.5 fold increase in Cmax; 1.7–17 fold increase in bioavailability; 1.3–2.3 fold reduction in tmax	[27]
Fenofibrate	Aqueous nanosuspension (340 nm)	1.67 fold increase in Cmax; 1.3 fold increase in bioavailability; 4.9 fold reduction in tmax	[13]
Aprepitant	Aqueous nanosuspension (120 nm)	No food effect at a dose of 2, 80, 125 mg/kg	[50]
Itraconazole	Aqueous nanosuspension (267 nm)	1.2–1.8 fold increase in Cmax; 1.2–1.8 fold increase in bioavailability; fsted/fed ratio of AUC markedly reduced	[33]

Many reports show that drug nanocrystals possess many positive effects on oral drug delivery of poorly soluble drugs. In particular, changes of pharmacokinetic parameters occurr after nanocrystals administration: these generally include increased maximum plasma concentration (C_{max}), reduced time to maximum plasma concentration (T_{max}), enhanced area under blood concentration–time curve (AUC) and reduced fasted/fed variability in comparison with conventional formulations as reported in Table 11.2. In particular, when drugs are administered as nanocrystals, a high drug concentration gradient takes place between GIT and blood vessel which will markedly improve absorption and result in a high bioavailability. This behavior is conceivably determined by the increased saturation solubility and dissolution rate of drug nanocrystals in digestive juice. One classic example is danazol, a poorly soluble gonadotropin inhibitor. The absolute bioavailability of marketed danazol conventional microsuspensions (200 mg, 10 μm) was only 5.2 % in beagle dogs. When administered as an aqueous nanosuspension (200 mg, 169 nm), an absolute bioavailability of 82.3 % could be achieved, with a 15 fold *Tmax* reduction and *Cmax* increase [30].

Jinno et al. [18] measured the cilostazol plasma kinetics after oral administration to beagle dogs, under fasted and fed conditions, of suspensions containing three different particle size distributions. The in vitro dissolution profiles in both water and biorelevant dissolution media showed an influence of particle size but the results could neither quantitatively predict the increase in bioavailability on decreasing particle size nor the food effect observed in vivo. Starting from these results, Willmann et al. [49] combined a physiologically based pharmacokinetic (PBPK) model for gastro-intestinal transit and absorption with a mechanistic dissolution model, previously developed by Johnson [19], to predict the influence of particle size on the plasma pharmacokinetics of cilostazol in beagle dogs.

The dissolution model was based on the Noyes–Whitney equation for spherical particles with a predefined particle size distribution, assuming the events illustrated

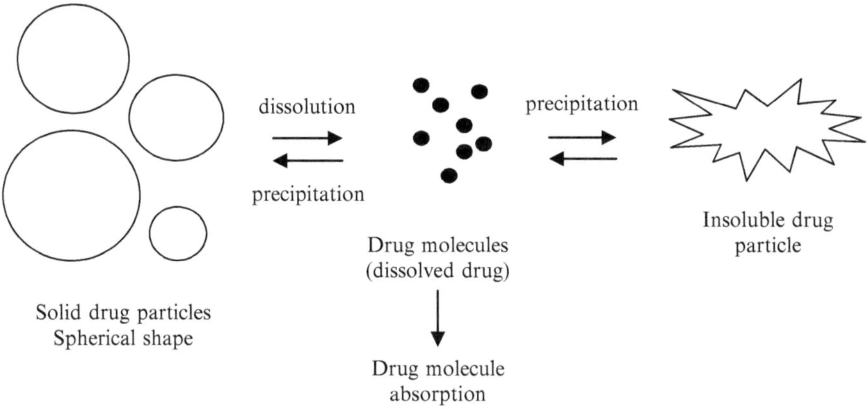

Fig. 11.4 Structure of the dissolution model: solid spherical particles of different diameter dissolved in gastrointestinal fluids (Modified from [19]). Precipitation can occur leading to particles similar to the original ones or to insoluble particles. Drug molecules can be absorbed

in Fig. 11.4. Three different cilostazol suspensions with different particle size distributions, previously described in the study of Jinno et al. [18], were used:

(a) Hammer-milled crystal sample with a median particle size of 13 μm,
(b) Jet-milled sample with a median particle size of 2.4 μm,
(c) Spray-dried powder with a median particle size of 0.22 μm.

The comparison between simulated curves and experimental data demonstrated that the dynamic dissolution model enabled an almost exact prediction of the decrease in rate and extent of absorption on increasing particle size, both in fasted and in fed conditions. In fact the bioavailability of cilostazol was increased by the reduction of particle size and consequently by the increase of dissolution rate. The model was also able to put in evidence that the rate-limiting step for absorption was dissolution in the case of the two micronized suspensions (obtained by hammer-milling and jet-milling), and it was the permeation across the intestinal epithelium for the nanocrystals. The use of the simulation approach in formulation development should aid the selection of the appropriate drug particle size for solid oral dosage forms, and could potentially support specification selection. In Fig. 11.5 the dissolution profiles of cilostazol from the suspensions in water at 37 °C are reported [18]. Dissolution study was performed at 50 rpm following USP Apparatus 2. About 5 mg of cilostazol was applied in 900 ml water. Results are expressed as the mean with the bar showing standard deviation values of six experiments and simulated curves (solid lines). The reduction of particle size dramatically increase cilostazol bioavailability (Fig. 11.5): in particular the highest enhancement with a minimum food effect was observed for the NanoCrystal cilostazol suspension, while the jet-milled suspension determined a moderate improvement of bioavailability in comparison with conventional hammer-milled crystals and showed a significant food effect. This means that food intake did not increase the

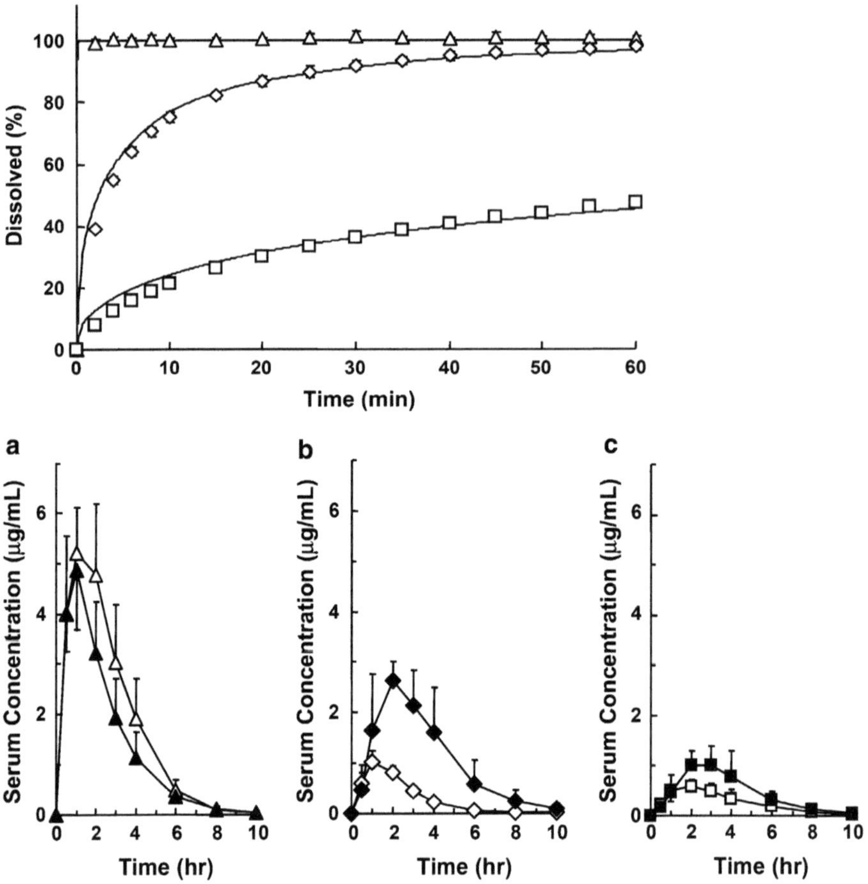

Fig. 11.5 Dissolution –time profiles for NanoCrystal® spray-dried cilostazol powder (median particle size 13 μm) (Δ) (**a**); jet-milled crystal (median particle size 2.4 μm) (◇) (**b**); hammer-milled crystal (median particle size 0.22 μm) (□) (**c**), reported in the upper part. In the lower part, plasmatic curves corresponding to the administration of each particle suspensions at a dose of 100 mg/body in beagle dogs are reported: the open symbols correspond to fed state, the closed ones to fasted state [18] (Copyright Elsevier Ltd.; reproduced with permission)

bioavailability of the NanoCrystal® cilostazol suspension. This occurs because dissolution rate of NanoCrystal® cilostazol is fast enough even under the fasted condition, where the absorption might be permeability-limited. Therefore, the further increase in the dissolution rate would not contribute to the improvement of the absorption.

Other papers that compare micronized particles and nanocrystals have been recently published: in particular they deal with the relationship between particle size and bioavailability of poorly soluble drug substances. Eg., Fu et al. [10] studied the relationship between dissolution and bioavailability of nimodipine microcrystals and nanocrystals aqueous dispersions, obtained by microprecipitation, with the aim of determining the critical size responsible for the improvement of oral bioavailability. Three size levels were considered in this study: 16.3, 4.1 and 0.83 μm, and the average volume-weighted particle sizes were determined by a laser diffraction method.

All three samples showed comparable aqueous equilibrium solubility, determined by shaking flasks, and the dissolution curves in purified water and 0.05 % SDS were also similar. However the pharmacokinetics data obtained in beagle dogs showed that AUC values were 1.69 and 2.59-fold higher for dispersions of 4.1 and 0.83 μm sizes than was for the 16.3 μm dispersion.

These results suggest that the aqueous solubility and the supersaturation dissolution were not effective indexes in evaluating the bioavailability of colloidal dispersions, and moreover they could not discriminate the in vivo performance. Fu et al. [10] evidenced that there is a critical particle size to improve the bioavailability of nimodipine and it ranges between 0.83 and 4.1 μm.

In another study, aprepitant (an antiemetic antagonist of NK1 receptor) release from micro-sized and nano-sized powders and its pharmacokinetics in fasted and fed humans is described (Shono et al. 2010) by means of Noyes–Withney equation coupled with STELLA® 9.0 software (Cognitus Ltd., North Yorkshire, UK). The change in particle size modifies the dissolution rates, and consequently the rate of drug appearance in plasma. The model was able to correctly simulate the experimental behavior, both in vitro and in vivo.

11.4 Final Remarks

The interest on particle size characterization and on its influence on bioavailability is renewed by regulatory requirements and by the introduction of the new nanonization technology. In particular several kinds of nanosized active ingredients are nowadays commercially available, and others are undergoing development studies. Nanonization to obtain a nanocrystal formulation is an approach well-adaptable to drugs having different chemical-physical properties. Moreover, the employment of excipients to stabilize nanocrystals, and in particular the use of surfactants for nanocrystal surface modification should further enhance drug bioavailability and could achieve prolonged release and targeted (site-specific) drug delivery.

In addition, nanotoxicological investigations of drug nanocrystals should be extensively carried out to better understand the fate of nanocrystals at a cellular level and to find new potential for nanocrystal applications for innovative treatment approaches.

11.5 Definitions, Abbreviations and Symbols

Absorption number	Ratio of the mean residence time to the absorption time.
Dissolution number	Ratio of mean residence time to mean dissolution time.
Dose number	Ratio of the total amount of drug added to the amount soluble in 250 ml water (being the standard volume of liquid available for dissolution in our gastro-intestinal system)
Fasted condition	Without eating food
Fed condition	After eating food
Food effect	Effect of the presence of food on drug fate in gastrointestinal tract
Initial saturation	Ratio of drug luminal concentration at the beginning of the intestine to drug solubility
Intrinsic dissolution	Dissolution rate of pure substances under the condition of constant surface area
Sink conditions	Conditions where C_t is negligible in comparison to C_s
AUC	Area under the blood plasma concentration-time curve
BCS	Biopharmaceutical Classification System
GIT	Gastro-intestinal tract
GL	Guideline
ICH	International conference on harmonization of technical requirements for registration of pharmaceuticals for human use
IR	Immediate release
NK1	Neurokinin 1
SDS	Sodium dodecylsulfate
A	Surface area
An	Absorption number
C_{max}	Maximum plasma concentration
C_s	Solubility of the substance
C_t	Concentration at time t
C_∞	Saturation solubility of an infinitely large drug crystal
D	Particle size (diameter of an equivalent sphere)
$D_{3,2}$	Area-weighted Sauter mean diameter
$D_{4,3}$	Volume-weighted mean size
Do	Dose number
Dn	Dissolution number
$\mathbb{D}$	Diffusion coefficient of a solute
dm/dt	Dissolution rate
G	Surface specific dissolution rate
Is	Initial saturation
M	Drug molecular weight,
m	Drug amount
R	Gas constant,
r	Particle radius
T	Temperature

(continued)

t	Time
t_{max}	Time to reach maximum plasma concentration
δ	Thickness of the diffusion boundary layer around a particle
λ	Crystal/medium interfacial tension
ρ	Particle density

References

1. Abdou, H.M.: Dissolution, Bioavailability and Bioequivalence. Mack, Easton (1989)
2. Amidon, G.L., Lennernas, H., Shah, V.P., Crison, J.R.: A theoretical basis for a biopharmaceutic drug classification: The correlation of in vitro drug product dissolution and in vivo bioavailability. Pharm. Res. **12**, 413–420 (1995)
3. Anderberg, E.K., Nyström, C., Bisrat, M.: Physico-chemical aspects of drug release. VII. The effect of surfactant concentration and drug particle size on solubility and dissolution rate of Felodipine, a sparingly soluble drug. Int. J. Pharm. **47**, 67–77 (1988)
4. Atkinson, R.M., Bedford, C., Child, K.J., Tomich, E.G.: Effect of particle size on blood griseofulvin levels in man. Nature **193**, 588–589 (1962)
5. Bisrat, M., Nyström, C.: Physicochemical aspects of drug release. VIII. The relation between particle size and surface specific dissolution rate in agitated suspensions. Int. J. Pharm. **47**, 223–231 (1988)
6. Brunner, E.: Reaktionsgeschwindigkeit in heterogenen Systemen. Z. Phys. Chem. **43**, 56–102 (1904)
7. Caramella, C.: Utilizzazione dei parametri granulometrici nella progettazione di forme farmaceutiche. Boll. Chim. Farm. **130**, 43–51 (1991)
8. Carstensen, J.T.: Solid Pharmaceutics: Mechanical Properties and Rate Phenomena. Academic, New York (1980)
9. Dressman, J.B., Reppas, C.: In vitro–in vivo correlations for lipophilic, poorly water-soluble drugs. Eur. J. Pharm. Sci. **11**, 73–80 (2000)
10. Fu, Q., Kou, L., Gong, C., Li, M., Sun, J., Zhang, D., Liu, M., Sui, X., Liu, K., Wang, S., He, Z.: Relationship between dissolution and bioavailability for nimodipine colloidal dispersions: The critical size in improving bioavailability. Int. J. Pharm. **427**, 358–364 (2012)
11. Gao, L., Liu, G., Ma, J., Wang, X., Zhou, L., Li, X.: Drug nanocrystals: In vivo performances. J. Control Release **160**, 418–430 (2012)
12. Gibaldi, M.: Biopharmaceutics and Clinical Pharmacokinetics, 3rd edn, pp. 44–63. Lea and Febiger, Philadelphia (1984)
13. Hanafy, A., Spahn-Langguth, H., Vergnault, G., Grenier, P., Tubic Grozdanis, M., Lenhardt, T., Langguth, P.: Pharmacokinetic evaluation of oral feno!brate nanosuspensions and SLN in comparison to conventional suspensions of micronized drug. Adv. Drug Deliv. Rev. **59**, 419–426 (2007)
14. Higuchi, T.: Mechanism of sustained-action medication. Theoretical analysis of rate of release of solid drugs dispersed in solid matrices. J. Pharm. Sci. **52**, 1145–1149 (1963)
15. Hintz, R.J., Johnson, K.C.: The effect of particle-size distribution on dissolution rate and oral absorption. Int. J. Pharm. **51**, 9–17 (1989)
16. Hixson, A.W., Crowell, J.H.: Dependence of reaction velocity upon surface and agitation. Ind. Eng. Chem. **23**, 923–931 (1931)
17. *ICH Guideline Q6A.*: Specifications: Test procedures and acceptance criteria for new drug substances and new drug products: Chemical substances; ICH (1999)
18. Jinno, J., Kamada, N., Miyake, M., Yamada, K., Mukai, T., Odomi, M., Toguchi, H., Liversidge, G.G., Higaki, K., Kimura, T.: Effect of particle size reduction on dissolution and

oral absorption of a poorly water-soluble drug, cilostazol, in beagle dogs. J. Control Release **111**, 56–64 (2006)
19. Johnson, K.C.: Dissolution and absorption modeling: Model expansion to simulate the effects of precipitation, water absorption, longitudinally changing intestinal permeability, and controlled release on drug absorption. Drug Dev. Ind. Pharm. **29**, 833–842 (2003)
20. Jounela, A.J., Pentikainen, P.J., Sothmann, A.: Effect of particle size on the bioavailability of digoxin. Eur. J. Clin. Pharmacol. **8**, 365–370 (1975)
21. Junemann, D., Dressman, J.B.: Analytical methods for dissolution testing of nanosized drugs. J. Pharm. Pharmacol. **64**, 931–943 (2012)
22. Kaneniwa, N., Watari, N.: Dissolution of slightly soluble drugs. I. Influence of particle size on dissolution behavior. Chem. Pharm. Bull. **22**, 1699–1705 (1974)
23. Kaneniwa, N., Watari, N., Iijima, H.: Dissolution of slightly soluble drugs. V. Effect of particle size on gastrointestinal drug absorption and its relation to solubility. Chem. Pharm. Bull. **26**, 2603–2614 (1978)
24. Kawabata, Y., Wada, K., Nakatani, M., Yamada, S., Onoue, S.: Formulation design for poorly water-soluble drugs based on biopharmaceutics classification system: Basic approaches and practical applications. Int. J. Pharm. **420**, 1–10 (2011)
25. Keck, C.M., Muller, R.H.: Drug nanocrystals of poorly soluble drugs produced by high pressure homogenization. Eur. J. Pharm. Biopharm. **62**, 3–16 (2006)
26. Kesisoglou, F., Panmai, S., Wu, Y.: Nanosizing – Oral formulation development and biopharmaceutical evaluation. Adv. Drug Deliv. Rev. **59**, 631–644 (2007)
27. Li, X., Gu, L., Xu, Y., Wang, Y.: Preparation of fenofibrate nanosuspension and study of its pharmacokinetic behavior in rats. Drug Dev. Ind. Pharm. **35**, 827–833 (2009)
28. Lipinski, C.: Poor aqueous solubility — An industry wide problem in drug discovery. Am. Pharm. Rev. **5**, 82–85 (2002)
29. Liversidge, G.G., Conzentino, P.: Drug particle size reduction for decreasing gastric irritancy and enhancing absorption of naproxen in rats. Int. J. Pharm. **125**, 309–313 (1995)
30. Liversidge, G.G., Cundy, K.C.: Particle size reduction for improvement of oral bioavailability of hydrophobic drugs: I. Absolute oral bioavailability of nanocrystal- line danazol in beagle dogs. Int. J. Pharm. **125**, 91–97 (1995)
31. Merisko-Liversidge, E., Liversidge, G.G., Cooper, E.R.: Nanosizing: A formulation approach for poorly-water-soluble compounds. Eur. J. Pharm. Sci. **18**, 113–120 (2003)
32. Mosharraf, M., Nystrom, C.: The effect of particle size and shape on the surface specific dissolution rate of microsized practically insoluble drugs. Int. J. Pharm. **122**, 35–47 (1995)
33. Mou, D., Chen, H., Wan, J., Xu, H., Yang, X.: Potent dried drug nanosuspensions for oral bioavailability enhancement of poorly soluble drugs with pH-dependent solubility. Int. J. Pharm. **413**, 237–244 (2011)
34. Muller, R.H., Jacobs, C., Kayser, O.: Nanosuspensions as particulate drug formulations in therapy. Rationale for development and what we can expect for the future. Adv. Drug Deliv. Rev. **47**, 3–19 (2001)
35. Muller, R.H., Peters, K.: Nanosuspensions for the formulation of poorly soluble drugs I. Preparation by a size-reduction technique. Int. J. Pharm. **160**, 229–237 (1998)
36. Nernst, W.: Theorie der Reaktionsgeschwindigkeit in heterogenen Systemen. Z. Phys. Chem. **47**, 52–55 (1904)
37. Neuvonen, P.J.: Bioavailability of phenytoin: Clinical pharmacokinetic and therapeutic implications. Clin. Pharmacokinet. **4**, 91–103 (1979)
38. Niebergall, P.J., Milosovich, G., Goyan, J.: E. Dissolution rate studies. 2. Dissolution of particles under conditions of rapid agitation. J. Pharm. Sci. **52**, 236–241 (1963)
39. Ning, X., Sun, J., Han, X., Wu, Y., Yan, Z., Han, J., He, Z.: Strategies to improve dissolution and oral absorption of glimepiride tablets: Solid dispersion versus micronization techniques. Drug Dev. Ind. Pharm. **37**, 727–736 (2011)
40. Noyes, A.A., Whitney, W.R.: The rate of solution of solid substances in their own solutions. J. Am. Chem. Soc. **19**, 930–934 (1897)

41. Nyström, C., Mazur, J., Barnett, M.I., Glazer, M.: Dissolution rate measurements of sparingly soluble compounds with the Coulter Counter model TAII. J. Pharm. Pharmacol. **37**, 217–221 (1985)
42. Oh, D.M., Curl, R.L., Amidon, G.L.: Estimating the fraction dose absorbed from suspensions of poorly soluble compounds in humens: A mathematical model. Pharm. Res. **10**, 264–270 (1993)
43. Oh, D.M., Curl, R.L., Yong, C.S., Amidon, G.L.: Effect of Micronization on the extent of drug absorption from suspensions in humans. Arch. Pharm. Res. **18**, 427–433 (1995)
44. Rabinow, B.E.: Nanosuspensions in drug delivery. Nat. Rev. Drug Discov. **3**, 785–796 (2004)
45. Shaw, T.R.D., Carless, J.E.: The effect of particle size on the absorption of digoxin. Eur. J. Clin. Pharmacol. **7**, 269–273 (1974)
46. Sheng, J.J., Sirois, P.J., Dressman, J.B., Amidon, G.L.: Particle diffusional layer thickness in a USP dissolution apparatus II: A combined function of particle size and paddle speed. J. Pharm. Sci. **97**, 4815–4829 (2008)
47. Simoes, S., Sousa, A., Figueiredo, M.: Dissolution rate studies of pharmaceutical multisized powders – A practical approach using the Coulter method. Int. J. Pharm. **127**(283), 291 (1996)
48. Watari, N., Hanano, M., Kaneniwa, N.: Dissolution of slightly soluble drugs. VI. Effect of particle size of sulfadimethoxine on the oral bioavailability. Chem. Pharm. Bull. **28**, 2221–2225 (1980)
49. Willmann, S., Thelen, K., Becker, C., Dressman, J.B., Lippert, J.: Mechanism-based prediction of particle size-dependent dissolution and absorption: Cilostazol pharmacokinetics in dogs. Eur. J. Pharm. Biopharm. **76**, 83–94 (2010)
50. Wu, Y., Loper, A., Landis, E., Hettrick, L., Novak, L., Lynn, K., Chen, C., Thompson, K., Higgins, R., Batra, U., Shelukar, S., Kwei, G., Storey, D.: The role of biopharmaceutics in the development of a clinical nanoparticle formulation of MK-0869: A beagle dog model predicts improved bioavailability and diminished food effect on absorption in human. Int. J. Pharm. **285**, 135–146 (2004)

Chapter 12
Pigments and Paint Dispersions

Henk G. Merkus

Abstract Signs of paint application have been found in caves, dating over 100,000 years ago. At that time, paint composition was based on natural resources. Nowadays, composition has changed considerably, because (1) many new constituents became available, (2) knowledge has increased and (3) new challenges and requirements have arisen. This chapter starts by listing the multitude of requirements for modern paints and overviews the relationship between composition and dispersion. The chapter deals with two paint types, viz. the conventional paint based on organic solvents and the modern water-based latex paint. Pigment particles provide the most important contribution to paint in terms of the desired optical properties (color, gloss and hiding power). These are explained and their relationships with refractive index, particle size distribution (PSD) and concentration of pigments and extenders are elucidated. The rheological, non-Newtonian behavior of paint is the second essential contribution to the performance of liquid paints. Adequate tuning of the viscosity at different shear rates is required for best performance at various stages during production and application. Especially where the concentration of solids comes close to their maximum packing density, attention should be paid to both concentration and packing density in relation to the PSD. Adequate de-agglomeration of pigments and extenders is necessary during their dispersion in view of both optical performance and minimum binder requirement. For water-based paints, both size and minimum film formation temperature of the colloidal polymer particles are extra points for attention. The quality of PSD measurement primarily depends on the presence of adequate analysis criteria and good sampling and dispersion methods that fulfill these criteria. Many measurement techniques are available, but they differ in requirements for concentration (dilution), precision, resolution and sensitivity.

H.G. Merkus (✉)
Retired Associate Professor, Delft University of Technology,
Park Berkenoord 30, Pijnacker 2641 CZ, The Netherlands
e-mail: henk.merkus@hetnet.nl

H.G. Merkus and G.M.H. Meesters (eds.), *Particulate Products: Tailoring Properties for Optimal Performance*, Particle Technology Series 19,
DOI 10.1007/978-3-319-00714-4_12,

12.1 Introduction

The earliest signs of paint application date back more than hundred thousand years, according to findings of some pigments and grinding equipment found in Twin Rivers cave, Zambia, Africa. Grinding equipment (abalone shells and bone pestles) and ochre pigment were also found in Blombos cave, South Africa. These also date to about 100,000 years ago. The earliest European paintings found date from about 32,000 years ago in the Chauvet cave in France and later in other caves. All have been found in rock caves where the people lived. They survived the extremely long period of time because the caves were later cut off from the surroundings. Some nice examples of such paintings are shown in Fig. 12.1.

Early paintings have been found in rock caves all over the world, in Africa, northern and southern America, Asia, Australia and Europe. The painters used pigments that were locally available, such as carbon black (charcoal) and materials like various iron and manganese oxides and salts. These were dispersed in water or animal fat after they had been ground to powders with primitive mortars (hollow stones or shells) and pestles (bones). The purpose of these paintings may have been merely artistic, but also educational and ritual. Also, the painters may have believed that the paintings gave them power over the prey that they painted.

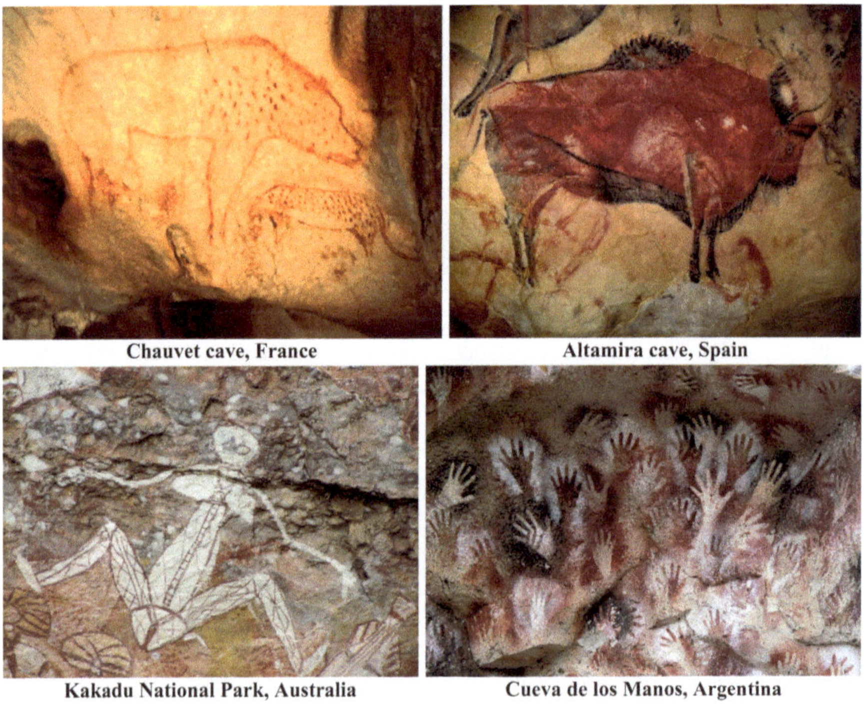

Fig. 12.1 Nice examples of prehistoric paintings [48]

Mesopotamian, Egyptian, Greek and Roman civilizations later on developed a wider color range of pigments as well as the art of paint making. Moreover, they extended paint application from artistic purposes to decoration and preservation of objects. Further development was promoted by the industrial revolution, when the increased application of iron and steel required delay or prevention of rusting and corrosion. This was further increased with scientific research, leading to both better insight in e.g. paint physics and physical chemistry, chemical structures and syntheses of new chemicals.

12.2 Functionalities and Composition

The main functions of paints have remained unaltered over the last centuries, including:

- Preservation of objects and structures
- Decoration of these objects and structures
- Artistic expressions.

Moreover, application has strongly widened, challenges have increased and composition improved. In relation to this, many functionalities may be required, viz.:

- Good adhesion to the surface of the object, for a multitude of objects and materials, e.g. wood, metal, concrete and plastic, all of which have quite different surface properties
- Providing a strong surface layer to the object to prevent contact with oxygen, salts and water, in order to give the object a prolonged life time
- Enhancing the aesthetic appeal of objects through provision of nice optical properties (color, gloss, opacity, etc.)
- Anti-corrosion properties
- Bactericidal properties
- Protection to irradiation of sun light to ensure long durability of both surface layer and color
- Good stability of paint during storage
- Easy application with optimum results
- Nice odor and minimum health hazards for humans during production and application
- Minimum detrimental effects for the environment
- Low costs.

The necessity for these functions depends upon the specific application. Thus, composition varies with application. For example, the application to cars poses different requirements from wood or concrete or boats. Transparent paints differ from opaque paints. Highly specific examples are the application of non-white pigments with high near-infrared reflectance for camouflage or to minimize solar heating [24].

Basically, liquid paints contain the following components:

- Binder, i.e. (precursor for) polymer or resin, which is often part of the continuous phase and provides the protective surface layer
- Solvent or diluents, forming the other part of the continuous phase, which gives paint the required rheological behavior during mixing and application; nowadays, water is most often used as a solvent in domestic applications whilst organic solvents are still used in challenging, industrial applications
- Pigments, to provide the required color, opacity and other optical or visual effects; they form the dispersed phase and are virtually insoluble in the continuous phase
- Extenders and fillers, which may assist in matting (reducing gloss) and better abrasion resistance of painted surfaces as well as in reducing pigment concentration and costs; they are also part of the dispersed phase and virtually insoluble in the continuous phase and, thus, contribute to paint viscosity.

In the past, paint systems were used, where the binder consisted of vegetable oils with a few double bonds that could be cross-linked with oxygen, in which the oils were dissolved in organic solvent/diluents mixtures (e.g. mineral spirits, alcohols or esters). Nowadays, most paints contain polymers or resins as binder, which are cross-linked or coalesce in the application. The function of the solvent/diluents is not only solvency but adds adequate rheological behavior and evaporation rate. Other important properties are odor, toxicity, flammability (flashpoint) and costs. In view of toxicity and flammability, organic solvents have been replaced in many applications by aqueous systems, which required significant adaptation of the composition (see Sect. 12.3.5). In organic, high-solids paint dispersions, the role of solvent is often fulfilled by monomers and/or oligomers of the binder.

A variety of further additives may be present, viz.:

- Surfactants to promote dispersion of pigments and extenders in the continuous phase, to stabilize the paint dispersion and to facilitate adhesion of protective layer to the surface of the object
- Initiators or catalysts to start or fasten cross-linking and/or polymerization (hardening; drying; curing) of the binder
- Thickening agents to optimize rheological behavior of the paint through formation of some kind of network
- De-foamers to counteract foam formation by lowering the surface tension of liquid in contact with air bubbles
- Corrosion inhibitors to prevent or slow down corrosion of metal substrates
- Antistatic additives to counteract building-up of static electricity and, thus, fire hazard
- Anti-skinning agents to prevent formation of a skin on the surface of stored paint
- Preservatives/bactericides to prevent growth of bacteria in stored paint

– Optical whiteners, which absorb UV light and re-emit it in the blue-violet region of visible light, thus boosting the white color impression.

Note: this chapter deals only with liquid dispersed paints containing pigments. Thus, powder paints (see Chap. 13) and lacquers, which do not contain pigments, are excluded.

12.3 Quality Requirements and Particle Size

The main requirements for paint have been mentioned in Sect. 12.2. A good deal of them is covered or dominated by binder, solvent and additives. Examples are adequate dispersion of pigments, formation and good adhesion of a protective surface layer on the substrate and rheological behavior. These aspects as well as various color aspects are treated in other books [2, 3, 18, 22, 23, 39, 40]. General test methods are described in the same books as well as in e.g. ASTM and ISO standards.

On the other hand, pigments and extenders are essential for producing the required aesthetic effects, like color, gloss and hiding power, as well as being relevant for their adsorptive properties for binder and contribute to the rheological behavior, especially at high solids contents. These aspects together with their related size distribution parameters are of particular interest in this book.

Attention should be dedicated to particle shape. For pigments and extenders, shape most often relates to the macroscopical shape, viz. the ratio of length, breadth and thickness, although sometimes surface roughness or particle porosity may play a role. Particle sizing techniques disregard particle shape through using some equivalent size based on assumed circles or spheres, viz. area equivalent size, volume equivalent size, Stokes' diameter, etc. On the other hand, if particle shape of pigments deviates significantly from spheres, for example in case of rods or plates, their properties will be quite different from expectations based on spheres.

12.3.1 Optical Properties

The optical properties of paints can be divided into color, gloss and hiding power (opacity). For some applications, special effects might be included, like fluorescence or texture. Color is related to the wavelength of light in the visual range of the spectrum (about 380–780 nm) (see also Sect. 2.7). In the world of paint this is usually described in terms of hue, saturation (or chroma) and lightness (or shade). Here, hue is the typical color, the human perception of which can be expressed in terms of two CIE parameters x and y; lightness indicates the position on a neutral gray scale ranging from black to white (without regard to hue) and saturation the color purity (its degree of departure from grayness) (see Fig. 12.2). In addition, many other words

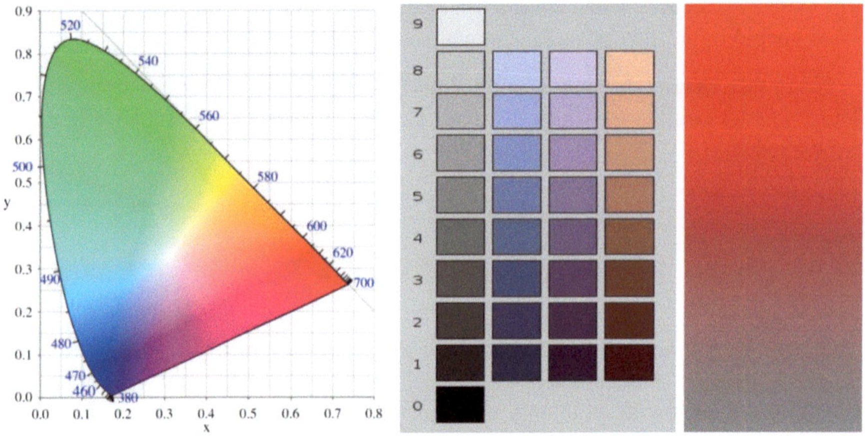

Fig. 12.2 Illustration of hue (CIE; *left*), lightness (*middle*) and saturation (*right*)

are being used (see Sect. 12.5 for some definitions) [18]. Moreover, the word color has a different interpretation in the world of science and that of human perception. In science, it relates to wavelengths. In normal life, it relates to feelings or appreciations, e.g. red may relate to warm, blue to cold and green to unripe.

There are three primary colors, viz. blue, green and red. All other colors can be formed by combinations of these three. The coloring of objects depends on their ability to selectively absorb light of some wavelengths and transmit or reflect the rest. This is easily demonstrated by color changes of objects when viewed under colored light instead of white light. For example, a red car viewed under blue light may appear black. Also, color shades may look false under sodium lighting commonly employed to light our streets. Pigments are primarily responsible for both the color effects and the hiding power of paint. Both effects relate to the degree of scattering and absorption of visible light by the pigment particles. When all incoming visible light is scattered, then the resulting color is white. When all incoming light is absorbed, then the result is a black color. And when visible light is scattered as well as selectively absorbed at some wavelength(s), a colored result is seen. A main goal of paint pigments is to provide aesthetic appeal to objects in the form of a nice color, thereby hiding unwanted surface effects and previous colors. Extenders may provide modifying effects on gloss.

Pigments are defined as solid particles that are insoluble in the continuous, liquid phase. They should be distinguished from dyes, which are soluble in the continuous phase. Pigments are either inorganic – e.g. titanium oxides, iron oxides, cadmium sulfide – or organic compounds, often azo-pigments or phthalocyanine pigments, containing multiple conjugated double bonds. Carbon black is somewhat intermediate, since it contains carbon but can not really be considered as an organic compound. Usually, pigments have a high refractive index (about 2–3). Typically, colored pigments are used in combination with white pigment, to provide for best opacity of the colored mixture.

Fig. 12.3 Pigment examples

Some nice, colorful pigment examples are presented in Fig. 12.3.

Extenders and fillers are much cheaper, insoluble particulate materials. Examples are chalk, China clay and talc. They contribute to paint viscosity, mat effects, mechanical properties of the paint film and improved weathering as well as to decreased paint costs, but have little or no color or opacity effect on the film. The reason is that their refractive index is typically about 1.5–1.8, close to that of the continuous phase of paint, which is about 1.3–1.6.

The basic properties of pigments are the result of chemical composition, degree of crystallinity and crystal structure. Besides solubility, they determine both the real value of the refractive index RI[1] (relevant to refraction and scattering of light at the particle's surface) and the imaginary RI value (relevant to absorption of light). Note that the same inorganic compound may have different crystal structures that may or may not have different colors (e.g. different iron oxides having different colors versus anatase and rutile titanium oxides, both being white). Also, minor amounts of other atoms (impurities) in crystal structures may have significant effects on color and refractive index. Higher real RI values mean larger amounts of scattered light and, thus, higher opacity of paint layers due to multiple scattering. Higher imaginary RI values mean larger light absorption and, thus, if the absorption is a function of wavelength of the visible light, a distinct color will result. For, when an object is illuminated by white light and selectively absorbs some of the wavelengths, then our eyes interpret the reflected or transmitted light as the resulting color.

Particle size, size distribution (PSD) and particle shape also influence light scattering by particles. Calculations for spheres using Mie theory for single scattering can exactly yield scattering properties versus particle size for given RI values. However, most particles are non-spherical and have a size distribution, whereas

[1] Note that the refractive index not only depends on composition and structure but sometimes also on the wavelength of light. Note further that refraction depends on the relative refractive index, that is the ratio of RI of pigment to that of continuous phase.

multiple scattering occurs in paint layers. The approximating equation of Weber (Eq. 12.1) allows estimation of maximum scattering power for spherical particles having a high RI [44]. It shows that this maximum is reached if particle size equals about half the dominant wavelength of the incident light.

$$D_{scmax} = \frac{2\lambda}{\pi(n_P - n_L)} \tag{12.1}$$

where:

D_{scmax} = particle size at maximum scattering power
λ = wavelength incident light
n_P = refractive index particle
n_L = refractive index liquid

Both color and opacity can be calculated from measured diffuse reflectance curves against wavelength through the Kubelka-Munk theory [14, 15, 17, 18, 20, 21]:

$$\left(\frac{K}{S}\right)_\lambda = \frac{(1 - R_{\lambda,\infty})^2}{2R_{\lambda,\infty}} = f(R) \tag{12.2}$$

where:

K = absorption coefficient of the K-M theory
S = scattering coefficient of the K-M theory
$R_{\lambda,\infty}$ = measured reflectance at given wavelength and infinitely thick paint layer
λ = wavelength

Although the K-M theory has a phenomenological origin, its absorption and scattering coefficients were proven later on to relate to the single-scattering and extinction coefficients, viz. ln$K$ is twice the extinction coefficient of the Beer-Lambert-Bouguer equation, even at high pigment concentrations; S has a more complex relationship with the Mie scattering coefficient, since it also depends on pigment concentration [17, 27, 31, 34].

The K-M equation is quite helpful since its results, set in Fig. 12.4, tell that reflectance will decrease at increasing K/S, for example if we add a strongly absorbing pigment such as carbon black to the system. On the other hand, reflectance increases if we increase S/K, e.g. by addition of a strongly scattering white pigment.

As said above, colored pigments are most often used in combinations to provide for the required color and hiding power. Duncan has demonstrated that the contributions of scattering and absorption by individual constituents in a mixture can be added [10, 15, 17, 18, 34]:

$$\left(\frac{K_{mix}}{S_{mix}}\right)_\lambda = \left(\frac{c_1K_1 + c_2K_2 + c_3K_3 + \ldots.}{c_1S_1 + c_2S_2 + c_3S_3 + \ldots.}\right)_\lambda \tag{12.3}$$

For color design, this equation is often used to calculate the composition of mixtures in relation to the desired color. For mixtures with white pigment, it can be

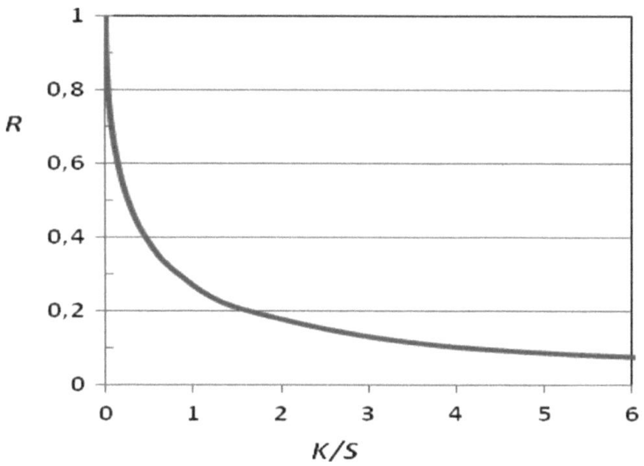

Fig. 12.4 Relationship between R and K/S

used in relation to the best opacity of the colored mixture. For such paints, which have a dominant, white scattering constituent, the equation is often further simplified (single-constant simplification) [4, 5]:

$$\left(\frac{K_{mix}}{S_{mix}}\right)_\lambda = \frac{c_1}{c_w}\left(\frac{K_1}{S_1}\right)_\lambda + \frac{c_2}{c_w}\left(\frac{K_2}{S_2}\right)_\lambda + \ldots\ldots + \left(\frac{K_w}{S_w}\right)_\lambda \tag{12.4}$$

where:

K_i = absorption coefficient of indicated constituent i (subscript)
S_i = scattering coefficient of indicated constituent i (subscript)
c_1, c_2, etc. = fractional concentration of indicated constituent
w = subscript for dominant, white scattering constituent
mix = pigment mixture

If one combines the colorant constituents (subscripts 1, 2, etc.) to a tinting constituent (subscript '*tint*'), then Eq. 12.4 further simplifies to:

$$\left(\frac{K_{tint}}{S_{tint}}\right)_\lambda = \frac{(K_{mix}/S_{mix})_\lambda - (K_w/S_w)_\lambda}{c_{tint}/c_w} \tag{12.5}$$

Now, only two samples are required to characterize the $(K/S)_\lambda$ of the tinting constituent, viz. the white constituent and a single mixture at an arbitrary concentration [4, 5].

Rutile TiO_2 is a white pigment that has highest maximum scattering power (about 5). This relates to its high refractive index (RI), which is much higher than that of older white pigments like ZnO (see Fig. 12.5).

The theoretically expected dependency for rutile scattering power on particle size and wavelength for blue, green and red light agrees well with measured values

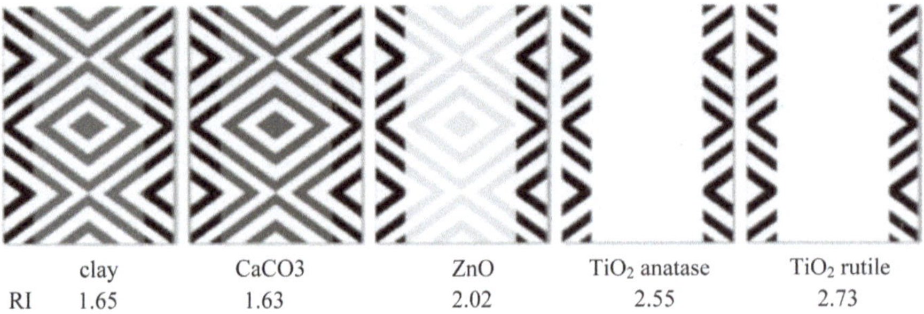

Fig. 12.5 Opacity depends on RI [11]

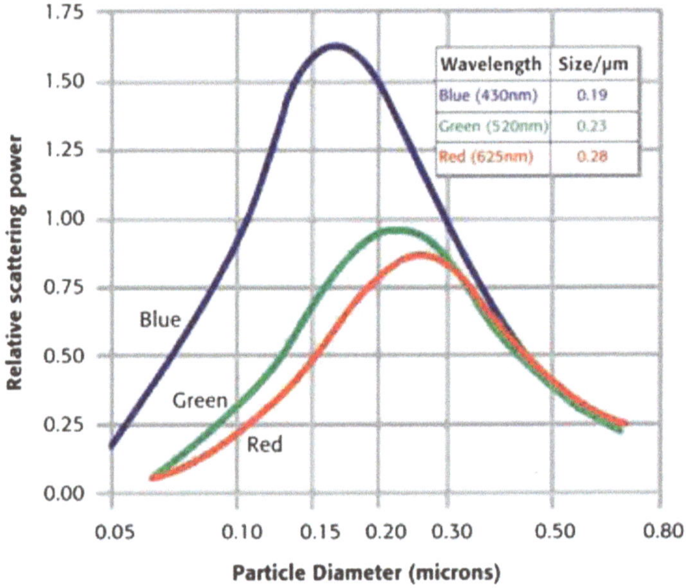

Fig. 12.6 Wavelength and particle size dependency of light scattering; (Predictions by Eq. 12.1 are given in insert; courtesy Malvern Instruments Ltd.)

(see Fig. 12.6). It clearly shows that highest scattering power is reached at particle sizes around 200 nm, and that blue light (smaller wavelength) is scattered more efficiently by smaller particles, green and red light (larger wavelength) by larger particles. Note that measured curves show less fluctuations with particle size than Mie calculations, due to the small particle size, presence of a size distribution, roughness of the particle's surface and/or non-spherical particle shapes in the TiO_2 samples. Note also that changes of the PSD may cause subtle undertone effects on the resulting colors [6]. Moreover, TiO_2 may act as an UV-activated oxidation catalyst for organic substances such as binder and organic pigments. This capability may promote weathering of the paint film, if not counteracted by coverage of the surface with another material (e.g. SiO_2).

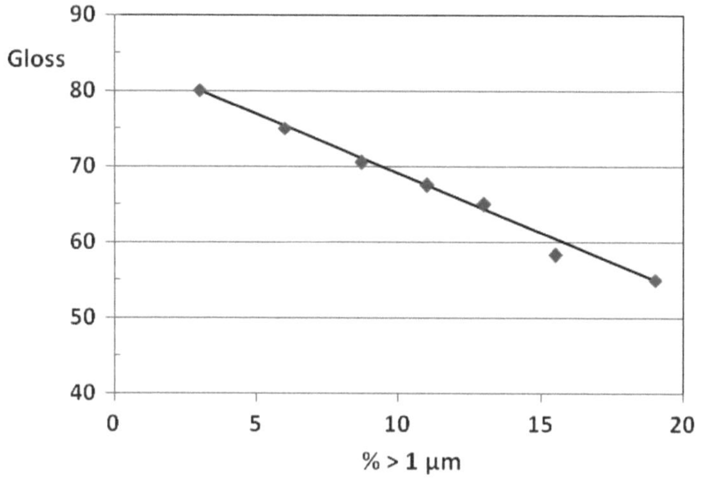

Fig. 12.7 Gloss versus amount of particles > 1 µm (Courtesy Horiba Instruments Ltd.)

Since visible light has wavelengths between about 380–780 nm, optimum particle size for pigments ranges from about 150–400 nm. If particles are smaller, scattering power decreases, below about 100 nm even in relation to the sixth power of size and, finally, they become transparent (Rayleigh regime). Above the optimum size, the decrease rate is smaller and depends upon the relative refractive index. Above a particle size of about 10–50 µm, only diffraction plays a role and RI is no longer of influence (Fraunhofer regime). Then, the scattering power of particles is 2. Note that secondary effects may come from particle shape. Especially plate-like particles may cause special effects. Moreover, rough particle surfaces may cause colors to become somewhat 'dirty'. The method of particulate production plays an essential role in the PSD. If the pigment or extender particles are synthesized by chemical reactions, like TiO_2, or produced by crystallization, like many organic compounds, then fairly narrow PSD's can be designed that meet the optimum size range reasonably. However, if they are produced by some kind of comminution of large particles (e.g. ores), then usually wide PSD's result. Then, some kind of classification is needed to reach an optimum PSD, with an acceptable maximum size and an acceptable amount of fines. Maximum particle size depends on the thickness of the ultimate paint layer on a surface as well as on the desired gloss and structural effects of the paint layer. Often, the dry paint film has a thickness in the range of 10–35 µm. Thus, if one does not want particles to stick out of the film layer, which decreases gloss, then this sets a maximum particle size. Figure 12.7 gives an example for the relationship between gloss and the amount of super-micrometre particles in a given pigment. Note that agglomerates resulting from poor pigment dispersion or aggregates have the same detrimental effect as large particles on gloss as well as on other paint performance aspects, since they behave more or less like single particles. Thus, adequate de-agglomeration during pigment dispersion is a must for optimum performance of paint.

Besides this maximum size, the most relevant particle size parameter in relation to both scattering behavior and costs is probably the volume-equivalent size of particles where the mean volume related size ($D_{4,3}$) is appropriate. A minimized amount of fines (smaller than about 200 nm) is required to achieve best scattering behavior (see above) at optimum costs. Note that caution is necessary in case of platelet particles, where the particle cross section is the most relevant size aspect in the paint film but costs still relate to volume related size.

In relation to optimum optical properties, most pigments have a PSD of their primary particles in the range of about 150–500 nm. Synthesized particles by gas-phase reactions or precipitation may have a much smaller primary particle size. For example, carbon black and TiO_2 particles may have a primary size starting at about 10 nm. However, such very small particles usually agglomerate or aggregate very easily. Agglomeration and aggregation as well as comminution processes for pigment preparation extend the upper particle size to much larger values, up to about 5 μm or even 50 μm. Note that the size of large clusters may deteriorate the performance of paints. The clusters may contain millions of primary particles, which make the dispersion/milling step essential for preparation of good paints. For, changing particle size and shape may influence hue as well as shade and tinting strength of pigments. Extenders and fillers often have slightly larger sizes than pigments [1, 11, 28, 33]. The upper size depends on application, a.o. gloss/matt requirements [15].

PSD results should be interpreted with due care. One reason is that results coming from different techniques may show large differences for non-spherical particles. Another reason for care is that large clusters may occur in dry powders, which are sometimes very difficult, if not impossible, to disperse into primary particles. Only image analysis techniques using images from optical microscopy, SEM or TEM are capable to measure the primary size through application of segmentation methods for overlapping particles. Most other PSD techniques, however, cannot distinguish between primary particles and clusters as they interpret the clusters as particles. Thus, the measurement results depend upon the dispersion method applied. Sonication and surfactants are often helpful to promote dispersion but they do not always lead to complete de-agglomeration or de-aggregation. In that case, these techniques will measure some combination of primary size and cluster size, which also depends on the applied sample concentration for measurement. This holds especially when the primary particle size is very small and complete de-agglomeration is virtually impossible, e.g. when primary particles are as small as a few nanometers. Moreover, different techniques often yield different PSD results when particle shape deviates from the assumed spheres (see also Sect. 12.3.5) [25]. Since both paint performance and PSD measurement depend upon the actual state of pigment dispersion, it is essential that this state is identical in both cases.

Close packing of the pigment particles in the paint film is essential for reaching optimum results for homogeneous colors and optimum hiding power at minimum costs. Densest packing can only be reached in size distributions where smaller particles can fill the voids left by the larger particles (see Chap. 2).

The main role of extenders and fillers is cost reduction by replacing expensive polymer binder and pigment by cheaper material. They play their role as particles in the rheological behavior of paint mixtures, influence the mechanical properties of the paint film and may have some effects on gloss reduction and paint structure.

12.3.2 Rheology

The rheological behavior of paints and especially its composition play an important role in several aspects, viz. [23, 28, 29, 30, 39] (see also Sect. 2.5):

- During mixing/milling. In this process, pigments and extenders should be mixed well with binder, solvent and additives as well as being de-agglomerated to an acceptable level. Often, two stages are applied. First, a 'mill base' is composed of solids and part of the liquid dispersion medium (optimum solids concentration), followed by blending of this base with the rest of the dispersion medium. Medium to strong shear forces, caused by pumping the viscous mixture through stirred beads or rolling balls or in between roller blades assist in the first de-agglomeration stage. Optimum viscosity of these mixtures and, thus, optimum particulate concentration is essential to reach good quality at minimum energy, time and costs. Note that the viscosity changes during this process due to de-agglomeration.
- To assist the shelf life time during storage. Modern paints are typically stabilized in some way through imparting electric and/or steric repulsion forces on the particles, which maximizes inter-particle distance forming some kind of 'structure' (see Sect. 12.3.3). Also the consistency of the liquid phase in the absence of shear helps, through slowing down Brownian movement of the particles.
- In the transfer stage from container via applicator to a (thin) film on the surface of an object. Several types of application systems are used, viz. brushes, hand-held rollers, sprayers, dipping or curtain roller coating. Hand-held brushes and rollers require the pickup of sufficient paint during dipping, which relates to both the viscosity of the paint and the quality of the applicator. Sprayers require low viscosity at fairly strong shear with relation to the pressure drop. Dipping, etc. also requires low viscosity to limit the thickness of the paint film. Each system has its own applicability potential and limitations, and market share.
- Finally, three aspects can be distinguished in the formation of a stable, strong and homogeneous film after leveling without sagging. The first aspect is good adhesion to the substrate's surface (wood, metal, concrete or polymer). This is mainly determined by the surface tensions of liquid and substrate (see Sect. 12.3.3). The leveling aspects relate to both viscosity at no-shear conditions and the solids content of the paint (see Sect. 12.3.4). Thirdly, evaporation of solvent and cross-linking/polymerization of binder are helpful to stabilize the paint film within reasonable times.

Table 12.1 Relevant shear rates during paint processing

Process step	Shear rate, s^{-1}
Leveling	10^{-3}–10^{-2}
Mixing, stirring	10^{1}–10^{2}
Brushing, spraying	10^{3}–10^{4}
Ball milling	10^{3}–10^{5}
High speed roller	10^{5}

Typical solvents show a Newtonian rheological behavior, where the viscosity is a material constant, independent of shear rate. In contrast, most paints have a complex, non-Newtonian rheological behavior, due to application of polymers, high solids contents and thickeners (see Chap. 2). Non-Newtonian rheological behavior means that the viscosity depends upon the shear rate applied. This behavior can be used to advantage in relation to the different properties required in the different shear regimes of the various stages indicated above. In view of this, some optimum rheological behavior has to be found. This requires good knowledge and/or understanding of the relationship between the rheological parameters and paint behavior, which comprises to different aspects, viz. (during preparation) de-agglomeration and mixing of pigments, extenders and binder, and (during application) paint transfer, film formation and leveling of the film [12, 28, 29, 30, 37, 38]. Note that the shear rate in these different stages of preparation and application of a paint may vary over a very wide range, up till about 10^5 s^{-1} (Table 12.1) [28, 39]. Note also that – after application – the solvent evaporation rate influences the possibilities for leveling and sagging.

Solvent, diluents and polymeric binder determine the basic rheological properties of paint. They define and control whether the paint is very viscous or not and what kind of non-Newtonian behavior it shows. However, the solid particles consisting of both pigments and extenders also may significantly contribute to such behavior, especially at high solids concentration in non-aqueous systems. Here, electric and steric interactions between particles are very strong and, even dominant near closest packing (see also Chap. 2.5). For such paints the viscosity may not only depend on shear rate but also on time of shearing and relaxation. The typical viscosity behavior against shear rate is – after removal of relaxation and time-dependency – schematically illustrated in Fig. 12.8 [12, 19].

For paints, often four regions can be recognized, for which the exact structure depends on product composition:

I region of zero-shear, Newtonian viscosity (viscosity often typified by η_0)
II region of shear-thinning (more or less power law, i.e. linear in double logarithmic plot)
III second region of Newtonian behavior (viscosity often typified by η_∞)
IV region of shear-thickening.

As stated above, most modern paints have a non-Newtonian rheological behavior. Typically, a minimum shear rate is required to make them flow (the yield stress), after which they show shear-thinning behavior. For many coatings, the

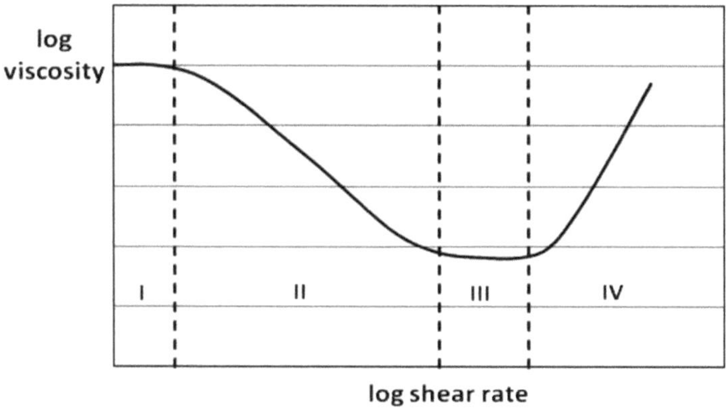

Fig. 12.8 Schematic viscosity – shear rate relationship (Relaxation and time-dependency removed)

Herschell-Bulkley equation fits the data for the shear-thinning region reasonably well (see also Chap. 8 on chocolate) [26, 35, 39]:

$$\tau = \tau_0 + \eta_{pl} \cdot (d\gamma/dt)^n \quad (12.6)$$

where:

τ = shear stress
τ_0 = yield stress
η_{pl} = apparent, plastic viscosity of this model
$d\gamma/dt$ = shear rate
n = flow viscosity index (related to shear thickening/thinning behavior).

For paint formulations where the particulate colloid concentration comes close to maximum packing, several equations relate the viscosity to the actual fraction of particulates. An equation that fits this situation well was formulated by Krieger-Dougherty [12, 19, 39]:

$$\eta_{rel} = \left(1 - \frac{\Phi_e}{\Phi_m}\right)^{-c\Phi_m} \quad (12.7)$$

where:

η_{rel} = relative viscosity
Φ_e = effective particulate fraction
Φ_m = particulate fraction of maximum packing
c = constant (intrinsic viscosity; often close to the Einstein constant of 2.5)

The effective fraction of particulates in dispersions is often larger than just the solids fraction. This is caused by adsorbed molecules (dispersant and liquid). It is represented in the hydrodynamic size, which includes adsorbed molecules within the layer of shear [25]. Particles in this size range may be estimated by the DLS technique.

A major difference between the rheological behavior of suspensions and emulsions is that the (liquid, mostly spherical) particles of the latter may deform at increased shear rates. At particulate concentrations smaller than about 50 % v/v, this effect is not visible and emulsions and suspensions show similar rheological equations [39]. However, at particulate concentrations larger than about 50 % v/v, it will result in relatively lower viscosities for emulsions than for suspensions, since this deformation leads to a larger packing density at maximum packing.

12.3.3 Adsorption and Dispersion of Pigments

Typically, pigments and extenders are produced by some company and then delivered as powders or granules to the paint manufacturer. In both forms, the primary particles are often contained in large clumps (mostly agglomerates, sometimes also aggregates), which originate in the filtration, drying and compaction stages of their production [7]. In agglomerates, the primary particles are held together by (medium) weak attractive forces between point contacts. These bonds can fairly easily be broken in liquid dispersions of particles having a primary size greater than about 1 μm. But decreasing particle size and the resulting stronger attractive forces make this increasingly difficult if not impossible. In aggregates, primary particles are typically in contact by their faces, often not only by physical forces but also by chemical bridges (formed during drying or sintering), which are much stronger. Therefore, breaking up aggregates is often impossible and aggregate formation should be avoided during production.

For paint production, the powders are mixed with the other paint components according to a given recipe and dispersed. This dispersion process is usually performed in roller, ball, pearl or bead mills. Therefore, it is often called milling, although it consists mainly of de-agglomeration of the clumps delivered. This is an essential step in the production process of paints, since it determines the ultimate performance quality of the paint dispersion, in relation to both homogeneity and degree of de-agglomeration. Another possibility is to disperse the pigments aided by surfactants in a little liquid (organic or water) to a paste (so-called 'mill-base'), which facilitates dosing as well as later mixing with other paint components.

Both surface area and degree of agglomeration of the solid particles are closely linked to the demand for binder, (formerly) expressed in the oil absorption value. Larger surface areas, i.e. smaller particles, have larger potential for adsorption as well as result in stronger attractive forces, thus promoting agglomeration if the dispersion is not properly stabilized. Moreover, the continuous phase present inside voids in clusters of primary particles, cannot contribute to the paints viscosity, which leads to a need for more liquid phase and, thus, lower solids content. The external surface area of particles inversely relates to the square power of particle size, their volume with the third power. Thus, both increase strongly with decreasing size, and vice versa. Note that the specific surface area, as determined by gas adsorption, always gives a nice impression of the primary particle size for

Table 12.2 Some typical mass and surface area values in relation to particle size

Particle size, nm	30	100	300	1,000	3,000
Volume, m^3	$1.4*10^{-23}$	$5.2*10^{-22}$	$1.4*10^{-20}$	$5.2*10^{-19}$	$1.4*10^{-17}$
Surface area, m^2	$2.8*10^{-15}$	$3.1*10^{-14}$	$2.8*10^{-13}$	$3.1*10^{-12}$	$2.8*10^{-11}$
Number of particles/g ($\rho_P = 2{,}000$ kg/m^3)	$35*10^{15}$	$9.6*10^{14}$	$35*10^{12}$	$9.6*10^{11}$	$35*10^{9}$
Number of particles/g ($\rho_P = 4{,}000$ kg/m^3)	$18*10^{15}$	$4.8*10^{14}$	$18*10^{12}$	$4.8*10^{11}$	$18*10^{9}$
Spec. surf. area, m^2/g ($\rho_P = 2{,}000$ kg/m^3)	100	30	10	3	1
Spec. surf. area, m^2/g ($\rho_P = 4{,}000$ kg/m^3)	50	15	5	1.5	0.5

non-porous particles, regardless of the degree of agglomeration, since the gas is capable of adsorbing to the surface of primary particles inside the agglomerates. For particle size distributions, the area weighted Sauter mean diameter $D_{3,2}$ is the relevant size parameter for relationships involving surface area. Note that the ratio of the specific surface area calculated from the PSD (which may include agglomerates) to that determined by gas adsorption (which mostly relates to the primary particles) gives an idea on the degree of de-agglomeration reached during dispersion. Some values for spheres showing typical relationships are given in Table 12.2.

The surface tension of both solid and liquid is another relevant factor for adsorption, wetting and dispersion (see also Chap. 3) [8, 25, 28, 30]. Wetting of the surface of solid particles is the first step in their dispersion. The driving force for wetting is the difference in surface tension between the solid-vapor interface (γ_{SV}) and the solid–liquid interface (γ_{SL}). The resisting force is the energy required for increasing the surface area of the liquid drop. The balance between these forces is represented by the equation of Young-Dupré:

$$\gamma_{SV} - \gamma_{SL} = \gamma_{LV} \cos \Theta \tag{12.8}$$

Adequate wetting is only possible if the contact angle Θ between solid and liquid is smaller than 90° ($\cos \Theta > 0$) and the driving force ($\gamma_{SV} - \gamma_{SL}$) approaches γ_{LV}. Compare, for example, water droplets on cars that are nicely waxed with the wet surface in absence of wax. Note, however, that adequate wetting of the particulate surface is not the only requirement for good dispersion/milling of clumps. In addition, the liquid should be able to easily penetrate into the pores of the clump of agglomerated primary particles [39].

Adequately wetted clumps require much less energy for de-agglomeration than dry clumps, since liquid reduces the attractive forces. Many pigments and extenders have an oxidic surface, which has a medium high surface tension, whereas binders and solvents usually have a fairly low surface tension. Then, wetting and dispersion is usually not problematic, except when dispersion in an organic liquid is the objective and the particles have an adsorbed water layer. In contrast, water has a very high surface tension, which makes wetting of surfaces having a lower surface tension very difficult. Often, surfactants that lower the surface tension of water, poly-electrolytes and/or steric stabilizers are used to promote dispersion of solids in aqueous liquids and to stabilize emulsions and suspensions [8, 25]. Here, stabilization results from adsorption of these components to the solid surfaces, which causes electric and/or steric repulsion between the particles. Thus, their required concentration relates to the specific surface area of the particles. Moreover, pigments may have undergone some surface treatment that allows for easier dispersion in liquids [32]. One way of treatment is by giving the particles an organic coating. Another way is through the addition of very small silica or alumina particles (typically up till about 1 %), which improves the flow of dry powders as well as their dispersion. These adsorb to the surface of larger particles and, thus, prevent direct contacts weakening the attractive forces between the particles. Note that their specific surface area may give a significant contribution to the total area.

Due to conflicting factors in the milling/dispersion process, the shortest dispersion time is often found at medium particulate concentrations and viscosity, where both agglomerated and primary particles have some space for free movement.

The ultimate behavior of pigment dispersions with time and the optimum concentration of dispersant aids can usually be seen from the sedimentation behavior. Slow sedimentation relates to optimum stability, very low sediment volumes to dense sediments (with fairly strong attractive forces between particles, which usually require considerable energy for re-dispersion) and relatively large sediment volumes to gelated or flocculated sediments (with relatively weak attractive forces between particles, which usually can easily be re-dispersed) [8].

12.3.4 Other Properties Related to Pigments and Extenders

The formation of good strong layers of paint depends upon both the close packing of the particles in the ultimate paint film (see Sect. 12.3.1 and Chap. 2) and the volumetric solids-binder ratio. The latter is typically expressed in pigment volume concentration (PVC). The pigment composition is important in aqueous systems at concentrations around about 20 % v/v, where adequate hiding power is required. Then, maximum hiding power can only be obtained by usage of TiO_2 without fillers [43]. At high PVC, fillers may be used in addition to pigment to reduce costs. A critical pigment volume concentration (CPVC) is defined, when the binder just fills the voids left by the pigment and filler particles [13, 28, 36]. It is often expressed as the oil absorption value of the powder, or derived from its bulk density. At this CPVC, films can have minimum thickness due to the densest particle packing. Use of less binder (pigment concentration larger than CPVC) causes the film to become porous and brittle. Too much binder will cause open spaces between particles unless compensated for by thicker films. Note that the word pigment here often includes extender. Note further that the CPVC is also influenced by the potential presence of agglomerates. The presence of agglomerates will increase the effective particle size as well as the amount of voids that will take up binder.

The stability of paint suspensions is another important item. Paints are often stored for long times (often more than a year), after which it should require only some stirring or shaking to make them ready for use. Two mechanisms are applied for stabilization, viz. electric and steric stabilization (see also Sect. 12.3.3). The zeta-potential of the particles is the essential characteristic in electric stabilization [25]. Both mechanisms prevent the particles from coming into close contact, while often some kind of 'structure' is built up. As a result, not only the paint dispersion is stabilized and has a yield stress value (see Sect. 12.3.2), but also they may form a loose, weak (flocculated) structure upon sedimentation, which can be broken up easily.

12.3.5 Water-Based Paint Dispersions

During application, the evaporating solvents of conventional organic solvent-based paints have negative effects on human health and the environment. Organic solvents are also a fire hazard. This lead – near the end of the twentieth century – to more strict rules for application of organic-based paints and to the development of alternatives. One alternative is the application of powder coatings (see Chap. 13). The other alternative is replacement of the organic solvents by water. This, however, was far from a simple job and only possible after extensive studies of all aspects of paint composition. Basically, water-based paints are stable mixtures of colloidal polymer particles and pigments in water. The polymer particles result from emulsion polymerization and generally have a particle size of about 100–300 nm; their concentration in latex paints is typically about 15–30 % w/w. These colloidal dispersions are only stable if sufficiently stabilized by electric and/or steric stabilization (see Sect. 12.3.3). Often organic thickeners are added to fulfill rheological requirements. At this moment, water-based paints have taken a major share of the household market and gain importance in industrial applications.

The required properties for the polymer particles mainly relate to film formation when a paint layer is brought upon a substrate. Three stages can be discriminated during film formation. They occur sequentially in local regions of the paint layer but simultaneously over the whole film. In the first stage, water evaporates from the dispersion layer resulting in a close packing of latex particles. In the next stage, water evaporates from the formed voids, which causes the voids to disappear through deformation of the packed, spherical particles, provided that these are sufficiently flexible. The strong attractive forces between the closely packed particles ideally lead to a perfect (honeycomb-type) packing. In the final stage, polymer molecules diffuse from one particle into another through the touching surfaces of the particles, and their coalescence starts. A requirement for adequate rates of deformation, diffusion and coalescence is that of the application temperature. This needs to be above the minimum film formation temperature (MFFT) of the polymer (which is typically close to its glass-transition temperature), or, in other words, that the polymer is soft enough to allow deformation and diffusion of polymer molecules [45, 47]. A sufficiently low MFFT is reached through adequate setting of the molecular weight of the polymer, through application of copolymers (polymerized mixtures of different monomers) or through addition of plasticizers to the polymer, or by increasing the application temperature. When the application temperature is below the MFFT, then brittle films result from insufficient possibilities for deformation, diffusion and/or coalescence.

The size of the voids in between the closely packed particles is also important in relation to deformation rate. For particles of about the same size, the void size is at maximum about the size of the particles. Thus, particle size (distribution) may play a role here as well. Firstly, larger voids take longer times to be filled up by particle deformation than smaller ones. And secondly, in wider PSD's small particles can partly fill the voids left by the larger particles, thus decreasing void volume. The specific surface area may be regarded as the relevant aspect for deformation (surface tension), inter-particle diffusion (contact area) and coalescence rate [45]. This relates more or less to the Sauter mean diameter $D_{3,2}$ of the particles. Note that the surface area increases during deformation, since spheres have the smallest surface area for a given volume. Also, the contact area between adjacent particles is strongly related to the degree of their deformation.

The progress of this film formation process can, for simple latex-water mixtures, be followed by transmission spectroscopy [41, 42]. Here, the changing wavelengths of characteristic absorption peaks correlate with the changing distance between the colloid particles, whereas their disappearance indicates coalescence to a homogeneous film. SEM and wide and small-angle neutron scattering have also been used for studying the stages in film formation [47].

12.4 Particle Size Measurement of Paint Dispersions

Adequate determination of PSD's require six basic elements, viz. stated quantitative analysis criteria, representative sampling, good dispersion/dilution of the measurement sample, well-trained analysts, good instrumentation and good interpretation of measurement results [25].

The *analysis criteria* relate to stated parameters of the PSD, particle shape and surface area, which in turn are related to stated aspects of the product performance during production and for the ultimate product, in a quantitative manner. This means that product performance characteristics have been translated into measurable parameters with a stated precision (repeatability and reproducibility) and resolution (see also Chap. 1). Note that size distributions and chosen parameters may be dependent on the measurement technique.

Good *sampling* requires a measurement sample that is representative for the total batch of product and contains sufficient particles for measuring the most critical parameter (e.g. largest size or few deviating particle shapes) at the desired precision.

For representative sampling of dry powders, it is usually necessary to take samples at different spots in heaps and/or at different times during product transport, to mix them carefully and, then, subdivide the collected sample in an adequate laboratory riffler to the measurement sample. For representative sampling of liquid dispersions, it is usually sufficient to mix them well and take the sample from the mixture while preventing sedimentation or other types of systematic product changes in the sampler (see further Chap. 3).

Good *dispersion/dilution* requires that the measured state of the particles agrees with the goals for measurement. Here, the dry powder dispersion must be distinguished from the dilution of paint dispersions. In the dispersion of dry powders, the emphasis most often lies in the PSD of the primary particles, although a first goal may be to find optimum dispersion conditions. For determination of the PSD of primary particles, an adequate liquid, surfactant type and concentration, as well as dispersion energy are required. Also, some criterion is necessary that the final goal is reached, for example through a microscopic examination of the particles after dispersion. For dispersions of prepared paints or pigment pastes, the goal is often to characterize the degree of agglomeration. Adequate care must be taken to ensure that no change of state of the dispersion takes place during the measurement procedure (see further Chap. 3).

Analysts who perform the measurements should be well trained. It means that they have a proven capability to perform measurements at the required quality, including laboratory sampling and dispersion/dilution, according to given procedures. Furthermore, they should be capable to report any deviation from the normal situation while doing their job. Their qualification should be regularly tested and approved. Testing of analysts can be done with in-house products that have PSD parameters of known quality. Their quality being proven, they can qualify instruments with standard reference materials or own products. If the procedures have been fully or partly robotized, then the equipment should be tested for repeatability and cleanliness, so that subsequent samples do not interfere.

Measurement techniques should be chosen that fulfill stated quality requirements in tests. The first quality requirement is to obtain sufficient repeatability and reproducibility (precision), but also stability, accuracy, resolution and sensitivity for measurement. Adequate product characterization means adequate discrimination between good and poor products. For size measurement of paint pigments and extenders, the Hegman gauge, microscopic techniques, laser diffraction, (centrifugal) sedimentation and dynamic light scattering are favored; in some industries, also ultrasound attenuation is applied (see further Sect. 12.4.1. For more techniques and more detailed information see Chap. 3 and [25, 43]). The different techniques have different capabilities and weak points. Thus, they can be applied in conjunction to yield additional information. Note that mixtures of different particles yield a combined PSD result, in which differences in density or refractive index may play a role as well. For surface area measurement of dry powders, gas adsorption using the BET theory is the best method.

Good *reporting of measurement results* means that reporting is done in accordance with given goals and instructions. Often, the basic, required parameters are given in a short report, eventually accompanied by signaled deviations from normal, while full data are stored in some memory. The same memory should contain data on instrument testing. All data should be accessible with authorization to both measurement personnel and users of the data.

12.4.1 Particle Size and Surface Area Measurement Techniques for Pigments and Paint

The Hegman gauge is an old, very simple device, which is mainly used to detect few large particles or clumps in paint pastes. It can also be applied to similar products, such as ink, peanut butter, chocolate paste or pharmaceutical pastes. It uses a steel body, containing a groove of linearly decreasing depth. A paste sample is brought up at the deep end of the groove and smeared with a specified scraper to the other end. As soon as particles have larger dimensions than the depth of the groove, they will show their presence by leaving speckles. The point where a specified number of speckles appears is read on the scale by the operator. A major advantage is that it allows high solids concentrations. A drawback is that results depend upon analyst capabilities.

The sieve residue test is also old. Typically, it is done with a 20, 32, 38 or 45 μm sieve, by means of wet or dry sieving. Note, however, that this test is much less sensitive than the Hegman gauge, since much larger numbers of large particles are required to yield a measurable residue mass (see Table 12.3). Moreover, the obtained residue results may differ from the amount of clusters present in the ultimate paint since dry sieving allows no de-agglomeration as in liquid suspensions and the solids concentration is often smaller in the wet-sieving test than in paint.

Microscopic techniques work with light in the visual range (optical microscopy) or with electron beams (electron microscopy), depending on size. In both systems, particle size is derived from the cross-sectional area of magnified particles. From the counting of particles of different size (typically related to cross-sectional area), a number-based PSD is obtained. Early determinations involved inspections of relatively few particles in view of tedious manual procedures. Nowadays, computers allow automation and, thus, analysis of tens of thousands of particles, for size as well as shape (by so-called Image Analysis). Low particle concentration is required to avoid statistical overlapping of particles in the image. Thus, many different fields of view have to be analyzed, which requires automated change of frames. Most often, the measurements are performed at a single magnification, which limits the measurement range to about a factor 30 (ratio of maximum to minimum size). The lower size limit of optical microscopy depends on magnification and is at minimum about 0.5 μm (while applying a 100 times objective with oil immersion; at this point, particles can only be seen but their size cannot be measured). The maximum magnification for electron microscopes is about 100,000 for scanning systems (SEM).

Table 12.3 Some typical particle mass and number data in relation to size

Particle size, μm	20	32	38	45
Volume, m^3	$4.2 \cdot 10^{-15}$	$1.7 \cdot 10^{-14}$	$2.9 \cdot 10^{-14}$	$4.8 \cdot 10^{-14}$
Mass, mg ($\rho_P = 2{,}000$ kg/m^3)	$8.4 \cdot 10^{-6}$	$3.4 \cdot 10^{-5}$	$5.7 \cdot 10^{-5}$	$9.5 \cdot 10^{-5}$
Number of particles/mg	119,000	29,000	17,000	10,000
Mass, mg ($\rho_P = 4{,}000$ kg/m^3)	$1.7 \cdot 10^{-5}$	$6.9 \cdot 10^{-5}$	$1.1 \cdot 10^{-4}$	$1.9 \cdot 10^{-4}$
Number of particles/mg	60,000	15,000	9,000	5,000

Laser Diffraction (LD; also named static light scattering) has become a popular technique, as it allows easy application at medium size class resolution. It yields volume-based sizing results and is based on a theoretical model to translate the measured angular light intensity pattern into a PSD. For this translation, a de-convolution procedure is applied to reach the PSD of spherical particles that yields the best fitting scattering pattern. The measurement typically takes place in dry state[2] or in suspensions after adequate particle dispersion and dilution to medium particulate concentration, and at stated refractive index values for both particles and dispersion medium. For de-convolution, two optical models are used, viz. the Fraunhofer model and the Mie model. The former is more simple as it only regards diffraction of light at the particles contours, the latter also takes refraction into account and, thus, requires refractive index (RI) parameters for both particles and dispersion medium. The Mie model is essential for particles larger than about 100 nm but smaller than about 2–50 μm; transparent particles require the lower size limit of this range, fully opaque particles allow the larger limit [16, 25]. Both models yield the same results above this upper limit. If the laser wavelength of an instrument lies within the strong absorption bands of a pigment, then it is not appropriate for size measurement. Another drawback is that instruments of different manufacture have different set-ups and use different software for de-convolution and, thus, often lead to different PSD results.

Sedimentation techniques can be distinguished in gravity and centrifugal systems. Instruments differ first in the force field that is applied for sedimentation, viz. gravity or a centrifugal field. Typical lower size of gravity sedimentation is about 0.5 μm, depending on particle density, which limits measurements mostly to some residue analysis. Centrifugal techniques allow a much lower minimum size, viz. about 20 nm depending on centrifuge speed and particle density. A second difference is that either a line start or a homogenous start is applied. And a third difference is the way of detection of particle amount after size separation. All techniques require low solids concentrations, a known liquid viscosity, a constant temperature and absence of sources of vibration. The terminal settling velocity of spherical particles in suspensions due to the force field is the basis for this technique.

[2] Note that adequate dispersion of dry powders containing sub-micrometre particles may be very difficult.

Viscous-flow conditions and dilute liquid dispersions are assumed (i.e. Reynolds number for settling particles < 0.25) and Stokes' law is applied to convert this velocity to particle size. Effective particle density, which includes the effects coming from gas or liquid present within potential closed and open pores of the particles and thus may have a smaller value than the true solid density, must be known for this conversion (in addition to liquid viscosity and liquid density). Absorption of X-rays or visible light is most often used to obtain remaining particle mass concentration after settling against time. Note that absorption of X-rays relate to mass concentration directly, provided that the atomic number of atoms is greater than about 20, whereas the relation between light absorption and mass concentration depends on size and refractive index of the particles as well.

In Dynamic Light Scattering (DLS; or Photon Correlation Spectroscopy, PCS), a collective of particles, dispersed in a transparent liquid, scatters light from a laser, similar to laser diffraction. But now, the variation of the scattered light intensity with time at some defined angle is measured[3]. The rate of change of this intensity is related to the diffusion coefficient of the particles by Brownian motion, which in turn is related to hydrodynamic particle size by the Stokes-Einstein equation. Several mathematical methods are used for conversion of the intensity-time relationship to a PSD. The fact that this conversion is ill-conditioned and has a poor signal-to-noise ratio limits the amount of PSD information that can be obtained. Usually, the polynomial 'cumulants' method is favored for data analysis. It leads only to an intensity weighted mean size and a value for PSD width. Other methods claim to lead to a PSD, but often results for broad PSD's are not stable. There are two versions for measurement. The conventional technique operates at very low particulate concentration, usually at an angle of 90° or another specified angle. New techniques, such as fiber-optics quasi-elastic light scattering (FOQELS) and diffusive wave spectroscopy, use back-scattered light and may operate at higher concentration. Then, particle-particle interactions often influence the particle movement and, thus, the sizing result. Still, however, a correlation with product quality (e.g. built 'structures') may be found provided that all other conditions including the particulate concentration are kept constant [46].

The technique can also be used for measurement of the zeta-potential (see Chap. 3).

Ultrasound attenuation shows similarities to laser diffraction in that it uses a source of radiation (in this case ultrasound), but now one detector and different frequencies are used instead of detectors at different scattering angles and one wavelength. Here too, de-convolution of detector signals to PSD's is necessary, on the basis of a theoretical or an empirical model. A drawback of the theoretical model is that it requires, especially at increased concentrations, many physical properties of the material system in addition to approximations for particle-particle interactions and overlap of acoustical fields of different particles (multiple

[3] A fairly new instrument measures Brownian motion of suspended particles directly, through microscopic inspection after illumination by a focused laser beam (particle tracking technique).

scattering). Empirical models are limited for similar reasons: they are specific for given products and, in view of non-linearity, valid only for a limited dynamic range of concentrations and other system variables. A major advantage of the technique is that it can be applied in concentrated dispersions (see Chap. 3 and [9, 25]).

Measurement of specific surface area is mostly done by gas adsorption techniques while the (evacuated) sample is at liquid nitrogen temperature. Typically nitrogen or argon is used as the adsorptive at stepwise increasing pressures; the amount adsorbed is calculated from the dosed amount and the equilibrium pressure after adsorption is completed [25]. Thus, an adsorption isotherm is created. The surface area is then calculated through the BET equation at the point where a complete monolayer is formed by using the number and cross sectional area of adsorbate molecules present in this monolayer.

12.5 Definitions, Abbreviations and Symbols

Agglomerate	Cluster of primary particles with (medium) loose bonds (point contacts; van der Waals type attractive forces)
Aggregate	Cluster of primary particles with strong bonds (physical attractive forces between particle faces, often accompanied by chemical bonds)
Antifoulant	Substance that inhibits growth of organisms on ship bottoms
Anti-skinning agent	Substance that helps to prevent the formation of a surface skin in paint
Chroma	Level of 'saturation' or intensity and richness of a color
Clump	Mostly dry cluster of primary particles (agglomerates and/or aggregates)
Cluster	Generic name for the total group of agglomerated, aggregated and flocculated particles
Color	Visual appearance of an object, for paints usually described in terms of hue, saturation and lightness
Color index	Standard way of identifying the essential chemical structure of pigments and dyes (by type, color and constitution number)
Critical PVC	Pigment volume fraction or concentration at which the binder just fills the voids left by the pigment
Effective density	Mass of particles divided by the volume of liquid they displace at optimum conditions, where the effects coming from gas or liquid present within potential closed and open pores of the particles, which decrease their true solid density, are taken into consideration
Floc	Cluster of primary particles with very loose bonds
Gloss	Ability of a surface to evenly reflect light at all angles, without any blurring
Hiding power	Ability of a paint film to complete mask any previous color or pattern on a surface (expressed as the number of square meters covered per kilogram of pigment, after it has been dispersed in paint and applied)
Hue	Pure color of an object, formed by a combination of the main wavelengths transmitted and the overtones of the other wavelengths (according to the CIE system the human color perception is expressed in the two parameters x and y)

(continued)

(continued)

Lightfastness	Stability of the color of a pigment when exposed to light (UV radiation may cause decomposition of molecules and, thus, change the ability of pigments to absorb light in the visible region of the spectrum)
Lightness	Variable that distinguishes black from white, without regard to hue, represented by a neutral gray scale ranging from black to white
Metamerism	Situation where two or more paint samples match at one set of viewing conditions, but mismatch when the conditions are changed
Oil absorption value	Weight of linseed oil in grams required to produce a smooth paste with 100 g of pigment
Opacity	Ability of a product such as paint pigments to scatter and absorb visible light
Pigment vol. conc.	Volume fraction or percentage of pigment in the dry paint film
Primary particle	Basic particle, which can not be separated unless by breakage
Refractive index	Optical property of a material, consisting of a real part (governing refraction at an interface with another medium) and an imaginary part (governing light absorption; typically the symbol 'i' together with a minus sign is added to this part)
Sagging	Excessive flow on a vertical surface resulting in drips and other imperfections of a paint layer
Saturation	Color purity or its degree of departure from grayness
Scattering power	Ratio of scattering particle size to actual size
Shade	Degree of darkness of a color (also called tone or undertone; mixture of a color with black)
Sheen	Gloss or flatness of a paint film when viewed at a small angle
Surfactant	Surface active agent, consisting of a polar part (head) and apolar tail, which improve contacts between substances and liquids by wetting and dispersion
Texture	Non-uniform surface of a paint film
Tinctorial strength	See tinting strength
Tint	Degree of lightness of a color (mixture of a color with white)
Tinting strength	Ability of a pigment to produce the same color shade in a mixture with white pigment, relative to a standard pigment of similar hue (when the standard requires less pigment, then the tinting strength of the alternative pigment is said to be lower)
Tone	Degree of darkness of a color (mixture of a color with gray; see also shade)
Undertone	See shade
Weathering	Degradation of a paint film or other material surface by exposure to nature, viz. sunlight, rain, frost, dust and wind.
ASTM	American Society for Testing and Materials
CI	Color index
CIE	Commission Internationale de l'Eclairage (International Commission on Illumination)
CPVC	Critical pigment volume concentration
DLS	Dynamic light scattering
ISO	International Standards Organisation
LD	Laser diffraction
MFFT	Minimum film formation temperature
PCS	Photon correlation spectroscopy

(continued)

(continued)

PSD	Particle size distribution
PVC	Pigment volume concentration
RI	Refractive index
SANS	Small-angle neutron scattering
SEM	Scanning electron microscopy
TEM	Transmission electron microscopy
$D_{3,2}$	Sauter mean diameter, area-weighted mean size, mean value of an area-based PSD
$D_{4,3}$	Volume-weighted mean size, mean value of a volume-based PSD
x	CIE parameter (1) for mapping of human color perception
y	CIE parameter (2) for mapping of human color perception
γ_{LV}	Surface tension of liquid
γ_{SL}	Surface tension of solid–liquid interface
γ_{SV}	Surface tension of solid-vapor interface
η	Viscosity
η_∞	Viscosity in second region of Newtonian behavior (see Fig. 13.7)
Θ	Contact angle Θ between solid and liquid
Φ_e	Effective fraction of particulate material
Φ_m	Fraction of particulate material at maximum packing
ρ_P	Density of particles

References

1. Abel, A.G.: Pigments for paint (Ch. 3), in [23]
2. ASTM Paint Testing Manual, ASTM International (1995)
3. Bentley, J., Turner, G.P.A.: Introduction to Paint Technology and Principles of Paint Technology, 4th edn. Chapman & Hall, London (1998)
4. Berns, R.S., Mohammadi, M.: Single-constant simplification of Kubelka-Munk turbid-media theory for paint systems - a review, Color Res. Appl. **32**, 201–207 (2007)
5. Billmayer, F.W., Adams, R.L.: Predicting reflectance and color of paint films by Kubelka-Munk analysis; I Turbid-medium theory, J. Paint Technol. **45**, 23–30 (1973)
6. Braun, J.H.: White Pigments; in [2]
7. Carr, W.: Pigment Powders and Their Applications (Ch. III-A-a), in [29]
8. Doroszkowski, A.: The Physical Chemistry of Dispersion (Ch. 6), in [23]
9. Dukhin, A.S., Goetz, P.J.: Ultrasound for Characterizing Colloids. Elsevier (2002)
10. Duncan, D.R.: The identification and estimation of pigments in pigmented compositions by reflectance spectrophotometry. J. Oil Color Chem. Assoc. **45**, 300–324 (1962)
11. DuPont, Titanium Dioxide for Coatings; CO_B_H_65969
12. Eley, R.R.: Rheology and Viscosimetry, in [2]
13. Graystone, J.A.: Coatings for Buildings (Ch. 9), in [23]
14. Hammond III, H.K., Kigle-Boeckler, G.: Gloss, in [2]
15. Howell, D.M.: In: Sanders, J.D. (ed.) The Technology, Formulation and Application of Powder Coatings. Surface Coatings Technology Series, vol. 1. Wiley, Chichester/New York (2000)
16. ISO 13320, Particle Size Analysis – Laser Diffraction Methods; International Standards Organisation
17. Johnston, R.M.: Color Theory (Ch. III-D-b), in [29]
18. Judd, D.B., Wyszecki, G.: Color in Business, Science and Industry. Wiley, New York (1975)
19. Krieger, I.M., Dougherty, T.J.: A mechanism for non-Newtonian flow in suspensions of rigid spheres. Trans. Soc. Rheol. **3**, 137–152 (1959)

20. Kubelka, P., Munk, F.: Ein Beitrag zur Optik der Farb Anstriche. Zeitschr. f. techn. Physik **12**, 593–601 (1931)
21. Kubelka, P.: New contributions to the optics of intensely light-scattering materials; Part I. J. Opt Soc. Am. **38**, 448–457 (1948)
22. Lambourne, R. (ed.): Paint and Surface Coatings – Theory and Practice, 1st edn. Wiley (1987)
23. Lambourne, R., Strivens, T.A. (eds.): Paint and Surface Coatings – Theory and Practice, 2nd edn. Woodhead Publishing (1999)
24. Levinson, R., Berdahl, P., Akbari, H.: Solar spectral optical properties of pigments - Part I: Model for deriving scattering and absorption coefficients from transmittance and reflectance measurements. Sol. Energy Mat. Sol. Cells **89**, 319–349 (2005)
25. Merkus, H.G.: Particle Size Measurements; Fundamentals, Practice, Quality. Springer, Dordrecht (2009)
26. Mohos, F.A.: Confectionery and Chocolate Engineering: Principles and Applications. Wiley-Blackwell, Chichester (2010)
27. Mudgett, P.S., Richards, L.W.: Multiple scattering calculations for technology. Appl. Optics **10**, 1485–1502 (1971)
28. Patton, T.C.: Paint Flow and Pigment Dispersion, vol. 2. Wiley, New York (1979)
29. Patton, T.C. (ed.).: Pigment Handbook Vol. III, Characterization and Physical Relationships. Wiley (1973).
30. Patton, T.C.: Paint Rheology and Pigment Dispersion (Ch. III-E-a), in [29]
31. Phillips, D.G., Billmeyer Jr., F.W.: Predicting reflectance and color of paint films IV; Kubelka-Munk scattering coefficients. J. Coating. Technol. **48**(616), 30–36 (1976)
32. Ritter, H.S.: Surface Properties of TiO_2 Pigments (Ch. III-B-d), in [29]
33. Rolinson, J.F.: Pigments for Paint (Ch. 3), in [22]
34. Schabbach, L.M., Bondioli, F., Ferrari, A.M., Petter, C.O., Fredel, M.C.: Colour in ceramic glazes: efficiency of the Kubelka-Munk model in glazes with a black pigment and opacifier. J. Eur. Ceram. Soc. **29**, 2685–2690 (2009)
35. Sokmen, A., Gunes, G.: Influence of some bulk sweeteners on rheological properties of chocolate. LWT-Food Sci. Technol. **39**, 1053–1058 (2006)
36. Stieg, F.B.: Pigment/Binder Geometry (Ch. III-C), in [29]
37. Strivens, T.A.: Introduction to Rheology (Ch. 14), in [23]
38. Strivens, T.A.: The Rheology of Paints (Ch. 15), in [23]
39. Tadros, T.F.: Colloids in Paints. Colloid and Interface Science Series, vol. 6. Wiley, Weinheim (2010)
40. Talbert, R.: Paint Technology Handbook. CRC Press/Taylor & Francis Group (2008)
41. van Tent, A., te Nijenhuis, K., Coll, J.: Turbidity study of the process of film formation of polymer particles in drying thin films of acrylic latices. Interface Sci. **150**, 97–114 (1992)
42. van Tent, A., te Nijenhuis, K.: Turbidity study of the process of film formation of thin films of aqueous acrylic dispersions. Prog. Org. Coat. **20**, 459–470 (1992)
43. Tiarks, F., Frechen, T., Kirsch, S., Leuniger, J., Melan, M., Pfau, A., Richter, F., Schuler, B., Zhao, C.-L.: Formulation effects in the distribution of pigment particles in paint. Prog. Org. Coat. **48**, 140–152 (2003)
44. Weber, H.H.: Lichtstreuung und Teilchengröszenverteilung Kugelförmiger Teilchen. Kolloid Zeitschrift u. Zeitschrift f. Polymere **188**, 40–44 (1962)
45. Wicks Jr., Z.W., Jones, F.N., Pappas, S.P., Wicks, D.A.: Organic Coatings; Science and Technology. Wiley, Hoboken (2007)
46. Willemse, A.W.: Optical measuring techniques for particulate systems at the fringes of concentration, Ph.D. thesis TUDelft (1998)
47. Winnik, M.A.: Latex film formation. Curr. Opin. Colloid Interface **2**, 192–199 (1997)
48. www.touropia.com/prehistoric-cave-paintings. Many more pictures available via Google search at prehistoric paintings cave.

Chapter 13
Powder Coatings and the Effects of Particle Size

Menno B. Claase, Paul Vercoulen, and Tosko A. Misev

Abstract Powder coatings are increasingly applied in various industrial coating applications. In this chapter the effects of particle size and particle size distribution on the manufacturing, application and coating properties of powder coatings are discussed. The chemical composition of thermoplast and thermoset resins and their corresponding powder coating formulations are addressed together with the premixing, extrusion, cooling, chipping, fine grinding, classification and collection steps in the manufacturing process. The resulting particle size and particle size distribution (PSD) of the powder coating have pronounced effects on the application process especially via the fluidized bed and electrostatic spray application. In many steps particle size influences the outcome: fluidization, charging, spray trajectories, layer build-up, adhesion are all affected.

In addition to the above effects, the PSD of coating components like pigments, fillers and additives has a large impact on the final properties of the coatings. This includes proper dispersion in the resin. This PSD is of prime importance since it directly affects properties like flow, degassing color, matting, adhesion and outdoor durability.

The final section addresses the effects of particle size on safety (explosions and worker safety) and sustainability (carbon footprint of powder coatings).

13.1 General Introduction into Powder Coatings

Although not very familiar to most people, the industrial use of powder coatings offers many advantages over the use of liquid paints. Advantages include:

- Ecology. Lack of solvents, emissions and better eco-footprint than most liquid systems.

M.B. Claase (✉) • P. Vercoulen • T.A. Misev
DSM Coating Resins, Zwolle, The Netherlands
e-mail: Menno.Claase@dsm.com

H.G. Merkus and G.M.H. Meesters (eds.), *Particulate Products: Tailoring Properties for Optimal Performance*, Particle Technology Series 19,
DOI 10.1007/978-3-319-00714-4_13,

- Efficiency. Powder paints can be recycled, increasing utilization rates up to 99 %, resulting in less waste, lower costs and efficient use of raw materials. Easy cleaning and low reject rates further add to the overall efficiency of the process.
- Ease of use. The application process usually requires less training of the work force and can be automated for a large part. Application of one layer is usually sufficient to obtain the desired properties and performance.
- Excellence of finish. With one layer of paint it is possible to obtain good coverage of the substrate and with the right choice of powder coating good appearance, mechanical properties, corrosion resistance, outdoor durability and/or chemical resistance can be obtained.

Therefore, industrial application of powder coatings has become popular in the last decades. The next paragraphs offer a short introduction into the field of powder coatings, their composition, manufacturing and application.

13.1.1 Thermoplasts and Thermosets

The first powder coatings produced were based on thermoplastic polymers which melt at the application temperature and solidify upon cooling. These developments were initiated in the 1950s when powdered polyethylene was successfully applied in a fluidized bed process on a preheated metal surface [12]. The simple method of manufacturing and application, no involvement of complicated curing mechanisms, raw materials that belong to commodity polymers and acceptable properties for many different applications boosted further the development of thermoplastic powder coatings. This resulted in the use of polyethylene, polypropylene, polyvinylidene fluoride, polyamides and polyesters as binders for thermoplastic powder coatings.

Properties of the thermoplastic powder coatings vary in a rather broad range and are exclusively determined by the polymer used as a binder: polyvinylidene fluoride (PVDF) based coatings are characterized with superior outdoor durability, polyolefines posses excellent solvent resistance, polyamides have exceptionally good wear resistance, polyesters have high aesthetic performance and polyvinylchloride thermoplastic powder coatings have very good price/performance ratio.

Thermoplastic powder coatings suffer, however, from number of weaknesses. Since they do not undergo curing during the film formation the initial molecular weight of the polymers must be high enough to ensure good mechanical properties of the film. Therefore, all polymers used for thermoplastic powder coatings have high melt viscosity and high modulus. The high melt viscosity affects adversely the pigment wetting during extrusion and significantly reduces the pigment loading resulting in a coating with low hiding power. To ensure satisfactory pigment wetting, big extruders typical for manufacturing of compounds for engineering plastics have to be used, which increases both the initial equipment investments and the exploitation costs due to higher energy consumption. Because of the high modulus of the polymers a cryogenic cooling is usually employed to take away the excessive heat developed during the grinding process. Thermoplastic powder coatings have poor adhesion on metal surface and require the use of primers, which adds to the expense of the coating process and is usually avoided unless special corrosion protection is required.

The problems inherent to the thermoplastic powder coatings were sufficiently overcome by the thermosetting powder coatings, which quickly took the largest part of the market. During stoving of the thermosetting powder coatings a cross-linking reaction takes place. Therefore, polymers with much lower molecular weights and lower melt viscosities can be used. As a result larger amounts of pigments and fillers can be successfully dispersed and incorporated into the coating during extrusion. The low molecular weight of the binders eliminates the necessity of cryogenic grinding. Due to the cross-linking during film formation the cured coating is exceptionally resistant to solvents. Because of the possibility to introduce polar groups in the used polymers or polar groups are formed during the curing of the coating, thermosetting powder coatings in general do not require primers as intercoat to improve the adhesion to metal.

First thermosetting epoxy powder coatings were developed in the Shell Chemical Laboratories in England and in The Netherlands in the beginning of the 1960s. Attempts to improve the outdoor durability and the yellowing resistance of the thermosetting powder coatings led to polyester powder coatings, which were introduced in 1970s by Scado in The Netherlands and UCB in Belgium [47, 53, 54]. In the 1980s acrylic [13, 18] and polyurethane [29] thermosetting powder coatings where commercialized followed by powder coatings based on unsaturated polyesters or acrylated acrylics. These coatings are cured by free radical mechanisms using peroxides as initiators or UV light to trigger the formation of free radicals [3, 7, 8, 27, 36].

Epoxy and polyester thermosetting powder coatings have a dominant position in the market. The cross-linkers used in epoxy powder coatings are solid aliphatic or aromatic amines, phenolic resins and polybasic organic acids or anhydrides. Epoxy powder coatings are generally characterized with excellent chemical and solvent resistance, very good salt spray resistance, very good adhesion to metal and good impact, abrasion and scratch resistance. In particular, epoxy powder coatings cross-linked with dicyandiamide show very good resistance towards industrial atmospheres containing sulphur dioxide. When cross-linked with phenolic resins, epoxy powder coatings can be cured at lower temperatures and have exceptionally high gloss and good flow. Polycarboxylic acids and anhydrides, which are often used, provide rapid curing, good electrical properties and good yellowing resistance. The drawback of epoxy powder coatings is the lack of outdoor durability, and tendency to yellowing specifically in the case of dicyandiamide based types.

Carboxyl functional polyesters in combination with epoxy resins as cross-linkers are used in the so called hybrid powder coatings, which benefit from the good yellowing resistance of the polyesters and the chemical and solvent resistance of the epoxy resins. Due to the inferior outdoor durability of the epoxy resins, hybrid powder coatings are only suitable for interior use. Thermosetting polyester powder coatings when cross-linked with triglycidyl isocyanurate (TGIC) or diglycidyl terephthalate (DGT) show very good resistance towards water, rain, snow, frost and solar UV radiation, which makes them suitable for outdoor use. Due to its mutagenicity, TGIC has been gradually replaced by β-hydroxyl alkyl amides in powder coatings for exterior use. This results in somewhat lower outdoor durability compared to TGIC and the additional drawback of a tendency to micro-pinholing due to release of water as a side product during curing.

In polyurethane powder coatings hydroxyl functional polyesters are cured with blocked isocyanates. The isocyanates are either internally blocked, so that no blocking agent is released during curing, or blocked with caprolactam as an external blocking agent. The release of the blocking agent during curing is not desirable from ecological point of view, but the blocking agent before leaving the film acts as a solvent providing exceptionally good flow approaching that of liquid coatings. In addition, polyurethane powder coatings have very good outdoor durability and excellent mechanical properties.

Acrylic powder coatings based on glycidyl methacrylate (GMA) copolymers with dodecane dicarboxylic acid (DDA) as cross-linker have excellent clarity, very good flow and excellent outdoor durability. Therefore these are the only powder coatings used as automotive clear top coats. A drawback of acrylic powder coatings is their inferior mechanical properties. For specific uses when excellent hardness and scratch/mar resistance are required carboxyl functional acrylic resins are combined with epoxy resin in formulating the so called acrylic hybrids, which are only suitable for interior use.

Unsaturated polyesters in combination with vinyl ether urethanes are used to formulate coatings that can be cured at low temperatures. The curing reaction proceeds via a free radical cross-linking mechanism that is initiated with organic peroxides or ultraviolet radiation. It is believed that this chemistry will open the market for coatings for heat sensitive substrates such as medium density fiber board, wood and plastics.

13.1.2 Composition

Powder coatings have five main constituents: resins, cross-linkers, pigments, fillers and additives. Resins and cross-linkers commonly used in powder coatings have been mentioned already in the previous section. In comparison with solvent based systems, most of the coating properties in the case of powder coatings are almost exclusively dependent on the resin. Main resin parameters such as molecular mass, functionality, glass transition temperature, and viscosity have to be carefully selected and tuned, because many of the desired properties of the coating are rather contradictory [32].

The proper ratio between the resin and the cross-linker is very important to obtain as high molecular weight as possible during curing which ensures optimal mechanical properties of the cured film. The highest molecular weight of the cured coating can be obtained when the resin and the cross-linker are in stoichiometric proportion [33].

To achieve uniform distribution in the coating, curing catalysts are often added to the resin or to the cross-linker during the manufacturing process. Powder paint manufacturers also add catalysts in the premix prior to extrusion. Catalysts used in thermosetting powder coatings include: tertiary amines, quaternary ammonium and phosphonium salts, imidazoles, organotin compounds, stannous octoate, zinc acetylacetonate and N,N,N-trisubstituted amidines. By choosing the right catalyst and its concentration one can reduce the curing temperature and the curing time. This will, however, increase the degree of pre-curing during extrusion, will shorten the time available for flow during film formation and will speed up the cross-linking during storage, i.e. will reduce the storage stability of the coating. The best way to get around this quandary is to select unimolecular first-order curing reactions [40]. Most of the thermosetting powder coatings, however, employ bimolecular curing processes. Therefore good compromise between the above contradictory requirements can be obtained by preceding the bimolecular curing reactions with rate-controlling unimolecular reactions utilizing latent catalysts [14, 33].

Up to 20 % (w/w) pigments and fillers are usually incorporated in thermoplastic powder coatings. Because of the much lower molecular weight and melt viscosity of the resins thermosetting powder coatings may contain up to 40 % (w/w) pigments and fillers. This corresponds to about 15 % pigment volume concentration (PVC). Higher concentrations of pigments dramatically degrade the flow of the coating and a PVC of 20 % is considered the maximum that can be used in practice [22].

Typical median sizes for the primary particles of TiO_2 pigments are in the range of 0.3–0.5 μm, whilst particles from often used fillers like talc, $BaSO_4$ or $CaCO_3$ are in the range of 0.7–15 μm. Note that in powders these particles usually are present in the form of clusters (agglomerates and/or aggregates).

As a consequence of the manufacturing process only solid additives, preferably with a glass transition temperature (Tg) above 50 °C, are suitable for use in powder coatings. Liquid additives can be used as a masterbatch in a resin or absorbed on silica carriers. Care should be taken, however, that when a masterbatch of liquid additive is used it will decrease the Tg of the overall system, which may adversely affect the storage stability of the coating. Additives used in powder coatings include: flow control agents, degassing additives, UV absorbers and light stabilizers, anti oxidants, pigment dispersants, antistatic and charge control additives, tribo charging additives, anti caking additives, mar resistance and slip improving additives, texturizing additives and matting agents. An excellent overview of the additives commonly used in powder coatings is given by De Lange [23].

13.1.3 Powder Coating Manufacture

The manufacturing process of powder coatings is much closer to plastic rather than paint technologies and consists of different processes connected in a logical sequence as presented in Fig. 13.1 [58].

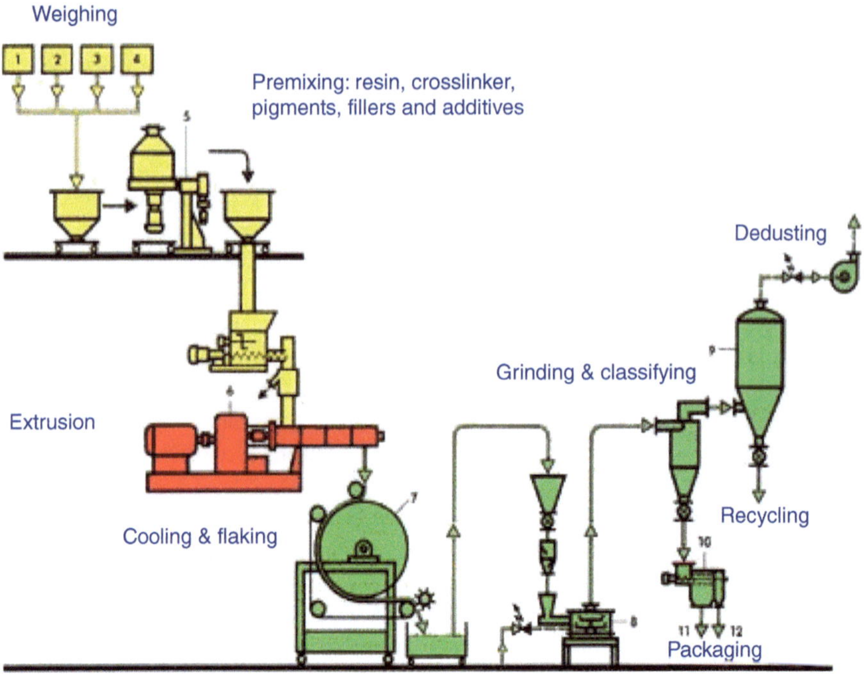

Fig. 13.1 Overview of the manufacturing process of powder coatings (Adapted from [58])

The first step in the manufacturing process is *premixing* of the weighed ingredients, viz. resin, cross-linker, curing agents, pigments, fillers and additives, if necessary after some size reduction. Premixing equipment can be chosen on basis of mixing intensity, color change frequency and whether the process is continuous or batch wise. For lines with a regular color change high-speed mixers are preferred, due to their easy cleaning and short mixing time of 2–3 min. For lines with very little color changes, ribbon blenders are more advantageous due to lower investment and larger batch sizes. Various mixers are used in practice such as tumbler mixers, double cone blenders, horizontal mixers, high speed blenders and conical mixers.

The second step is *extrusion* of the mixture. The extruder is maintained at a sufficiently high temperature to reach an acceptable viscosity of the mixture so that the solid ingredients can be adequately wetted and dispersed by the liquefied resin. Since molten thermoplasts have a higher viscosity than thermoset mixtures, the latter allows easier and faster dispersion. Three types of extruders are used for continuous compounding of powder coatings. The screw of the Buss "Ko-Kneader" executes not only a radial movement but also an axial movement. In this way the kneading flights positioned on the shaft knead the mixture axially against the three rows of stationary kneading pins in the barrel casing. This allows very homogeneous mixing despite an extremely short processing length.

Twin screw extruders have two co-rotating screws which are designed according to the block building principle. The barrel consists of separate sections bolted together to provide the required processing length. The material is continuously wiped from the agitator screw and the barrel wall and transferred from one zone created between the wall and the kneading element to another. These are the places where very effective dispersing and homogenizing takes place.

Planetary extruders use a helically grooved central shaft, surrounded by smaller planetary shafts. Upon rotation of the central shaft the planetary shafts are rolling on the central one and the outer cylindrical housing of the extruder, which also carries helical grooves, and rotate as planets. During extrusion the mixing action takes place at the interfaces of all of the shafts.

Single screw and twin screw extruders dominate the market, the latter consuming somewhat more energy per unit produced powder, but having an advantage of short residence time of the material in the extruder, which allows safe production of powder coatings with rather high reactivity.

The third step in the process is *cooling* of the molten product and *chipping* it after solidification into chips of about 1–10 mm.

The fourth step is *fine grinding* of the chips, often combined with the fifth, *classification* step. Hammer mills as first devices are still in use. They operate with a stepless speed variation of the rotor that carries rotating hammers, which allows adjustment of the output of the mill as well as the quality of the ground powder.

Classifying pin disc mills are widely used at present. They combine both the milling and the classifying action in one piece of equipment and offer compact design that saves floor space. Pin disc mills are characterized by steep particle size distributions, sharp classifying cut points, easy cleaning and maintenance, and low energy requirements.

The grinding mechanism of the opposed jet mill is based on inter-particle collision between opposed suspension jets in the grinding chamber. In most cases this is combined with a pneumatic micro-classifier in a closed circuit so that the milling and classifying are united into one operation. Opposed jet mills are specifically suitable for fine grinding where particle sizes below 15 μm are required.

The resulting particle size and particle size distribution are controlled by various elements in this grinding and classification process. The maximum particle size is determined by the sharpness of the classification inside the classifier mill [43]. A steep curve at the coarse end of the particle size distribution characterizes a highly efficient classification inside the impact classifier mill. The parameters of influence on the maximum particle size are the dimensions and geometry of the housing, design of the milling chamber, speed of air and amount of product inside the classifier mill, type of classifier wheel and design of the outlet. A highly efficient classification process avoids deflected particles or particles that are too coarse, which will have a negative influence on the quality of the powder coated surface.

Particle size is decisive for the quality of powder coatings. Particle size distribution affects the optimal application as well as the surface properties of the powder coating. Powder flow, and hence ease of handling during production and application, are directly influenced not only by particle size and size distribution, but also by particle shape [6].

Normally the majority of particles should be between 40 and 50 μm, so that best reproducibility, optimal flowability and good fluidization qualities are achieved. Particles smaller than 8 μm should only be present in very small proportions since they can enter the lungs and pose a health risk. Furthermore these so-called fines can cause problems in the electrostatic application process creating poorer fluidization behavior, low deposition efficiency on the substrate and clogging of the spray gun.

Although the amount of fines can be influenced by various measures in a certain range, it is not possible to reduce it independently from the overall particle size distribution [56]. This fact has been confirmed by single impact grinding tests which showed that the actual grinding technology is very close to the physical limits. This makes it inevitable to have an additional classifier downstream of the mill, if the amount of fines in the ground material is higher than acceptable.

Therefore in the final step of the manufacturing process, *air classifiers* are applied to give the powder coating particles the required size distribution for further application. Centrifugal air classifiers are common equipment used in continuous processes or when high output of the plant is required. These classifiers are usually combined with the grinding mills into one compact unit, which simplifies the manufacturing process. For batch operations or in plants with lower capacities, a number of different sieving machines is used such as tumbler sifters, vibratory tumbler sifters, pneumatic tumbler sifters and centrifugal sifters.

Powder *collection* and *de-dusting* are two sequences in the manufacturing process preceding the packing of the powder coatings. The collection of the fine particles leaving the outlet of the air classifier is usually done by side removal filters. Often, these filters follow a cyclone, where preliminary removal of the less fine particles facilitates more efficient use of the dust removal filters.

In most cases after the grinding and classification steps powder coatings are ready for use. In some cases, however, further mixing takes place. In the case of so-called dry-blend systems two or more powders are mixed together for instance in double cone blenders. Especially for powder coatings that are cured with β-hydroxy alkyl amides the resulting mixtures typically give matt coating appearances in gloss ranges that usually cannot be achieved by use of matting agents and fillers (10–40 gloss units).

In the last decade several investigations have been performed to modify powder properties by coating powders with other powders by using newly developed mixing techniques like dry particle coating [39, 55]. These techniques, however, have not been implemented in the powder coating industry, most likely because of the associated costs of this additional step in the manufacturing process. Although these techniques significantly add to the costs of the final powder they could be useful in the future for modifying fine particles or introducing properties not available via other routes.

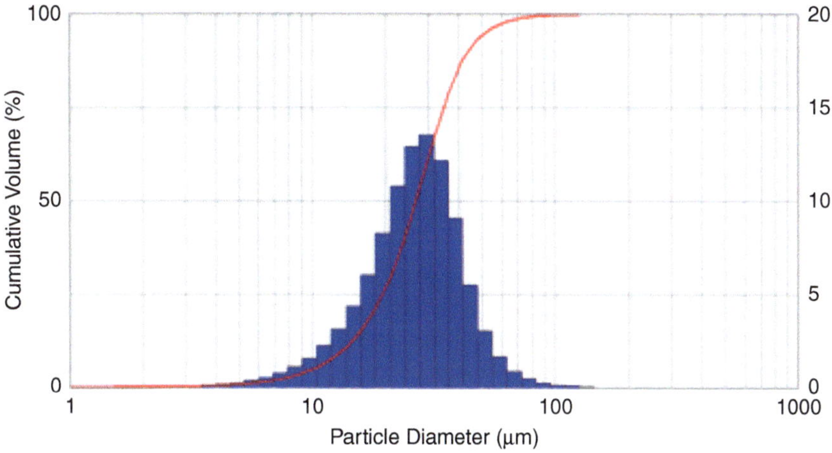

Fig. 13.2 Typical particle size distribution of a thermoset powder coating; $D_{50;3}$ = 38–40 μm (laser diffraction) [26] (Reproduced with permission of Malvern Instruments Ltd)

13.1.4 Powder Coating Application

In principle, three different techniques can be used for the application of powder coatings, viz. fluidized beds, electrostatic spray guns or flame spray guns.

The oldest technology for application of powder coatings is the fluidized bed process where the preheated workpiece enters a space in which powder particles are kept in a fluidized bed by means of an air stream. The particles coming in contact with the surface adhere to the substrate and melt. In this way the powder particles are confined to an enclosure without the need to recycle and re-blend the overspray, which makes the fluidized bed installations very compact compared to the spray gun/booth combinations. Layer thicknesses can be controlled by varying the number of dip coats, the dip coat time and the preheat temperature of the substrates. Layers of well over 250 μm can easily be obtained. To ensure proper fluidization the median particle sizes of the powder coatings are usually around 100 μm [26].

Electrostatic spraying is the most common process used in application of powder coatings. The basic principle of the process is propulsion of powder particles by means of compressed air through a spray gun, in which they become electrically charged. After leaving the spray gun, the movement of the particles to the earthed substrate is governed by a combination of mechanical forces derived from the air that blows the powder through the gun and electrical forces being a result of interaction between the charged particles and the electric field between the substrate and the gun. Powder particles as electrically insulating material retain their charge and adhere to the substrate, even after the electric field is removed. Median particle sizes of powder coatings for electrostatic spraying are usually in the range of 30–50 μm. A typical particle size distribution is shown in Fig. 13.2 [26].

Two types of guns are used for electrostatic spraying of powder coatings. The corona gun in its simplest design is a tube with a charging electrode inside, which has the form of a sharp pointed needle or fine gauge wire. By applying a high voltage to the electrode a local electric field of about 3 MV/m is created. At this field strength an electrical breakdown or discharge of the air occurs in the vicinity of the pointed needle. As a result of the local discharge the air is ionized producing negatively charged ions, which are picked up by the powder particles passing this space.

Tribo charging guns make use of the principle of frictional charging. The construction of the tribo gun is very simple. It is in most cases a Teflon-made tube through which the powder particles are propelled by means of compressed air. The tube walls are earthed, so that the charge generated on the walls by the friction with the powder particles is permanently drained to the earth. In this way, the powder particles leaving the gun are positively charged.

A combination of fluidized bed and electrostatic charge of powder particles is effected in the electrostatic fluidized bed process. The installation comprises the same chamber like in the fluidized bed process separated from the plenum by a porous membrane underneath which charging electrodes are located. The air passes the plenum, gets charged by the electrodes and via the porous membrane enters the upper chamber where the charge is transferred to the powder particles. Being identically charged, powder particles repel each other and move upwards forming a powder cloud. The earthed workpiece entering the cloud is covered by the powder particles due to the difference in the electrical potential.

In the step that follows the electrostatic spray or fluidized bed application, the film formation and in the case of thermosets curing of the powder coating takes place. The articles covered with powder are brought into a batch oven, or if the production line is a continuous one transported into half open oven by means of a conveyor belt. Typical curing temperatures for powder coatings for metal substrates are 160–200 °C for a period of 10–15 min. After melting, sintering in a continuous film, flowing, and curing of the coating the coated objects are cooled under ambient conditions or undergo forced cooling by using air blowers or water baths before they are ready for further packaging, storage and/or transport.

The flame-spray technique is used only for thermoplastic powder coatings and allows application of powder on practically any substrate since the coated article does not undergo additional heating to ensure film formation. The technique is relatively simple. A powder coating is fed into the gun by compressed air and the particles are injected at high velocity in a flame of burning propane. The residence time in the flame is short enough to prevent burning of the particles, but long enough to allow complete melting. The molten particles deposit on the substrate in the form of high viscous droplets and form a high-build film upon solidification.

Electrostatic spraying is a very suitable technique for application of powder on large conductive workpieces, while the fluidized bed is very suitable for application on objects of smaller size, which are not necessarily conductive. Coatings with high aesthetic appearance are usually applied by electrostatic spraying. In this respect fluidized bed technology is less forgiving, but on the other hand it ensures application of thick films with very good uniformity. Flame spray is very suitable for field applications on workpieces which are large or permanently fixed thus not able to fit inside an oven. Capital investments decrease going from electrostatic spraying to fluidized bed and flame spray, whilst the labor costs increase.

13.1.5 Application Areas

The metal furniture industry was one of the first that largely converted from solvent borne to powder coatings and is still a big application area. The industry of domestic appliances fully switched to powder coatings and at present represents probably the largest single application area. Both applications are dominated by so called hybrid polyester/epoxy powder coatings. Other important applications for the polyester/epoxy hybrids include kitchen equipment, ceiling panels for the building industry, lawn and garden products, automotive parts, electric components and housings, functional coatings for appliance parts and the like.

Thermoplastic polyolefinic powder coatings are widely used for dishwasher racks, refrigerator racks, washer tubes, metal containers, drums and pipes. Being non-toxic, odor and taste free and not facilitating bacterial growth, thermoplastic nylon powder coatings are used for coating machine parts and pipings for the food processing industry as well for coating containers for potable water. Due to the excellent abrasion resistance, low coefficient of friction, good solvent resistance and resistance to acids and weak alkalis, nylon powder coatings are used for coating car wheels, motorcycle frames, luggage trolleys, farm machinery, pumps housings, sport equipment, handles of all types, steering wheels and the like.

Because of their excellent chemical, solvent and corrosion resistance, pure epoxy powders are used as functional coatings, as primers for polyolefines powder pipe coatings for subsurface pipelines, primers for nylon powder coatings, coatings for drums, tanks and pipes, etc. Epoxy powder coatings cover a specific application area of coatings for concrete reinforcing bars, or the so called rebar coatings, where good resistance towards alkali environment is required.

Next to the excellent chemical and abrasion resistance, epoxy/phenolic powder coatings have good decorative properties and are used for appliances, wire goods, bicycle frames, hot heater radiators, fire extinguishers, toys and tools, electronic components, heavy duty electrical cabinets, control boards, automotive armatures, etc.

Polyester, acrylic and PVDF powder coatings are used for exterior applications. PVDF powder coatings have superior outdoor durability, and despite their high price are used as architectural coatings for monumental-type structures, where the maintenance costs exceed by far the initial investments. Acrylic powder coatings have very good combination of exterior durability and high aesthetic appearance, and they are primarily used as automotive clear top coats. The main market for polyester powder coatings cured with triglycidyl isocyanurate, diglycidyl terephthalate, β-hydroxyalkyl amides and polyisocyanates is for the protection of outdoor architectural elements, where outdoor durability of the coating and the aesthetic appearance are very important. Other applications include metal furniture for outdoor use, automotive trim parts such as wheel rims, fenders, bumpers, window frames, bicycle frames, motor-cycles etc. These coatings are also used as automotive primer surface that precedes the automotive top coat.

Due to their excellent performance, powder coatings replaced 20 % of the liquid coatings for the industrial metal market. Outside metal there are niche markets for powder coatings such as in mould powder coatings and powder coatings for medium density fiber board and plastics. The main challenge for the future remains lowering of the curing temperature so that powder coatings can enter the market for heat sensitive substrates, where liquid coatings play a dominant role at present.

13.2 The Effects of Particle Size on Powder Coatings

The powder coating field is multidisciplinary where different aspects of science and technology play an important role in obtaining a high quality coated object [28]. Among others these are polymer science, powder manufacturing, fluidization and transport of powder, particle charging and dispersion, electrostatic spraying and coating, curing and melt rheology, and process modeling and optimization.

In many of these areas particle size and particle size distribution are parameters that have significant effects on the final properties. The following paragraphs will focus on the role of particle size in the application of powder coatings as well as the final properties of the cured coating.

13.2.1 Electrostatic Spraying and Particle Size

Electrostatic spraying is at present the most versatile and widespread application technique for powder coatings. Therefore, it is treated in more detail in this section.

The electrostatic spraying process can basically be divided into five stages [1]:

1. Charging of the powder;
2. Transport of charged particles to a conducting earthed substrate;
3. Powder deposition and adhesion;
4. Fusion of powder to form a coating;
5. Recycle of unused powder (overspray) to powder feeder.

Particle size and mass have a strong influence on the aerodynamic, gravitational and charging behavior of the particles that play a role in these stages.

Before powder enters the spray gun, powder is fluidized in a reservoir by blowing air through it from a perforated base and this fluidized powder is then blown along a feed pipe to the gun [2].

Two types of gun are available: the tribo-electric and the corona gun.

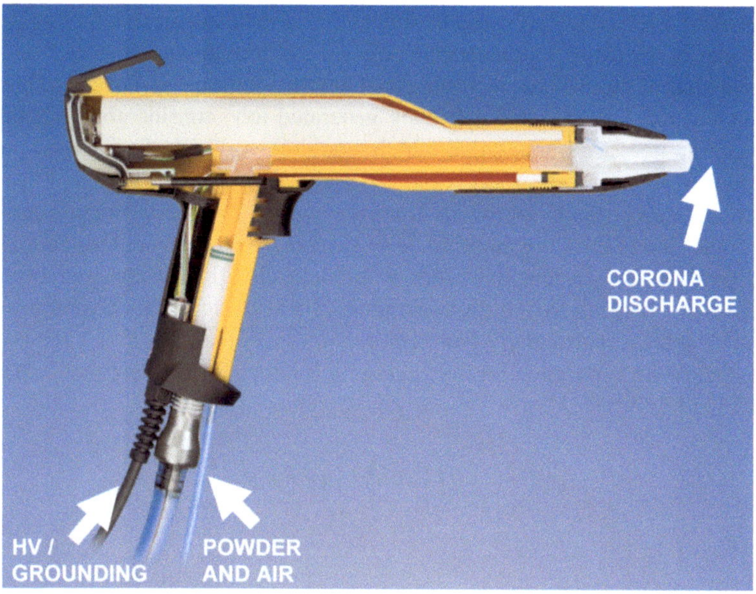

Fig. 13.3 Schematic representation of a conventional corona gun from ITW-GEMA. Corona discharge happens at the end of the electric filament running through the center of the spray nozzle (Reproduced with permission of GEMA GmbH)

In the tribo-electric gun fluidized powder by means of compressed air is blown vigorously through the barrel, which consists of a carefully chosen material, mostly Teflon, to maximize tribo-electric charging. Tribo charging guns make use of friction to generate charge. The tube walls are earthed, so that the charge generated on the walls by the friction with the powder particles is permanently drained to the earth. The charge exchange processes between powder particles and pipe walls and that due to particle/particle collisions is exceedingly complex. Particles of various sizes and shapes are present and impact conditions vary with particle size and mass. The tribo charging process results in a linear increase in particle charge with the surface area of particles and consequently the charge-to-mass ratio of particles decreases inversely with respect to aerodynamic diameter. The powder particles leaving the gun are positively charged and are blown towards the substrate to be coated.

In the alternative and more common corona gun, the powder is charged by passing it through a corona discharge in the area near the gun tip (Fig. 13.3). The charging electrode inside has the form of a sharp pointed needle or fine gauge wire and is maintained at a high negative potential of 60–100 kV. The intense voltage gradient in the vicinity of the electrode tip sets up a stable corona discharge through which powder particles pass. Charging of these particles by negative ions occurs according to ion attachment.

Both types of powder coating gun produce a charged powder cloud in the transport region between the gun and the substrate which results in a space-charge electric field directed towards, and ending at the earthed substrate, ensuring that the trajectories of the charged particles are directed towards and onto the substrate. The high potential of the corona electrode generates an additional electric field between the gun and the substrate and the many generated ions considerably enhance the space-charge field well above that which would exist due to the charged powder cloud alone.

The ions move rapidly outwards from the source, such as the region around a sharp point, due to the Coulombic force of the intense electric field, directing them to the particle surface and thus charge it. Field charging is considerably more efficient than charging by ion diffusion alone. Field charging has been considered by several authors, particularly Pauthenier.

The following expression for field charging is known as the Pauthenier equation (Eq. 13.1) [42, 59]:

$$q = \left[1 + 2\left(\frac{\varepsilon_r - 1}{\varepsilon_r + 1}\right)\right] 4\pi a^2 \varepsilon_0 E \frac{t}{t + \tau} \tag{13.1}$$

where:

q = the charge acquired by the spherical dielectric particle
a = particle radius
ε_r = relative permittivity
E = electrical field strength
τ = charging time constant (ion flux is implicit here)
t = time

Often, especially in numerical calculations, it is being assumed that particles are approximately spherical and monodisperse. In reality all powder coatings are however characterized by particle size distributions since they consist of particles of various shapes and sizes.

As particles are transported to and from the powder gun by air, particle trajectories are also governed by aerodynamic forces, which depend on particle aerodynamic diameter. The exact trajectories followed by particles therefore depend on the balance between electrostatic and aerodynamic forces and are particle size dependent.

The dynamics of the charged particles in the region between a spray gun and a substrate are complex [2]. The air flow from the spray gun, usually at a velocity of a few meters per second, diverges and may be turbulent. In addition ventilation is used in powder coating installations that imposes an overall flow on the system, but this is often at air speeds less than 1 m/s and usually can be ignored in particle trajectory calculations.

Experimentally, it was determined that without corona voltage the velocity in the central region between gun and substrate is approximately 2 m/s [51]. With an applied corona voltage of 60 kV it was found that the particle velocity, v_E increased to about 3 m/s and it was estimated that the mean electric field strength, E, between gun and substrate was approximately 2.4×10^5 V/m. This increase, of approximately 1 m/s, is caused by Coulombic forces and was shown to be in accordance with the viscous drag, as described in Eq. 13.2:

$$\upsilon_E = \frac{Eq}{6\pi D\eta} \tag{13.2}$$

where:

ν_E = particle velocity
E = electric field strength
q = particle charge
D = particle diameter
η = viscosity of air

Aerodynamic forces are mainly responsible for conveying the powder particles towards the substrate but electrostatic forces dominate in close proximity to the substrate. For further descriptions of other types of spray guns and the fundamentals of electrostatic powder spraying, transport and coating the reader is further referred to an excellent review on this topic by A.G. Bailey [2].

Next to practical experiments where one can identify the effects of different particle sizes on processing, application and coating properties, numerical simulations offer a great deal of insight in the underlying physical reasons why particle sizes influences the application process and the final properties so much. Several numerical simulations have been reported in literature [24, 48, 49], with insightful perspectives on the spraying process.

It was shown that the particle motion inside the transport region between gun and substrate depends on the size of the particles. Larger particles tend to fall towards the bottom of the booth but smaller particles stay in a higher position inside the booth [48]. This is because larger particles with bigger mass are under higher gravitational forces and tend to move downwards while smaller particles tend to follow more the gas flow path and remain suspended for a longer time.

This effect can be clearly visualized in a numerical simulation on 15 and 35 μm powder as shown in Fig. 13.4 [49].

This behavior of fine particles shows that the spray flow of fine powders can be better controlled than that of larger particles with properly selected airflow rate. Variation in airflow rate does, however, influence the transfer efficiency of the powder.

The First Pass Transfer Efficiency (FPTE) represents the ratio between the amount of powder deposited on the substrate during one passage and the amount of powder coming from the spray gun. Low efficiencies mean longer spraying times and larger amounts of recirculation, and, thus, higher costs. The FPTE can be influenced by both particle size and air flow rate.

For powders of size 25, 35 and 45 μm, the FPTE initially increases and then decreases when the airflow rate increases. For these powders, the airflow rate initially helps the particles to reach the substrate, but a further increase in the airflow rate eventually decreases the FPTE because the increased turbulence makes the particles more dispersed near the substrate and causes particles to move past the substrate. An increase in the FPTE is more noticeable with larger size particles.

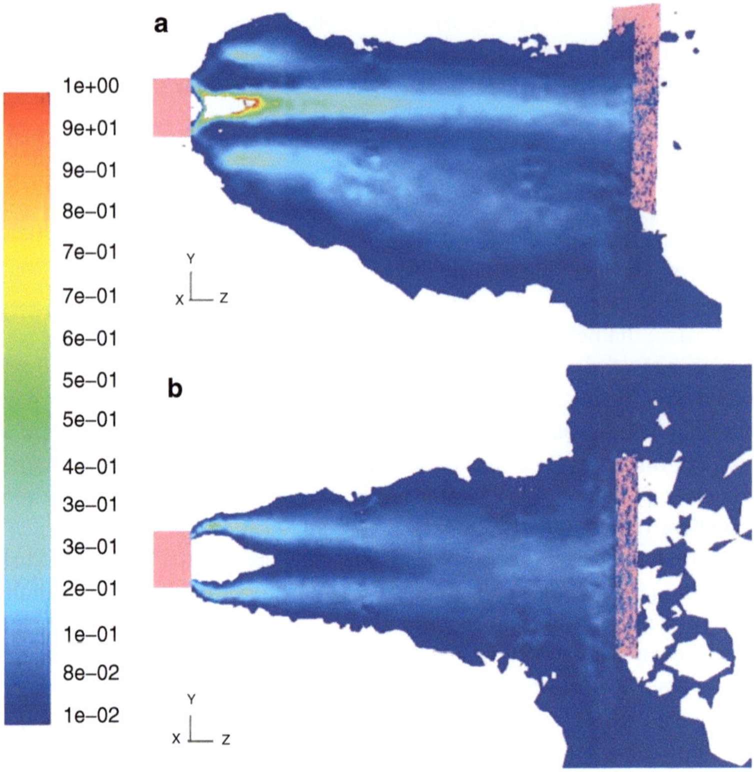

Fig. 13.4 Side view of predicted solid particle concentration (kg/m3):(**a**) 35 μm powder; (**b**) 15 μm powder [49] (Reproduced with permission by Elsevier)

It is important to notice that finer powder can result in better coating quality. Calculations show that the 15 μm size powder provides a very uniform coating thickness compared to the larger sized powders.

Further numerical simulations of 15 μm fine powder coating systems (in this case compared to a coarser powder of 45 μm) showed that the previously described strong effect on the particle transfer efficiency is different for different positions of the pattern adjust sleeve (PAS) of the spray gun compared to the corona electrode [24]. When the PAS is moved to a more forward position, the particle transfer efficiency increases. When the particle size decreases, the particle transfer efficiency tends to decrease. In general, however, since only about half the quantity is required for ultrafine powders to cover the same surface area, there is actually a net increase of particle deposition for finer powders.

Several experimental studies have also been conducted to study the effect of particle size on particle trajectories. The trajectories of charged epoxy particles, with size range 45–125 μm, in an electrostatic powder coating system were studied using a photographic technique [1].

Epoxy particles of three sieved size ranges: 45–53, 75–90 and 105–125 μm, were used in the experiments. As seen in the previously discussed numerical experiments the presence of an electric field effectively helped to overcome the gravitational effect and confined the particles within the air stream. Agreement is reasonable for particles emerging near the centerline but away from it, larger deviations were observed. This difference becomes more severe with increasing angle of emergence of the particles.

It was shown that the air flow from the spraying device was responsible for the initial particle transport, with increasing dominance of electrostatic forces near the substrate mainly due to the field enhancement effect of the space charge, again in line with the previously discussed numerical experiments.

In an experimental study, a non-spherical black polyester paint powder with an average size of 35 μm was used and particle profiles without and with electrostatic charge were determined [57].

It was noticed that the overall particle sizes measured under the electrostatic charge are generally smaller than those without the charge. Since the particles are charged with the same polarity, they repel each other in the process of traveling towards the substrate. As a result, agglomerates may be broken up and smaller particle sizes are recorded.

For the particle size profile without electrostatic charge it was again seen that the portion of fine particles distributed near the top is significantly higher than near the bottom.

Additional studies on the particle velocities clearly showed that particle velocity changes with spray time. The particle velocity drops while the spraying is progressing until it reaches a steady value. The falling off of the particle velocity indicates the occurrence of so-called back ionization in the first 20–40 s. These experimental results clearly indicate that the initial deposition period is a significant part of the electrostatic spray coating process. During this period, particles experience the stronger electrostatic forces and have higher velocities than they do after a prolonged spray. In fact, when the particle velocity starts to level off, it is the first indication that the bulk of the powder deposition has occurred. Further spraying of powder to the substrate will cause unnecessary powder overspray, lower transfer efficiency and a decrease of the coating quality.

It was also revealed that locally deposited particles have narrower size distributions than the original particles [31]. It was demonstrated that small particles are more prone to adhere to the substrate in coarse powder coating processes while large particles take the priority to deposit in ultrafine powder coating processes, due to different dominant factors.

In more detail it was found for black polyurethane paint with a volume median particle diameter (D_{50}) of 35 μm (coarse), and another with a D_{50} of 12 μm (fine) that the particle size distributions of locally deposited particles were narrowed with respect to their original size distributions. Results indicated that the median particle size changes in radial direction. The overall size decrease across the substrate is up to 12 μm for the coarse system but less than 4 μm for the ultrafine one. The observed size change is not only depending upon the original particle size of the powder but also upon radial distance and operating conditions such as charging voltage, flow velocity and spraying duration.

For the coarse powder, the particles deposited in the central region are larger while the particles in the middle and the fringe region are smaller than the original particles. For the ultra fine powder, all the particles deposited on the substrate are larger than the original particles.

Thus, large particles are liable to be eliminated from the powder cloud by electro-separation, and consequently the overall deposited particles are smaller than the original coarse particles. In contrast, for the ultrafine system electro-separation takes the secondary place, due to a sharp reduction in particle size and mass. Small particles are very likely to stay in the airflow and to be reclaimed as the overspray in the ultrafine powder coating. This contributes to the overall deposited particles to be larger than the original ultrafine particles. Also it was observed that the median particle size decreases with spraying duration in the two powder coating systems. It was believed that this phenomenon is mainly attributed to the influence of the intensifying back corona.

Corona charging of a dispersed powder cloud is very inefficient, with only about 0.5 % of all the available ions actually attaching to the particles. The remaining 99.5 % remain as free-ions and alight on the deposited paint layer during the coating sequence. Rapid onset of back-ionization disrupts the coating, leading to severe 'orange peel' effects. In most cases, back-ionization will be initiated as soon as the first monolayer of powder is deposited [17].

Differences in size also result in differences in deposition efficiencies with a higher efficiency for the coarse powder [30]. The coarse powder is superior to the ultrafine powder in charging in-flight particles, which directly contributes to its higher deposition efficiencies. Furthermore, it was observed that the two powders exhibit distinct characteristics in charging deposited particles. Compared to the coarse powder, the ultrafine powder is more uniform in charging deposited particles, mainly owing to its greater particle number and higher specific surface area but less mass. Both powders behave differently in the amount of charge per mass (*Q*/*M*) with the increased charging voltage; an increasing tendency is seen for the coarse powder while a decreasing tendency with the ultrafine one. Likely the coarse powder forms a dilute powder cloud and induces weaker corona quenching compared to the ultrafine powder. As a result, the deposited layer can receive more ions by the secondary charging as a whole and thereby increase its *Q*/*M*. By comparison, the ultrafine powder forms a denser powder cloud (due to its greater particle number and higher specific area), which incurs stronger corona quenching and thereby a more severe reduction. In particular, the denser cloud of the ultrafine powder inhibits the ions from penetration.

Q/*M* of the sprayed powder is one of the most critical parameters in powder coating and has been studied in detail for three clear coat powder samples [50]. A high *Q*/*M* yields high transfer efficiency, but the onset of back corona reduces the transfer efficiency as a function of time and disrupts the uniformity of the surface, thereby causing a poor appearance. Onset of back corona is a function of the product of the thickness and the charge density of the powder layer.

With a gun-to-target distance of 30 cm it was found experimentally that the gun voltage should be 40 kV or higher for consistent corona discharge with minimum ion current. The ion current is proportional to the square of the applied voltage (V) whereas Q/M is directly proportional to the applied voltage. However, transfer efficiency is proportional to both the applied voltage and Q/M. Therefore, both reduction of positive tribo charging and optimization of the applied voltage are necessary for high transfer efficiency and excellent appearance.

The appearance is influenced by Q/M, the particle size distribution, the thickness of the powder layer, the powder mass flow rate, the molecular weight distribution, melt viscosity of the powder, and the target current. Each of these parameters affects the final appearance of the film.

The uniformity of powder flow and dispersion of the powder sprayed from the powder gun are critically dependent upon coarse sieving and loosening of the powders. Powder fresh from the bag had agglomerates, but when sieved through a 400 μm opening sieve, both flowability and dispersion of powder improved greatly.

Several researchers have shown that when the mean diameter of the powders used in the coating process is decreased from 25 to 10 μm, the smoothness of the films improved and the surface irregularities decreased by almost 50 %. This improvement can be explained by the formation of a smooth powder layer and a decrease in the void spaces in between the particles. Also the melting time (t_m) is directly proportional to the particle radius [4, 38]:

$$t_m = f\left(\frac{a - \eta}{\gamma}\right) \tag{13.3}$$

where:

f = not further specified function

t_m = melting time

η = viscosity of the molten powder

γ = surface tension

a = radius of the particle.

For powders with a very narrow particle size distribution, the particles will melt at the same time and the melted material will flow more uniformly, depending mostly upon the surface tension.

For a given Q/M of the powder and thickness of the powder layer, there is a minimum particle size that can overcome the net repulsive force. Particles smaller than this minimum size are not attracted for deposition, and if already deposited, they are loosely bound to the powder layer.

For a given particle size distribution of the powder, there is a self-limiting powder layer thickness, above which the film build-up rate diminishes and the final top layer of the powder contains large particles and packs loosely. Therefore the film thickness needs to be tightly controlled to obtain the desired appearance. A wide particle size distribution will produce a film surface with large short and long wave structures. Agglomeration of fine particles caused by storage,

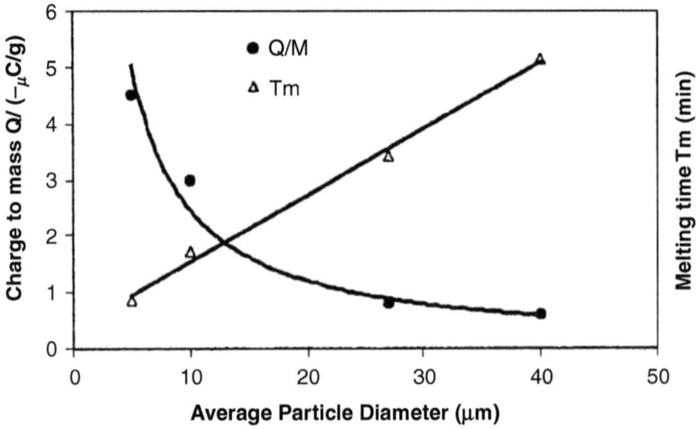

Fig. 13.5 Dependence of the melting time and charge-to-mass ratio with the average particle diameter [38] (Reproduced with permission by Taylor & Francis)

fluidization, and transport will also produce an uneven surface profile of the deposited powder layer.

A wide particle size distribution associated with a wide distribution in the molecular weight of the polymer chains will accentuate the orange peel effect and reduce the glossiness of the cured films.

The melting point of the particles during the curing cycle depends upon both the molecular weight and the particle size. The curing cycle is adjusted so that the flow rate of the powder melt is controlled. By the time that the small chains of small particles are melted, the larger ones are still in a semisolid state. These variations in the melting properties generate differences in the melt flow patterns with direct implications on the film surface appearance.

Fine powders, with $D_{50} < 10$ μm in diameter, are very cohesive and do not fluidize or disperse well. The powder agglomerates easily and deposits as aggregates rather than fine dispersed powder. Since Q/M of the powder increases as diameter decreases (Fig. 13.5), back corona sets in at a very low powder layer thickness. Dispersion, delivery, and controlled charging of fine powders remain as major impediments to obtaining thin film with excellent appearance.

Choosing an optimum particle size distribution for the best appearance of film is a compromise between many opposing factors. Particles with D_{50} larger than 20 μm will have low Q/M of bulk powder and much better deposition and fluidization properties compared to those for fine powders. The advantage of fine powders with D_{50} close to 10 μm for producing a smoother top layer of powder is difficult to realize in practice. A practical approach is to have a narrow particle size distribution with a minimum value of D_{50} set by the dispersion and flowability of the powder.

The thicker the powder films are, the smoother the surface finish is. This decrease in surface roughness is due to the much higher leveling velocity of the powder melt for thick films. If the surface of the workpiece has scratches, the film thickness required to cover the 20 μm deep surface asperity is approximately 40 μm. The optimum film thickness depends upon the functional properties and appearance of the film and the application costs.

The "critical coating thickness" is reached when the rise in surface potential, as charged powder is deposited, becomes just equal to the dielectric strength of the layer. Beyond this point, incoming particles are discharged as soon as they land. As a result, the powder adhesion, deposition efficiency and smoothness of the coating suddenly decrease [16].

To study this effect preparing plots of surface potential versus coating thickness can prove useful. Theoretical curves can be calculated by using published values of the breakdown voltage across small air gaps, whose widths equal the thickness of the powder layer. To predict the behavior of the surface potential in the breakdown region, one only needs the value of the bulk density of the powder. The value where the experimentally obtained surface potentials cross the calculated values is the critical thickness, which marks the point beyond which one will not ordinarily wish to build up the coating thickness.

Under typical conditions, the image force[1] will be about 100 times as strong as the force of gravity. The depth of the layer is not part of the relation describing powder adhesion, so a coating of any thickness, no matter how great, would tend to adhere to the substrate as long as the particles retained their charge. This is why the critical thickness is an important property of the powder layer. Coating particles can also lose their charge slowly via several electrical conduction mechanisms through the powder layer to the grounded substrate. Experimental data on several types of powder coatings (epoxy, acrylic and polyester) show that none of the tested powders lose more than 10 % of their initial charge during a period of 10 min, meaning that the force of adhesion on the outer particles has decreased by no more that 20 %. Thus a coating should normally adhere well enough to be transferred to the curing oven even after considerable delay.

The critical thickness rises with decreasing gun voltage, increasing particle diameter and density, with increasing bulk density and dielectric constant of the powder and with decreasing particle charge. Yet powder adhesion is affected adversely by most of these same factors. The adhesive force is proportional to the square of the particle charge and rises with increasing bulk density, but falls with increasing particle diameter, dielectric constant and density. Adhesion however will usually be adequate under all but the most extreme conditions. Generally speaking, the same fundamental principles on critical thickness and powder adhesion apply both to powder spraying and to fluidized bed coating.

Considering the large impact particle size can have on various aspects of the application, the commonly used practice of recycling the overspray of the powder needs to be considered carefully from case to case.

---

[1] The image force is the electrostatic force calculated through application of an image field. This field is used to simplify calculation and visualization of the electric field distribution near a conducting surface. The image field is constructed by an imaginary charge, which is the mirror image of the real charge in respect to the surface but has the opposite sign.

In a study on fresh and reused polyester resins, the particles of the reused polyester resin were smaller than fresh polyester resin and somewhat agglomerated [25]. Mean particle sizes of fresh and reused polyester resins were 23.6 and 14.6 μm, respectively.

In the case of Q/M, the value of the reused polyester resin was higher than that of fresh resin due to the greater surface area. Powder with high Q/M easily causes back ionization.

Aerate and packed bulk densities of the reused polyester resin were almost the same as those of fresh resin. The flowability index of the reused polyester resin, particularly the fluidity, were much lower than the values of fresh resin, again related to more cohesion and agglomeration. The flowability index and fluidity influence the movement of powder particles from the spray gun to the substrate.

The suitable weight ratio of fresh and reused polyester resins was achieved here at 3:1 and gave the on specification coating film thickness.

In the process of electrostatic application particle size has an effect on particle charging, particle trajectory, deposition behavior, deposition efficiency and layer thickness, which makes it one of the most important parameters in powder coatings to take into account when coating an object.

Besides the area of powder coating of mainly metal substrates several authors have also dedicated time and effort in investigating the effects of electrostatic spray deposition of powders on food. Also here topics like corona charging vs. tribo charging, charge-to-mass ratio, transfer efficiency and the effects of particle size have been thoroughly studied. References are provided for the interested reader, further discussion of this topic is beyond the scope of this overview [15, 52, 60].

13.2.2 Particle Size and Coating Properties

As discussed in the previous section on application, particle size and particle size distribution have an effect on layer thickness and coating appearance. One of the well-known effects in powder coatings is that with increasing layer thickness the leveling of the paint film improves resulting in a smoother coating. Besides this (indirect) effect there are various other aspects of coating properties where particle size plays a role in a direct or indirect way.

Pigment Dispersion

As in liquid paints, pigment dispersion affects many quality aspects of powder coatings, a.o. optical properties such as gloss, color shade and hue, and hiding power of the paint [41], but also other properties like viscosity and anti-corrosion protection. Adequate dispersion, i.e. de-agglomeration of clusters to primary particles, is of fundamental importance for optimum paint quality. Typically, the dispersion process takes place in the extruder, with typical residence times of less than 30 s. For strong clusters, this may be too short to reach full de-agglomeration. In that case the type of dispersant or binder has to be changed or some extra processing step is required, like masterbatching of pigment and binder.

Next to processing conditions, the nature of the pigment surface and the cohesive strength of the pigment agglomerates are known to influence the dispersing process [10].

The dispersing process can be divided in three steps: wetting of pigment clusters, de-agglomeration, and stabilization of the resulting primary particles. The driving force for this wetting process is governed by surface tension differences between the pigment surface and the resin. For most resins, typical surface tensions are 30–40 mN/m at 150 °C and for TiO_2 pigments typical values are 45–60 mN/m, depending on the nature of the surface treatment. For low-viscous liquids, good wetting is therefore expected, but for highly viscous liquids, such as polymer melts, the wetting process is dramatically retarded and may be incomplete in practice. For resins and polymers the rate of de-agglomeration of pigment clusters depends largely on the rate of polymer infiltration and on the mechanical stresses supplied by the extruder.

It was observed that TiO_2 pigments are almost completely dispersed without further use of dispersants [10]. Immediately after extrusion, the pigment particles do not have enough time to re-agglomerate because of the high viscosity of the powder coating after cooling. In addition, re-agglomeration during cross-linking may be retarded because of a smaller drop in viscosity, resulting from the cross-linking reactions.

De-agglomeration of pigment clusters can take place by rupture and by erosion. Rupture is an abrupt and fast process which requires a specific viscous stress in the medium in order to overcome the agglomerate's cohesive strength. Smaller particles require a larger stress to be ruptured. Erosion, on the other hand, is a slow and gradual process, in which small fragments are disrupted from a large agglomerate. This requires a smaller viscous stress, compared to rupture. During rupture, the agglomerates are broken up into a relatively small number of large particles; erosion yields a large amount of small particles. Although the rate of de-agglomeration due to erosion is highly dependent of the rate of wetting, de-agglomeration due to rupture is mainly influenced by the magnitude of the extrusion forces imposed onto the agglomerates. Since the mean residence time of the powder coating material in an extruder is very short (typically less than 30 s), rupture is supposed to be the dominant mechanism of pigment dispersing in the preparation of a powder coating. De-agglomeration of TiO_2 pigments most likely proceeds by rupture and the de-agglomeration of carbon black predominantly by erosion.

The state of pigment dispersion in powder coatings can easily be determined by applying scanning electron microscopy (SEM) on coating surfaces that have been etched using an oxygen plasma [21]. This technique has therefore been used to analyze the influence of input power of the extrusion equipment on the pigment dispersion in powder coatings [19]. The relative amount of pigment aggregates and agglomerates affects gloss and haze of the final coating and may also change the viscosity of the melted powder. A twin-screw extruder ZSK 50 with three different input powers was tested with 20 %, 40 % and 60 % of the maximum power available (43.2 kW).

The results show that the degree of dispersion clearly depends on the power input of the extruder. The size distribution of the dispersed pigment shifts towards smaller particle sizes when higher input power was applied during the extrusion process, indicating the more extensive break down of the pigment agglomerates and aggregates. The degree of pigment dispersion is the lowest at the 20 % input power. The corresponding distribution graph shows the largest particle sizes, exceeding the 2 μm size. The results for 40 % and 60 % input power are relatively similar. They differ mostly in the medium-sized particle range (0.3–0.5 μm). The state of pigment dispersion in powder coatings with the same formulation therefore depends on the applied input power during the extrusion process.

Appearance

One of the most important challenges in powder coatings is how to improve the appearance of the films and how to decrease the film thickness [37]. A decrease in glass transition temperature or molecular weight of resins may improve the leveling properties, but by this measure the storage stability is usually decreased and the physical properties become worse.

In powder coatings, particle size is one of the important factors associated with the wavelength and initial amplitude of irregularities. If the particle size and the surface tension, σ, are considered to be constant in powder systems, the flow is proportional to the reciprocal viscosity, $h^3 t/\eta$ (with t the time, h the mean film thickness and η the viscosity).

Next to viscosity, yield stress also needs to be considered. Even assuming the severe conditions corresponding to distinct orange peel, i.e. a film thickness of 73.5 μm, a surface tension of $3 \cdot 10^{-2}$ N/m, a roughness of 2.33 μm and a wavelength of 1.45 mm, this stress is only $\approx$ 0.45 μN/mm^2. The existence of any yield stress of this magnitude would prevent any further leveling, so it is necessary to consider the yield stress in paint formulations, especially when pigmented.

As previously discussed, the thickness of a charged powder layer is limited. Inside the powder coating, the electric field induces an increased electrostatic pressure on the powder grains at the substrate–coating interface and a vanishing pressure on the grains at the coating–air interface [5]. This internal field is amplified into air bubbles and it may be responsible for the back ionization process and for the formation of moon craters via the ionization of the air molecules followed by the pressure exerted by the ions pushed in the direction of the free surface. The result is eventually a bubble explosion via the inner pressure exerted by the ion bombardment.

A way to influence the particle size distribution and therefore the appearance is by using other ways than the regular extrusion and grinding to prepare the powder. Clear powder coatings that consist of spherical particles with a narrow particle size distribution were produced via a suspension method that involves the agglomeration and unification of smaller particles [46]. These powder coatings have the advantages of excellent powder flowability and high transfer efficiency. This behavior is attributable to the spherical particle shape and absence of finer particles.

Prepared by agglomeration and later unification to produce thermosetting resin particles having a narrow particle size distribution a 10 μm spherical powder coating was prepared. Also, low-*Tg* powder coatings that were encapsulated by high-*Tg* submicrometre resin particles were produced.

Film surface roughness was found to reduce dramatically with a decrease in weight average particle size for each film thickness. This improvement was due to the absence of coarse particles producing large irregularities in the film that cause it to coalesce and degas slowly, resulting in a poor appearance.

Despite these powder coatings having a low *Tg*, blocking resistance of suspension powder coating using 2 wt% ultra fine powder was equal to that of high-*Tg* pulverized powder coating. At a film thickness of 50 μm, the film surface smoothness level of encapsulated powder coating has reached the same level as a solvent-based 40 μm film thickness acrylic clear coating. Thus, it was established that low *Tg* powder coatings which were uniformly encapsulated with small amount of high Tg ultra fine powder improved both blocking resistance and film appearance for thinner films. This encapsulated powder coating had excellent film surface smoothness at a thinner film thickness, and could be applied as an automotive clear topcoat.

Matting

Several techniques are known to produce coatings with lower gloss finishes [20]. They can be divided into two basic categories, whether they act due to the presence of non-soluble particles or whether they are effective due to the addition of resinous materials with different compatibility behavior. Gloss reduction using non-soluble particles is widely used in liquid coatings. During drying processes, the loss of solvents and condensation products results in a fairly high shrinkage of the coating film. Consequently, a diffusely reflecting micro-textured surface is formed. This matting technique is less effective in coatings that undergo little shrinkage during film formation like powder coatings and some UV curing systems. Here, resinous materials are frequently applied that show limited compatibility with the other resinous materials that are present in the system. This causes the formation of a coating that has an imperfectly smooth surface. In addition, fillers with larger particle sizes are frequently applied to reduce the surface smoothness.

Three different types of matt carboxy polyester based powder coating systems have been investigated by scanning electron microscopy after oxygen plasma etching: (a) matted with wax, (b) with wax and fillers, and (c) with hardener. Each resulting in its own typical sample surface. Circular features were seen for all coating surfaces in which waxes were applied as matting additives and fillers appeared as larger particles of irregular shape. No typical features were observed for the sample where a resinous material was used as additional hardener. The structures that are responsible for the matt effect were seen on micrographs that were scanned at an approximately 65 ° angle. The wax particles lie on the top of the surface and their incompatibility with the binder causes sharp edges that are clearly seen on the SEM micrographs. This diminishes the surface smoothness and forms a vertical surface profile. The hardener does not cause such effects and, therefore, the resulting surface of the coating is generally smooth.

Wax particles of irregular shape can be seen on surfaces even at the lowest curing temperature (80 °C). At much higher temperatures they form circular structures. The lowest temperature, where wax particles appear as circles on the coating surface, corresponds well to the corresponding softening temperature of the wax.

The average size of wax particles, D_{wax}, is a significant parameter when no fillers or hardeners are used as matting additives in the coating. Coatings with larger D_{wax} have a lower gloss. One filler does not influence this relation to a large extent, but the use of two fillers results in a considerable deviation of this trend.

Adhesion

Studies have been undertaken to examine how the particle size of epoxy phenol novolac powder coatings may affect adhesion to metal substrates [44]. Powders with particle sizes of 21 and 83 μm diameter were used for these studies. DSC analysis showed that the activation energy of cross-linking with 41 kJ/mol for the 21 μm particle size powder is lower than the 58 kJ/mol for the 83 μm particle size one.

Additional pull-off adhesion measurements on aluminum substrates showed that when the 21 μm particle sized powder is cross-linked for 10 min at 110 °C, 140 °C, and 170 °C, adhesion is consistently higher than for the same coating system at 83 μm particle size. This difference is attributed to the time required for powder particles to reach a proper melt viscosity, followed by reactions of functional groups leading to cross-linking. Longer curing times up to 120 min for the 83 μm particles resulted in adhesion similar to the 21 μm particles.

DSC measurements showed that the 21 μm powder exhibits a higher extent of reaction at temperatures up to 155 °C, as compared to that of the 83 μm. As the temperature increases above 155 °C, the extent of cross-linking difference between the two particle sizes narrows, and at 185 °C, the extent of cross-linking between the two powder coatings is similar.

Powder packing plays a key role in which the melting point of a specimen will depend on the thermal conductivity of the heated surface. Since the thermal conductivity coefficient of air is much lower than that of organic polymers, the presence of voids will extend the time required for a specimen to melt due to the lower contact area between neighboring particles. In the case of epoxyphenol novolac powders, thermal conductivity of powders based on 21 μm particles will be approximately four times larger than that based on 83 μm particles. Smaller particle size powders are expected to reach a molten state in 25 % shorter times than larger sizes. Therefore the extent of cross-linking at a given temperature will be greater for smaller particles. As the temperature increases, the magnitude of thermal transfer will also increase, and thus the extent of reactions for larger particle sizes will become similar to those with a smaller particle size.

The faster melting of the finer powder results in improved wetting of the substrate in a molten state, thus promoting chemical and mechanical adhesion. Smaller particles provide shorter cure times, thus powder coalescence plays a very important role in achieving cross-linking and proficient adhesion.

Outdoor Durability

Pigment particle size has been found to be of influence on the performance of UV powders as shown in weathering tests on three UV powder formulations on coil-coated aluminum test panels [34]. The coatings were melted in a convection oven at an air temperature of 130 °C and, while still liquid, exposed under two 80 W/cm medium pressure mercury lamps. After cooling to room temperature, the degree of cure was evaluated.

At a given pigment volume concentration next to the hiding power of the coating the pigment particle size and shape also determine light penetration and through cure in such systems. In liquid UV curable coatings it has been found that both UV penetration depth and hiding power increase with pigment particle size. With the use of iron oxide pigment a considerable increase in through-cure performance was found with increasing pigment particle size. This is an indication that reflection properties of pigment particles can be as important as their absorption characteristics in determining the penetration depth of UV light in these coatings.

13.2.3 Particle Size and Safety, Health and Environment

Although powder coatings offer a safer and more environmentally friendly alternative to solvent-borne coatings there are several safety, health and environmental issues to be considered.

When working with fine particles of organic matter one should always be aware of the risk of explosions (see also Chap. 4). Since the energy for ignition can already be provided by the electrostatic discharge between an earthed object and a spray gun, manufacturers of spray guns have attempted to counteract this by equipping guns with a protective device that reduces the amount of current if the electric field between the gun and the object becomes too strong [11].

Besides reducing the amount of potential ignition sources, thorough knowledge of the explosion behavior is paramount to improve the safety of factories and workers. Therefore, also in the area of powder coating, studies have been performed to investigate the effects of composition and particle size on important explosion parameters like the minimal ignition energy (MIE) and maximum explosion pressure (P_{max}).

In an extensive study, the MIEs of more than 40 coating powders have been investigated. The MIE was shown to depend on the particle size distribution or the specific surface area of the powder calculated from it, the powder agglomeration, the chemical composition of the resin used and the resin content [45].

The influence of the particle size distribution has been investigated, using transparent polyester powder coatings. Due to the different portions of fine powder in the samples, there was only a poor correlation between the MIE values and the D_{50} of the powders. Satisfactory correlation was obtained for the specific surface area, with lower MIE values for higher specific surface areas. For some of the finest sieve fractions with a calculated specific surface area of more than 500 cm^2/cm^3 considerable deviations towards higher MIEs were found due to agglomeration of these fine powders. To study the effect of resin composition, transparent powder coatings based on epoxy, epoxy-polyester, polyamide and polyurethane resins were investigated. The dependence of the MIE of these powders on the calculated specific surface area is basically the same as for the transparent polyester powder coatings. Starting from the coarse powders, the MIE rapidly decreases with decreasing particle size and tends towards a lower limiting value at a D_{50} of 32 μm, corresponding to a specific surface area of about 300 cm^2/cm^3. This lower limiting value depends on the chemical composition of the resin. The limiting values are about 1.7 mJ for epoxies, about 2 mJ for polyurethanes, 2–3 mJ for epoxy-polyester combinations, about 2.9 mJ for polyester powders and about 4 mJ for polyamides.

Pigmented coating powders differ from the respective transparent coating powders by their pigment content which can generally be regarded as inert substance (often mainly TiO_2). The values of the MIE are therefore generally higher than those of the respective transparent coating powders or equal to these at the most. Higher MIE values of pigmented powder coatings are observed when the pigment content is higher (i.e. lower resin content) and in the case of coarser powders (i.e. lower specific surface areas).

When recycling powder in powder coating production lines, the powder sprayed can easily reach the lower limiting value of the MIE even if relatively coarse basic powder is used as the initial powder for spraying. Reducing overspray besides improving the efficiency of the spraying process also improves the safety of the process.

Results from Hartmann bomb tests on 11 epoxy-polyester systems show, with one exception, that the maximum explosion pressures are within the range 5.4–6.4 bar(g) for all the powders [11].

The inorganic pigments act as heat sinks, and thus contribute to the lowering of the explosion pressure, whereas the organic pigment burns and contributes to the production of heat. The specific effect of particle size is demonstrated by two chemically identical powders. The coarser powder, with a D_{50} of 59 μm, has a maximum rate of pressure rise of 200 bar/s, whereas the value for the finer powder, with a D_{50} of 23 μm, is 410 bar/s.

There is a clear tendency that the minimum explosible dust concentration increases with increasing content of non-combustible material, i.e. increasing pigment content. The lowest minimum explosible concentrations of 33–35 g/m^3 were obtained for the powders containing combustible material only. There is an indication that the minimum explosible dust concentration increases slightly with increasing particle size.

In view of the very low MIEs of some powder coatings and the significant explosion potential in terms of maximum explosion pressure and maximum rate of pressure rise, one may wish to maintain the existing, fairly strict requirements to ventilation of coating chambers in terms of the maximum permissible quantities of fine powder fractions in recirculation.

Proper ventilation besides reducing the explosion risks also reduces the exposure of the work force to powder. As with other materials consisting of fine particulates, diseases of the respiratory tracts should be a concern for any employer and employee. The specific type of personal protection equipment used is partly influenced by the particle size of the powder.

Specific health concerns further depend on the chemical composition of the powder coating. Epoxy-amine systems are known to cause sensitization among workers. Powders based on the cross-linker TGIC are registered as known carcinogens in the EU area.

Next to health and safety, where particle size plays an important role, one of the major focus points of the powder coating industry will continue to be sustainability. It is a common understanding that from environmental point of view powder coatings are better than solvent borne coatings, but this was never confirmed in a quantitative way. Recently, Morris [35] presented a study where the carbon footprint (CF) of powder coatings is compared to that of other industrial coatings applied on metal. The CF was calculated using the IPCC 2007 GWP 100a assessment method and the results were expressed as kg CO_2 equivalents; using the associated characterization factors for the relevant greenhouse gases. Figure 13.6 represents the results for solvent borne coatings (conventional and high solids polyester and acrylic coatings in red), water soluble alkyd coatings (in blue), and eight different types of powder coatings (in green).

The study shows that solvent borne coatings have 2–2.5 times higher CF than either water borne coatings or powder coatings. Although powder coatings have a more favorable CF than solvent borne coatings there are still specific areas where further

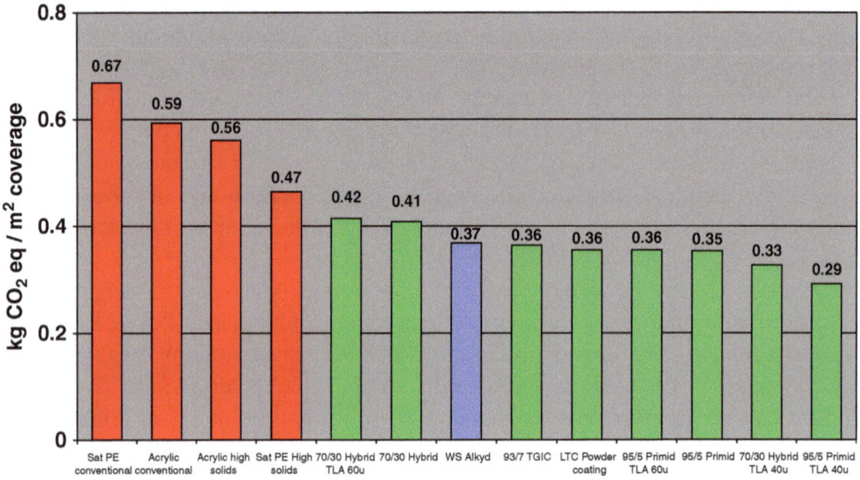

Fig. 13.6 Carbon Footprint Comparison – Coatings applied to 1 m^2 steel substrate. TLA stands for thin layer application, WS for water soluble and LTC for low temperature cure [35]

improvements can be made. Since the coating itself is responsible for approximately 85 % of the CF, reducing the layer thickness will have a significant impact.

The study also shows that hybrid powder coatings containing epoxy resins have a higher CF than the alternative types. Epoxy and hybrid replacement programs could help companies to reduce the CF of these coatings.

The third potential improvement is the reduction of the energy required to cure powder coatings. While the CF reduction impact of this improvement on the entire system studied is much smaller than those previously mentioned, curing temperature reduction could reduce application energy related CF impacts for powder coatings by at least 20 %.

13.3 Outlook

Powder coating has taken a good position on the market over the last decades, based on thorough investigations and good insight in the processes involved. Still, many challenges remain that require further research. Some of these challenges are [28]:

1. The problems that arise from back corona by minimizing the presence of free ions.
2. Minimizing or getting around the Faraday cage effect, enabling the coating of complex objects with hollow shapes and crevices.
3. Development of a process for applying thin films.
4. Deposition of charged powder on non-conducting and heat-sensitive substrates such as MDF, wood and plastic.
5. Coating of large objects such as aircrafts.

6. Variation of charging levels and inconsistent operation of tribo charging guns.
7. Long time required for color changes.

Focusing on small(er) particles, several issues also need to be resolved. For powder coating with film thickness smaller than 25 μm, using fine powders in the size range of 5–20 μm in diameter has many fundamental problems. Unless these problems are well resolved, applications of powder coating may be limited to thicker films. The research needs in the thin film powder coating application are:

1. Dry dispersion of fine powders: fine powders are not easily fluidized or dispersed. The possibility of using an ordered mixture of coarse and fine powders in fluidized bed applications should be investigated.
2. Preventing agglomeration during storage: fine polymer powders, even in small quantities, form agglomerates during storage, making it extremely difficult to get the primary size fraction after dispersion. Research is needed to study different surface treatment processes, such as plasma treatment, plasma polymerization, microencapsulation (of the powder with ultrafine powder in the sub-micron range), to minimize inter-particle attraction forces such that even when the agglomerates are formed, they can be easily separated. Coating of particles with other finer particles for instance by means of a mechanofusion process could hold promise here [61].

In the area of non-conductive and heat-sensitive substrates like MDF, wood and plastics the development of UV curable powder coatings and low temperature curing powder coatings is a big step forward in coating these materials without excessive damage to the substrate in the form of cracking (MDF, wood), melting (plastics) or warping (applicable to all three materials classes).

In the UV curing of powder coatings, further research work is needed to apply a thin layer of powder, to produce thin films that can be applied both on metals and nonmetals [28]. Since solvent-based application of acrylates, followed by UV curing, has been successfully applied to develop scratch-free, thin film coatings, it should be possible to implement a similar process for powder coatings. Application of UV and electron-beam curing at room temperature with appropriate inorganic nanoparticle additives, such as silicone dioxide, could be used for the preparation of powder coatings with superior resistance to chemical corrosion, scratches, and UV radiation.

Further improvement lies in the fact that powder coatings for heat-sensitive substrates do not need UV or electron-beam for curing but that they can cure thermally at low (80–100 °C) temperatures. This development will greatly increase the design freedom of the substrates going from (almost) flat panels for UV coatings to complex 3D geometrical shapes as is the case for regular powder coatings that cure at high temperatures. The limited process window between storage stability, extrusion temperature and cure temperature offers a great challenge in obtaining suitable binder systems for these kinds of low temperature cure applications. Resin manufacturers like DSM Coating Resins are putting substantial research effort in achieving this goal [9].

13.4 Definitions, Abbreviations and Symbols

Mar resistance	Resistance of the coating surface towards scratching or scuffing
Modulus (elastic)	Slope of the stress–strain curve in the elastic deformation region, measure for the elasticity of a material
Orange peel	Uneven surface appearance of a coating reminiscent to the skin of an orange
CF	Carbon footprint
DDA	Dodecane dicarboxylic acid
DGT	Diglycidyl terephthalate
DSC	Differential scanning calorimetry
FPTE	First pass transfer efficiency
GMA	Glycidyl methacrylate
MIE	Minimum ignition energy
PAS	Pattern adjust sleeve
PSD	Particle size distribution
PVC	Pigment volume concentration
PVDF	Polyvinylidene fluoride
SEM	Scanning electron microscopy
TGIC	Triglycidyl isocyanurate
a	Particle radius
D	Particle size/diameter
E	Electrical field strength
H	Film thickness
M, m	Particle mass
P_{max}	Maximum explosion pressure
Q, q	Particle charge
Tg	Glass transition temperature
V	Voltage
ε_r	Relative permittivity
ν_E	Viscous drag
η	Viscosity
σ	Surface tension
τ	Charging time constant

References

1. Ang, M.L., Lloyd, P.J.: Int. J. Multiphas. Flow **13**, 823–836 (1987)
2. Bailey, A.G.: J. Electrostat. **45**, 85–120 (1998)
3. BASF EPA 650978 (1994)
4. Biris, A.S., Mazumder, M.K., Yurteri, C.U., Sims, R.A., Snodgrass, J., De, S.: Particul. Sci. Technol. **19**, 199–217 (2001)
5. Cazaux, J.: J. Electrostat. **65**, 764–774 (2007)
6. Crompton, C.: Eur. Coat. J. **4**, 56 (2006)

7. DSM Resins BV, EO 0.188.840 (1984)
8. DSM Resins, WO 93/25596 (1993)
9. DSM IP Assets BV, WO 2010052290 (2010)
10. Duivenvoorde, F.L., Van Nostrum, C.F., Laven, J., van der Linde, R.J.: Coat. Technol. **72**, 145–152 (2000)
11. Eckhoff, R.K., Pedersen, G.H., Arvisson, T.: J. Hazard Mater. **19**, 1–16 (1988)
12. Gemmer, E.: Industrie-Anzeiger **75**, 1095 (1953)
13. The Glidden Company, EP 0.256.369 (1986)
14. Greidanus, P.J.: 12th Congress of Federation of Scandinarian Paint and Varnish Technologist, Helsinki, 8–11 May 1988
15. Halim, F., Barringer, S.A.: J. Electrost. **65**, 168–173 (2007)
16. Hardy, G.F.: J. Paint. Technol. **46**, 73–82 (1974)
17. Hughes, J.F.: J. Electrostat. **23**, 3–23 (1989)
18. Kansai Paint Co., US Pat. 3.876.578 (1975)
19. Klanjsek-Gunde, M., Kunaver, M., Mozetic, M., Hrovat, A.: Powder Technol. **148**, 64–66 (2004)
20. Klanjsek-Gunde, M., Kunaver, M., Cekada, M.: Dyes Pigments **74**, 202–207 (2007)
21. Kunaver, M., Klanjsek-Gunde, M., Mozetic, M., Kunaver, M., Hrovat, A.: Surf. Coat. Int. Part B Coat. Trans. **86**, 175–179 (2003)
22. de Lange, P.G.: J. Coat. Technol. **56**, 23 (1984)
23. de Lange, P.G.: Powder Coatings Chemistry and Technology, Vincentz Network, p. 202 (2004)
24. Li, Z., Zhu, J., Zhang, C.: Powder Technol. **150**, 155–167 (2005)
25. Lothongkum, A.W., Nonthapone, R., Seangkiatiyuth, K., Tanthapanichkoon, W.: Adv. Powder Technol. **18**, 175–186 (2007)
26. Malvern Instruments Ltd., Powder Coatings Industry Overview; www.malvern.com (2012)
27. Maetens, D.: Proc. Radtech'98, Chicago, p. 170 (1998)
28. Mazumder, M.K., Sims, R.A., Biris, A.S., Srirama, P.K., Saini, D., Yurteri, C.U., Trigwell, S., De, S., Sharma, R.: Chem. Eng. Sci. **61**, 2192–2211 (2006)
29. McLafferty, J.J., Figlioti, P.A., Camilleri, L.T.: J. Coat. Technol. **58**, 23 (1986)
30. Meng, X., Zhu, J., Zhang, H.: J. Electrostat. **67**, 663–671 (2009)
31. Meng, X., Zhang, H., Zhu, J.: Powder Technol. **195**, 264–270 (2009)
32. Misev, T., Belder, E.: JOCCA **9**, 363 (1989)
33. Misev, T.A.: Powder Coatings Chemistry and Technology. Wiley, Chichester (1991)). 196 and 203
34. Misev, L., Schmid, O., Udding-Louwrier, S., de Jong, E.S., Bayards, R.: J. Coat. Technol. **71**, 37–44 (1999)
35. Morris, D.: Comparative Footprint Study on Industrial Coatings for Metal, European Coating Show. Nurnberg (2011)
36. Morton Thiokol, EP 309 088 (1987)
37. Nakamichi, T.: Prog. Org. Coat. **8**, 19–46 (1980)
38. Nix, V.G., Dodge, J.S.: J. Paint. Technol. **45**, 59–63 (1973)
39. Ouabbas, Y., Dodds, J., Galet, L., Chamayou, A., Baron, M.: Powder Technol. **189**, 245–252 (2009)
40. Pappas, S.P., Hill, L.W.: J. Coat. Technol. **53**, 43 (1981)
41. Patton, T.C.: Paint Flow and Pigment Dispersion, p. 510. Wiley (1973)
42. Pauthenier, M.M., de Moreau-Hanot, M.J.: Physique Radium **7**, 590–613 (1932)
43. PCI "Efficient Grinding and Classifying of Powder Coatings ", PCI Magazine **5**(1) (2003)
44. Pennington, B.D., Grunlan, J.C., Urban, M.W.: J. Coat. Technol. **71**, 135–142 (1999)
45. von Pidoll, U., Krämer, H.: J. Electrostat. **30**, 103–114 (1993)
46. Satoh, H., Harada, Y., Libke, S.: Prog. Org. Coat. **34**, 193–199 (1998)
47. Scado, B.V.: US Pat. 3.624.232 (1970)
48. Shah, U., Zhu, J., Zhang, C., Nother Sr., J.H.: Powder Technol. **164**, 22–32 (2006)

49. Shah, U., Zhang, C., Zhu, J.: J. Electrostat. **64**, 345–354 (2006)
50. Sims, R.A., Mazumder, M.K., Chok, W., Wankum, D.L., Guo, W., Tebbets, G.: Part. Sci. Technol. **18**, 239–248 (1998)
51. Singh, S., Bright, A.W.: IEEE IAS Proc. 1–4 (1977)
52. Sumonsiri, N., Barringer, S.A.: J. Food Sci. **75**, E537–E543 (2010)
53. UCB SA, DE Pat. 2.352.467 (1972)
54. Unilever N.V., DE Pat. 2.163.962 (1971)
55. Vilela, A., Concepcion, L., Accart, P., Chamayou, A., Baron, M., Dodds, J.A.: Part. Syst. Charact. **23**, 127–132 (2006)
56. Wadenpohl, C.: Int. J. Miner. Process. **74S**, S155–S164 (2004)
57. Wang, F., Martinuzzi, R., Zhu, J.: Powder Technol. **150**, 20–29 (2005)
58. Werner & Pfleiderer GmbH, Brochure WER 05 066/3-1.5-VIII (1986)
59. White, H.J.: Industrial Electrostatic Precipitation. Addison-Wesley, Reading (1963)
60. Yousuf, S., Barringer, S.A.: J. Food Eng. **83**, 550–561 (2007)
61. Zhou, Q.T., Qu, L., Larson, I., Stewart, P.J., Morton, D.A.V.: Int. J. Pharm. **394**, 50–59 (2010)

Chapter 14
An Overview of Physical (Particulate) Sunscreens

David Fairhurst

Abstract Skin is constantly assaulted by sun exposure and terrestrial solar ultraviolet radiation (UVR) is a major factor deleterious to our health. The most reliable approach to sun protection is to cover up the skin by using a thin film of a topical sunscreen formulation containing active chemical ingredients that absorb, scatter or reflect the incident UVR to reduce the direct penetration and effect of UV exposure. Physical (particulate) actives are very stable and benign. From a regulatory perspective, zinc oxide and titanium dioxide are the only ones universally allowed as sunscreen actives. The development of sunscreen formulations containing ZnO and TiO_2 requires knowledge of their optical properties as these dictate both the resulting aesthetics and the efficacy of the final sunscreen product and, thereby, acceptance by the user; these factors critically impact the economics of commercial sunscreen manufacturing and sales. The optical properties are uniquely related to the particle size (distribution) and particle morphology (e.g., surface area) of the two oxides. Precise measurement of these parameters is therefore a critical metric in formulating sunscreen products. X-ray disc centrifugation, acoustic attenuation spectroscopy and nuclear magnetic resonance relaxation are useful techniques that can be used to characterize fundamental properties of particulate suspensions at concentrations that are commercially relevant.

14.1 Introduction

For those not familiar with the topic under discussion in this chapter, it is perhaps instructive to first briefly review the issue of sun protection and the formulation of sunscreen products containing what are termed physical (particulate) sunscreen actives.

D. Fairhurst (✉)
Particle Sciences Inc, Bethlehem, PA18017, USA
e-mail: colloidman@atlanticbb.net

H.G. Merkus and G.M.H. Meesters (eds.), *Particulate Products: Tailoring Properties for Optimal Performance*, Particle Technology Series 19,
DOI 10.1007/978-3-319-00714-4_14, © Springer International Publishing Switzerland 2014

14.1.1 Ultraviolet Radiation and Sun Protection

Skin, the outermost defense against noxious stimuli, is constantly assaulted by environmental insults, most prominently sun exposure; sunlight is a part of our everyday life. However, while generally perceived as "good" and beneficial to our well-being, with the advent of our knowledge in cutaneous photobiology and photodermatology within the last 50 years, it has become evident that terrestrial solar ultraviolet radiation (UVR) has become a major factor deleterious to our health [3]. For convenience, the solar UV spectrum is arbitrarily divided into three bands; the UVA from 400 to 320 nm, the UVB from 320 to 290 nm and the UVC from 290 to 100 nm.

UVC

The UVC spectral band, is referred to as the germicidal band because it is erythemogenic, mutagenic and carcinogenic but is absent at the surface of the earth because it is absorbed and filtered by the stratospheric ozone layer. Hence, the heightened concern regarding depletion of this layer by chlorofluorocarbons, nitric oxides, halons, etc., since this will, inevitably, contribute to an increase in effects harmful to human life [70]. Although solar UVC does not currently constitute a risk to the general population, UVC is emitted from artificial sources [7] and can be hazardous in certain occupations such as arc welding.

UVB

UVB rays do pass through the ozone layer and are responsible for most of the effects induced by sun exposure such as sunburn (erythema), photoaging, skin cancer and immunosuppression [20]. UVB radiation values have increased, on average, more than 6 % annually since the early 1980's [31]. Until fairly recently, the active ingredients used in the vast majority of commercial sunscreen products (see Sect. 14.1.2) absorbed primarily UVB radiation; they efficiently provided, thereby, high *sun protection factor* (SPF) products. Erythema (sunburn) is the most familiar manifestation of UVB exposure and the *minimal erythema dose* (MED) is used as a metric in the determination of the SPF number, as approved by the US Food and Drug Administration (FDA) [25].

The SPF is defined as the ratio of the time of exposure necessary to produce MED in sun-protected skin to that time for unprotected skin. However, the question of the appropriate *methodology* to determine the effectiveness of sunscreen products, as expressed by the SPF number, is of considerable international industry concern. European, American, Australian and Japanese sunscreen products compete in world markets but national marketing concerns, tradition and less than unanimous scientific opinion have hampered agreement on many aspects of significant testing details. Further, in the USA, sunscreen products are regulated as drugs; in the EU they are considered to be cosmetics. Thus, there is a difference not only in the legal status of sunscreen products between the US and the EU but also in the number of chemical ingredients approved for use as sunscreen actives.

Although UVB-blocking sunscreens decrease the risk for development of non-melanoma skin cancer [59], paradoxically, since high-SPF sunscreen products prevent sunburn, their use allows for prolonged UVA exposure without a warning sign thus encouraging susceptible individuals to spend more time in the sun [27] leading to an increased risk of the development of melanoma skin cancer. Further, issues of photo-and nonphoto-induced toxicity and allergy/irritancy have been raised with some of the UVB sunscreen actives [66].

UVA

Unfortunately, until only fairly recently the UVA region was considered to be innocuous and of little concern to dermatologists. The importance of UVA radiation, which is much more abundant – at approximately 99 % of UVR – than UVB in natural light, is now apparent [13]. UVA radiation, although of lesser energy, penetrates the skin more efficiently and deeply than UVB [29] and we know now that, in contrast to UVB, UVA radiation constitutes a severe oxidative stress [69] that can lead to DNA mutations [8]. UVA is especially efficient for generating free radicals and many of the harmful effects produced by UVB radiation are induced by UVA radiation. Thus, it is clear that providing complete broad-spectrum protection (i.e., both UVB and UVA radiation) is paramount [16].

Sadly, the action spectra for UVA-induced phenomena (including photosensitivity dermatitis and melanoma) have yet to be specifically defined and so, currently, there is no consensus of an *in-vivo* assay to assess protection against UVA. A variety of methods has been described [43] and the FDA has recently made some proposals to address this very important issue [26].

14.1.2 Sunscreen Actives

The most reliable approach to sun protection *is to cover up the skin* by using a thin film of *topical sunscreen* formulations (with SPF values of 15 or higher) and they have become an essential armament in providing protection to human skin against acute and chronic adverse effects of UVR. Topical sunscreens contain *active chemical ingredients* that absorb, scatter or reflect the incident UVR; they reduce the direct penetration and effect of UV exposure. These actives are broadly classified as organic (chemical/soluble) or inorganic (physical/particulate).

Organic (Chemical/Soluble) Actives

This constitutes by far the largest category of commercially available sunscreen actives. These chemicals are generally aromatic compounds conjugated with a carbonyl group and with typically an amine, or methoxyl, group substituted in the *ortho*- or *para*-position of the aromatic ring. They are (as mentioned earlier) primarily UVB-*absorbing* chemicals [55] and the vast majority are oil-soluble. Hence, the choice of the oily vehicle (solvent), when formulating a sunscreen product, is important; because of its polarity, it may modify the action of the sunscreening agent by shifting the absorption spectrum toward, or away from, the desired range [42, 58].

Since these actives absorb UVR, the absorbed radiation must be dissipated as either heat or light, or else be used in some chemical reaction. With the organic sunscreens this may result in the creation of reactive oxygen species (ROS) or photoproducts and this has led to questions regarding their photostability [6]. In most cases, the radiation simply is re-emitted at a longer wavelength. The skin, itself contains many compounds capable of such behavior; under radiation melanin, for example, produces free radicals [38]. Indeed, a key reason to use a sunscreen product is to "sacrifice" the extrinsic molecule (the sunscreen active) to spare the body's own chromophores.

Currently, there are very few UVA-absorbing chemicals allowed for general use in cosmetics. Only one, avobenzone, has been approved for use in the USA. It absorbs in the longer UVA wavelengths (from 350 to 390 nm) but studies have shown that it rapidly degrades under solar radiation [12] and, moreover, as avobenzone degrades other sunscreen actives in the formula may also be destroyed [61]!

Because they have been solubilized in the carrier vehicle, the organic actives can be systemically absorbed into the skin. Long-term use and increased daily exposure to these chemicals will likely result in an increased incidence of adverse reactions such as allergenicity and toxicity [57].

Inorganic (Physical/Particulate) Actives

The incorporation of sunscreen actives into daily-wear products is beginning to push the existing organic (soluble) actives to their limit. Physical (particulate) actives offer additional flexibility in the formulation of broad-spectrum, high SPF products without the concerns associated with the organic actives.

Physical sunscreen agents comprise particles of a size that scatter, reflect and absorb solar radiation in the UV, visible and IR ranges. *Particulate* is generally accepted to describe any material that is insoluble in the chosen medium. Commonly recognized particulates include zinc oxide, titanium dioxide, talc, iron oxide, kaolin, bentonite, silica and mica. These physical sunscreens are very stable and benign [10, 30]. As insoluble particulates they lay on top of the skin and are not absorbed systemically; for example, application of a 40 % zinc oxide (ZnO) ointment has not been found to raise serum zinc level [14]. An important consequence of this is the lack of opportunity for direct contact with any meaningful cellular structures. For this reason, they are perceived as safer than the organic sunscreen actives. They are especially useful in UVR-sensitive patients and they do not cause photosensitization nor any photoallergic or contact dermatitis.

Ironically, particulates were amongst the first actives to be recognized; in an article in the first volume of the Journal of the Society of Cosmetic Chemists in 1947 [28] the author discloses the UV absorption spectrum for several metal oxides including ZnO and titanium dioxide (TiO_2). From a regulatory perspective, these latter two oxides are the only ones universally allowed as sunscreen actives and they account for almost all of the physical sunscreen actives used today because they are the two most effective materials in terms of UVR attenuation efficiency and are widely available; consequently, what is discussed in the section on characteristics is focused on them. The other particulates tend to be used as adjuncts to help increase the optical path length and so "boost" the SPF number, or as fillers, especially in colored products such as face powders and lipsticks.

14.2 Sunscreen Formulations

Although the cosmetic chemist has many types of delivery vehicle from which to choose – emulsions, oils, gels, sticks, mousses, aerosols and ointments – by far the most popular of all for sunscreens, the emulsion, offers almost unlimited versatility in meeting the primary market objectives of efficacy, aesthetics and cost parameters [62]. Although emulsions are the most popular delivery type, they are also the most difficult to stabilize, particularly at elevated temperatures [5].

There are two basic "varieties" of emulsions used to formulate sunscreen products: oil-in-water (O/W) and water-in-oil (W/O). The former are used for daily-wear products such as moisturizers, while the latter are virtually exclusively used for beach-wear products. The proper oil vehicle can enhance a sunscreen's *substantitivity* or ability to remain on the skin and be effective. Substantivity is of enormous importance because sunscreen products are used outdoors and many times in settings where abundant sweating and repeated immersion in water is common [35].

There are two basic "styles" of emulsions: lotions and creams. Lotions are more popular than creams owing to their easier spreadability on the skin and dispensability (from tubes, bottles, etc.). An emulsion is termed a cream or a lotion on the basis of its viscosity (resistance to flow). The exact point at which a lotion becomes a cream is quite arbitrary but lotions typically have viscosities below 50 Pa.s.

In order to produce a true broad-spectrum product, and especially one also having a high SPF value, a combination of two or more sunscreen actives with overlapping or complimentary attenuation spectra is necessary, no matter what variety or style of emulsion is used. Indeed, the design of a sunscreen formulation is quite complex and, in addition to the sunscreen actives, it typically can comprise 15 or more components (emollients, humectants, thickeners, etc.). Thus, it often requires several iterations to produce a stable, efficacious, safe, cost effective and elegant product.

14.3 Characteristics of Physical (Particulate) Sunscreens

The development of sunscreen products containing ZnO and TiO_2 requires knowledge of both their optical properties and their surface chemistry relative to the dispersion medium.

14.3.1 Optical Properties

Here, two separate aspects need to be considered:

1. The color strength (whiteness/opacity/hiding power) of the "film" applied to the skin. What is of concern here is light in the visible part of the spectrum (from about 400 to 720 nm). The particulate active should impart as little opacity as possible. Ideally, it should be transparent to the human eye.
2. The ability of this film to act as a sunscreen by attenuating (stopping) UV radiation (specifically from 300 to 400 nm) from reaching the skin.

The first is an aesthetic concern, the second is one of performance (e.g. SPF) but they are equally important for, without proper aesthetics (minimal whitening) no level of sunscreen performance is high enough because most persons will simply not use it. Both of these aspects are uniquely related to the particle size (distribution) and particle morphology (e.g., surface area) of the material.

14.3.2 Particle Size

Both ZnO and TiO_2 are commercially available in *pigmentary* and *microfine* grades. The pigmentary grades are used to provide opacity in, for example, paints and "whiteness" in paper. Microfine is also termed micronized, ultrafine or nanosize. By definition, micronized refers to a pigment that has been physically reduced to its given particle size by grinding or milling. Microfine is a term without a universally agreed meaning; it is used commercially to describe particles with claimed particle size (PS) anywhere from 20 nm to a few micrometers. This loose terminology has led to a great deal of confusion. It is the microfine grade that concerns us when using ZnO and TiO_2 as sunscreen actives. Figure 14.1 is a schematic illustrating the typical difference in particle size distribution (PSD) between *pigmentary* and *microfine* grades of TiO_2.

Light Absorption in the Visible Spectrum

The parameter that describes the scattering-absorption characteristic of a material is its *refractive index* (RI); though an oversimplification, this single number does

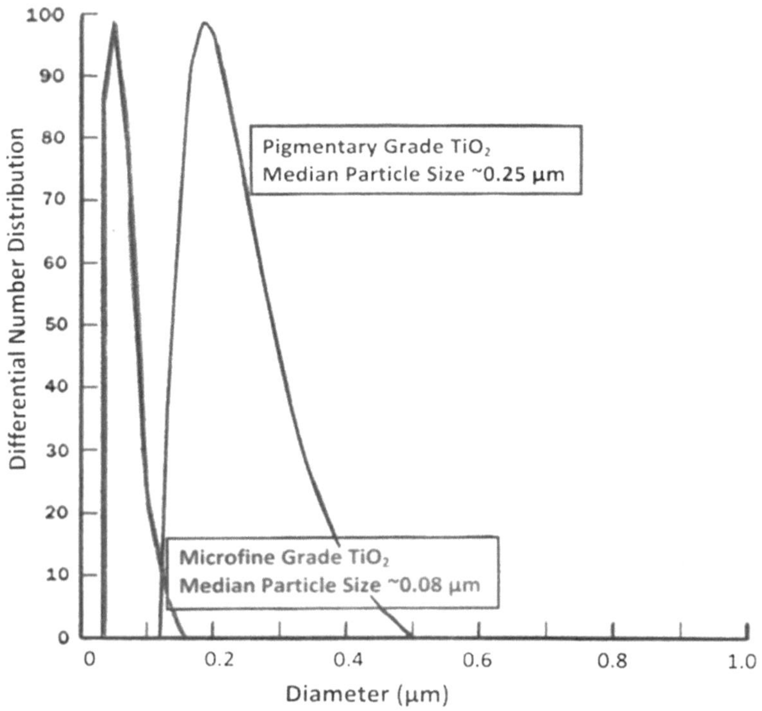

Fig. 14.1 Schematic of comparison of PSD for Microfine and Pigmentary grades of TiO_2

allow reasonable real-world predictions. Both ZnO and TiO_2 are white pigments. The actual "whiteness" of a suspension is a function of the ratio of the RI values for the material and its dispersion medium. The organic actives have RI values similar in magnitude (typically <1.5) to those of the oily vehicles in which they are solubilized. Although an emulsion will appear white in bulk, when it is spread as a layer onto skin and the water phase evaporates what is left is a thin film of the oil phase. Since the RI values are virtually matched, no matter what the concentration of organic active (and it can be as high as 30 % wt) the film will appear transparent. It is a much different situation for the physical particulates TiO_2 and ZnO; these materials have RI values of ca 2.4–2.8 and 2.01–2.03 respectively. The RI is an intrinsic property of a material and is, therefore, fixed. The particle size, however, can be chosen. As a rule of thumb, the scattering power of a material is highest when its PS is equal to about half the wavelength of the incident light. For monochromatic light this is straightforward. The problem is the visible spectrum which constitutes "white" light; it is not a single wavelength. Thus, there is no single optimum PS for a white pigment. Interestingly, the human eye has a peak response at about 560 nm (in the green portions of the spectrum).

The relation between particle size and scattering can be estimated using Lorenz-Mie theory [36] and is graphically illustrated in Fig. 14.2 for TiO_2 and ZnO using an integrated value for the wavelength of "white" light. The maxima in the two plots

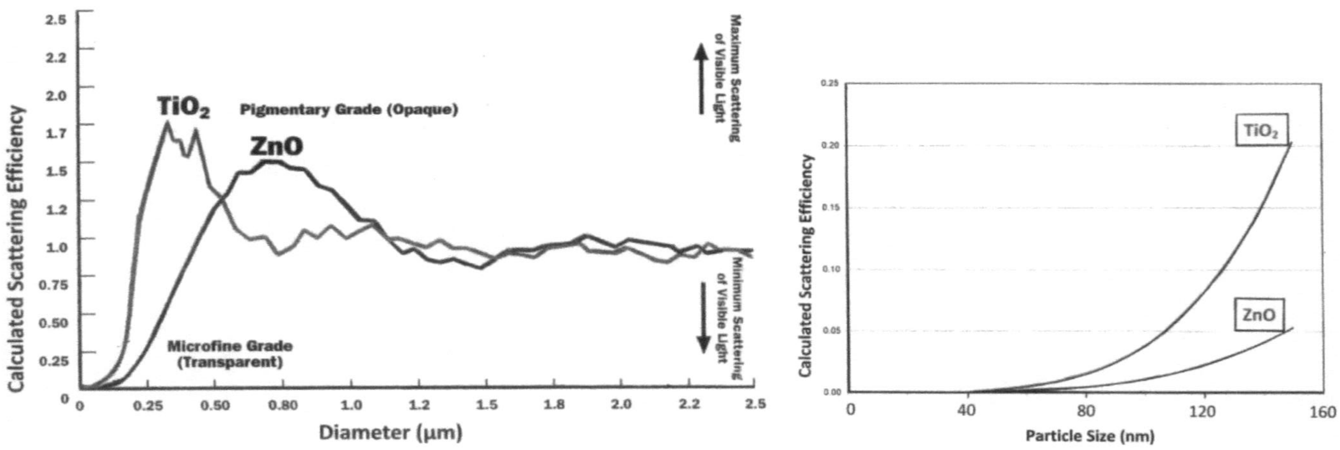

Fig. 14.2 Extinction efficiency against wavelength for aqueous suspensions of TiO_2 and ZnO; *at right*: graph expanded for small particle sizes

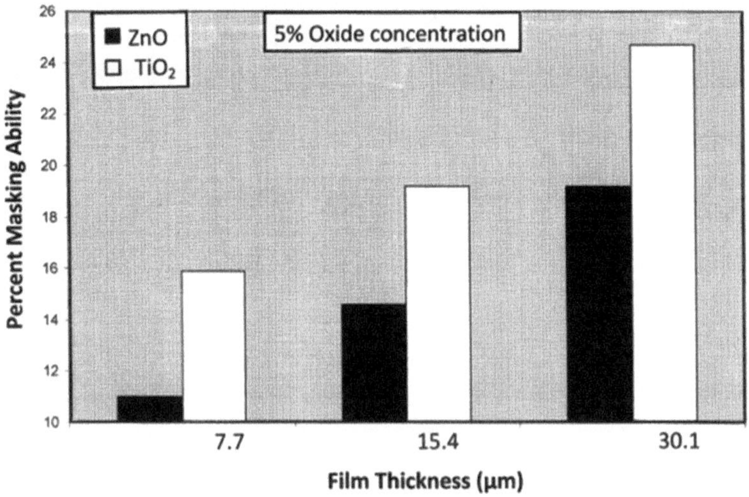

Fig. 14.3 Masking efficiency against film thickness for TiO_2 and ZnO

are the "sizes" where maximum scattering (i.e. highest opacity) will be obtained for the two materials. In other words, for the pigmentary grades the greatest hiding power occurs at something around mean sizes of 250 nm for TiO_2 and 750 nm for ZnO (see also Chaps. 2 and 12).

These theoretical calculations are based on oversimplified assumptions and it is not possible to derive a mathematical treatment in any rigorous form for the optical properties of an actual film such as that from a sunscreen emulsion, although attempts have been made for paint films [52]. What is clear is that for all PS sizes less than 500 nm, TiO_2 is a much more efficient light scatterer than ZnO and so will be more difficult to "hide" in a topical product. The cosmctic clegancc of thcsc sunscreen actives is limited by the degree of whiteness they impart. Figure 14.2, *right*, expands the plot below 200 nm from which it can be seen that a microfine grade of ZnO or TiO_2 with a PS less than about 150 and 100 nm respectively should be effectively transparent to the human eye.

Film Thickness

The thickness of a film and the number of particles in a unit volume of that film will both impact the degree of scattering. This is illustrated in Fig. 14.3 in which the relative masking ability of microfine ZnO and TiO_2 is compared at 5%wt concentration as a function of film thicknesses. Although both ZnO and TiO_2 have allowed use levels of up to 25 % [25], typical concentrations in commercial sunscreen products are from 2 % to 8 %. The procedure followed to obtain the data in Fig. 14.3 was adapted from an ASTM standard method for determining the hiding power of paints [4]. It can be seen that, at a given film thickness, TiO_2 has almost twice the hiding power of ZnO. The practical implication of this in the formulation of a sunscreen product is that a much smaller TiO_2 particle size (PS) (typically ca 50–90 nm) is needed for "transparency". Furthermore, maintenance of this size throughout formulation is critical. Should agglomeration occur the TiO_2 reverts to behaving as a pigmentary material thereby increasing interaction in the visible and, hence, appearing white. In this respect, ZnO is a much more forgiving pigment.

The amount of sunscreen product mandated by the US FDA in the determination of SPF *in-vivo* [25] is 2 mg/cm^2 and this would represent a film thickness of approximately 20 μm. It is most important that the sunscreen application be very uniform; large variations of SPF value are obtained with irregularly applied amounts of sunscreen. Hence, technicians in testing laboratories are carefully trained in this *in-vivo* technique. Studies suggest that, in "real-life" use, the average individual typically applies < 2 mg/cm^2 [60, 65]; this could result in less protection than the product label SPF value suggests.

Light Absorption in the UV Spectrum

Scattering and adsorption are not restricted to the visible spectrum. However, at shorter wavelengths the optical behavior of ZnO and TiO_2 becomes more complex. Not only the PS but also the crystal structure and the energy gap between valence and conduction bands have to be taken into consideration. For all crystals there is some characteristic wavelength that will excite an electron from the valence band to the conduction band. This is the process through which these metal oxides owe their semiconductor properties. Wavelengths that are shorter than this characteristic wavelength and that fall on the crystal will, therefore, be dissipated (i.e. absorbed by the crystal structure).

Being crystalline semiconductors, they have the ability to contain, internally, the production of ROS. As with the organic actives, most of the radiation can be re-emitted at a longer wavelength (e.g., heat). However, they may also produce oxygen free radicals at their crystal surface; this *photoreactivity* has been extensively studied [71]; generally TiO_2 is much more photoactive than ZnO [50]. Although it has been suggested that photoactive metal oxides may initiate deleterious events in the skin, to do so a particle would have to traverse the *stratum corneum*; TEM studies have concluded that there is no evidence that this occurs [18]. Furthermore, for a variety of reasons, manufacturers of ZnO and, especially, TiO_2 apply coatings (both inorganic and organic) [74] and these result in the reduction or even elimination of photoreactivity [37]. In the US, owing to regulatory concerns, virtually no uncoated ZnO or TiO_2 is used as a sunscreen active.

The ZnO crystal structure efficiently absorbs radiation over virtually the whole UV spectrum (380 nm and shorter); the attenuation efficiency varies very little with PS from about 80 to about 150 nm. In contrast TiO_2 is predominantly a very efficient *UVB absorber* but a *UVA scatterer*. For example, at 360 nm about 90 % of TiO_2's attenuation is via scattering, whereas for ZnO about 90 % is by absorbance. Thus, when using ZnO and TiO_2 as sunscreen actives for human applications

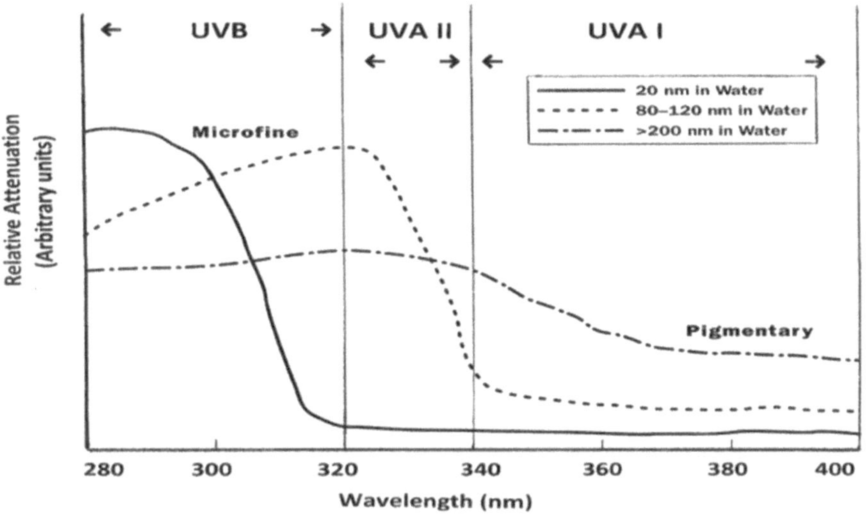

Fig. 14.4 Particle size dependence of UVR attenuation for TiO_2

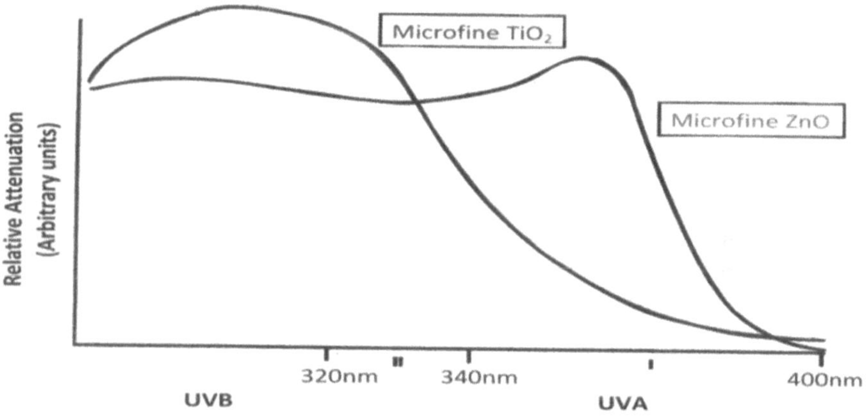

Fig. 14.5 Schematic of comparison of UVR attenuation for microfine TiO_2 and ZnO

this distinction has some important consequences. Figure 14.4 shows the efficacy of attenuation for TiO_2. Very small (ca 20 – 60 nm PS) TiO_2 particles that are non-whitening are very efficient in attenuating in the UVB but are very poor at attenuating UVA radiation; larger 80 – 100 nm TiO_2 particles have good UVA attenuation but lose their ability to attenuate UVB. Microfine TiO_2 and ZnO can, therefore, be considered complimentary in terms of their sun protective effects (Fig. 14.5); TiO_2 is primarily used for UVB protection and ZnO for UVA protection. Thus, they are often used in combination to obtain true broad-spectrum coverage.

14.4 Measurement of Particle Size

Although both uncoated ZnO and TiO_2 are hydrophilic and could be dispersed in the water phase of either an O/W or a W/O emulsion, more often than not these particulates are dispersed into the oil phase because this provides a continuous film on the skin after dry-down and also there are a wide variety of primary oil vehicles from which to choose – including mineral oils, fatty esters, triglycerides and silicone fluids [73] – depending upon the aesthetics required, such as skin feel. In this case, coated versions of the ZnO or TiO_2 are used in which the coating is chosen to match as closely as possible the vehicle – for example a dimethicone coated ZnO in a silicone fluid. It is critical that an optimized stable dispersion be prepared before the other components are added and the whole then emulsified. Indeed, "pre-dispersions" can be obtained from cosmetic ingredient suppliers eliminating this initial step in the formulation. In either case the particle size needs to be determined.

14.4.1 Techniques

Particle size analysis techniques are often applied inappropriately, primarily because of a lack of understanding of the underlying principles of size analysis and confusion arising from claims and counter-claims of the analytical ability of the various size determination techniques and the wide choice of commercially available instrumentation. Further, techniques will *weight* the raw data differently. An excellent comprehensive treatise on the whole subject of particle size measurement has recently appeared [48].

Most data, particularly in pharmaceutical applications, is useful when quoted as volume-weighted since drugs are delivered by volume or mass (proportional to volume for a fixed density). In sunscreen applications, it is number-weighted data that is preferred because it more closely correlates with coating and SPF efficiency.

14.4.2 Dynamic Light Scattering/Photon Correlation Spectroscopy

The size of these microfine pigments would seem to make them candidates for measurement by dynamic light scattering (also termed photon correlation spectroscopy). However, DLS is a technique in which the signal, from which the size distribution is calculated, is a sum over all the signals from all the particles during the entire measurement [72]. Thus, the results are an average over an ensemble of particles. In general the resolution is low since the particles are neither counted nor physically separated. Instead, over the duration of the entire measurement, particles of all sizes contribute to the final signals. Then the signals are mathematically deconvoluted to produce size distribution information. As these signals are not simple linear functions of particle size, but often complicated functions of size, it is not surprising that information is lost. Also, DLS requires that the dispersion under investigation be extremely dilute; as mentioned earlier typical use concentrations of microfine pigments are from 2 % to 8 % wt. The viscosity of the oily vehicles can be >10X that of water. This slows diffusion increasing the length of time to obtain a useful autocorrelation function necessitating repeat measurement and lengthening the total time to obtain reliable data. A complication is that both ZnO and TiO_2 are dense materials ($\rho_P = 5.6$ and 4.2 respectively); even at the microfine size settling occurs and this impacts reproducibility.

Finally, in DLS measurement of very small particles the intensity of the light scattered is weighted by the 6th power of the diameter. Hence in relatively broad PSD as found in microfine pigments the data tends to be skewed towards the larger end of the PSD. Further, most data, particularly in pharmaceutical applications, is most useful when quoted as either volume- or number-weighted values. Although it is a simple matter to write equations for converting one type of weighting to another, the results calculated this way are often erroneous, especially in the case of intensity-weighting – in part because of the assumptions that need to be made. A fundamental tenet of particle size analysis is, if at all possible, to choose a technique whose fundamental weighting is closest to that which is needed, or used, in practice [2].

14.4.3 X-ray Disc Centrifugation

Given the above, the preferred method for PSD measurement of ZnO and TiO_2 pigments tends to be X-ray disc centrifugation [47]. In any particle sizing techniques based on an optical system to monitor the sample, the extinction efficiency of particles of different sizes must be taken into account [68]. Such corrections become unnecessary when the detection is based on X-rays rather than visible radiation. X-rays are absorbed in direct proportion to the total mass of particles present. X-ray detection is therefore ideal for metal oxides in general. By using both combining gravitational sedimentation with centrifugal sedimentation an extremely wide range of sizes can be analyzed. Importantly, for an instrument that weights by volume (such as an XDC device), even if a few of the larger particles are missed, the volume-weighted diameter is not much affected. The maximum particle concentration allowed is 2 % vol, thus the typical use concentrations of ZnO and TiO_2 can be measured directly, without dilution.

An example is shown in Fig. 14.6 that compares the PSD (measured using an XDC device) for two commercial samples of supposedly equivalent microfine ZnO supplied by different manufacturers; in this case, 5 % wt suspensions of the ZnO materials were prepared in isopropyl myristate (IPM), an ester commonly used in sunscreen formulations [73].

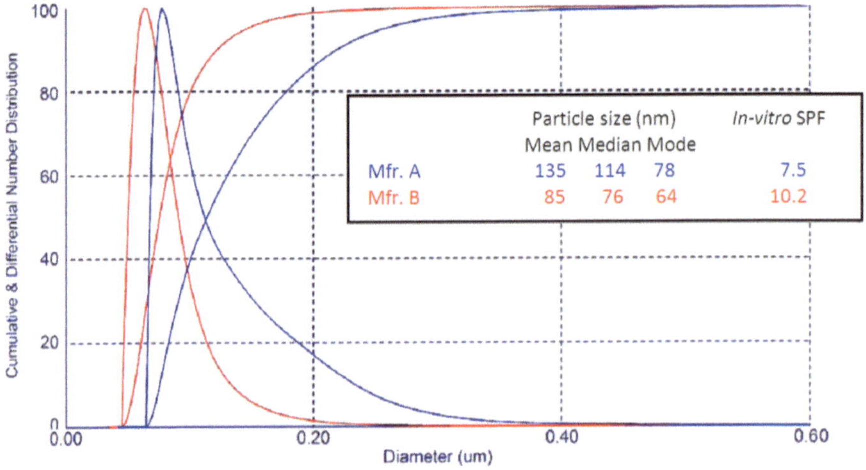

Fig. 14.6 Comparison of PSD for two microfine ZnO samples

The particle size data is displayed as number-weighted values. The resolution obtained is unlikely to be achieved with DLS. Although the modal values (the majority fraction comprising any PSD) for the two pigments are very similar (at 64 and 78 nm), the "tails" of each PSD (made up of massive single particles, aggregates and agglomerates) differ enormously and this critically affects the resultant SPF values that would be obtained. The SPF values shown in Fig. 14.6 were determined from *in-vitro* measurements [15] using the 5 % ZnO dispersion in IPM and the data shows that the ZnO from manufacturer B is clearly superior. While not allowed for labeling purposes, *in-vitro* SPF measurements are routinely used to screen formulations before submission for *in-vivo* measurements which are extremely expensive, using as they do human volunteers. Since the microfine ZnO tends to be the most expensive component in a sunscreen product, the economic impact of choice of this material can be considerable.

14.4.4 Acoustic Attenuation Spectroscopy

Ideally, it would be preferable to be able to measure the PSD of microfine ZnO and TiO_2 in an actual finished sunscreen product. Alas, cosmetic (sunscreen) emulsions are multi-component systems that contain, in addition to the microfine pigments, emulsified oil (or water) droplets and solid particles such as waxes and fillers that have particle sizes well into the micrometer range.

It has been long known that ultrasound based techniques open an opportunity to eliminate dilution [46]. However, instrumentation hardware was too complex and suffered from weaknesses in practical application of the technique to systems of high internal phase concentration. In addition, the theoretical basis of the

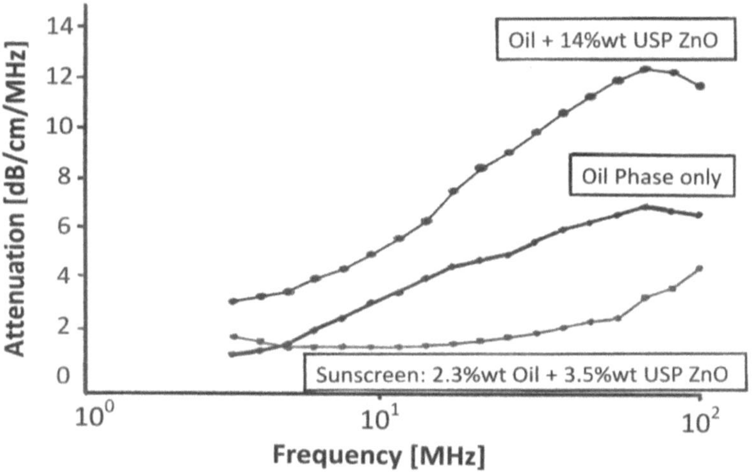

Fig. 14.7 Ultrasound attenuation spectra of an oil phase alone, with predispersed ZnO and diluted to the ultimate sunscreen

deconvolution of the raw data was not capable of dealing with particle-particle interaction – an essential feature of concentrated dispersions. Recently, these problems have been addressed [17] and initial results, from studies of a series of commercially formulated cosmetic (O/W and W/O) emulsions, suggest that acoustic attenuation can be used be used as a formulation aid to "fingerprint" emulsion composition [23].

Figure 14.7 compares the attenuation spectra of three different "components" that would typically be produced during the course of the formulation, or manufacture, of a sunscreen product. These were an "oil phase", the same oil phase containing predispersed 14 % wt USP ZnO and the final sunscreen following addition of water and homogenization. The spectra are clearly different and track the compositional nature. There is an increase in attenuation of the oil phase upon addition of the particulate ZnO because, as a solid, it has a larger attenuation than any liquid. However, the final sunscreen has the smallest attenuation; it is considerably less than the oil phase even though it contains ZnO. This is primarily because a large concentration (ca 60 % wt) of water was added (to create the final emulsion) and water has the smallest attenuation (almost zero) of any liquid. In addition, the oil/ZnO attenuation is also significantly reduced because the concentration of that component is now only about 35 % wt; hence, the actual ZnO concentration in the sunscreen is only 3.5 % wt. Thus, the use of the attenuation spectrum as a measure of reproducibility and repeatability at any stage of formulation could lead to improvement in the consistency of the final product and, overall, result in a cost benefit.

The raw attenuation spectra for two examples of non-ionically stabilized O/W moisturizers are shown in Fig. 14.8. Primarily they differ in that one also contains 3.5 % wt of a microfine "hydrophobically coated" ZnO; they are clearly

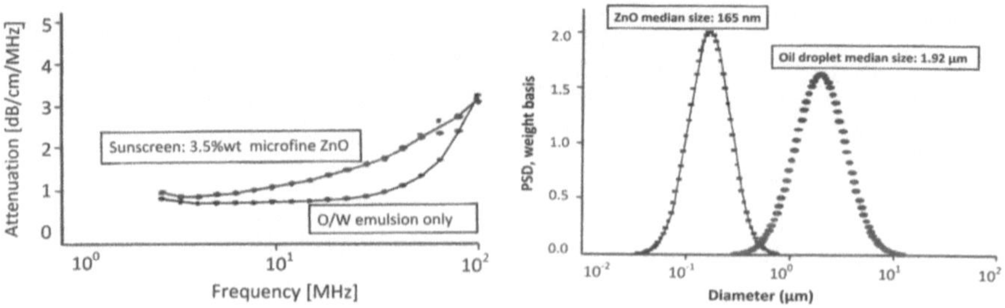

Fig. 14.8 Particle size distributions (*right*) and ultrasound attenuation spectra (*left*) of a ZnO sunscreen and its O/W emulsion basis

distinguishable from one another and it is also apparent that the shapes of the two spectra are dissimilar. Note also, the good reproducibility of the technique. While not straightforward, it is possible to obtain the PSD of both the oil droplets and the ZnO, dispersed therein from the two spectra. In contrast to light scattering methods, the raw data so obtained from AAS measurements is *volume*-weighted [17]. The PSD is calculated assuming a log-normal distribution (Fig. 14.8, *right*); the median values of 1.92 μm and 165 nm are typical for emulsion oil droplets and microfine ZnO particles respectively. Indeed, the ZnO size is in very good agreement with the mean volume-weighted PS value of 148 nm, measured for a predispersion of the ZnO in the oil phase prior to emulsification using an XDC.

14.5 Particle Shape

Particle shape is a factor of considerable importance in determining the overall properties (especially those of a physico-mechanical nature) of colloidal systems and this includes dispersions of microfine ZnO and TiO_2. The activity of catalysts is a consequence not just of their surface-to-volume ratio but also the particle shape; different crystal facets result in different selectivity [1]. The difference in crystal shape is also a reason why uncoated TiO_2 is much more photoreactive than uncoated ZnO.

It has long been recognized that the shape of TiO_2 and ZnO pigment particles affects the way they pack in a paint film [40]. In sunscreens it is critical for optimum efficacy that the active particles be evenly distributed within the oil phase so that a uniform coating can be applied to the skin. TiO_2 has a compact tetragonal crystal lattice; that for ZnO is hexagonal. Because of their different symmetry they will pack differently; in uniform films, at equal concentrations, microfine TiO_2 is a much more efficient UVB absorber than ZnO.

While it is known that particle shape can be manipulated during the manufacturing formation [9] or altered during subsequent processing [49], for two important reasons there has been no impetus for formulators to use anything other than standard production material. The first relates to cost. Contrary to what the consumer might think when looking at the price paid at a retail store for a 120 mL (4 oz) tube, or bottle, of sunscreen product, the actual total raw ingredient cost is quite low; typical figures (including manufacturing) for a W/O beach-wear product and an O/W daily-wear (moisturizer) product, both containing microfine ZnO, are about $8.50/kg and about $7.00/kg respectively. The profit margin on manufacturing is very small and manufacturers are reluctant to pay even a few cents more for an ingredient. The second relates to formulation. As mentioned earlier, it is critical that an optimized stable dispersion initially be prepared and then kept so throughout the whole formulation process. Even starting with a material of optimum morphology and a finely controlled PSD, poor formulation will always result in aggregates and agglomerates that result in inferior aesthetics and mediocre performance.

14.6 Particle Surface Area

It is clear that the particle size of particulate sunscreen actives is important, determining as we have seen the aesthetics and performance of sunscreen products formulated with them. It is also evident that as the size is reduced during the dispersion process the particle surface area will increase (as $1/D^2$, for spheres). However, what is less obvious is the exceptionally large surface area-to-volume ratio per given mass for the particles involved – an essential characteristic common to all colloidal dispersions and of critical import to microfine pigments and especially so with nanoparticulate systems; it becomes orders of magnitude larger than it is for particles of even only a few micrometers in size [21].

In practical terms, the total *wetted* surface area of a given suspension is a consequence of the concentration and size of particles. The latter derives from the specific dispersion process used, *viz* the choice/type and concentration of dispersion aids (i.e., wetting agent, de-agglomerating agent and stabilizing agent) and the type/duration of the mechanical dispersing treatment [22].

An approximate amount of surfactant or dispersant required for surface coverage can be estimated using the Sauter mean diameter (SMD; $D_{3,2}$) [48]. SMD is an average of the particle size traditionally used in applications where active surface area is important (e.g., in catalysis). NMR relaxation measurements can be used to study adsorption at interfaces [24] and so offers the possibility of determining optimum coverage directly and, moreover, using industrially relevant particle suspension concentrations.

The practical significance to product performance and the economics of manufacture of the use of excessive mechanical attrition during the formulation of wet suspensions that results in "fines" can be considerable [11]. This is especially so for crystalline materials such as ZnO and TiO_2 in which fracture can occur at defect crystal planes. Monitoring the presence of nanoparticles in suspensions that have a broad particle size distribution is a general problem for many current particle sizing instruments.

In addition, the increase that occurs in particle surface area can dramatically affect not just adsorption of surfactants/dispersants and other moieties onto the particle surface but also the interaction between particles and system properties such as suspension rheology, coating and adhesion. All of these performance factors impact sun products formulated to contain microfine ZnO and TiO_2.

14.6.1 Measurement of Particle Surface Area

BET Method

The most common method of particle surface area determination is BET nitrogen (N_2) gas adsorption [44]. This technique requires the material under test to be a dry powder. However, drying wet suspensions inevitably results in unintentional (and unwanted) aggregates and agglomerates. As a consequence, the subsequent surface area results will be skewed. Indeed, it has long been accepted that any procedure which requires a dry sample (such as BET or Knudsen flow of a gas at low pressure) may give an area which, for all its accuracy, has little relation to the effective area in suspension [34].

It should be obvious that the state of dispersion will critically affect the particle surface area and, to be useful, it is essential that a technique appropriate to the application be used. While it is possible to calculate surface area from measurements of the particle size this assumes spherical particles and a monodisperse size distribution, a condition clearly not met by microfine pigment dispersions; any calculated surface area is, at best, only a crude approximation and it is well-recognized that particle shape, surface irregularities and porosity will inevitably lead to estimated values significantly less than the true value [2].

NMR Relaxation

It is clear that it is the *wetted* particle surface area of any suspension (and especially for colloidal dispersions and nanoparticulate systems) that is the vital metric in quantifying the performance of such materials and, ideally, measurement should be made directly on suspensions as formulated.

High resolution nuclear magnetic (NMR) spectroscopy is one of the most powerful analytical tools used to probe details of molecular structure and dynamics. The determination of wetted surface area from NMR relaxation measurements is well-established [45] and, with the advent of small powerful permanent magnets, a low field (13 MHz) NMR bench-top device is now available to measure the wetted particle surface area of suspensions [24, 56]. The technique is based on the fact that liquid in contact with, or "bound" to, the particle surface behaves differently than the bulk or "free" liquid. The NMR relaxation time of bound versus bulk liquid is markedly different: the relaxation time of the latter is much longer. There are no

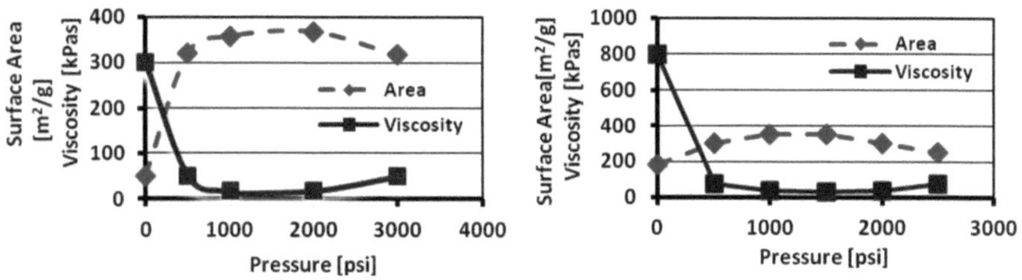

Fig. 14.9 Surface area and viscosity as a function of mill pressure; (*left*) without dispersant; (*right*) with 0.1 % dispersant

assumptions made about the particle size (distribution) or shape in the determination of the wetted surface area. Measurement of virtually any particle in any fluid is accessible and the technique does not require dilution of the sample.

Figure 14.9 (*left*) shows the effect of milling a pre-mix high solids (52 % wt) suspension of a microfine grade ZnO in C_{12}-C_{15} alkyl benzoate (widely used as an emollient in cosmetic formulations for the skin such as sunscreens and facial moisturizers). No dispersing agent was used; the pre-mix was initially prepared using high shear (rotor/stator) mixing and then further processed using a cavitation device run at increasing processing pressures. The initial wetted surface area of pre-mix was determined to be 50 m^2g^{-1} and the optimum (milled) value was 365 m^2g^{-1}; for this particular ZnO material, milling at pressures over 2,000 psi ($1.4 \cdot 10^4$ kPa = 140 bar) is, clearly, counter-productive.

Figure 14.9 (*right*) then demonstrates the effect of using a dispersing agent. The same ZnO was used; a dispersing agent – polyhydroxystearic acid (PHSA) – was added at a concentration of 0.1 % wt (determined empirically) and it was found possible in the pre-mix to significantly increase the ZnO solids loading (to 60 % wt) without any detrimental change in pre-mix viscosity. Importantly, with the use of only 0.1 % wt PHSA, the initial pre-mix wetted surface area increases 50 – 180 m^2g^{-1} and an optimum (milled) value of was 355 m^2g^{-1} at only 1,500 psi (10^4 kPa = 100 bar).

In both experiments the dispersions were all measured at full concentration, without dilution. The correlation of milling with surface area measurement is quite clear and since each surface area measurement took less than five minutes this suggests that NMR relaxation could prove useful in monitoring a milling process, almost in real-time, to obtain optimum conditions thus saving time and money – clear economic advantage to manufacturer. The data further suggests that NMR relaxation could be used to optimize the amount of dispersant used.

Although the flow behavior of real commercial suspensions is complex – the systems are visco-elastic (VE) – it is directly related to the particle size distribution (PSD) of the dispersed phase [51]. Thus, milling should result in changes in viscosity and this is evident also from Fig. 14.9; milling dramatically decreases the dispersion viscosity. A lower viscosity means that the ZnO dispersions are easier to pour making them more “processable”, an advantage since these materials will subsequently be used in the preparation of emulsions.

14.7 Particle Surface Chemistry

Whatever the so-called primary PS of ZnO or TiO_2 particles, it is the state of dispersion of the final product that determines the "effective" PS and thus the optical properties. However, the preparation of wet suspensions having a defined dispersed state, with specific properties, is often difficult to achieve because complete knowledge of the dispersed system is usually lacking. It is probably fair to say that most industrial (cosmetic) methods of formulation have been developed and optimized empirically, or on the basis of working hypotheses that provide adequate guidance to the formulation chemist, but are insufficient as a basis for understanding how and why suspensions of particles in liquids behave as they do and, many times, not as the formulator wants.

And this is never more so true than in the formulation of particulate-based sunscreen products where a stable (non-agglomerating) reproducible PSD needs to be achieved. There is a natural tendency to treat all particulates as being alike. This makes no more sense than treating all the organic (soluble) sunscreens as being the same. All oxides have quite distinct surface chemistries [33]. For example, the point of zero charge (PZC) for pure, uncoated TiO_2 and ZnO are found at ca pH6 and ca pH9 respectively [39]. This means that in neutral water (pH7, 25 °C) TiO_2 will be mildly acidic and have a small negative charge whereas ZnO will be basic and positively charged. The actual PZC for a specific material is influenced by its source or preparation method, pretreatment and presence of trace impurities [53, 64]; and should be determined batch-to-batch and lot-to-lot; there is probably greater manufacture-to manufacture variation with the particulates than with the organic (soluble) sunscreen actives. Add to this the wide variety of surface treatments and the result is that sunscreen formulation is more of an art than a science. And it is this, more than anything else, which to-date has limited the wider adoption of particulate actives.

The subjects of surface chemistry and the dispersion of particulates, though important, are beyond the scope of this chapter. There are a variety of references to which the interested reader is directed [22, 41, 54, 67].

14.8 Regulatory Issues

Because of perceived concerns regarding the safety of so-called nanoparticles (NP), especially because of a less than well-informed public and organizations with their own agendas, regulatory bodies both in the USA [75] and the EU [19] have recently presented "guidelines" regarding the manufacture and the use, and the application, of "nanotechnology" as it relates to cosmetics. In 2007, an FDA Task Force issued a report in which nanotechnology was defined as "the ability to measure, see, manipulate and manufacture *things* usually between one and 100 nm". In 2011, the European Commission defined a "nanomaterial" as "anything made up of natural, or manufactured, particles that are *unbound or aggregated/agglomerated*. At least 50 % of particles *must have one or more dimensions* between 1 nm and 100 nm and/or have a *specific surface per unit volume* of greater than 60 m^2cm^{-3}". Although presently just "guidelines", this author has little doubt that the regulatory agencies will move to make them "regulations".

The Hepatitis B virus (42 nm in diameter) is biologically active and penetrates the skin. Thus, as long as the minimum particle size of a pigment is not less than 42 nm it should be physically impossible for such an "inert" particle to be absorbed through the skin; however, the oily components of a sunscreen formulation may impact potential penetration. Penetration of NP can be aided by damage of the skin, e.g. by removing hairs. Although studies of TiO_2 and ZnO NP suggest that they are mainly localized in hair follicle openings and on the stratum corneum surface, there are indications for cytotoxicity (TiO_2) and genotoxicity (ZnO) [63]. The picture is less certain for particles whose primary size is <40 nm. A complicating factor is that NP are often agglomerated in formulation so that the "effective" PS is much larger. A recent overview of the pathology and toxicology of nanoparticulates concludes that significant data gaps exist on skin penetration and the toxicology of true NP [32].

Notwithstanding the fact that "fines" have been present for decades in commercially available pigments, that ZnO, for example, has been used a skin therapeutic agent for at least 300 years and that a search of available literature has not revealed a single adverse reaction, from this authors perspective these guidelines now impose an unnecessary need to measure, or evaluate, not just the particle size of particulate actives (and now as a **number**-weighted value and not on the traditional volume-weighted basis) but also the surface area of materials (and the guidelines do not specify whether it is to be determined on a sample of the dry powder or on a suspension….) and, potentially, an added requirement to determine particle characteristics such as the aspect ratio.

All this will certainly have an impact, as yet unknown, on the use of microfine grade pigments as sunscreen actives.

14.9 Definitions, Abbreviations and Symbols

BET equation	Linearized equation to extract the specific surface area from gas adsorption results at different gas pressures (see Sect. 3.6.3)
Emulsion	(Stable) dispersion of small liquid droplets in another, immiscible, liquid
Substantivity	Ability of a sunscreen to remain on the skin and be effective
BET	Brunauer-Emmett-Teller
DLS	Dynamic light scattering
ESD	Equivalent spherical diameter: the diameter of a sphere of equivalent volume of an irregular-shaped particle, measured according to a defined technique

(continued)

(continued)

EU	European Union
FDA	Food and Drug Administration (USA)
IPM	Isopropyl myristate
MED	Minimal erythema dose
NMR	Nuclear magnetic resonance
NP	Nanoparticle
O/W emulsion	Oil-in-water emulsion
PHSA	Polyhydroxystearic acid
PS	Particle size
PSD	Particle size distribution
PZC	Point of zero charge: the pH at which oxide particles dispersed in water have no effective surface charges
RI	Refractive index
ROS	Reactive oxygen species
SPF	Sun protection factor
TEM	Transmission electron microscopy
US(A)	United States of America
UVA	Ultraviolet radiation in 320 – 400 nm range
UVB	Ultraviolet radiation in 290 – 320 nm range
UVC	Ultraviolet radiation in 100 – 290 nm range
UVR	Ultraviolet radiation
VE	Visco-elastic
W/O emulsion	Water-in-oil emulsion
XDC	X-ray disc centrifuge
D	Particle size
% vol	Percentage by volume
% wt	Percentage by weight
ρ_P	Particle density

References

1. Ahmadi, T.S., et al.: Science **272**, 1924 (1996)
2. Allen, T.: Particle Size Measurement, 4th edn. Chapman and Hall, London (1990)
3. Armstrong, B.K., Kricker, A.: *Cancer Surveys – Skin Cancer*, pp. 133–153. Imperial Cancer Research, Oxford (1996)
4. ASTM D2805-11: Standard Test Method for Hiding Power of Paints by Reflectometry. ASTM, Philadelphia (2011)
5. Becher, P.: Encyclopedia of Emulsion Technology, 2nd edn. Marcel Dekker, New York (1985)
6. Beck, I., et al.: Int. J. Cosmet. Sci. **3**, 139–152 (1981)
7. Birmingham, D.J.: Prog. Dermatol. **3**, 1–8 (1968)
8. Black, H.S., et al.: Photochem. Photobiol. **40**, 29–47 (1997)
9. Buckley, H.E.: Crystal Growth. Wiley, New York (1951)
10. Butler, H.: Poucher's Perfumes, Cosmetics and Soaps, 10th edn. Kluwer Publisher, Dordrecht (2000)

11. Daniel, F.K.: Natl. Paint, Varn. Lacquer Assoc. Sci. Sect. Circ. No.**744**, (1950)
12. Deflandre, A., Lang, G.: Int. J. Cosmet. Sci. **10**, 53–62 (1988)
13. de Laat, J.M., de Gruijl, F.R.: Cancer Surv. **26**, 173–191 (1996)
14. Derry, J.M., McLean, W.M., Freeman, J.B.: J. Parenter. Enteral Nutr. **7**(2), 131–135 (1982)
15. Diffey, B.L., Robson, J.: J. Soc. Cosmet. Chem. **40**, 127–133 (1989)
16. Diffey, B.L.: Int. J. Cosmet. Sci. **16**, 47–52 (1994)
17. Dukhin, A.S., Goetz, P.J.: Ultrasound for Characterizing Colloids. Elsevier, Amsterdam (2002)
18. Dussert, A.S., Goori, E., Hemmerle, J.: Int. J. Cosmet. Sci. **19**, 119–129 (1997)
19. ec.europa.eu/nanotechnology/
20. Epstein, J.H.: In: Smith, K.C. (ed.) The Science of Photobiology, 2nd edn, pp. 155–192. CRC Press, Boca Raton (1989)
21. Everett, D.H.: Basic Principles of Colloid Science. RSC Publications, London (1988)
22. Fairhurst, D., Mitchnick, M.A.: In: Lowe, N.J., Shaath, N.A., Pathak, M.A. (eds.) Sunscreens: Development, Evaluation and Regulatory Aspects, 2nd edn. Marcel Dekker, New York (1997)
23. Fairhurst, D., Dukhin, A.S., Klein, K.: ACS Symposium Series 472, Chapter **16** 249 (2004)
24. Fairhurst, D., Prescott, S.: Spectrosc. Eur. **23**(4), 13–16 (2011)
25. Federal Register: Sunscreen Drug Products for Over-the-Counter Human Use; Tentative Final Monographs: Proposed Rule, The Code of Federal Regulations, 21 CFR Part 352, pp. 28–33. FDA, Washington, DC (1998)
26. Federal Register: Final Rule: Testing and Labeling Requirements, 21 CFR 201.327, 17 June 2011
27. Garland, C.F., et al.: AEP **3**, 103–110 (1993)
28. Grady, L.D.: J. Soc. Cosmet. Chem. **1**(1), (1947)
29. Groves, G.A.: Drug Cosmet. Ind. **37**(2), 155 (1994)
30. Harry, R.G.: The Principles and Practice of Modern Cosmetics, vol. 2. Leonard Hill, London (1963)
31. Herman, J.R., et al.: Geolphys. Res. Letter **23**, 2117 (1996)
32. Hubbs, A.F., et al.: Toxicol. Pathol. **39**(2), 301–324 (2011)
33. Hunter, R.J.: Zeta Potential in Colloid Science. Academic, London (1981)
34. James, R.O., Parkes, G.A.: In: Matijevic, E. (ed.) Surface and Colloid Science, vol. II. Wiley, New York (1980)
35. Kaidby, K.H.: In: Lowe, N.J., Shaath, N.A. (eds.) Sunscreens: Development, Evaluation and Regulatory Aspects. Marcel Dekker, New York (1990)
36. Kerker, M.: The Scattering of Light and Other Electromagnetic Radiation. Academic, New York (1969)
37. Kobayashi, M., Kalreis, W.: Sun Products: Protection and Tanning, pp. 142–146. Allured Publishing, Carol Stream (1998)
38. Kochevar, I.E.: In: Gilchrist, B.A. (ed.) Photodamage, pp. 201–220. Blackwell Science, Cambridge (1995)
39. Kosmulski, M.: Surface Charging and Points of Zero Charge. CRC Press, Boca Raton (2009)
40. Lambourne, R. (ed.): Paint and Surface Coatings; Theory and Practice. Ellis Horwood, Chichester (1987)
41. Laskowski, J.S., Ralston, J.D. (eds.): Colloid Chemistry in Mineral Processing. Elsevier, New York (1992)
42. Lowe, N.J., Weingarten, D., Wortzman, M.: J. Dermatol. Surg. Oncol. **17**, 744–746 (1991)
43. Lowe, N.J., Friedlander, J.: In: Lowe, N.J., Shaath, N.A., Pathak, M.A. (eds.) Sunscreens: Development, Evaluation and Regulatory Aspects, 2nd edn. Marcel Dekker, New York (1997)
44. Lowell, S., Shields, J.E.: Powder Surface Area and Porosity. Chapman Hall, London (1984)
45. Maciel, G.E.: In: Grant, D.M., Harris, R.K. (eds.) Encyclopedia of Nuclear Magnetic Resonance. Wiley, New York (1996)
46. Marlow, B.J., Fairhurst, D., Pendse, H.P.: Langmuir **4**, 611–626 (1983)
47. McFadyen, P.: In: Freer, R. (ed.) Nanoceramics. Institute of Materials, London (1993)

48. Merkus, H.G.: Particle Size Measurements; Fundamentals, Practice, Quality. Springer, Netherlands (2009).
49. Mitchnick, M.A.: Sun Products: Protection and Tanning. C&T Ingredient Resource Series. Allured Publishing, Carol Stream (1998)
50. Mitchnick, M.A., Fairhurst, D., Pinnell, S.R.: J. Am. Acad. Dermatol. **40**, 85–90 (1999)
51. Mollet, H., Grubenmann, A.: Formulation Technology. Wiley-VCH, Weinheim (2001)
52. Orchard, S.R.: Oil Color Chem. Assoc. **51**, 44 (1968)
53. Parfitt, G.D., Sing, K.S.W. (eds.): Characterization of Powder Surfaces. Academic, London (1976)
54. Parfitt, G.D. (ed.): Dispersion of Powders in Liquids. Applied Science, New York (1981)
55. Pathak, M.A., Fitzpatrick, T.B.: In: Fitzpatrick, T.B. (ed.) Dermatology in General Medicine, 4th edn, pp. 1689–1717. McGraw-Hill, New York (1993)
56. Race, S., Fairhurst, D., Brozel, M.: US Patent, 7,417,426, 26 Aug 2008
57. Reiger, M.M.: Cosmet. Toil. **108**(12), 43 (1993)
58. Riegelman, S., Penna, R.: J. Soc. Cosmet. Chem. **11**, 280–291 (1960)
59. Roberts, J.: In: Auerbach, P. (ed.) Management of Wilderness and Environmental Emergencies, 2nd edn. Mosby, St. Louis (1989)
60. Sayre, R.M.: Arch. Dermatol. **12**(2), 745–746 (1986)
61. Sayre, R.M., Dowdy, J.C.: Photodermatol. Photoimmunaol. Photoderm. **38,** (1998)
62. Schlossman, M.L. (ed.): The Chemistry and Manufacture of Cosmetics, Vol. II, Formulating, 2nd edn. Allured Publishing, Carol Stream (2000)
63. Smijs, T.G.M., Bouwstra, J.A.: in J. Biomed. Nanotechnology **6**(5), 469–484 (2010)
64. Solomon, H., Hawthorne, G.: Chemistry of Pigments and Fillers. Wiley, New York (1983)
65. Stenberg, C., Larko, O.: Arch. Dermatol. **121**, 1400–1402 (1985)
66. Szczurko, C., et al.: Photodermatol. Photoimmunol. Photomed. **10**, 144–147 (1994)
67. Tadros, T.F.: Solid–Liquid Dispersions. Academic, New York (1987)
68. Tscharnuter, W.W., Weiner, B.B., Fairhurst, D.: ACS Symposium Series 472, Chapter **14**, 184 (1991)
69. Tyrell, R.M.: Photochem. Photobiol. **63**, 380–386 (1996)
70. Urbach, F.: In: Urbach, F. (ed.) Biological Responses to Ultraviolet Radiation, pp. 1–6. Valdenmar Publishing, Overland Park (1992)
71. Wamer, W.G., Yin, J.J., Wei, R.R.: Free Radic. Biol. Med. **23**, 851–858 (1997)
72. Weiner, B.B.: In: Barth, H. (ed.) Modern Methods of Particle Sizing, pp. 93–116. Wiley-Interscience, New York (1984)
73. Wenniger, J.A., McEwen, G.N.: Cosmetic Ingredient Handbook, 2nd edn. CTFA, Washington, DC (1992)
74. Wiseman, T.J.: In: Parfitt, G.D., Sing, K.S.W. (eds.) Characterization of Powder Surfaces. Academic, London (1976)
75. www.fda.gov/nanotechnology/

Chapter 15
Increased Antimicrobial Activity of Cheese Coatings Through Particle Size Reduction

Gabriel M.H. Meesters and Stephen L.A. Hennart

Abstract This chapter discusses the results of reducing the particle size distribution of an anti-microbial solid to facilitate a better use of the material. The product, an antifungal agent is size reduced using a wet-stirred media mill. The obtained product has the size of around 180 nm. The way of grinding is described and the used sizing techniques discussed. In the last part the product is characterized and tested in its application. Here it is shown that the reduced size gives a better performance against fungal anti-microbial infections than a product containing larger particles. This is explained by discussing the coverage, as well as the diffusion of the product through a layer of material, in this case a cheese surface.

15.1 Introduction

This chapter focuses on food coatings such as cheese coatings. A preservative additive is often incorporated in the coating to prevent microbial growth in the food products. The shelf life of the food product is defined as the time without any microbial growth. The preservative being poorly water soluble is mainly present in the form of particles. These slowly dissolve and the dissolved molecules diffuse into the coating.

The approach reviewed in this chapter is to reduce the particle size of the preservative particles in the coating and to prove that the size of the particles and the efficiency of the antimicrobial agent are correlated. At a given concentration, both the number of particles and the total surface area of product available increase with decreasing particle size. This enables one to reduce the distances between the

G.M.H. Meesters (✉) • S.L.A. Hennart
DSM Food Specialities, Alexander Fleminglaan 1, 2613 AX Delft 1, Delft 2600 MA, The Netherlands
e-mail: gabrie.meesters@DSM.com; Stephen.Hennart@qiagen.com

H.G. Merkus and G.M.H. Meesters (eds.), *Particulate Products: Tailoring Properties for Optimal Performance*, Particle Technology Series 19,
DOI 10.1007/978-3-319-00714-4_15,

particles. The increase in specific surface area improves the product dissolution rate. At a given concentration in the food coating, the product diffuses along shorter distances when decreasing product particle size. This will improve the shelf life of the food product by a better antimicrobial action.

The example taken in this chapter is a product named Plasticoat®, produced by DSM Food Specialties, The Netherlands. Plasticoat® is used as cheese coating and contains a poorly soluble antifungal active product. The antifungal compound is active only in dissolved state.

15.1.1 Antimicrobial Protection of Food Surfaces

Gustavsson et al. [18] reported in 2011 that "roughly one-third of food produced for human consumption is lost or wasted globally, which amounts to about 1.3 billion tons per year". One of the causes of food waste is fungal growth that shortens shelf life.

In the food industry several research groups are working on the protection of food and food surfaces against mold growth. The challenges are optimization of the protection systems to enhance the shelf life of food products. In various studies reported by Teerakarn et al. [43], Gill et al. [13], Vojdani et al. [46], Ouattara et al. [36], Warin et al. [47], Ozdemir and Floros [37, 38] the food surface is protected by a coating containing antimicrobial products. The antimicrobial product is often a poorly soluble solid crystalline compound. So a large part of the active molecules is in a particulate form. The antimicrobial particles are dispersed in the coating.

From the previous research cited above it is known that the different aspects regulating the concentration of the product in the coating are product diffusion, product dissolution, particle distribution, product degradation and the sensitivity of the microbe to the product (minimum concentration necessary to inhibit the growth of the organism).

The present work is based on the idea that the particle size has an influence on the shelf life of a food product protected by a film which contains the antimicrobial particles. Indeed, a reduction in particle size should increase the product specific surface area and the number of particles per unit of coating surface area. The particle surface area has an influence upon the dissolution properties and upon the product diffusion. The number of particles has influence on the distribution of particles in the coating. The smaller the particles the more influence of the parameters is to be expected.

15.1.2 Production of Sub-Micrometer Particles

Different methods are available to produce small particles. Particles can be obtained by bottom up approaches like controlled precipitation or cryogenic spraying for example. The inverse approach – top-down – consists in particle size reduction by milling.

Published research shows a large potential of wet ball milling to produce sub-micrometer particles. Inorganic material, like aluminum particles and some organic pigments particles are, for example, reduced to the sub-micrometer size range. Very fine powders of inorganic products, for example silica particles or metals, are produced by grinding in wet ball mills. Grinding enables, in such applications, a particle size reduction to less than 50 nm. Inorganic surfactants are used to prevent agglomeration of nano-particles.

Grinding of organic products is studied in the pharmaceutical industry. In most cases, grinding is carried out using additives such as polymers or cyclodextrines [14, 22]. Those additives enable finer grinding, better physical stability and/or chemical stability. An example is given by Horter and Dressman [23]. When the dissolution process is a rate-determining step, an improvement in the dissolution rate of poorly water-soluble drugs is required for enhancement of the gastrointestinal absorption. Grinding with polymers is known to be one pharmaceutical approach to improve the dissolution rate of drugs [34, 42, 45, 48] and may be useful for enhancement of the gastrointestinal absorption of poorly water-soluble drugs. The polymer acts as a surfactant to enhance the physical stability of the ground particles.

Stirred media mills are therefore expected to be capable of reducing the particle size of antimicrobial products to be used in the food coatings. Many polymers used in the pharmaceutical industry do not have a food grade status. Where grinding without additives is possible, costs are limited and no toxicological issues are raised.

15.1.3 Characterization of the Particle Stability

As shown by Mende et al., after grinding the particles can form agglomerates, which is undesirable [11, 26, 33]. Focus is required on the prevention of agglomeration after wet grinding of poorly soluble crystalline organic compounds.

When two colloidal particles approach one another, attractive and repulsive forces come into play. Particles can approach each other because of Brownian motion and convective transport. The main attraction and repulsion forces are the van der Waals force and the electrostatic interaction force, respectively.

The research by Mende et al. on sub-micron grinding of inorganic products shows that the stability of the product and its final particle size are a function of the electrostatic properties of the particles (zeta-potential) and therefore of the pH of the solution [33].

The detection of agglomerates is, in practice, not always straightforward. Different particle sizing techniques can give different results depending on the measurement conditions. Imaging techniques enable a visual representation of the agglomerates in suspension. Sedimentation measurements are another approach to the characterization of the agglomerates.

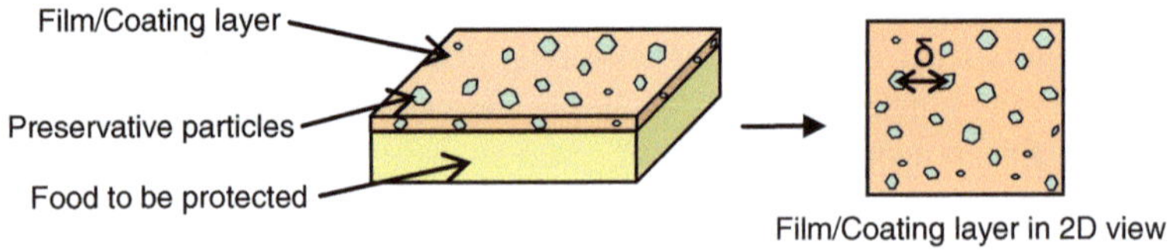

Fig. 15.1 Schematic view of the studied particle dispersion in a coating and its equivalent 2D representation (δ is the variable distance between the particles in the coating distance between particles)

15.1.4 Application of Particles in Coatings on Food Surfaces

As mentioned the different aspects regulating the concentration of the preservative in the coating are diffusion, degradation, dissolution, the particle distribution and the sensitivity of the mold to the preservative. Accelerated shelf life tests are used to experimentally investigate the influence of particle size on shelf life.

The present chapter explains the behavior of a poorly water soluble preservative coated onto food surfaces and describes the effect of product particle size on shelf life of a food product. Figure 15.1 illustrates the studied system.

15.2 Production of Sub-Micrometer Particles

15.2.1 Grinding Process

This part focuses on the grinding of the poorly water soluble antifungal crystalline compound. The material has a starting mean product particle size ($D_{4,3}$) of 15 μm; its density is of 1,300 kg.m^{-3}. Grinding was performed using a ball mill (Dynomill, Bachofen AG, Switzerland). These grinding experiments have been performed in the absence of additives. Particle size distributions have been measured with a laser diffraction size analyzer (LS230 equipment from Beckman Coulter).

The ball mill was operated in a recirculation mode. The poorly water soluble organic product was suspended in water (4.8 % w/w, 500 mL solution). Figure 15.2 shows the circulation of the product suspension through the grinding chamber (volume of the chamber: 300 mL, pump speed: 2 mL.s^{-1}) that was filled at 80 % (bulk volume) with grinding beads (Zirconium oxide, Yttrium stabilized beads from Tosoh, Japan). The product suspension was extracted through a 0.1 mm gap to prevent the grinding medium exiting the milling chamber. The system was cooled to keep a constant temperature throughout the entire system.

The particle size distribution was measured as a function of time throughout the grinding process. Figure 15.3 shows an example of data that were obtained during a grinding experiment.

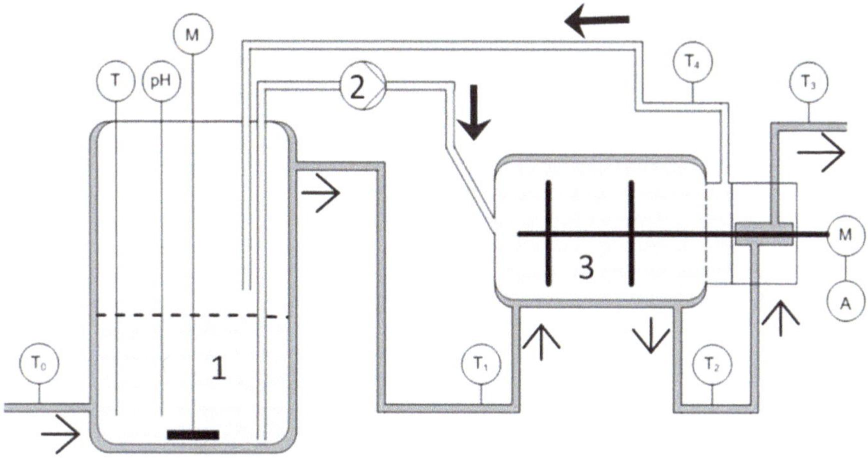

Fig. 15.2 Grinding set-up and controls, where *T* is a temperature measurement and pH is a pH control point. Arrows show the circulation of the cooling water (→) and product (➔). The product is pumped from the stirred tank marked as 1 by the pump 2 to the grinding chamber 3. M stands for motor and A for Amp meter

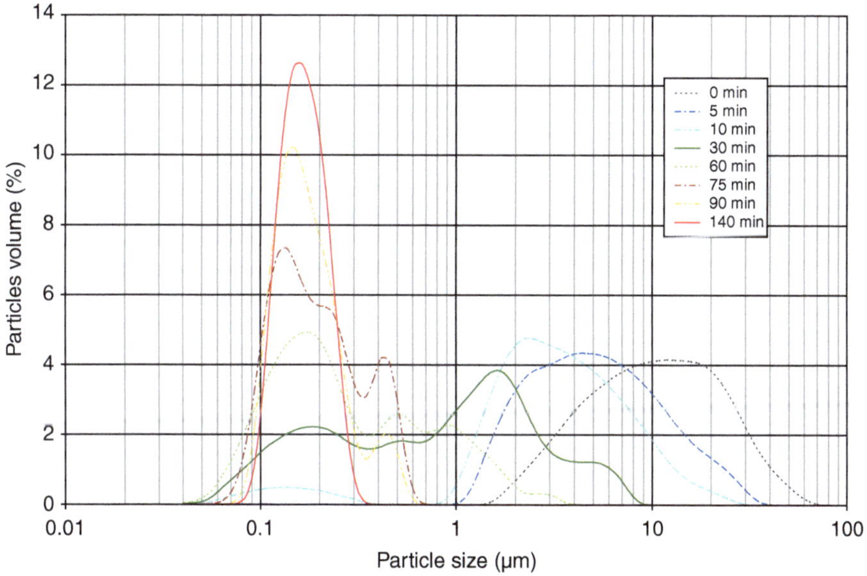

Fig. 15.3 Evolution of the particle size distribution as function of the grinding time (process parameters: rotation speed: 2,000 rpm, grinding media size: 0.5 mm)

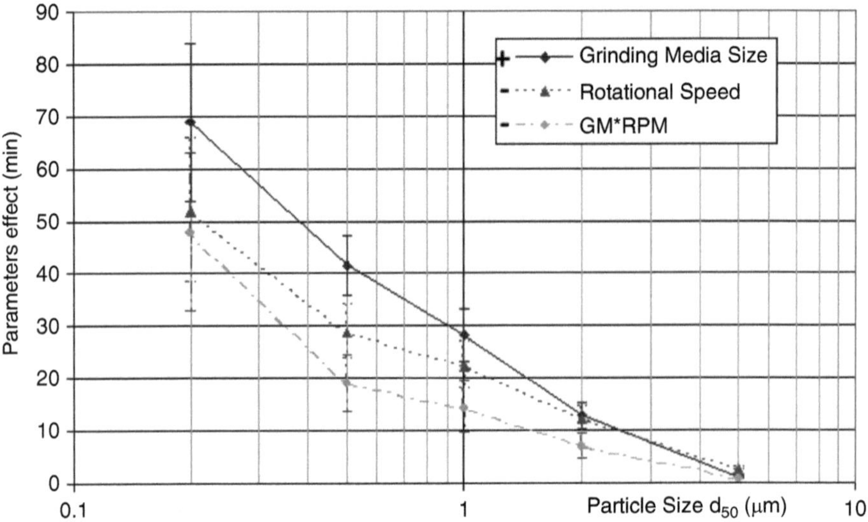

Fig. 15.4 Effect of grinding medium bead size, rotational speed and final product particle size (d_{50}) on grinding time. Effects can be positive (+) or negative (−). As example, the effect of grinding media size is negative (−): the bigger the grinding media size, the longer the grinding time

15.2.2 The Effect of Grinding Medium Bead Size and Rotational Speed on Grinding Time

The first objective was to analyze the effect of grinding medium bead size and rotational speed on grinding time on the basis of factorial design. The calculations were done using Statgraphics 5.1 software ANOVA (ANalysis Of VAriance). The effect was measured as the average response at a high parameter value minus the average response at a low parameter value. Figure 15.4 shows that the influence of grinding medium bead size and rotational speed on grinding time increases exponentially as the final product particle size decreases. The influence of these parameters on grinding time is thus very strong, especially when sub-micrometer product particles are desired.

The influence of rotational speed alone, and the product of rotational speed and grinding medium bead size is negative (negative in this case implies shorter grinding time). Thus, at higher rotational speed, a shorter grinding time is required. Opposite to this the grinding medium bead size has a positive effect on grinding time, i.e. smaller grinding beads give a shorter grinding time to reach a certain product particle size.

The data from Fig. 15.4 can be normalized. The normalized effect of one parameter is the ratio of the effect of that parameter and the sum of the effects from all parameters. Figure 15.5 shows the normalized data in order to compare the

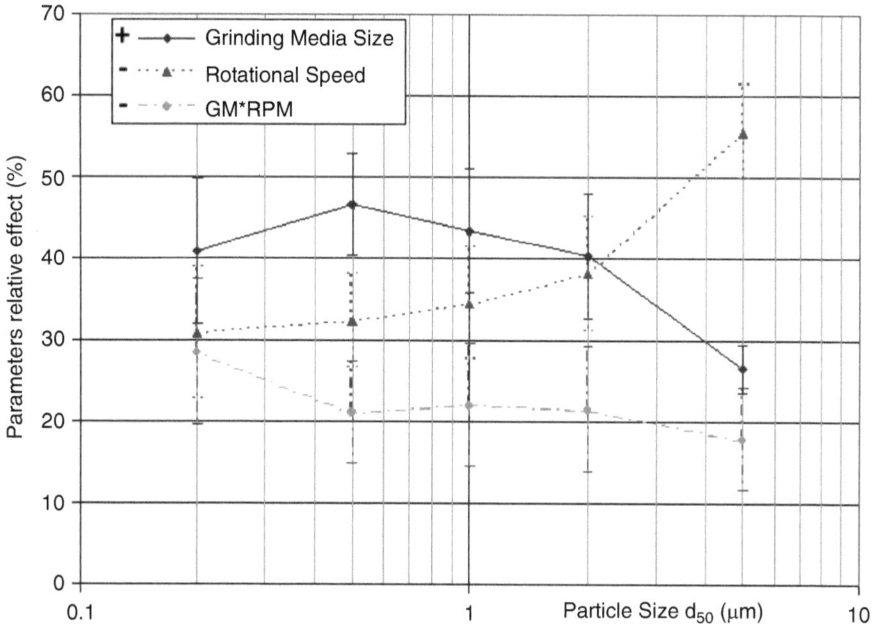

Fig. 15.5 Relative effect of the different parameters and combination of parameters on the time of grinding to reach different particle sizes

relative effects of the grinding parameters. The rotation speed is the dominant factor for grinding in the median particle size range above about 2 um, but its influence decreases in the sub-micrometer range. The bead size of the grinding medium becomes dominant for reaching median particle sizes in the sub-micrometer range.

The strongest effect of the rotation speed is observed in the coarse product size range and the finer product size range is less influenced by the speed. This is confirmed by the observations from Jankovic [27] and Hou et al. [24], the relative effect of the rotation speed is decreasing with the product particle size.

15.2.3 The Effect of Grinding Medium Bead Size and Rotational Speed on Heavy Metal Contamination

The forces applied in the grinding chamber are of great magnitude and high frequency. The stainless steel grinding chamber and the grinding medium (zirconium oxide stabilized with yttrium) undergo wear. Heavy metals can therefore enter the final suspension. Stainless steel is typically composed of iron, chromium, nickel and carbon. Carbon contamination cannot be detected since the ground product is organic and contains carbon atoms. The focus was therefore on the heavy metals. The content of iron, chromium, nickel and zirconium was determined using an atomic absorption

Table 15.1 Heavy metal content after 3 h grinding (ppm, ± 0.05 ppm)

d_{GM} (mm)	v_{tip} (rpm)	Zr	Cr	Fe	Ni
0.8	6,000	103	80	295	59
0.8	2,000	9.9	1.7	6.8	1.5
0.3	6,000	56	20	73	15
0.3	2,000	7.9	0.29	1.8	0.6

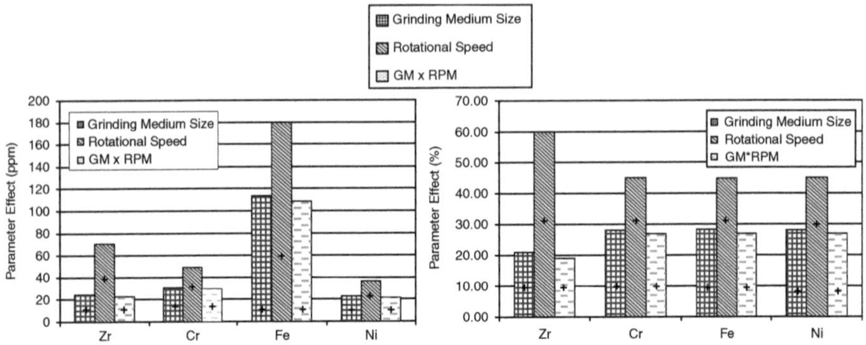

Fig. 15.6 Parameter effect (*left*) and relative effect (*right*) of grinding medium size and rotational speed on the heavy metal contamination after 3 h grinding

spectrophotometer. Table 15.1 shows the influence of the rotational speed, v_{tip}, and the grinding medium bead size, d_{GM}, on the heavy metal content.

The contamination by heavy metals is strongly dependent on the grinding conditions and the material hardness. The contamination by zirconium and iron ranges from a few ppm to a couple of hundreds ppm; the chromium and nickel contents are somewhat smaller. The contamination by zirconium is limited because the grinding medium is relatively hard in comparison to the grinding chamber that is composed of stainless steel. Table 15.1 and Fig. 15.6 present the heavy metal contamination after 3 h of grinding. The required grinding time to reach a specific product particle size may be shorter than 3 h. Contamination levels can therefore be reduced in some cases. An optimum balance needs to be found between heavy metal contamination and the product particle size that is reached. The biggest impact is found for the rotation speed. At higher speeds the heavy metal concentrations increase. The effect of the grinding medium bead size is less important. The number of impacts between the grinding beads and the grinding chamber wall thus seems to have a smaller impact than the collision force.

The results from Breitung-Faes et al. [5] showed that the bigger the grinding media the more contamination would occur. This effect is seen here and it is very limited. The hardness of the product material as mentioned by Becker and Schwedes [1] determines the bead wear as well. The present organic product milled is presumed to be much softer than a metal or ceramics (as used in the references) and thus the wear of the grinding media is very limited.

The conclusions from Howorth [25] are consistent with the present observation. Howorth indicated that the harder the medium, the less contamination would be

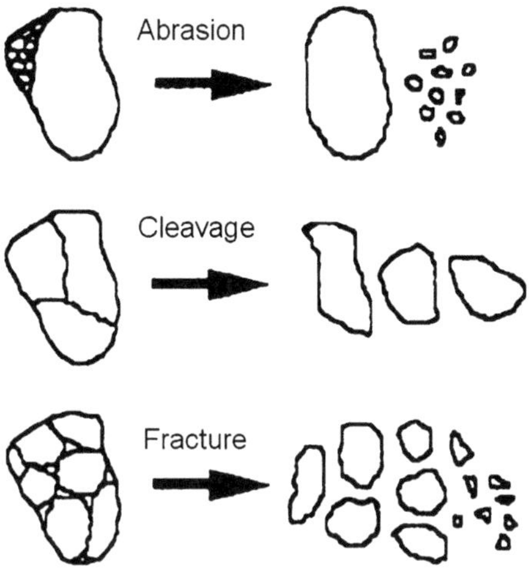

Fig. 15.7 Fragmentation mechanisms (Adopted from Varinot et al. [44])

detected. The present experiments are carried out with zirconium oxide, which is one of the hardest materials used as grinding media. Its wear is according to Howorth and, as observed here, is very limited.

The stainless steel composition is constant. Wear should therefore be similar for all metals that are present in stainless steel. This is confirmed by Fig. 15.6.

15.2.4 Breakage Mechanisms

The nature and the intensity of the applied stresses on the particles affect the particle size reduction process. In comminution the following three main breakage mechanisms have been identified (Fig. 15.7) [2, 4, 19, 41]:

- Abrasion occurs when stress is applied on particles along the tangential axis (shear). Particle breakage in this case gives a bimodal particle size distribution comprising fine particles that are released from the surface of the initial particles, in addition to particles with a size close to that of the initial particles.
- Cleavage of particles occurs when intense stresses are slowly applied on a particle (compression). This produces fragments of sizes 50–80 % v/v smaller than the initial particles.
- Fracture occurs through rapidly applying intense stresses (impact). The broken particle size distribution in this case will range between 20 and 70 % v/v of the size of the initial particle.

The grinding mechanisms can be studied by comparing the experimental data with the characteristic behavior of the grinding mechanisms as illustrated in Fig. 15.3.

The initial particle size distribution of the material slowly shifts to smaller particle sizes during the first 5 min of grinding. The first breakage events only

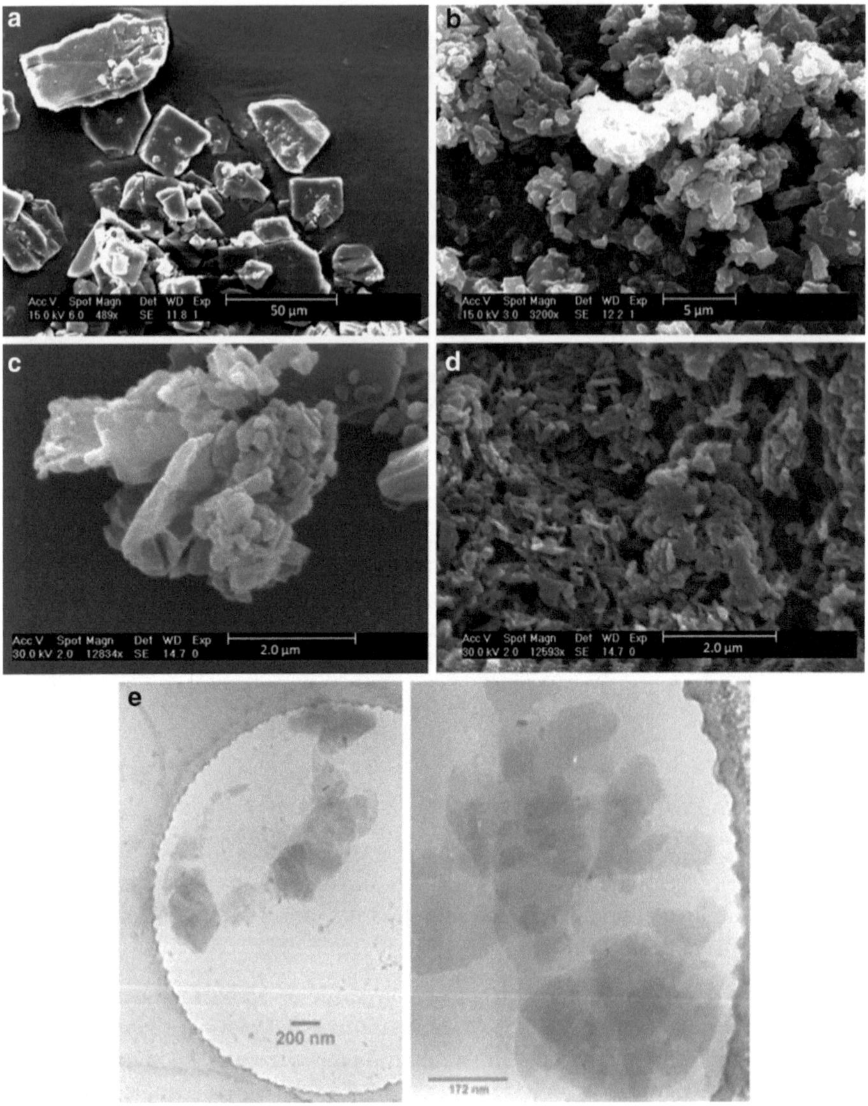

Fig. 15.8 Scanning electron microscopy imaging of particles after different grinding times: (**a**) starting material; (**b**) 10 min grinding; (**c**) 30 min grinding and (**d**) end product after freeze drying—140 min grinding; (**e**) Cryo-TEM picture after 20 min grinding

require a small amount of energy. This breakage leads to only a few daughter particles and therefore cleavage is the dominant grinding mechanism.

In a later stage during grinding, smaller particles appear and the distribution becomes bimodal. The particle size measurements have been complemented with scanning electron microscopy (SEM) pictures taken from samples dried at room temperature at different grinding times. Figure 15.8 shows that the decrease in

Table 15.2 Particle size and grinding time after start of the circuit operation as function of the grinding medium bead size and rotation speed. (time t_Z required to reach the minimum reachable (final) $D_{4,3}$)

		Mean diameter grinding media (mm)		
		0.300	0.500	0.800
		Final mass mean diameter $D_{4,3}$ (Time of grinding to final product t_Z)		
Rotation speed (rpm)	2,000	0.181 μm (53 min)	0.173 μm (116 min)	0.175 μm (260 min)
	3,000	0.180 μm (47 min)	0.188 μm (120 min)	0.179 μm (175 min)
	4,500	0.184 μm (29 min)	0.183 μm (69 min)	0.175 μm (145 min)
	6,000	0.177 μm (26 min)	0.183 μm (50 min)	0.176 μm (112 min)

particle size between 5 and 10 min of grinding leads to coarse particles covered by many fine particles. This behavior is typical for the abrasion breakage mechanism.

The formation of intermediate sized particles is observed after circa 30 min of grinding. This broad range of particles shows that the breakage is mainly resulting from a complex breaking process which includes abrasion, cleavage and/or fracture. The SEM image in Fig. 15.8c shows a large amount of fines, intermediate sized particles, coarse particles and agglomerates/aggregates ('clusters'). The presence of fines is characteristic of the abrasion mechanism. It is not easily visible on the picture if the other broken particles are resulting from a cleavage or fracture process. The breakage of intermediate sized particles is mainly resulting from a double breaking process which includes abrasion and cleavage or fracture.

Finally, all intermediate sized particles break down to the final particle size. The difference between those particle sizes is not big and breakage can therefore be classified as a cleavage process. SEM picture (Fig. 15.8d) was made of a freeze dried sample. Small particles appear to be present as clusters but their presence can still be observed.

15.2.5 Milling in Absence of Additives

A stirred media mill is a complex system with many parameters that influence the process of grinding. The more important parameters are rotation speed, grinding medium size hardness and density, filling ratio, product concentration, etc. This study focuses on grinding medium size and rotation speed. The grinding medium used is made of Zirconium oxide stabilized with Yttrium. The sizes of the grinding beads (D_{GM}) are of 0.3, 0.5 and 0.8 mm. The rotation speed ranges from 2,000 to 6,000 rpm. Filling ratio of the mill is 80 %. Considering the size of the impellor blades, the tip speed ranges from 6.8 to 20.4 m/s. The results are presented in Table 15.2.

In all experiments a similar final particle size (about 180 nm) was reached regardless of grinding medium size and rotation speed. The grinding time was, however, strongly dependent upon both parameters. The model to be developed should therefore allow definition of the time required for grinding as a function of its parameters, which themselves should be defined as a function of the process parameters.

15.2.6 Milling in Presence of Surfactants

The free energy of particles increases with an increase in specific surface area. When particles form agglomerates the total surface area decreases. As a result the free energy is lowered and the system is more stable. Agglomerates are formed when particles overcome the inter-particle energy barrier [33].

The driving forces for agglomeration are Brownian motion and fluid motion. In a stirred media mill the particles can be pressed together between two grinding beads or between grinding beads and the wall of the mill. When particles are pressed together they can agglomerate or aggregate. The resulting clusters can again be broken up into smaller particles in the stirred media mill.

In the final coating application, at a given concentration, clustered particles are more spaced apart from each other than non-clustered particles would be. One objective of grinding is to obtain smaller particles to reduce the distance between them in the coating application. The benefits from grinding would thus not be optimized in a case clustering occurred.

Clustering can be prevented with additives that give electrostatic and/or steric repulsion between the particles. The choice of additives is dependent on the product and its application(s). Griffin [15, 16] and Davies [6] calculated the hydrophilic-lipophilic balance (HLB) number for surfactants on the basis of the molecular weight of the hydrophilic and hydrophobic parts of the molecule. The HLB number helps to make a first selection of a stabilization system.

The grinding experiments from Sect. 15.2.5 were repeated with CTAB (Cetyl-trimetyl-ammonium bromide), a cationic surfactant to stabilize the particle suspension and to prevent agglomeration of small particles. A cationic surfactant is used as the particles have a negative zeta potential in the condition of grinding (pH ~6; Zeta potential in Fig. 15.11). As described for example by Bernhardt [3], the use of an additive will help to prevent broken particles from agglomerating. Therefore the final particle size reachable could be smaller when surfactants are used.

Figure 15.9 shows that an increase in surfactant concentration reduces the mean particle size after grinding. The observed effect is most likely due to stabilization of the particle suspension by the surfactant, which prevents/reduces particle agglomeration. The surfactant was present as micelles at the concentrations that were used (critical micelle concentration CTAB $= 1$ mmol.L^{-1}). These micelles could affect the particle size measurement. As a result a lower average particle size could be measured in the presence of surfactant micelles. The surfactant was added prior to grinding. The surfactant thus might affect all particle size measurements

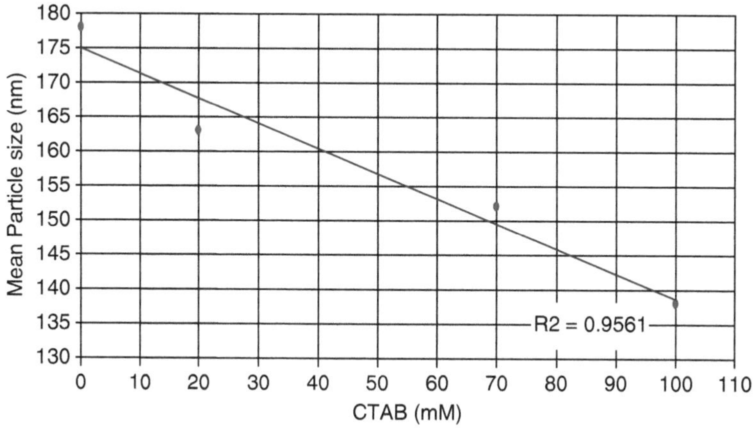

Fig. 15.9 Final particle size in ball milling as function of CTAB surfactant concentration

in one grinding experiment. The particle size before grinding was the same in all experiments regardless of the use of surfactant. It may therefore be concluded that the micelles did not affect the particle size measurement in any of the experiments.

The addition of surfactant can change the solubility and degradation rate of the product and thereby induce a particle size reduction. Solubility could be measured using filtration of dispersed suspensions (0.02 μm filters) and HPLC techniques. Tests were performed to verify if the particle size decrease was related to a solubility increase. For CTAB concentrations of 0 and 100 $mmol.L^{-1}$ the particle size changed from 0.18 to 0.14 μm after grinding, which implies an average single particle mass reduction of 47 %. The concentration of total product in the liquid phase was 0.5 % w/w in all experiments. In absence of surfactants the solubility was limited to 50 ppm. In the presence of surfactants (100 $mmol.L^{-1}$) the solubility reached 170 ppm. The decrease of total particle mass due to solubilization therefore was 2 % w/w. This is much less than the decrease of 47 % that is observed as the result of grinding. This proves that the surfactant prevents agglomeration of particles.

Another experiment was performed to determine the influence of the surfactant on the product solubility and the particle size analysis. 100 mM CTAB was added to a 180 nm particle suspension that did not contain surfactant. The particle size was measured before and after surfactant addition. Upon the addition of surfactant the particle size decreased from 180 to 170 nm after several hours of magnetic stirring. The decrease in particle size could be related to solubilization or de-agglomeration of the particles. The particle size in presence of surfactant was thus much higher than 140 nm, which was reached when surfactant was added prior to grinding. This confirms that the surfactant had a small effect on the particle size measurement.

15.3 Characterization of Extra Fine Milled Product: Particle Size and Particle Stability

15.3.1 Particle Size Characterization of Extra Fine Milled Product

In this part the particle size distribution of sub micrometer particles suspended in a liquid is described. The particles milled are an organic poorly water soluble crystalline product. To characterize the size of these particles, different techniques have been tested: Imaging techniques (SEM, Cryo-TEM), static light scattering techniques, dynamic light scattering techniques, centrifugation and flow field flow fractionation.

The results indicate that the studied milled particles have a primary particle size close to 180 nm and there is strong evidence of larger particles which are very likely clusters. This is clearly seen from the Cryo-TEM results (Fig. 15.8e).

All the above mentioned techniques should in principle be able to measure samples of dispersion containing particles of ca 180 nm but several are disturbed by the presence of clusters. It is difficult to estimate the amount of cluster present, but most of the time one is interested in what the primary particle size distribution is.

It is clear that no single piece of equipment is capable of exactly determining the particle size distribution of our samples, but the static light scattering with low shear on mixing does give a good representation of what is seen with the image analysis by Cryo-TEM.

In this work frame, it is important to determine the final particle size distribution of the milled particles. This enables the evaluation of the efficiency of the grinding process.

It is also important to evaluate the stability versus aggregation of the milled particles. It has been shown in previous studies that milled particles can quickly aggregate [21, 33]. The presence of aggregated milled particles can be seen as large particles in the particle size distribution.

Several commercially available techniques have been tested. All the tested techniques are capable to analyze sub micrometer particles dispersed in a liquid. For each technique, particles milled and stored at the same conditions have been used. The size distribution has been measured with each sizing technique.

The different techniques available on the market can be drawn together in several groups [17, 31]:

Image analysis techniques
Static light scattering based techniques using:

- Polarization Intensity Differential Scattering (PIDS)
- High resolution CCD detectors
- Blue wave technology
- Backscattering detection

Techniques based on the Brownian motion of the particles. Those techniques uses:

- Photon correlation spectroscopy (PCS)
- Photon Cross Correlation Spectroscopy (PCCS)
- The Heterodyne principle
- Particle Tracking

Centrifugation
Flow field flow fractionation techniques
Other techniques available commercially which are not tested in this work are:

- Particle separation by size exclusion chromatography
- Acoustics and electro-acoustics

The results of the various techniques used to measure the particle size distributions are summarized in Table 15.3.

A detailed description of each technique is presented in the next section.

Image analysis of SEM pictures shows primary particles of around 180–200 nm. The accuracy of the SEM results depends on how representative the sample in the picture is and how many particles are counted. Measuring 100 particles does not pretend to give an accurate result but is an indication of an order of size. The product suspension is dried and gold coated before being introduced in the SEM. Therefore the samples are not systematically representative for the original particles in suspension. The clustered structure could be created during the drying step. An answer to this point is possible using the Cryo-TEM technology.

The sample preparation for the Cryo-TEM imaging includes a process of cryonization that is quick enough (ca 1 ms) to freeze the water in a glassy state and to fix the particles as they are in the suspension. The particles seen in the frozen matrix are thus expected to be in the same as in the suspension. Figure 15.8 shows clearly that the ground particles are clustered. The size of the primary particles corresponds to the size measured by the SEM (0.18–0.20 μm) and the clusters are in the micrometer range. From the Cryo-TEM (Fig 15.8e) analysis it can be concluded that the aggregates are present in suspension. Similar clusters are seen in the SEM pictures as well. Thus, the clustered structures observed in the SEM pictures were probably not created during the drying step but are present in the suspension of ground particles.

Cryo-TEM also shows presence of larger clusters. Several techniques do see the small particles of the same size as well as some larger particles. The size of these larger particles could be measured by some techniques to range mainly between 1 and 50 μm.

Some of the static light scattering instruments show both small and larger particles but the Beckman Coulter LS 13320 and the Horiba LA950 only see the small particles. There is very little difference between the processing of the particles during the size measurement. They are all measured at similar

Table 15.3 Overview of the results given by the different particle size measurement techniques

Technique		Time of measur ement	Mean primary particle size $D_{4,3}$ (nm)	Large particles	Remarks
Imaging techniques	Imaging techniques SEM	1day	194[a]	Yes	The SEM requires drying and therefore aggregates cannot be differentiated from the smaller particles. The technique enables a good evaluation of the shape of the particles
	Imaging techniques Cryo-TEM	20 min	180–200[a]	Yes (2–3 μm)	Cryo-TEM enables an "in solution" look to the particles and enables to see the aggregates and estimate their size as well as the one of the primary particles
Static light scattering	+ PIDS – Beckman coulter LS13320	~ 5 min	181	No	
	High Res. Detect. Saturn DigiSizer 5,200	<5 min	354	Yes (1–10 μm)	
	+ Blue Light Malvern Mastersizer	<5 min	187	Yes (1–50 μm)	
	+ Back scattering Horiba LA950	<5 min	180	No	
Brownian motion based techniques	PCS	~ 10 min	–	"Yes"	No measurement was possible because of the high influence of the concentration and the presence of large particles
	PCCS	~ 5 min	188	Yes (>1 μm)	The light aggregates are detected and are above the detection maximum of the machine (1 μm)
	Heterodyne	~ 5 min	853	Yes	Only the big particles are detected, they most probably disturb the measurement

Particle tracking	~ 5 min	277	No	The few number of particle actually analyzed in the present case do not enable an accurate quantitative analysis. The coarse particles are not detected; they might have sediment before the measurement
Disc Centrifugation	20–30 min	381	No	Streaming effects and high Reynolds number might be responsible of the higher values of the particle size measured. High gravity and drag forces might break aggregates
Flow field flow fractionation	60–90 min	170[b]	Yes	The aggregate particles elute after a very long time. The total distribution could not be therefore detected

[a]average length of particle (not $D_{4,3}$)
[b]median particle size

concentrations and all in demineralized water without surfactant. All samples are pumped through the measurement cell by using a centrifugal pump. Slight differences are seen in the design of this pump and the shear induced on the product. Should the Horiba LA 950 and the Beckman Coulter LS13320 devices induce higher shear on the clustered particles than the Malvern Mastersizer, those clusters might break in these first two instruments and remain intact in the Mastersizer. Differences might also be due to the de-convolution procedure or to filtering of the data when small amounts of larger particles are detected.

The Digisizer shows larger particles than the other instruments based on static light scattering. Note that the measured scattering pattern includes signals of both primary particles and clusters. The resulting scattering pattern is therefore very complex. The CCD detectors are capable to detect that scattering pattern at higher resolution but not all acquired data can be processed because of lack of data processing power. That inaccuracy in the calculation might explain the difference in the particle size measured. No quantification of the error made on the value of the particle size could be made to confirm this hypothesis.

Dynamic light scattering techniques are difficult to use because they are strongly influenced by the presence of larger particles. When they settle they give rise to a high motion which results in a shift of the particle size distribution.

Only photon cross-correlation spectroscopy shows about the same primary particle size as the image analysis. In addition the particle size distribution indicates that larger particles may be present. But their size cannot be obtained, because the upper limit of detection is around 1 μm.

Since the sample is neither stirred nor undergoing any shear, clusters would not be broken. The large particles observed are thus most probably the clusters observed on the Cryo-TEM and SEM pictures. No measurement can be done while stirring or applying shear. That would induce a movement of the product particles in addition to Brownian motion, and the measurement results would be wrong.

The measurement in the case using the heterodyne technique is not reliable enough to clearly determine the cluster size. Nevertheless, the presence of large particles is clear. Larger particles move because of sedimentation and not only Brownian motion. The signal emitted by big particles is influencing the correlation function a lot. The fitting of the correlation function is extremely sensitive and results in huge differences in the obtained particle size distributions.

The mean volume-based particle diameter $D_{4,3}$ obtained using particle tracking technique is 277 nm, which is in the order of magnitude as the results given by image analysis and static light scattering techniques.

No coarse particles are detected; it is possible that the coarse particle sediment out from the focal plane before or during the measurement. They would then not be analyzed.

Centrifugal sedimentation is capable to see smaller particles but larger particles are expected to sediment that fast that they are not detectable. The mean particle size measured is 381 nm; slightly bigger than the one observed by image analysis (180–200 nm) but the results remain in the order of magnitude [30]. Two aspects can be responsible for a larger mean particle size as expected (Laidlaw and Steinmetz):

(a) At the beginning of the analysis, the entire sample is contained in a thin fluid layer near the surface. A streaming phenomenon enables the particles to sediment faster which leads to a broad initial band. The sedimentation follows then a normal pattern but the calculation of the particle size from the slightly shorter sedimentation time would give slightly bigger results of the particle size.
(b) At very high rotation speed as in the present case (24,000 rpm, ca 2,900 G), the Reynolds number of the particle flow becomes higher and the Stokes law does not describe the sedimentation process accurately. That error could induce a bigger particle size than expected.

Flow field flow fractionation shows primary particles of 170 nm. The results are similar to the results by image analysis. The particle size as detected by a multiple angle scattering device shows a wide distribution. It is not possible from the data to determine the exact particle size distribution of the tested product, a.o. because the elution was not totally finished and no calibration was executed with a reference material. Larger particles are detected by the UV detector but the time of analysis is then very long.

15.3.2 *Stability of Particle Suspension After Fine Grinding*

The physical stability of sub-micrometer particle suspensions of organic crystalline food compounds after grinding will be described next. A Dynomill ball mill was used in combination with zirconium oxide grinding beads. The organic product was a poorly water soluble product. During grinding the average particle size of the particulate product was reduced to a minimum value in the sub-micrometer range. Forward light scattering was used to analyze the particle size distribution. Dynamic light scattering measurements, on the other hand, showed that clusters were present after grinding. The difference in the obtained particle size distributions using both techniques was related to the shear in the measurement device, i.e. in the laser diffraction measurement the shear was higher than in the dynamic light scattering device. Thus in the laser diffraction measurement the clusters were broken up by shear, while this was not the case in the dynamic light scattering measurements. The difference in the measurements showed that the particles formed clusters at low to zero shear.

The clustering of the particles was studied by measuring the sedimentation behavior of the particles suspension at various pH values. The impact of the pH on the clustering rate can be explained by the zeta-potential of the particles.

Sedimentation Measurements

Freshly ground particle suspensions (1.0 % solution with 0.01 mol.L^{-1} NaCl) were divided over several bottles. The pH was adjusted to values between 3 and 9 using sodium hydroxide (0.01 mol.L^{-1}) and hydrogen chloride (0.01 mol.L^{-1}). The ionic strength was thus constant at all pH values. At pH 10, to be able to reach the right pH, the ion concentration was increased to 0.05 mol.L^{-1}. The amount of acid and base that was required was low compared to the sample volume. The dilution was thus negligible. The pH was set over a period of 48 h by incremental acid or base dosage until the desired pH was reached. During the process the samples were continuously stirred. After pH adjustment the samples were put in graduated pipettes of 10 mL that were closed with a rubber stop to avoid evaporation of liquid. The sedimentation front was recorded over a period of 15 days.

Zeta Potential Measurements

The measurements were carried out with freshly prepared suspensions of ground particles. These suspensions were diluted to 0.1 % prior to measurement in the ZetaSizer (Malvern, UK). The ionic strength was fixed to 1 mM using potassium chloride. The zeta potential was determined as function of pH by automatic titration from pH 3 to 10 using a 4 N sodium hydroxide solution.

Sedimentation Tests

The sedimentation data is presented in Fig. 15.10. The initial sedimentation velocity can be used to determine the cluster size at t = 0, i.e. directly after mixing and setting the pH. Particle sedimentation in these experiments occurred under laminar flow, because the Reynolds number for particle sedimentation was very low in all cases. The suspensions contained 0.76 % v/v of organic particles. It is expected that this concentration is low enough to avoid swarm effects in sedimentation. The sedimentation rate of particle swarms is defined by the Richardson-Zaki equation [7, 28]:

$$v_{swarm} = v_s \cdot c_V{}^n \tag{15.1}$$

where v_{swarm} is the sedimentation rate of a swarm, v_s the settling rate of one particle, c_V the volume fraction liquid ($c_V = 99.24$ %), n is a factor dependent on the Reynolds number (Re < 0.2 in any case), and can be calculated by:

$$n = 4.6 + 20 \cdot \frac{D_p}{D_v} \tag{15.2}$$

where the particle diameter being D_p ($D_p = 1.64$ μm in the worse case) and the vessel diameter being D_v (1.0 cm).

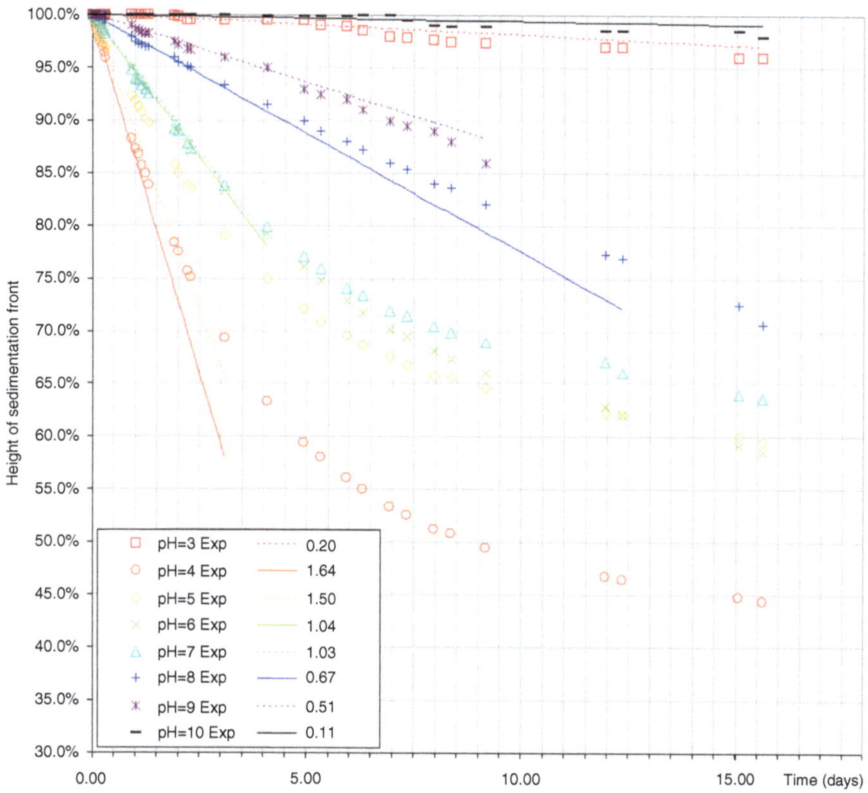

Fig. 15.10 Sedimentation profile of ground particles as function of pH. Symbols relate to experimental values, lines to fitting of Stokes sedimentation theory with indicated cluster size (in μm)

For the system that is considered in this work $n = 4.60$ and the swarm speed is 97 % of the settling velocity of the single particles. This confirms that swam effects are negligible in this work. The settling velocity of the particles can thus be expressed by Stokes law:

$$v_s = \frac{(\rho_p - \rho_f) \cdot D_s^2 \cdot g}{18 \cdot \eta_f} \tag{15.3}$$

with η_f is the fluid viscosity, D_s the Stokes particle diameter, ρ_f the fluid density (1,000 kg/m^3), ρ_p the particle density and g is the gravitational acceleration. The particle density was measured by sucrose gradient centrifugation [32]. This density was approx. 1,300 kg/m^3.

Figure 15.10 depicts sedimentation data and fits of the initial sedimentation rates with Stokes law. The graph shows that extreme pH values give higher suspension stability than intermediate pH values. This observation can be explained by the zeta

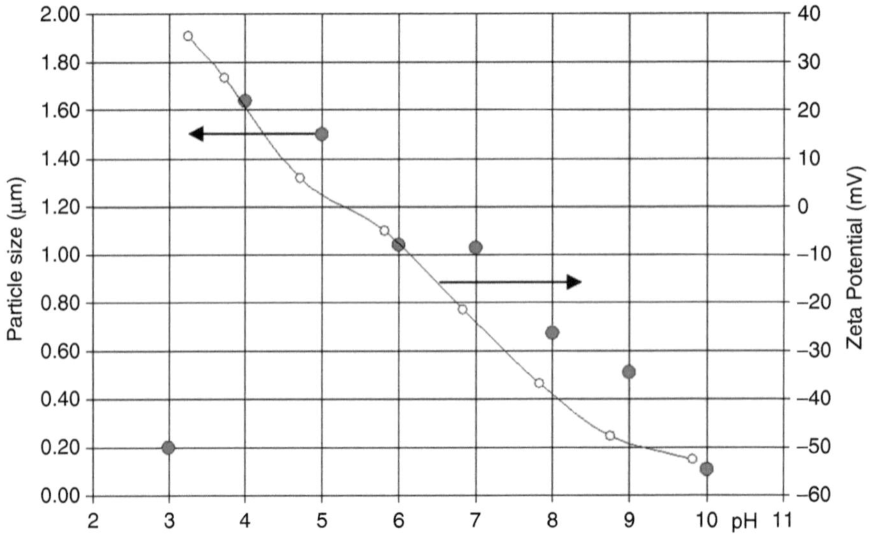

Fig. 15.11 Zeta-potential (*line*) and calculated Stokes equivalent particle size (*dots*) as function of applied pH

potential measurements that are presented in Fig. 15.11. The particle size determined by sedimentation measurements corresponds reasonably well to the Cryo-TEM picture in case of unstable suspensions, i.e. highest sedimentation rates. The particle size determined by sedimentation measurements in case of stable suspension, i.e. lowest sedimentation rate, corresponds reasonably well to the forward light scattering measurements.

The iso-electric point of the ground organic product particles is approx. 5.3. The degree of agglomeration increases when the pH approaches the iso-electric point, which can be explained by the reduction in the electrostatic repulsion between the particles.

At pH 4 the sedimentation seems to be higher than at pH 5 or 6. This is contradictory to the expectation on the basis of the zeta-potential measurements. This contradiction can be explained on the basis of the experimental protocol. The pH of the particle suspensions was set during a period of 48 h. The pH after grinding was 5.7. The iso-electric point was thus passed for suspensions with a pH below 5.7. The stability during pH control was thus different for samples above pH 5.7 and below pH 5.7. The clusters that were formed during pH control could be broken up at different rates depending on the zeta-potential and the applied shear. The shear was constant in all cases. It is therefore expected that cluster break up rate during pH control was higher for more extreme zeta potentials.

An alternative approach would have been to grind the product at the desired pH values instead of setting the pH after grinding in order to avoid passing the iso-electric point for stability testing. Grinding experiment were carried out at various pH values for the study of the grinding mechanism. It was shown that the grinding kinetics are independent of pH.

15.4 Distribution of Poorly Soluble Particles in Films and Coatings

In Sect. 15.2, sub-micrometer particles are produced. In Sect. 15.3, the particles are characterized and there stability investigated. In this part, the ground particles are applied in an example of food stuff: a cheese coating.

Several research groups in the food industry have been studying antimicrobial agents in food and on food surfaces (among others [13, 36–38, 43, 46, 47]). The aim of these studies was to obtain optimal antimicrobial protection of the food stuff. In all the cited papers it is highlighted that the shelf life is systematically dependent upon an optimum balance between:

1. The molecular diffusivity of the ingredient
2. The consumption or degradation rate of antimicrobial ingredient
3. The dissolution behavior
4. The homogeneity of the distribution of the particles of the antimicrobial compound.
5. The minimum inhibitory concentration (MIC) of the antimicrobial compound. Microorganisms are able to grow in or on the food stuff when the concentration of the antimicrobial compound in or at the surface of the food stuff is below the MIC. The microbial shelf life of the food stuff is thus limited by the time the dissolved preservative concentration remains above the MIC.

Optimization of the microbial shelf life of the coated food stuff requires a study on the interaction between all five parameters that were mentioned above. Diffusivity depends amongst others on the nature of the matrix, temperature, and the size of the antimicrobial compound. Increasing the diffusivity enables a better coverage of the food stuff and may thus lead to an increase of the microbial shelf life of the food stuff when diffusion of the preservative is limiting [10, 12, 29, 39]. Consumption and/or degradation of the preservative are dependent upon the chemical stability and the presence of microorganisms that are attacked by the antimicrobial compound. Dissolution of the preservative particles depends on the properties of the particles and the surrounding medium and the available surface area for dissolution. The preservative particle distribution is dependent on the preservative dosage and particle size. The critical dissolved concentration of the preservative is given by the MIC for the microorganisms from which the food stuff needs to be protected.

In this section the objective is to determine the influence of the preservative particle size on the shelf life of the coated food stuff. In a first step, a model is used to evaluate the impact of the above mentioned parameters. All aspects of the model (diffusion, degradation, dissolution and distribution) are presented before a theoretical model is calculated. In a second part the experimental method for the determination of each aspect is presented. Finally, the experimental results are given and an accelerated shelf life test is used to experimentally verify the model calculations.

Different sizes of the preservative particles were obtained by wet grinding. Particle sizes of 15 μm and approx. 0.2 μm were tested.

15.4.1 Modeling of the Distribution of Particles and Diffusion of Preservative Molecules in a Coating

The MIC for the microorganisms is assumed to be constant at one third of the solubility of the preservative, C_S. The shelf life t_c of the coated food stuff is reached when the preservative concentration at the weakest location in the coating is below the MIC. The weakest location in the coating is the farthest point from the preservative particles. The distance between two particles is δ. The farthest point between the two particles is thus $\delta/2$. The shelf life t_c of the coated food stuff is thus resembled by Eq. 15.4:

$$C\left(\frac{\delta}{2}, t_c\right) = \frac{C_S}{3} \tag{15.4}$$

The concentration C of preservative molecules is defined by Fick's [8, 9] second law of diffusion and a degradation term $B(x,t)$.

$$\frac{\partial C(x,t)}{\partial t} = \mathbb{D} \cdot \frac{\partial^2 C(x,t)}{\partial x^2} + B(x,t) \tag{15.5}$$

where C is the dissolved preservative concentration (ppm), x the position between two preservative particles (m), t the time (s), $\mathbb{D}$ the diffusion coefficient ($m^2.s^{-1}$).

Diffusion The diffusion coefficient $\mathbb{D}$ for a molecule can be described by Eq. 15.6.

$$\mathbb{D} = \frac{k \cdot T}{3 \cdot \pi \cdot \eta_f \cdot d_m} \tag{15.6}$$

where k is the Boltzmann constant, d_m the diameter of the molecule assuming a spherical molecule, η_f the fluid (or matrix) viscosity and T the temperature. The apparent diffusivity will be lower when the molecule is not spherical [40]. The characteristic time for diffusion (τ_{diff}) is defined as the time required for a molecule to diffuse over the distance δ (Eq. 15.7):

$$\tau_{diff} = \frac{\delta^2}{\mathbb{D}} \tag{15.7}$$

Dissolution and solubility The dissolved preservative concentration directly after application of the coating to the food stuff is assumed to be equal to the solubility (C_S) of the preservative. This is the starting point for the model calculations, i.e. $C(x,0) = C_S$. The second assumption for the model calculations is that the concentration of dissolved preservative at the preservative particle surface is equal to the solubility of the preservative, i.e. the dissolution rate is much faster than the diffusion and degradation rates. The boundary condition for the model calculations thus is: $C(0,t) = C_S$ and $C(\delta,t) = C_S$. The validity of these assumptions will be proven further on in this chapter (Sect. 15.4.3).

The dissolution rate constant (k_{diss}) for the preservative is defined from Noyes and Whitney [35] as:

$$\frac{dC(t)}{dt} = k_{diss} \cdot (C_S - C(t)) \tag{15.8}$$

With $C(t)$ the concentration of the preservative in time and C_S the product solubility.

The value of $k_{diss,}$ degradation rate constant (only the dissolved material will degrade, not the solid material present), was determined in the coating prior to application at room temperature in absence of light. The preservative concentration was measured as function of the storage time. The initial preservative concentration was 6.5 times the value of the solubility C_S. First order degradation reaction kinetics (Eq. 15.8 with $n = 1$) were fitted to the experimental data.

In application the preservative-containing coating will be exposed to another environment. The preservative will undergo more potential degradation mechanisms such as oxidation due to contact with air and higher microbial contamination which may lead to enhanced consumption of the preservative. The degradation rate constant that has been measured in this work is thus expected to be an underestimation of the degradation rate constant that can be encountered in application.

From the fit of the experimental data using Eq. 15.8 $k_{diss} = 0{,}95\ \mathrm{s}^{-1}$ was obtained for 0.2 μm particles. For 15 μm particles this was $0.05\ \mathrm{s}^{-1}$.

The characteristic time for the preservative dissolution process (τ_{diss}) is inversely proportional to the dissolution rate constant k_{diss}.

$$\tau_{diss} = \frac{1}{k_{diss}} \tag{15.9}$$

Distribution: The shelf life of the coated food stuff is assumed to be equal to the time that is required to reach the MIC for the dissolved preservative at any position in the coating. The most critical point in this respect is the position in the coating where the distance between the preservative particles is the largest. At this position the distance that needs to be traveled by the dissolved preservative molecules from the preservative particle surface, where dissolution takes place, to a position in the middle between two preservative particles is the largest. The biggest gaps in the coating with regard to preservative particle coverage are thus most critical for the shelf life of the coated food stuff.

The determination of this critical gap size in the preservative particle coverage is discussed in Sect. 15.4.3, results are given in Sect. 15.4.4.

Equation 15.5 becomes thus Eq. 15.10:

$$\frac{\partial C(x,t)}{\partial t} = \mathbb{D} \cdot \frac{\partial^2 C(x,t)}{\partial x^2} - k_d \cdot C(x,t) \tag{15.10}$$

The initial (IC) and boundary (BC) conditions as described above are:

$$\begin{cases} IC : C(x,0) = C_s \\ BC : C(0,t) = C(\delta,t) = C_s \end{cases} \tag{15.11}$$

With C_S being the solubility of the product in the matrix.

The model has been analytically solved using a Fourier series. Equation 15.12 shows the solution of Eq. 15.10 using the initial and boundary conditions:

$$C(x,t) = C_s + \sum_{n=0}^{\infty} \frac{4 \cdot k_d \cdot \delta^2 \cdot C_s}{\mathbb{D} \cdot ((2 \cdot n+1) \cdot \pi)^3 + (2 \cdot n+1) \cdot \pi \cdot k_d \cdot \delta^2} \cdot \left(e^{-\left(\mathbb{D} \cdot \left(\frac{(2 \cdot n+1) \cdot \pi}{\delta}\right)^2 + k_d\right) \cdot t} - 1 \right) \sin\left(\frac{(2 \cdot n+1) \cdot \pi \cdot x}{\delta}\right) \tag{15.12}$$

Equation 15.12 is valid until the preservative particles are fully dissolved in the coating matrix. At the time noted $t_\varnothing$ the preservative particles are fully dissolved and the boundary conditions change into $BC_\varnothing$:

$$BC_\varnothing : \left.\frac{\partial C(x,t)}{\partial x}\right|_{x=0} = \left.\frac{\partial C(x,t)}{\partial x}\right|_{x=\delta} = -k_d \cdot C(x,t) \tag{15.13}$$

The time $t_\varnothing$ at which the preservative particles are fully dissolved is given by Eq. 15.14:

$$\left. m_{part} \right|_{t=0} = \mathbb{D} \cdot S_{part} \cdot \int_{t=0}^{t_\varnothing} \left.\frac{\partial C(x,t)}{\partial x}\right|_{x=0} .dt \tag{15.14}$$

With $\left.\frac{\partial C(x,t)}{\partial x}\right|_{x=0}$ being the flux of product from the particle surface into solution, m_{part} being the mass of the particle and S_{part} the surface of the product particle.

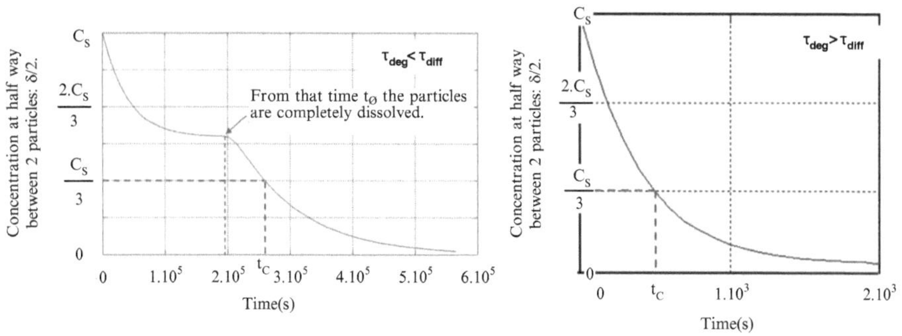

Fig. 15.12 Concentration profile of active product at equal distance between two particles: $C\ (\delta/2, t)$ *Top*: The shelf life is reached after the preservative particles are fully dissolved $(t_C > t_{\varnothing})$. *Bottom*: The shelf life is reached before the preservative particles are fully dissolved $(t_C < t_{\varnothing})$

Numerical calculations were performed in Matlab.

The critical time was determined for a range of values of the characteristic times of diffusion (τ_{diff}) and degradation (τ_{deg}). The characteristic time for dissolution was assumed to be much smaller than the characteristic times for degradation and diffusion.

An example of the concentration profiles of the preservative at $\delta/2$ is given in Fig. 15.12 for conditions when diffusion is faster than degradation and vice versa. As illustrated, the initial concentration at $t = 0$ is the product solubility C_S. The concentration decreases in time because of the degradation process that is partly compensated by the diffusion from the product particle. The graph shows that when the particles are completely dissolved, at $t = t_{\varnothing}$, only degradation takes place and a second drop in concentration is observed. The critical time (the shelf life t_C) is reached when the concentration at $\delta/2$ is of one third of the product solubility C_S. When the characteristic diffusion time for degradation is higher than the characteristic time for diffusion ($\tau_{deg} > \tau_{diff}$) then $t_C > t_{\varnothing}$, and vice versa.

The numerical calculations show that a small characteristic diffusion time and a high characteristic degradation time are the best for a long shelf life, which is logical because under these conditions diffusion of dissolved preservative molecules is much faster than degradation of dissolved preservative molecules. Under these conditions the diffusion process is fast enough to supply preservative throughout the coating without degradation causing local dips in the dissolved preservative concentration. The opposite, a short characteristic degradation time and a long characteristic diffusion time, gives a short shelf life because the diffusion process is not fast enough to allow for rapid replacement of degraded preservative in the coating.

15.4.2 Characterization of the Distribution of the Particles in the Matrix

Inhomogeneities in the preservative particle distribution are seen as a Normal and broad distribution of particles in the matrix. Those inhomogeneities are unfavorable for the shelf life of the coated food stuff because inhomogeneities result in larger inter-particle distances locally in the coating matrix. An increase in inter-particle distance gives an increase in the diffusion distance and thus shifts the balance between the diffusion rate and the degradation rate in favor of the degradation rate. The process of dispersing the particles in the matrix should thus be optimized to achieve a homogeneous preservative particle distribution. In the present case study the particles were well mixed throughout the coating matrix. The coating was applied by dipping the food stuff into the coating. The applied coating was dried at ambient conditions. This procedure is expected to give a random distribution of the preservative particles in the applied coating.

15.4.3 Calculation of the Inter-Particle Distance Distribution

In Fig. 15.1 the preservative particles are projected onto a 2D surface. Assuming a random particle distribution in the coating matrix, the distribution of squares that do not contain any preservative particles can be calculated analytically using the Bose-Einstein statistics. The number of empty squares M_{empty} is calculated as function of the total number of squares M and the number of particles N according to the following equation:

$$M_{empty} = \frac{M \cdot (M-1)}{M+N-1} \tag{15.15}$$

The probability P that a square surface area is empty is equal to the fraction of empty squares:

$$P = \frac{M_{empty}}{M} = \frac{(M-1)}{M+N-1} \tag{15.16}$$

Equation 15.16 gives the probability distribution for the empty square size. Given a confidence interval (probability P) and a concentration (number of particles N), the value of the number of cells (M) can be calculated (Eq. 15.16). The surface area of a cell is calculated as the ratio between the surface area of coating and the corresponding number of cells (M) (Eq. 15.17):

$$Surface\ of\ 1\ cell = \frac{1}{M} = \frac{1}{\frac{P}{1-P} \cdot N_{part} + 1} \tag{15.17}$$

The inter-particle distance distribution δ_p is taken as the diagonal size of the empty squares and can be calculated for a given confidence interval P (Eq. 15.18).

$$\delta_p = \sqrt{2} \cdot \sqrt{\frac{1}{\frac{P}{1-P} \cdot N_{part} + 1}} \tag{15.18}$$

With P being the confidence interval ($P = 0.90$ corresponds to δ_{90}, which indicates that 90 % of the empty squares are smaller than the given value) and N_{part} being the number of product particles per unit surface area. The number of particles per unit surface area is calculated from the preservative concentration, the coating thickness and the preservative particle size.

15.4.4 Accelerated Shelf Life Test; Determination of Antimicrobial Activity of the Coating

Freshly brined Gouda cheeses were challenged with Penicillium discolor and coated with various coating formulations to determine the relation between shelf life and coating formulation. Coatings with various preservative particle sizes were applied. The methods that were used to obtain these preservative particle sizes are reported by Van Hee et al. [20]. The number of mould spores that was able to germinate on the treated cheese surfaces is indicative for the anti-fungal activity of the preservative in the coating. The cheeses (10 per coating type) were considered to be no longer protected by the product when 50 % of the cheeses analyzed had at least one growing colony on their surface.

Plasticoat (DSM Food Specialties, The Netherlands) was used as a coating. Several preservative concentrations with various particle sizes were added to this coating. The cheeses were contaminated with Penicillium discolor PED1, CR1A and Penicillium discolor PED74, L1 spores at a concentration of 1.1×10^3 cm^2 after application of the coating. The cheeses were visually checked for fungal growth on a daily basis for 3 weeks. The experiment was carried out with two different preservative particle sizes ($D_{4,3}$), i.e. 15 μm and 0.2 μm

The rate of dissolution is an important aspect. The characteristic dissolution time (τ_{diss}) is much smaller than the characteristic diffusion time (τ_{diff}) and the characteristic degradation time (τ_{deg}).

The model is accurate if the inter-particle distance is at least 10 times larger than the particle size x_i. Fig. 15.13 can be used to determine when this is the case:

For 15 μm particle $\delta_{50} > 150$ μm $\rightarrow C_T/C_S < 70$
For 15 μm particle $\delta_{90} > 150$ μm $\rightarrow C_T/C_S < 700$
For 0.2 μm particle $\delta_{50} > 2$ μm $\rightarrow C_T/C_S < 2$
For 0.2 μm particle $\delta_{90} > 2$ μm $\rightarrow C_T/C_S < 9$

The model as plotted in Fig. 15.14 will be valid in case the concentrations considered are above the limits calculated. The corresponding τ_{diff} can be calculated with Eq. 15.17.

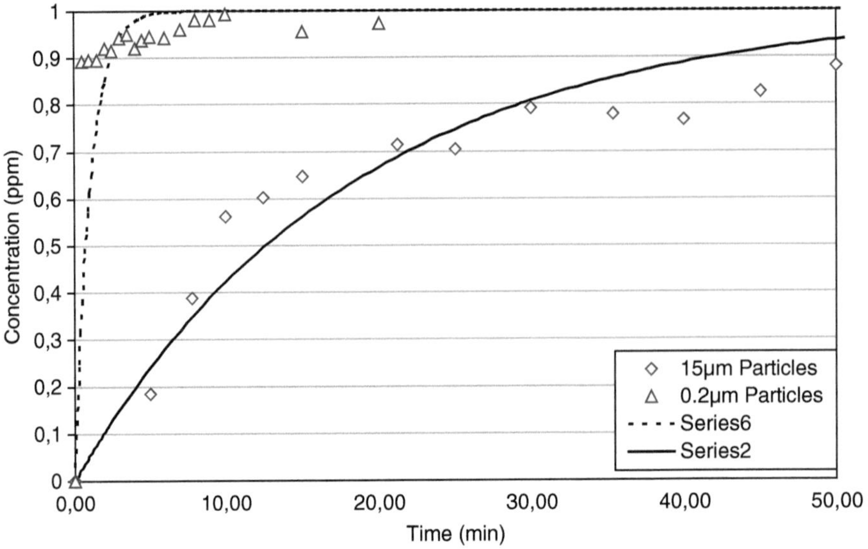

Fig. 15.13 Concentration of dissolved preservative in water as function of time (data points represent measurement data and lines are fits)

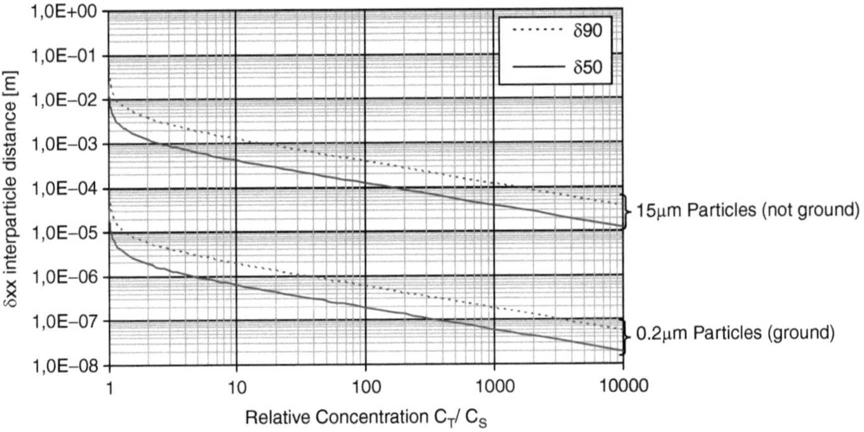

Fig. 15.14 The inter-particle distance distribution (δ_{50} and δ_{90}) for random distributed particles in a matrix as function of the total particle concentration C_T (coating thickness 10^{-4} m)

15.4.5 Experimental Results of the Accelerated Shelf Life Test

Ten cheeses were tested per preservative particle size. As a reference a coating without preservative was tested. The results are presented in Table 15.4.

Table 15.4 Microbial shelf life of coated cheese (the time indicates when 50 % of the cheeses had at least one mould colony; the standard deviation indicates the measurement variation for each set of 10 cheeses)

	Product relative concentration (C_T/C_S)			
Particle size	0.0	3.0	5.0	8.3
15 μm	4 days	–	10.7 ± 2.1 days	11.3 ± 3.0 days
0.2 μm		19.6 ± 4.5 days	20.1 ± 9.6 days	–

The results show that the protection against fungal growth increases when the preservative concentration is increased and/or smaller preservative particles are used.

15.4.6 Conclusions with Comparison of Model and Experimental Results

Model calculations on the distribution of a preservative in a coating were used to determine the microbial shelf life of a coated food product. A case study was performed with a cheese coating containing an anti-fungal compound with low solubility. The preservative was dosed to the coating at concentrations above its saturation concentration. Preservative particles were thus dispersed in the coating. The anti-fungal compound is active only in dissolved state. The dissolution, diffusion and degradation behavior of the anti-fungal compound thus affect the anti-microbial protection that is given by the coating. In addition, the inter-particle distance is of importance. The weakest points in the coating are in the middle between the preservative particles.

The model has shown that an optimum shelf life can be obtained when the characteristic time for diffusion is much shorter than the characteristic time for degradation. The characteristic time for diffusion can be controlled by optimizing the inter-particle distance, i.e. a reduction of the inter-particle distance gives a reduction of the characteristic time for diffusion. At constant preservative concentration the inter-particle distance can be reduced by reduction of the preservative particle size.

The impact of the preservative concentration decreases dramatically when the characteristic time for diffusion is much larger than the characteristic time for degradation. This is due to the fact that the diffusion process is too slow to replenish the degraded preservative in between the preservative particles. Under these conditions the shelf life is mainly determined by the dissolved preservative concentration at $t = 0$ and preservative particles will be present in the coating when the shelf life is reached. This implies that the preservative is not used very efficiently, because not all the preservative that is present in the coating is used for the anti-microbial action of the coating.

Experimental results show that the microbial shelf life of coated cheeses increases with an increase in preservative concentration and/or with a reduction in preservative particle size at constant preservative dosage. These findings correspond to the model calculations. Thus the model gives a reasonable prediction of the accelerated shelf life test. The model shows that the distribution of the preservative particles throughout the coating is of great importance. Such a model can be used to find the optimum formulation for food coatings including the optimum particle size and concentration of active ingredient.

15.5 Definitions, Abbreviations and Symbols

Cluster	Agglomerate and/or aggregate of primary particles
Primary particle	Basic particle that cannot be separated other than by breakage
CTAB	Cetyl-trimethyl-ammonium bromide
PSD	Particle size distribution
SEM	Scanning electron microscopy
TEM	Transmission electron microscopy
$B(x,t)$	Birth/death term [ppm.s^{-1}]
$C(x,t)$	Concentration of product [ppm]
C_S	Solubility of product particles [ppm]
d_m	Diameter of a molecule [m]
d_{50}	Median particle size
$D_{4,3}$	Volume-weighted mean size, mean value of a volume-based PSD
D_p	Particle size
D_v	Vessel diameter
$\mathbb{D}$	Diffusion coefficient or diffusivity [m^2.s^{-1}]
k	Boltzmann constant [m^2.kg.s^{-2}.K^{-1}]
k_{degr}	Constant of degradation [first order: s^{-1}]
k_{diss}	Constant of dissolution [first order: s^{-1}]
m_{part}	Mass of one particles [kg]
$N_{particles}$	Number of particles [-]
S_{part}	Surface area of a particle [m^2]
t	Time [s]
t_C	Microbial shelf life of coated food stuff [s]
$t_ø$	Time at which particles are dissolved [s]
T	Temperature [K]
$V_{suspen.}$	Volume of suspension in the system [m^3]
x	Position [m]
x_i	Particle size [m]
α	Proportionality factor [m.s^{-1}]
δ	Distance between two particles [m]
η_f	Fluid viscosity [Pa.s]
τ_{diff}	Characteristic time of diffusion [s]
τ_{deg}	Characteristic time of degradation [s]
τ_{diss}	Characteristic time of dissolution [s]

References

1. Becker, M., Schwedes, J.: Comminution of ceramics in stirred media mills and wear of grinding beads. Powder Technol. **105**, 374–381 (1999)
2. Bel, F.H., Frances, C., Mamourian, A.: Investigations on ultra-fine grinding of titanium dioxide in a stirred media mill. Powder Technol. **105**(1–3), 362–373 (1999)
3. Bernhardt, C., Reinsch, E., Husemann, K.: The influence of suspension properties on ultra-fine grinding in stirred ball mills. Powder Technol. **105**(1–3), 357–361 (1999)
4. Bilgili, E., Hamey, R., Scarlett, B.: Nano-milling of pigment agglomerates using a wet stirred media mill: Elucidation of the kinetics and breakage mechanisms. Chem. Eng. Sci. **61**(1), 149–157 (2006)
5. Breitung-Faes, S., Kwade, A.: Nano particle production in high-power density mills. Chem. Eng. Res. Design. **86**, 390–394 (2008)
6. Davies J.T.: A quantitative kinetic theory of emulsion type, I. Physical chemistry of the emulsifying agent, Gas/Liquid and Liquid/Liquid Interface; Proceedings of International Congress Surface Activity, pp. 426–438 (1957)
7. Davies, L., Dollimore, D., Sharp, J.H.: Sedimentation of supsensions: Implications of theories of hindered settling. Powder Technol. **13**, 123–132 (1975)
8. Fick, A.: Poggendorff's annal. Physik **94**, 59 (1855)
9. Fick, A.: Phil. Mag. **10**, 30 (1855)
10. Fox, J.B.: Diffusion of chloride, nitrite, and nitrate in beef and pork. J. Food Sci. **45**(860), 1740–1744 (1980)
11. Gao, M., Forssberg, E.: Prediction of product size distributions for a stirred ball mill. Powder Technol. **84**(2), 101–106 (1985)
12. Gennadios, A., Weller, C.L., Testin, R.F.: Temperature effect on oxygen permeability of edible protein-based film. J. Food Sci. **58**, 212–219 (1993)
13. Gill, C.O.: A review, intrinsic bacteria in meat. J. Appl. Bacteriol. **47**, 367–378 (1979)
14. Gontard, N., Guilbert, S., Cuq, J.: Edible wheat gluten film: Influence of the main process variables on film properties using response surface methodology. J. Food Sci. **57**, 190–199 (1992)
15. Griffin, W.C.: Classification of surface-active agents by 'HLB'. J. Soc. Cosmetic Chem. **1**, 311 (1949)
16. Griffin, W.C.: Calculation of HLB values of non-ionic surfactants. J. Soc. Cosmet. Chem. **5**, 259 (1954)
17. Grubenmann, A.: Particle size distribution and aspect ratio of organic pigments. Part. Part. Syst. Charact. **3**(4), 179–186 (1986)
18. Gustavsson, J., Cederberg, C., Sonesson, U., van Otterdijk, R., Meybeck, A.: Global food losses and food waste. Study conducted for the Internat. Congress at Interpack Düsseldorf, Germany Published by the FAO. (http://www.fao.org/docrep/014/mb060e/mb060e00.pdf) (2011)
19. He, M., Wang, Y., Forssberg, E.: Parameter effects on wet ultrafine grinding of limestone through slurry rheology in a stirred media mill. Powder Technol. **161**, 10–21 (2006)
20. Hee, P. Van, Meesters, G.M.H., Wildeboer, W.J., Hennart, S.L.A, Vis, A.J.: Stabilized micronized particles; WO Patent # 2008/110626
21. Hennart, S.L.A., Wildeboer, W.J., Meesters, G.M.H.: Study of the process of stirred ball milling of poorly water soluble organic products using factorial design. Powder Technol. **198**(1), 56–60 (2009)
22. Herbst, J.A., Sepulveda, J.L.: Fundamentals of fine and ultra-fine grinding in a stirred ball mill. International Powder and Bulk Solids Handling Conference, Chicago, pp. 452–470 (1978)
23. Horter, D., Dressman, J.B.: Influence of physicochemical properties on dissolution of drugs in the gastrointestinal tract. Ad. Drug Deliv. Rev. **46**, 75–87 (2001)

24. Hou, T.-H., Su, C.-H., Liu, W.-L.: Parameters optimization of a nano-particle wet milling process using the taguchi method, response surface method and genetic algorithm. Powder Technol. **173**, 153–162 (2007)
25. Howorth, C.J., Lee, W.E., Rainforth, W.M., Messer, P.F.: Contamination rates from Ce and Y-TZP ball milling media. Br. Ceram. Trans. J. **90**, 18–21 (1990)
26. Hu, J., Johnston, K.P., Williams, R.O.: Nanoparticles engineering process for enhancing the dissolution rates of poorly water soluble drugs. Drug Devel. Ind. Pharm **30**(3), 233–245 (2004)
27. Jankovic, A.: Variables affecting the fine grinding of minerals using stirred mills. Minerals Eng. **16**, 337–345 (2003)
28. Janssen, L., Warmoeskerken, M.: Transport Phenomena Data Companion. Edward Arnold Ltd, London (1987). ISBN 0-7131-3618-9
29. Kärger, J., Grinberg, F., Heitjans, P.: Diffusion Fundamentals. Leipziger Universitätsverlag, Leipzig (2005). ISBN ISBN: 3-86583-073-0
30. Laidlaw, I., Steinmetz, M.: Introduction to differential sedimentation. In: Scott, D.J., Harding, S.E., Rowe, A.J. (eds.) Analytical Ultracentrifugation Techniques and Methods, pp. 270–290. Royal Society of Chemistry, Cambridge (2005)
31. Lange, H.: Comparative test of methods to determine particle size and particle size distribution in the submicron range. Part. Part. Syst. Charact. **12**(3), 148–157 (1995)
32. Lett, J.T.: Measurement of single strand breaks by sedimentation in alkaline sucrose gradients. In: Friedberg, E.C., Hanawalt, P.C. (eds.) DNA repair – A Laboratory Manual of Research Procedures, vol. 1, pp. 363–378. Marcel Dekker, New York (1981)
33. Mende, S., Stenger, F., Peukert, W., Schwedes, J.: Mechanical production and stabilization of submicron particles in stirred media mills. Powder Technol. **132**, 64–73 (2003)
34. Mura, P., Faucci, M.T., Parrini, P.L.: Effects of grinding with micro-crystalline cellulose and cyclodextrins on the ketoprofen physicochemical properties. Drug Dev. Ind. Pharm. **27**, 119–128 (2001)
35. Noyes, A.A., Whitney, W.R.: The rate of solution of solid substances in their own solutions. J. Am. Chem. Soc. **19**, 930–936 (1897)
36. Ouattara, B., Simard, R.E., Piette, G., Begin, A., Holley, R.A.: Diffusion of acetic and propionic acids from chitosan-based antimicrobial packaging films. J. Food Sci. **65**, 768–773 (2000)
37. Ozdemir, J.D.: Floros, analysis and modeling of potassium sorbate diffusion through edible whey protein films. J. Food Sci. **65**, 149–155 (2001)
38. Ozdemir, M., Floros, J.D.: Film composition effect on diffusion of potassium sorbate through whey protein films. J. Food Eng. **68**, 511–516 (2003)
39. Perez-Gago, M.B., Krochta, J.M.: Water vapor permeability of whey protein emulsion films as affected by pH. J. Food Sci. **64**, 695–698 (1999)
40. Redl, A., Gontard, N., Guilbert, S.: Determination of sorbic acid diffusivity in edible wheat gluten and lipid based films. J. Food Sci. **61**, 116–120 (1996)
41. Redner, S.: Statistical model for the fracture disordered media. In: Hulin, J.P. (ed.) Fragmentation, pp. 321–328. Elsevier, London (1990)
42. Saito, M., Ugajin, T., Nozawa, Y., Sadzuka, Y., Miyagishima, A., Sonobe, T.: Preparation and dissolution characteristics of griseofulvin solid dispersions with saccharides. Int. J. Pharm. **249**, 71–79 (2002)
43. Teerakarn, A., Hirt, D.E., Action, J.C., Rieck, J.R., Dawson, P.L.: Nisin diffusion in protein films: Effect of film type and temperature. J. Food Sci. **67**(8), 3019–3025 (2002)
44. Varinot, C., Hiltgun, S., Pons, M.-N., Dodds, J.: Identification of the fragmentation mechanisms in wet-phase fine grinding in a stirred bead mill. Chem. Eng. Sci. **52**(20), 3605–3612 (1997)
45. Vippagunta, S.R., Maul, K.A., Tallavajhala, S., Grant, D.J.: Solid-state characterization of nifedipine solid dispersions. Int. J. Pharm. **236**, 111–123 (2002)

46. Vojdani, F., Torres, J.A.: Potassium sorbate permeability of methyl-cellulose and hydroxypropyl methyl-cellulose coatings: Effect of fatty acids. J. Food Sci. **55**, 841–846 (1990)
47. Warin, F., Gekas, V., Voirin, A., Dejmek, P.: Sugar diffusivity in agar gel/milk bilayer systems. J. Food Sci. **62**, 454–456 (1997)
48. Yamada, T., Saito, N., Imai, T., Otagiri, M.: Effect of grinding with hydroxypropyl cellulose on the dissolution and particle size of a poorly water-soluble drug. Chem. Pharm. Bulletin **47**, 1311–1313 (1999)

Subject Index

H.G. Merkus and G.M.H. Meesters (eds.), *Particulate Products: Tailoring Properties for Optimal Performance*, Particle Technology Series 19, DOI 10.1007/978-3-319-00714-4,

Author Index

H.G. Merkus and G.M.H. Meesters (eds.), *Particulate Products: Tailoring Properties for Optimal Performance*, Particle Technology Series 19,
DOI 10.1007/978-3-319-00714-4, © Springer International Publishing Switzerland 2014

MIX
Papier aus verantwortungsvollen Quellen
Paper from responsible sources
FSC® C105338

If you have any concerns about our products,
you can contact us on
ProductSafety@springernature.com

In case Publisher is established outside the EU,
the EU authorized representative is:
Springer Nature Customer Service Center GmbH
Europaplatz 3, 69115 Heidelberg, Germany

Printed by Libri Plureos GmbH
in Hamburg, Germany